ELECTROGENERATED CHEMILUMINESCENCE

MONOGRAPHS IN ELECTROANALYTICAL CHEMISTRY AND ELECTROCHEMISTRY

Consulting editor
Allen J. Bard
Department of Chemistry
University of Texas
Austin, Texas

Electrochemistry at Solid Electrodes, *Ralph N. Adams*

Electrochemical Reactions in Nonaqueous Systems, *Charles K. Mann and Karen Barnes*

Electrochemistry of Metals and Semiconductors: The Application of Solid State Science to Electrochemical Phenomena, *Ashok K. Vijh*

Modern Polarographic Methods In Analytical Chemistry, *A. M. Bond*

Laboratory Techniques in Electroanalytical Chemistry, *edited by Peter T. Kissinger and William R. Heineman*

Standard Potentials in Aqueous Solution, *edited by Allen J. Bard, Roger Parsons, and Joseph Jordan*

Physical Electrochemistry: Principles, Methods, and Applications, *edited by Israel Rubinstein*

Scanning Electrochemical Microscopy, *edited by Allen J. Bard and Michael B. Mirkin*

Electrogenerated Chemiluminescence, *edited by Allen J. Bard*

ADDITIONAL VOLUMES IN PREPARATION

ELECTROGENERATED CHEMILUMINESCENCE

EDITED BY

ALLEN J. BARD

The University of Texas at Austin
Austin, Texas, U.S.A.

MARCEL DEKKER, INC.　　　　　　　　　　　　NEW YORK

Although great care has been taken to provide accurate and current information, neither the author(s) nor the publisher, nor anyone else associated with this publication, shall be liable for any loss, damage, or liability directly or indirectly caused or alleged to be caused by this book. The material contained herein is not intended to provide specific advice or recommendations for any specific situation.

Trademark notice: Product or corporate names may be trademarks or registered trademarks and are used only for identification and explanation without intent to infringe.

Library of Congress Cataloging-in-Publication Data
A catalog record for this book is available from the Library of Congress.

ISBN: 0-8247-5347-X

This book is printed on acid-free paper.

Headquarters
Marcel Dekker, Inc., 270 Madison Avenue, New York, NY 10016, U.S.A.
tel: 212-696-9000; fax: 212-685-4540

Distribution and Customer Service
Marcel Dekker, Inc., Cimarron Road, Monticello, New York 12701, U.S.A.
tel: 800-228-1160; fax: 845-796-1772

Eastern Hemisphere Distribution
Marcel Dekker AG, Hutgasse 4, Postfach 812, CH-4001 Basel, Switzerland
tel: 41-61-260-6300; fax: 41-61-260-6333

World Wide Web
http://www.dekker.com

The publisher offers discounts on this book when ordered in bulk quantities. For more information, write to Special Sales/Professional Marketing at the headquarters address above.

PRINTED IN THE UNITED STATES OF AMERICA

Preface

Although the first experiments in electrogenerated chemiluminescence (ECL) were carried out about 40 years ago, no comprehensive monograph devoted to this field has appreared. This is very surprising, since ECL has entered the mainstream of research and has found a number of applications. Numerous organic and inorganic compounds have been shown to be capable of producing ECL and the study of ECL reactions has provided useful insight into the mechanisms and energetics of electron transfer reactions and electrogenerated intermediates. The use of coreactants in ECL allowed practical analytical applications and also provided information about short-lived intermediates of the coreactant species. In terms of applications, ECL is widely used in immunoassay, thanks largely to the initial development efforts at IGEN, Inc., and now accounts for hundreds of millions of dollars in sales per year. Moreover, light emission in polymer and solid-state electrochemical cells is actively being investigated and developed (often without recognition of the ECL roots of such studies), with the promise of useful light sources and displays. It is clearly time to publish this first monograph, which provides comprehensive reviews of different aspects of ECL.

Chapters 1–7 of this book present an overview and brief history of ECL (Chapter 1), both the experimental (Chapter 2) and theoretical aspects of ECL (Chapters 3 and 4), a review of the behavior of coreactants (Chapter 5), organic molecules (Chapter 6) and metal chelates (Chapter 7). The other chapters are dedicated to applications: immunoassay (Chapter 8), other analytical uses, e.g. in chromatography (Chapter 9), and devices, such as systems for light production and displays (Chapter 10). Finally there is a brief chapter containing several additional topics and conclusions. Although some knowledge of electrochemistry and physical chemistry is assumed, the key ideas are introduced in Chapter 1 and discussed at a level suitable for beginning graduate students.

I would like to thank my many students, postdoctoral fellows, and colleagues, and the contributors, who have done so much to develop ECL. I also thank Drs. David Hercules and Ed Chandross, two pioneers in this field, for their comments on Chapter 1. The future for ECL remains bright. It is an interesting and aesthetically pleasing phenomenon. It is a shame it has not yet found its way into elementary general chemistry texts!

Allen J. Bard

Contents

Contributors

Allen J. Bard   The University of Texas at Austin, Austin, Texas, U.S.A.

Mihai Buda*   The University of Texas at Austin, Austin, Texas, U.S.A.

Jai-Pil Choi   The University of Texas at Austin, Austin, Texas, U.S.A.

Neil D. Danielson   Miami University, Oxford, Ohio, U.S.A.

Jeff D. Debad   Meso Scale Discovery, Gaithersburg, Maryland, U.S.A.

Fu-Ren F. Fan   The University of Texas at Austin, Austin, Texas, U.S.A.

Samuel P. Forry   University of North Carolina at Chapel Hill, Chapel Hill, North Carolina, U.S.A.

Eli N. Glezer   Meso Scale Discovery, Gaithersburg, Maryland, U.S.A.

Andrezej Kapturkiewicz   Insitute of Physical Chemistry, Polish Academy of Sciences, Warsaw, Poland

Jonathan K. Leland   IGEN International, Inc., Gaithersburg, Maryland, U.S.A.

Joseph T. Maloy   Seton Hall University, South Orange, New Jersey, U.S.A.

Wujian Miao   The University of Texas at Austin, Austin, Texas, U.S.A.

Mark M. Richter   Southwest Missouri State University, Springfield, Missouri, U.S.A.

George B. Sigal   Meso Scale Discovery, Gaithersburg, Maryland, U.S.A.

R. Mark Wightman   University of North Carolina at Chapel Hill, Chapel Hill, North Carolina, U.S.A.

Jacob Wohlstadter   Meso Scale Discovery, Gaithersburg, Maryland, U.S.A.

*Present address: Politehnica University of Bucharest, Bucharest, Romania

1

Introduction

Allen J. Bard
The University of Texas at Austin, Austin, Texas, U.S.A.

I. BACKGROUND AND SCOPE

This monograph deals with light emission at electrodes in electrochemical cells caused by energetic electron transfer (redox) reactions of electrogenerated species in solution. A typical system would involve a solution containing reactants A and D in a solution with supporting electrolyte, for example, MeCN with tetra-*n*-butylammonium perchlorate (TBAP), in an electrochemical cell with a platinum working electrode. The reaction sequence to generate an excited state and light emission is:

$$A + e^- \rightarrow A^{-\bullet} \qquad \text{(reduction at electrode)} \qquad (1)$$

$$D - e^- \rightarrow D^{+\bullet} \qquad \text{(oxidation at electrode)} \qquad (2)$$

$$A^{-\bullet} + D^{+\bullet} \rightarrow {}^1A^* + D \qquad \text{(excited state formation)} \qquad (3)$$

$${}^1A^* \rightarrow A + h\nu \qquad \text{(light emission)} \qquad (4)$$

where A and D could be the same species, e.g., an aromatic hydrocarbon such as rubrene. The term *electrogenerated chemiluminescence* (ECL) was coined to describe this class of reaction [1,2] and to distinguish it from other systems where light is emitted from an electrode in electrochemical cells.

As pointed out in earlier reviews [3,4], light can be emitted during an electrochemical reaction in a number of different ways, and reports concerning such emission date back at least to the 1880s. These types of reactions are briefly described below but are not considered within the class of ECL reactions and are not dealt with further in this monograph.

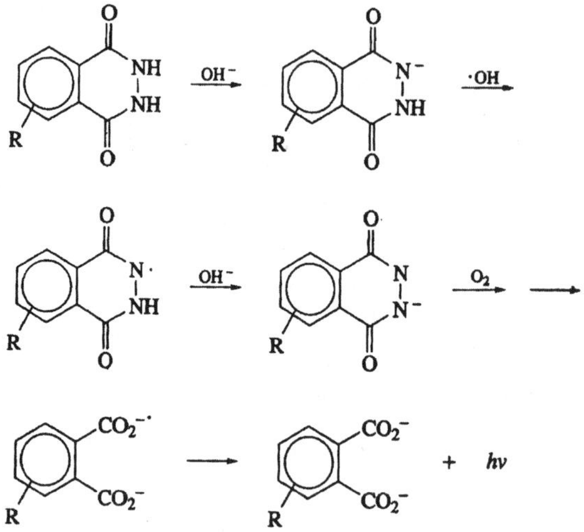

Scheme 1

A. Chemiluminescence Reactions

There are numerous chemical reactions where the decomposition of a species, often triggered by reaction with oxygen or peroxide, results in the production of excited states and emission [5]. The best known, e.g., those involving luminol, luciginen, and oxalate esters, generate intense light and are the basis for consumer products such as "light sticks." Consider the luminol (5-amino-2,3-dihydro-1,4-phthalazinedione) chemiluminescent reaction (R = NH_2) shown in Scheme 1. This chemiluminescent reaction is generally carried out in alkaline solutions by the addition of hydrogen peroxide in the presence of an ion such as Fe^{2+}, which generates hydroxyl radical. The mechanism is complicated and leads to destruction of the luminol starting material, with the emitter, the excited phthalate, completely different than the starting material. One can generate chemiluminescence in an electrochemical cell, for example by generating peroxide by the reduction of oxygen or oxidizing luminol [6,7]. Although this reaction can be useful, for example in imaging reactions at electrodes [8], it is not the type of reaction considered in this volume. Perhaps it would be useful to classify these in a more general class as *electrochemiluminescence*.

B. Glow Discharge Electrolysis

If electrolysis is carried out with a small electrode, e.g. the anode, and very high current densities (and high voltages) are applied, the evolution of gas (e.g., oxygen at the anode) forms a gas layer that separates solution from the electrode. At higher applied voltages, breakdown occurs in the gas film, leading to a glow discharge and emission of light [9]. This phenomenon is closely related to gas discharge and also occurs when one of the electrodes in the cell is moved into the gas phase above the liquid [10]. The emission of light results from ion–radical and electron–radical reactions that have been widely studied in gas-phase chemiluminescence [5].

C. Electroluminescence

There are several mechanisms by which light is emitted in solids, usually semiconductors or insulators [11]. Dielectric breakdown, similar to the glow discharge phenomenon that occurs in gases, can produce light emission. This involves rather high electric fields. Another high-field form of luminescence in solids is related to cathodoluminesence, which occurs when cathode rays (beams of electrons) strike a phosphor, as in cathode ray tubes in oscilloscopes. In this case the electron beam behaves as an excitation source, like the ultraviolet excitation beam in fluorescence, and causes excitation via electronic transitions in the solid material. This mechanism, sometimes called *acceleration–collision electroluminescence* [11], occurs in many electroluminescence (EL) devices, such as those with ZnS, where electrons injected into the conduction band are accelerated by the applied electric field and cause excitation of activator centers produced by doping of the ZnS.

An EL process that is closer to ECL is one that involves electron–hole recombination in the solid following injection of carriers (sometimes called *injection electroluminescence*). This is the process responsible for light emission at suitable *p-n* junctions in light-emitting diodes (LEDs), for example, those of GaAs. This process can also occur at semiconductor electrodes in electrochemical cells. For example, when an electrode of ZnS doped with Mn immersed in a solution containing peroxydisulfate ($S_2O_8^{2-}$) passes cathodic current, fairly intense light emission centered at about 580 nm is observed [12]. This emission results from injection of electrons into the conduction band of the ZnS:Mn, as shown schematically in Figure 1. The passage of current leads to the reduction of the $S_2O_8^{2-}$ and in turn the production of the strong oxidant, $SO_4^{-\bullet}$. The latter can inject a hole into the valence band of the ZnS:Mn, resulting in electron–hole recombination and emission. This process is formally the same as the use of $S_2O_8^{2-}$ as a coreactant in ECL systems as described below and in Chapter 5. This same EL process occurs with many semiconductor electrodes and also with thin oxide films on metals such as Ta and Al. These are considered briefly in Chapter 11.

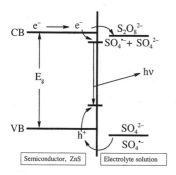

Figure 1   Electroluminescence of ZnS electrode immersed in a 0.2 M Na$_2$SO$_4$ solution containing 0.2 M ammonium peroxydisulfate. (See Ref. 12).

II. HISTORY OF ECL

A. Background

To understand how ECL experiments arose in the middle 1960s, it is useful to consider the scientific environment that existed at the time, especially with respect to the electrochemistry of organic compounds and to studies of radical ions. The state of organic electrochemistry in the 1950s was quite confused. The problem was that most studies were carried out in aqueous and partially aqueous solutions (e.g., mixtures of water with ethanol or dioxane to improve solubility) [13]. Water has the disadvantage of having a limited range of available potentials before it is oxidized to oxygen or reduced to hydrogen (i.e., it has a small *potential window*). Moreover, water shows reasonable acidic and basic properties so that proton transfer reactions are frequently coupled to electron transfers in this solvent. The result was that many of the electrode reactions showed waves resulting from the addition of two or more electrons. For example, the reduction of anthracene was described as a two-electron, two-proton reduction leading to 9,10-dihydroanthracene as the product. Moreover, many oxidation reactions of organic compounds were described in terms of adsorbed intermediates, with reactions like the Kolbe reaction being very popular [14]. Radical ions were not thought of as the usual intermediates, because the prevailing concepts of organic chemistry at the time were based largely on "pushing" electron pairs, in contrast to the currently accepted idea of electrode reactions generally proceeding in elementary steps involving single electron transfer.

The situation began to change with the introduction of the use of aprotic solvents such as MeCN [15]. MeCN has a large potential window, which allows difficult oxidations and reductions to be examined. Many organic compounds show reasonable solubility in MeCN, and the solvent itself has negligible acid and base properties. A problem with the early studies with MeCN, however, was that the solvent was often contaminated with small amounts of water (and sometimes oxygen), so that even with this solvent, electrogeneration of radical ions was not noticed [16]. However, with the advent of electrochemical cells that could be used on vacuum lines or in glove boxes, with highly purified and degassed solvents and electrolytes, electrochemical experiments showed that reduction of many aromatic hydrocarbons led to rather stable radical anions and that oxidations produced radical cations. (For reviews of this early work, see, e.g., Refs 17a–c.) The application of electron spin resonance to the study of electrogenerated radical ions, which showed that they were the same as those prepared chemically, was also important in proving the ready production of these species at electrodes [18]. Thus, by the early 1960s, there was a good deal of interest in the electrogeneration of aromatic hydrocarbon radical ions in aprotic solvents such as MeCN, DMF, and DMSO and in understanding their behavior. At the same time there was increased interest by the Office of Naval Research and the Army Research Office in chemiluminescence for possible military applications such as wide area markers. Conferences involving some of this work were held at Duke University in 1965 and at the University of Georgia in 1972 [5].

B. Early ECL Experiments

A time line showing various events in the development of ECL is given in Figure 2. A good historical perspective of the discovery of radical ion annihilation ECL and the first experiments in this area can be found in several early reviews by the key groups in the area at the time [4,19–21]. The earliest chemiluminescence experiments involving radical ions used radical anions prepared chemically and reacted with different oxidants [22]. The experiments involved light emission when the radical anion of 9,10-diphenylanthracene (DPA), produced by reduction with potassium in THF, reacted with the 9,10-dichloride of DPA (DPACl$_2$), Cl$_2$, benzoyl peroxide, and several other species. Chandross and Sonntag [22] point to a desire to study the cation radical of DPA as an oxidant but their inability to prepare solutions of this chemically. They also refer to a private communication that DPA cations readily undergo internal cyclization (which later work would demonstrate to be untrue). This work also mentions unpublished experiments by G. J. Hoytink at the University of Amsterdam in attempts to study the reaction between chemically generated anthracene radical anions and cations. The instability of the radical cation of anthracene, as shown in later ECL experiments [23], made the results of these experiments "very doubtful" [19], and they were never published. Then, within the next six

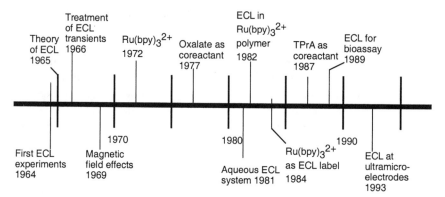

Figure 2 Time line of ECL.
1964–1965: First experiments [24–26]
1965: Theory [30].
1966: Transients [31].
1969: Magnetic field effects [29].
1972: $Ru(bpy)_3^{2+}$ [38].
1977: Oxalate [39].
1981: Aqueous [40].
1982: $Ru(bpy)_3^{2+}$ polymer [46].
1984: $Ru(bpy)_3^{2+}$ label [43].
1987: TPrA [41–42].
1989: Bioassay [44,45].
1993: Ultramicroelectrodes [34].

months, three papers describing the electrochemical generation of radical ion species and light emission appeared [24–26]. Hercules [24] used Pt electrodes in MeCN or DMF solutions and showed emission of light with a number of hydrocarbons, including DPA, anthracene, and rubrene when the electrode potential was cycled at frequencies up to 10 Hz and also when a dc current was applied to two closely spaced Pt electrodes. No detailed electrochemical or spectroscopic results were reported with visual detection of the emitted light. The radical ion annihilation reaction was suggested, but considered not consistent with all of the experimental results reported. In Refs 25 and 26 cyclic voltammetric evidence was reported in an attempt to understand better the reaction mechanism with DPA. The results suggested that either the radical cation or other oxidants could be generated during cyclic electrochemical steps and result in light emission. The potentials needed to generate the light were shown to be those at which cathodic and anodic processes occur at potentials sufficient to form the singlet excited state. Patents on ECL were applied for on July 13, 1964 [27] and August 18, 1964 [28].

Within the next few years the list of ECL emitters was extended to include many more hydrocarbons and heterocyclic molecules, such as substituted isobenzofurans and carbazoles. From the beginning, the energetics of the annihilation reaction compared to the energy of the emitting singlet state were considered. Those cases where the energy was significantly smaller were attributed to the formation of triplets followed by triplet–triplet annihilation (TTA), a process known to occur with optical excitation (delayed fluorescence):

$$A^{-\bullet} + D^{+\bullet} \rightarrow {}^3A^* + D \qquad \text{(excited triplet formation)} \qquad (5)$$

$$^3A^* + {}^3A^* \rightarrow {}^1A^* \; A \qquad \text{(triplet–triplet annihilation)} \qquad (6)$$

$$^1A^* \rightarrow A \; h\nu \qquad \text{(light emission)} \qquad (7)$$

In studies of these processes, species were added to solutions of the aromatic hydrocarbon to produce a cation radical that was significantly less oxidizing than the aromatic hydrocarbon cation radical. Similar experiments were carried out with species with less reducing anion radicals reacting with aromatic hydrocarbon cation radicals. The mechanism was also probed by studying magnetic field effects [29]. The generation of excimers (A_2^*) and exciplexes (AD*) during the ion radical reaction was also studied.

$$A^{-\bullet} \; A^{\bullet} \rightarrow A_2^* \qquad \text{(excimer formation)} \qquad (8)$$

$$A^{-\bullet} \; D^{\bullet} \rightarrow AD^* \qquad \text{(exciplex formation)} \qquad (9)$$

Details about this work and later investigations of organic species in ECL are considered in Chapter 6.

Other issues involved "preannihilation ECL," with which weak light was observed in solutions of the aromatic hydrocarbon even when the applied potentials were significantly smaller than those needed to generate both ion radicals, and the question of whether direct excitation at an electrode, for example by electrochemical oxidation of the anion radical, was possible.

C. Theoretical Studies

Approaches to a theoretical understanding of the ECL response, in terms of the reasons both for excited state formation and for the shape of the electrochemical ECL transients, followed soon after the first experiments. ECL actually provided the first evidence of the Marcus inverted region, where generation of the excited state rather than the energetically more favored ground state was seen [30]. This work and later developments are described in Chapter 4. A digital simulation of the diffusion and kinetic processes involved in the generation of reactants at a single electrode by potential steps and the presentation of an analysis to distinguish between direct formation of the singlet state and triplet–triplet annihilation appeared [31]. This work is discussed in Chapter 3. Several experimental studies that used this approach appeared. One major motivation in these kinds of simulation studies was the desire to measure the rate constant of the radical ion annihilation reaction. However, the rate of this reaction was found to be too fast (compared to the rate of diffusion) to be measured by this technique with the electrodes and instrumentation available at that time. Similarly, studies with the rotating ring-disk electrode (RRDE), which allow steady-state, rather than transient, measurements under faster mass transfer (convective) conditions, were not successful in determining this rate constant [32,33]. This rate constant $(2 \times 10^9 \text{ M}^{-1} \text{ s}^{-1})$ was finally measured electrochemically by use of an ultramicroelectrode (radius 5 μm) and a frequency of 20–30 kHz for generation of DPA ECL [34].

D. Metal Chelate ECL

The photoluminescence of $Ru(bpy)_3^{2+}$ was reported in 1959 [35] and its spectroscopy was studied rather extensively in the late 1960s [36]. $Ru(bpy)_3^{3+}$ in aqueous solution (formed by treatment of the 2+ form with PbO_2) chemiluminesced when a solution in 9 N H_2SO_4 was treated with 9 N NaOH [37]. The ECL of this species in MeCN solution was reported in 1972 [38] and ascribed to the reaction sequence

$$Ru(bpy)_3^{2+} - e \rightarrow Ru(bpy)_3^{3+} \quad \text{(oxidation during anodic step)} \qquad (10)$$

$$Ru(bpy)_3^{2+} + e \rightarrow Ru(bpy)_3^{+} \quad \text{(reduction during cathodic step)} \qquad (11)$$

$$Ru(bpy)_3^{+} + Ru(bpy)_3^{3+} \rightarrow Ru(bpy)_3^{2+}* + Ru(bpy)_3^{2+} \qquad (12)$$

$$Ru(bpy)_3^{2+}* \rightarrow Ru(bpy)_3^{2} + h\nu \qquad (13)$$

Details about this reaction system, which has been studied extensively, and other metal chelate systems, are discussed in Chapter 7.

E. Aqueous ECL Systems

The discovery of ECL with the water-soluble $Ru(bpy)_3^{2+}$ suggested that aqueous ECL was a possibility, especially because it was also known that $Ru(bpy)_3^{3+}$ was reasonably stable in aqueous solution. The problem, however, was that the potential window of water is rather small compared to that in aprotic solvents such as MeCN [thermodynamically 1.23 V, but practically somewhat larger, depending on the electrode material, because of kinetic (overpotential) effects], so that it is difficult to generate cathodically a reductant suitable for ECL. What was needed was a water-soluble species that could generate the reductant for reaction with

$Ru(bpy)_3^{3+}$ by *oxidation* rather than by reduction. Such a species has become known as a *coreactant*. Although species that could serve as coreactants in nonaqueous solvents, such as benzoyl peroxide, were known from earlier work, they weren't suitable for aqueous solutions. The discovery that oxalate anion would behave as a coreactant via the reaction sequence

$$C_2O_4^{2-} - e \rightarrow CO_2^{-\bullet} + CO_2 \tag{14}$$

$$Ru(bpy)_3^{3+} + CO_2^{-\bullet} \rightarrow Ru(bpy)_3^{2+}* + CO_2 \tag{15}$$

meant that ECL could be generated simply by oxidizing $Ru(bpy)_3^{2+}$ and oxalate in a mixture [39]. Coreactants in ECL for both nonaqueous and aqueous solutions are discussed in Chapter 5.

The aqueous $Ru(bpy)_3^{2+}$/oxalate system, with intense emission (maximum at pH 6) that was the same as that for photoexcitation, was described in 1981 [40]. Since that study, many other Ru- and Os-based chelates have been studied (see Chapter 7) and many other coreactants used. A more efficient coreactant for $Ru(bpy)_3^{2+}$, tri-*n*-propylamine (TPrA), at pH values near 7 was found a few years later [41,42].

F. Analytical Applications

From the beginning there was interest in using ECL as an analytical method, with demonstrations of the linearity in emission and concentration [1]. However, there was not much interest in determination of hydrocarbons under the restricted solution requirements needed for stable and reproducible ECL. The advent of aqueous ECL, however, especially the discovery of coreactants, changed the picture. Because different coreactants such as oxalate, pyruvate, and other organic acids would all react with $Ru(bpy)_3^{3+}$ to produce emission [40], analytical applications of ECL, e.g., in chromatography, now became possible. The $Ru(bpy)_3^{2+}$ system became the basis, for example, of sensitive chromatographic determination of amines [41]. These kinds of applications are discussed in Chapter 9.

A different type of analytical application uses $Ru(bpy)_3^{2+}$ or other ECL active emitter as a label (or tag), usually on a biologically interesting molecule such as an antibody (replacing radioactive or fluorescent labels). The sensitivity of $Ru(bpy)_3^{2+}$ ECL determinations is high, down to picomolar levels, and the linearity of response with concentration is good [43]. ECL has an advantage over fluorescent labels in that it does not require a light source and is free from effects of scattered light and fluorescent impurities in the sample. ECL has thus been used commercially for immunoassay and DNA analysis [44,45], as discussed in Chapter 8.

G. Other Applications

From the earliest days of ECL there was interest in using this phenomenon for lighting, displays, lasers, and up-conversion devices. Work in this area was carried out at several laboratories, including Bell Canada-Northern Electric, Westinghouse, Battelle-Geneva, and Philips. Much of this work was motivated by the possibility of producing an active (light-emitting) display, because it predated the appearance of solid-state LEDs. This early work did not lead to commercial devices, perhaps because the operating life was not adequate and also it was difficult at the time to encapsulate liquids. However, ECL was also observed in polymer films, such as a film of polymerized $Ru(bpy)_3^{2+}$ and poly(vinyl-9,10-diphenylanthracene [46,47]. Polymeric and solid-state light-emitting electrochemical cells (LECs) have been investigated in recent years and are discussed in Chapter 10.

III. ELECTROCHEMISTRY FUNDAMENTALS FOR ECL

The following is a brief overview of the electrochemical concepts needed to understand the basics of ECL. For more complete treatments of electrochemical methods, a number of textbooks and monographs are available [48].

A. Electrode Potential

The standard potential, $E°$ $(A, A^{-\bullet})$, of the electrode reaction

$$A + e \rightleftharpoons A^{-\bullet} \tag{16}$$

is measured with respect to a given reference electrode, based on a redox pair O/R

$$O + e \rightleftharpoons R \tag{17}$$

and represents the Gibbs free energy, $\Delta G°$, of the reaction

$$A + R \rightleftharpoons A^{-\bullet} + O \tag{18}$$

For aqueous solutions, the standard reference electrode is the H^+, H_2 couple, with all reactants at unit activity. For practical purposes, other reference electrodes such as AgCl, Ag, or the saturated calomel electrode (SCE) are usually employed [48]. For nonaqueous solvents such as MeCN and DMF, the reference redox couple is usually the ferrocene (Fc), ferrocenium (Fc^+) couple. When measuring the difference of potential between two redox couples, as discussed below, the nature of the reference electrode is unimportant (as long as

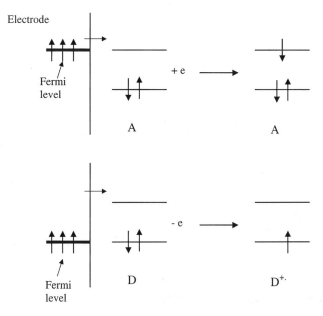

Figure 3 Molecular orbital representation of reduction and oxidation.

its potential remains constant), so that a quasi reference electrode, such as a silver wire immersed in the solution, can sometimes be used.

The potential of the couple represents the energy needed at equilibrium to add an electron to A or to remove an electron from $A^{-\bullet}$. The more negative the potential, the more difficult it is to reduce A or the easier it is to oxidize $A^{-\bullet}$. A molecular orbital (MO) picture of the process is shown in Figure 3, where one can identify the energy as that needed to place an electron in the lowest unoccupied MO (LUMO) of A.

Similarly, the standard potential, $E°(D^{+\bullet},D)$, of the electrode reaction

$$D^{+\bullet} + e \rightleftharpoons D \tag{19}$$

represents $\Delta G°$ for the reaction

$$D^{+\bullet} + R \rightleftharpoons D + O \tag{20}$$

with respect to the reference electrode O, R. The potential represents the energy needed to remove an electron from D or to add an electron to $D^{+\bullet}$. The more positive the potential, the more difficult it is to oxidize D or the easier it is to reduce $D^{+\bullet}$. The MO picture of this oxidation shown in Figure 3 identifies this potential as that needed to remove an electron from the highest occupied MO (HOMO) of D.

Of interest in ECL is the $\Delta G°$ of the reaction

$$D^{+\bullet} + A^{-\bullet} \rightleftharpoons D + A \tag{21}$$

which is given by

$$\Delta G° = -FE°_{rxn} \tag{22}$$

$$E°_{rxn} = E° (D^{+\bullet},D) - E° (A, A^{-\bullet}) \tag{23}$$

The reaction entropy, $\Delta S°$, can be calculated from the temperature coefficient of the reaction potential:

$$\Delta S° = F \left(\frac{\partial E°_{rxn}}{\partial T} \right)_P \tag{24}$$

For ECL reactions, $\Delta G°$ is typically in the range of 2–3 eV (193–289 kJ/mol) and $T \Delta S°$ is small, ≤ 0.1 eV.

B. Cylic Voltammetry (Potential Sweeps)

As a prelude to ECL studies, one usually investigates the electrochemical characteristics of the system, and cyclic voltammetry (CV) is a powerful tool for a fast overview [48]. The experimental setup for these experiments is discussed in Chapter 3. Typically one obtains a current–potential (i vs. E) curve in a solution of the compound(s) of interest in a solvent and supporting electrolyte of choice at an inert electrode, typically Pt, as the potential of the electrode is swept with time at a scan rate v. The resulting voltammogram shows characteristic peaks showing the reduction and oxidation processes in the system (Fig. 4). In a well-behaved system, i.e., one where the electron transfer reactions at the electrode are rapid (nernstian systems) and the products of the reactions, e.g., the ion radicals, are stable, an analysis of the waves yields the $E°$ values. Typically, for a one-electron reaction ($n = 1$) where the reactant and products have essentially the same diffusion coefficients (D),

$$E° = E_p \pm 1.109 \frac{RT}{F} = E_p \pm 28.5 \text{ mV at } 25°C. \tag{25}$$

where E_p is the peak potential and the positive sign applies to the cathodic (reduction) process and the negative sign to the anodic one (i.e., $E°$ is slightly smaller in magnitude than E_p. One can also obtain the D value (in square centimeter per second) from the peak current i_p (in amperes), when the concentration of reactant, C^* (in moles per cubic centimeter), is known, from the equation

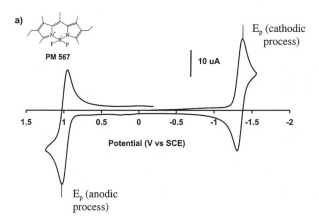

Figure 4 Typical cyclic voltammogram for a species (PM 567 – structure shown) that involves formation of stable reduced and oxidized forms. (From Ref. 66.)

$$i_p = (2.69 \times 10^5) \, ADC^* v^{1/2} \tag{26}$$

where A is the electrode area (cm^2) and v is in volts per second. The stability of the products can be assessed from the CV waves. The existence of waves on scan reversal demonstrates that the product, e.g., A$^{-\bullet}$, formed by reduction of A, is reoxidized on the reverse scan. If A$^{-\bullet}$ were unstable in the time it takes to traverse the wave at v, the reverse wave would be absent (and the cathodic wave would be shifted to less negative potentials).

C. Potential Steps

During ECL experiments, the potential is often stepped (rather than swept) between potentials where the redox processes occur. A first potential step from a potential where no reaction occurs to one well beyond E_p, e.g., where reaction (1) occurs, produces a current i that decays with time t and follows the Cottrell equation,

$$i = FAC^* D^{1/2} / (\pi t)^{1/2} \tag{27}$$

The current decay with $t^{-1/2}$ represents a growing diffusion layer of formed product, A$^{-\bullet}$. A step toward positive potentials to the region where reaction (2) occurs is more complicated, because it involves oxidation of both A$^{-\bullet}$ and D and the occurrence of reaction (3), and has been treated by digital simulation [31].

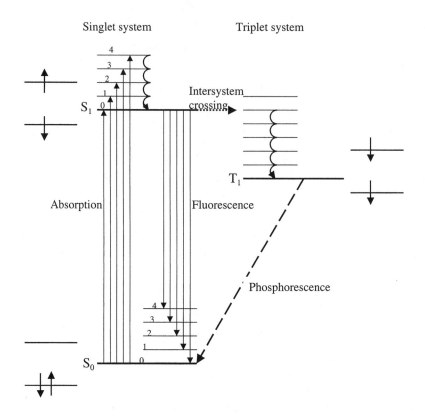

Figure 5 Representation of energy levels and molecular orbitals during the absorption and emission of radiation.

IV. SPECTROSCOPY/PHOTOCHEMISTRY FUNDAMENTALS FOR ECL

The following is a brief overview of some of the principles needed to understand the luminescent processes of importance in ECL. A number of monographs cover these in detail [49,50].

A. Energy Levels

One usually represents the various energy states available to a molecule by an energy level diagram like that in Figure 5, which shows the electronic and vibrational (but not rotational) states. The electronic energy levels can be related

c)

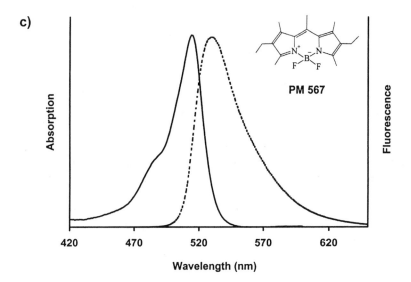

Figure 6 The absorption (solid line) and emission (dashed line) spectra of PM 567. (From Ref. 66)

to the molecular orbitals (HOMO and LUMO) of the molecule. Shown in this diagram is the lowest, singlet, electronic state of the molecule (the ground state), S_0, where the electrons are paired in the same HOMO. The multiplicity, singlet or triplet, of a state is given by $2s + 1$, where s is the spin quantum number, so when $s = 0$ (electrons paired), the multiplicity is 1. The first excited singlet state, S_1, and the lowest triplet state, T_1 (with two unpaired electrons, $s = 1$), are also shown on the diagram. Not shown are higher electronic excited states, S_2, T_2, ..., that are usually not of importance in ECL processes. The vibrational levels for each of the states are numbered $v = 0, 1, 2,....$

It is of interest to consider the lifetime of a molecule in different states. The lifetime of an excited vibrational state in solution is in the picosecond regime, because energy can be easily transferred to the solvent. Thus excited vibrational states, as illustrated in the S_1 level, rapidly decay to the $v = 0$ vibrational level. The radiative lifetime of the excited singlet state, S_1, for organic molecules is typically of the order of nano seconds, but that of a triplet state is much longer, often in the millisecond to second regime, because radiational transitions to the ground S_0 state are "forbidden." However, the radiative lifetimes of triplet states of metal chelates are much shorter, because the presence of the heavy nucleus can promote spin–orbit coupling. Thus the triplet charge transfer excited state of $Ru(bpy)_3^{2+}$ emits with a lifetime of less than 1 μs.

B. Photoluminescence

Emission of light results from transitions between excited states and the ground states. The absorption from S_0 to S_1 involves transitions from the ground vibrational state, $v = 0$, of S_0 to various vibrational levels in S_1 (Fig. 5), resulting in the observed absorption band (Fig. 6). Emission occurs from the $v = 0$ state of S_1 to the different vibrational levels of S_0 (Fig. 5), resulting in the emission band (Fig. 6). The energies of the emission band are lower (shifted to longer wavelengths or red-shifted) compared to the absorption band as shown in Figure 6. There is a difference in energy between the absorption and emission bands for the transition between the $v = 0$ states, the so-called 0,0 band. This difference, known as the Stokes shift, is the result of the optical transitions being essentially instantaneous ($\sim 10^{-15}$ s), so that no structural or solvation changes occur during the transition. Thus during excitation, the transition is from a ground-state structure to an excited state structure of the same configuration. However, the excited state, because of the different electronic configuration, can undergo some structural changes, for example, reorientation of solvent, to relax to a lower energy state. Emission from this state then occurs to a ground state that has the configuration of the excited state (which then relaxes to the original ground state). The energy of the singlet transition can be estimated from the spectra, for example, as the average of the energies of the 0,0 bands for excitation and emission, if they are discernible, or sometimes from the wavelength, λ where the excitation and emission bands cross, by the formula

$$E_S \text{ (in eV)} = \frac{1239.81}{\lambda \text{ (in nm)}} \tag{28}$$

Emission from the triplet state results in phosphorescence. This is rarely seen for organic molecules in solution, because these molecules are readily quenched by solution species before emission. Oxygen, because it is a triplet molecule, is an effective triplet quencher. Emission from triplet states is found, however, with metal chelates, which have much shorter emission lifetimes than organic molecules. The triplet states can produce excited singlets, a type of energy up-conversion, by triplet–triplet annihilation (TTA), Eq. (6). This is observed spectroscopically as delayed fluorescence. In such an experiment a species is excited by pulse irradiation to the S_1 state, but observation of the emission is delayed until all of the S_1 states have decayed either by radiating to the ground state or by radiationless processes such as intersystem crossing to the T_1 state. Observation of S_1 emission after such a delay results from conversion of the triplets to excited singlets. Note that in spectroscopy some singlet states, for example, 9,10-diphenylanthracene, have close to 100%

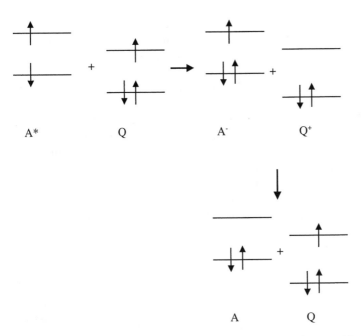

Figure 7   Molecular orbital representation of electron transfer quenching.

efficiency for radiation to the ground state, with very few triplets formed by intersystem crossing. In this case delayed fluorescence is not observed. However, with such systems, triplets can be formed in ECL by adjustment of the energetics of the annihilation reaction, where TTA is then observed. Thus states are sometimes accessible in ECL that are not available spectroscopically.

C. Quenching

An excited state can be quenched by another molecule (a quencher, Q) to produce the ground state, i.e.,

$$D^* + Q \rightarrow D + Q^* \tag{27}$$

where Q^* can often decay to the ground state without emission. Quenching can occur by either energy transfer or electron transfer. Energy transfer, sometimes called Förster transfer, is favored by the electronic energy of D^* being greater than that of Q^* and a large overlap of the emission band of D^* with the absorption band of Q. The energy transfer occurs by a direct electrodynamic interaction between D^* and Q and occurs at short distances between the reactants

[49]. Electron transfer quenching occurs because the excited state is easier to oxidize and easier to reduce than the corresponding ground state of the same molecule, by an amount essentially equal to the excitation energy of the molecule. Thus, if the reduction of A, Eq. (16), occurs with $E°$ (A,A$^{-•}$), the reduction of A* occurs at $E°$ (A,A$^{-•}$) + $E*$, where $E*$ is the excitation energy of A. For example, if the ground state is reduced at -1.0 V and the excitation energy is 2.5 eV, the excited state would be reduced at $+1.5$ V. This can be rationalized by the greater ease of introducing an electron into the unfilled HOMO of the excited state rather than into the LUMO of the ground state. Similarly, oxidation of the excited state occurs at $E°$ (D$^{+•}$, D) $-$ $E*$. Thus, excited states can be quenched rather easily by an electron transfer process, as shown in Figure 7, so that the radical ions are effective quenchers. The kinetics of a quenching reaction, Eq. (27), is governed by the Stern–Volmer equation

$$\frac{\phi^0}{\phi-1} = \frac{R^0}{R-1} = \kappa_Q \tau_0 [Q] \tag{28}$$

where ϕ^0 and ϕ are the fluorescence efficiencies and R^0 and R are the fluorescence responses in the absence and presence of a quencher at concentration [Q], respectively, k_Q is the rate constant for quenching, and τ_0 is the lifetime of the excited state in the absence of a quencher.

Metal electrodes can similarly act as quenchers by either energy transfer or electron transfer processes. The energy transfer mode is analogous to Förster transfer and results from coupling of the oscillating field of the excited dipole to the surface plasmon modes of the metal [51,52]. Quenching by electron transfer follows the same arguments as those given above for redox quenching by a solution species [53]. The effective quenching by metals has been invoked as an argument against direct formation of excited states at metal electrode surfaces, e.g., by reduction of a cation radical at a sufficiently negative potential. Semiconductor electrodes are less effective quenchers than metals, so direct production of excited states is possible at these (Chapter 11).

V. FUNDAMENTALS OF ECL

The main ideas and reactions in ECL have been described in the preceding sections, e.g., in Eqs. (1)–(8). The basic requirements for efficient annihilation ECL to occur are (1) stable radical ions of the precursor molecules in the electrolyte of interest (as seen in the CV response), (2) good photoluminescence efficiency of a product of the electron transfer reaction, and (3) sufficient energy in the electron transfer reaction to produce the excited state. When these criteria are met, ECL will usually be observed, although the efficiency often depends upon details of the kinetics of the reaction (Chapter 4).

A. Energy Requirements

The energy of the electron transfer reaction governs which excited states are produced. Because the excitation energy is fundamentally closest to a thermodynamic internal energy, the usual energy criterion for production of an excited singlet state of energy E_S (in electronvolts) is

$$-\Delta H^0 = E° (D^{+•},D) - E° (A,A^{-•}) - T \Delta S^0 > E_S \qquad (29)$$

The value of $T \Delta S^0$ is usually estimated as 0.1 (± 0.1) eV [54–60]. Frequently the criterion is given based on CV peak potentials (in volts) at $T = 298$ K as [54]

$$E_p (D^{+•},D) - E_p (A,A^{-•}) - 0.16 > E_S \qquad (30)$$

[see Eq. (25)]. Such reactions are sometimes called "energy sufficient," and the reaction is said to follow the S route.

When $-\Delta H^0 > E_T$, the reaction energy can produce the triplet state at energy E_T, with excited singlet production occurring through TTA. Such reactions are said to follow the T route. Sometimes, both excited singlets and triplets are formed in significant amounts in the annihilation, in a reaction said to follow the ST route. When the reaction produces excimers or exciplexes, Eqs. (8) and (9), it is said to proceed by the E route.

B. Coreactants

Establishing the energetics of ECL reactions that occur with coreactants is more difficult, because the energies of the reactive intermediates, such as $CO_2^{-•}$, may not be known. However the fundamental principles are the same as with annihilation reactions in terms of the energy of the electron transfer step, e.g., Eq. (15). A number of coreactants for ECL have been investigated. For example, tri-n-propylamine (TPrA) has been used in the oxidative ECL of $Ru(bpy)_3^{2+}$ [41,42]. The proposed mechanism of this species involves ultimate generation of a radical that acts as the reductant with the electrogenerated $Ru(bpy)_3^{2+}$.

$$Ru(bpy)_3^{2+} - e^- \rightarrow Ru(bpy)_3^{3+} \qquad (31)$$

$$Pr_2NCH_2CH_2CH_3 - e^- \rightarrow Pr_2NCH_2CH_2CH_3^{+} \qquad (32)$$

$$Pr_2NCH_2CH_2CH_3^{+•} \rightarrow Pr_2NC^•HCH_2CH_3 + H^+ \qquad (33)$$

$$Ru(bpy)_3^{3+} + Pr_2NC^•HCH_2CH_3 \rightarrow Ru(bpy)_3^{2+*}$$
$$+ Pr_2NC^+H_2CH_2CH_3 \qquad (34)$$

The oxidation of TPrA can occur by reaction directly at the electrode as shown in Eq. (32) or by reaction with $Ru(bpy)_3^{3+}$. The overall path of the TPrA coreactant system depends upon the electrode material and solution concentrations and can be quite complicated, involving the generation of oxidants as well as reductants (See Chapter 5) [61].

Species that form strong oxidants upon reduction can also serve as coreactants. For example, the reduction of $S_2O_8^{2-}$ proceeds as follows:

$$S_2O_8^{2-} + e \rightarrow SO_4^{2-} + SO_4^{-\bullet} \tag{35}$$

where $SO_4^{-\bullet}$ is a strong oxidant that can react with $Ru(bpy)_3^+$ or anion radicals to produce the excited state [62].

C. Monolayers and Films

In addition to studies of solution-phase species, ECL has also been observed in monolayers on electrode surfaces and in various types of films. For example, a monolayer of a surfactant derivative of $Ru(bpy)_3^{2+}$ (i.e., one containing a C_{18} chain linked through an amide link to a bipyridine) on Pt, Au, or ITO electrodes showed ECL when oxidized in the presence of oxalate as a coreactant [63]. Similarly, a film of a $Ru(bpy)_3^{2+}$ surfactant can be formed at the air/water interface and contacted (transferred) by the horizontal touch method with an ITO or HOPG electrode and the ECL observed on the Langmuir trough [64,65]. ECL active layers on the surface of beads or electrodes form the basis of the ECL immunoassays discussed in Chapter 8. Films of polymers or solids can also show ECL and are the basis of light-emitting devices. This work is discussed in Chapter 10.

REFERENCES

1. Cruser, S.A., Bard, A.J. Anal. Lett. **1967**, *1*, 11.
2. Faulkner, L.R., Bard, A.J. J. Am. Chem. Soc. **1968**, *90*, 6284.
3. Kuwana, T. In *Electroanalytical Chemistry*, Bard, A.J., Ed.; New York: Marcel Dekker, 1966; Vol. 1. p. 197.
4. Bard, A.J., Santhanam, K.S.V., Cruser, S.A., Faulkner, L.R. In *Fluorescence*, Guilbault, G.G. Ed.; Marcel Dekker: New York, 1967; p. 627
5. Cormier, M.J., Hercules, D.M., Lee, J. Eds., *Chemiluminescence and Bioluminescence*, Plenum Press: New York, 1973.
6. Vojir, V. Coll. Czech. Chem. Commun. **1954**, *19*, 862.
7. Kuwana, T., Epstein, B., Seo, E.T. J. Phys. Chem., **1963**, *67*, 2243.
8. Bowling, R.J., McCreery, R.L., Pharr, C.M., Engstrom, R.C. Anal. Chem. **1989**, *61*, 2763.

9. Hickling, A. Ingram, M.D. Trans. Faraday Soc. **1964**, *60*, 783.
10. Hickling, A. Ingram, M.D. J. Electroanal. Chem. **1964**, *8*, 65.
11. Ivey, F. *Electroluminescence and Related Effects*; Academic Press: New York, 1963.
12. Ouyang, J., Fan, F.R.F., Bard, A.J. J. Electrochem. Soc. **1989**, *136*, 1033.
13. Kolthoff, I.M., Lingane, J.J. *Polarography*, 2nd ed., Interscience: New York, 1952; p. 634.
14. Piersma, B.J. Gileadi, E. Mod. Aspects Electrochem. **1966** *4*, 47.
15. (a) Kolthoff, I.M., Coetzee, J.F. J. Am. Chem. Soc. **1957**, *79*, 870; (b) Popov, A.I., Geske, D.H. J. Am. Chem. Soc. **1958**, *80*, 1340.
16. Lund, H. Acta Chem. Scand. **1957**, *11*, 1323.
17. (a) Bard, A.J., Santhanam, K.S.V., Maloy, J.T. Phelps, J., Wheeler, L.O. Disc. Faraday Soc. **1968**, *45*, 167; (b) Peover, M.E. *Electroanalytical Chemistry*, Bard, A.J. Ed.; Marcel Dekker: New York, 1967; Vol. 2, p. 1; (c) Adams, R.N. *Electrochemistry at Solid Electrodes*, Marcel Dekker: New York, 1969; p. 309.
18. (a) Geske, D.H., Maki, A.H. J. Am. Chem. Soc. **1960**, *82*, 2671; (b) Wheeler, L.O., Santhanam, K.S.V., Bard, A.J. J. Phys. Chem. **1967**, *71*, 2223.
19. Hercules, D.M. In *Physical Methods of Chemistry*, Weissberger, A., Rossiter, B., Eds.; Wiley: New York, 1971; Vol. 1, Part IIB, p. 257.
20. Chandross, E.A. Trans. NY Acad. Sci. Ser. II, **1969**, *31*, 571.
21. Zweig, A. Adv. Photochem. **1968**, *6*, 425.
22. Chandross, E.A., Sonntag, F.I. J. Am. Chem. Soc. **1964**, *86*, 3179.
23. Faulkner, L.R., Bard, A.J. J. Am. Chem. Soc. **1968**, *90*, 6284.
24. Hercules, D.M. Science. **1964**, *145*, 808.
25. Visco, R.E., Chandross, E.A. J. Am. Chem. Soc. **1964**, *86*, 5350.
26. Santhanam, K.S.V., Bard, A.J. J. Am. Chem. Soc. **1965**, *87* 139.
27. Maricle, D.L., Rauhut, M.M. US Patent 3,654,525, April 4, 1972.
28. Chandross, E.A., Visco, R.E. US Patent 3,319,132, May 9, 1967.
29. Faulkner, L.R., Bard, A.J. J. Am. Chem. Soc. **1969**, *91*, 209.
30. Marcus, R.A. J. Chem. Phys. **1965**, *43*, 2654.
31. Feldberg, S.W. J. Am. Chem. Soc. **1966**, *88*, 390; J. Phys. Chem. **1966**, *70*, 3928.
32. Maloy, J.T., Prater, K.B., Bard, A.J. J. Am. Chem. Soc. **1971**, *93*, 5959.
33. Maloy, J.T., Bard, A.J. J. Am. Chem. Soc. **1971**, *93*, 5968.
34. Collinson, M.M., Wightman, R.M. Anal. Chem. **1993**, *65*, 2576.
35. Paris, J.P., Brandt, W.W. J. Am. Chem. Soc. **1959**, *81*, 5001.
36. Crosby, G.A. Acc. Chem. Res. **1975**, *8*, 231.
37. Lytle, F.E., Hercules, D.M. J. Am. Chem. Soc. **1966**, *88*, 4745.
38. Tokel, N.E., Bard, A.J. J. Am. Chem. Soc. **1972**, *94*, 2862.
39. Chang, M. Saji, T. Bard, A.J. J. Am. Chem. Soc. **1977**, *99* 5399.
40. Rubinstein, I. Bard, A.J. J. Am. Chem. Soc. **1981**, *103*, 512.
41. Noffsinger, J.B., Danielson, N.D. Anal. Chem. **1987**, *59*, 865.
42. Leland, J.K., Powell, M.J. J. Electrochem. Soc. **1990**, *137*, 3127.
43. Ege, D., Becker, W.G., Bard, A.J. Anal. Chem. **1984**, *56*, 2413.
44. Bard, A.J., Whitesides, G.M. US Patent 5,221,605, June 22, 1993.
45. Blackburn, G.F. Shah, H.P., Kenten, J.H. Leland, J. Kamin, R.A. Link, J. Peteran, J. Powell, M.J. Shah, A. Talley, D.B., Tyuagi, S.K., Wilkins, E., Wu, T.G., Massey, R.J. Clin. Chem. **1991**, *37*, 1534.

46. Abruña, H.D., Bard, A.J. J. Am. Chem. Soc. **1982**, *104*, 2641.
47. Fan, F.R., Mau, A. Bard, A.J. Chem. Phys. Lett. **1985**, *116*, 400.
48. Bard, A.J., Faulkner, L.R. *Electochemical Methods*; Wiley: New York, 2001 and references therein (pp. 39–41).
49. Parker, C.A. *Photoluminescence of Solutions*, Elsevier: Amsterdam, 1968.
50. Birks, J.B. *Photophysics of Aromatic Molecules*, Wiley-Interscience: New York, 1970.
51. Kuhn, H.J. Chem. Phys. **1970**, *53*, 101.
52. Chance, R.R. Prock, A. Silbey, R. J. Chem. Phys. **1975**, *62*, 2245.
53. Chandross, E.A., Visco, R.E. J. Phys Chem. **1968**, *72*, 378.
54. Faulkner, L.R., Bard, A.J. In *Electroanalytical Chemistry*, Bard, A.J., Ed.; Marcel Dekker: New York, Vol. 10, p. 1.
55. Weller, A., Zachariasse, K. J. Chem. Phys. **1967**, *46*, 4984.
56. Faulkner, L.R., Tachikawa, H., Bard, A.J. J. Am. Chem. Soc. **1972**, *94*, 691.
57. Hoytink, G.J. Disc. Faraday Soc. **1968**, *45*, 14.
58. Van Duyne, R.P., Reilley, C.N. Anal. Chem. **1972**, *44*, 142.
59. Visco, R.E., Chandross, E.A. Electrochim. Acta, **1968**, *13*, 1187.
60. Pighin, A. Can. J. Chem. **1973**, *51*, 3467.
61. Miao, W., Choi, J−P., Bard, A.J. J. Am. Chem. Soc. **2002**, *124*, 14478.
62. White, H.S., Bard, A.J. J. Am. Chem. Soc. **1982**, *104*, 6891.
63. Zhang, X., Bard, A.J. J. Phys. Chem. **1988**, *92*, 5566.
64. Miller, C.J., McCord, P., Bard, A.J., Langmuir **1991**, *7*, 2781.
65. Miller, C.J., Bard, A.J. Anal. Chem. **1991**, *63*, 1707.
66. Lai, R.Y., Bard, A.J. J. Phys. Chem. B **2003**, *107*, 5036.

2
Experimental Techniques of Electrogenerated Chemiluminescence

Fu-Ren F. Fan
The University of Texas at Austin, Austin, Texas, U.S.A.

I. INTRODUCTION

This chapter discusses experimental techniques used in electrogenerated chemiluminescence (ECL). It merely introduces some mechanistic pictures and chemical aspects required for illustration, so that the goal and the nature of experimentation can be appreciated. The topic has been extensively studied and thoroughly reviewed [1–3], and there are also a few commercial instruments available on the market, (e.g., IGEN, Boehringer, etc.; also see the discussion in Chapter 8), so we mainly focus on some of the experimental aspects encountered in conventional ECL measurements. Readers who are interested in detailed discussion on the early works are encouraged to refer to Refs. 1–3. Also included are some combined techniques, including magnetic field effects, interception techniques, and an up-to-date account of the newly developed combined techniques, thus, interested individual investigators may appreciate the design of ECL-related techniques for their specific applications.

II. APPARATUS AND TECHNIQUES

Several questions are often raised in ECL experiments: What is the efficiency of the luminescence emission? How efficiently does excitation occur in charge transfer? How do efficiencies change from system to system or from one medium to another? Where do key energy losses occur, and how can they be circumvented? Quantitative questions like these require careful and quantitative experimental approaches. Much of the difficulty in experimentation arises from the fact that one can rarely find a solvent in which both anion and cation radicals of any system are completely stable even over a short time domain (e.g., a few minutes). Anions are generally stable in DMF, but cations are often less stable; the opposite is true for acetonitrile. One thus must seek techniques that permit generation and reaction of the radicals on a time scale that is short compared to ion decay times. Electrochemical (EC) techniques are frequently used because they allow sequential or simultaneous generation of reactants on time scales ranging from nanoseconds to hours and because theoretical treatment of diffusive or convective transport permits analysis of reaction rates. In addition, careful control of the generating electrode's potential gives excellent selectivity in ion production compared to other techniques. In a sense, ECL is similar to a photoexcitation (e.g., fluorescence) method; however, it has the advantage that a light source is not used, so that scattered light and interferences from emission by impurities are not problems. The price for electrochemical control and versatility is the acceptance of complex reactant distributions and, sometimes, low light levels. However, with currently available techniques for the detection of low light intensity on the time scale of nanoseconds (e.g., by single-photon counting methods) and powerful digital simulation methods, these difficulties seem not intractable, and electrochemical methods have been established as valuable quantitative, as well as qualitative, tools.

A. Electrochemical Media and Cell Design

The solvent-supporting electrolyte systems should be chosen for their lack of reactivity with the electrogenerated species, their wide potential limits, their high solubility for the compound to be studied, and their good conductivities. Typical systems used and the approximated useful potential ranges are shown in Table 1. The purity of the medium is important in determining both the quality of the electrolytic background and the stabilities of the ECL reactants, hence special attention must be paid to it.

The usual supporting electrolytes, TBAP or $TBABF_4$, can be purchased from SACHEM as "electrometric grade" reagent. They can be used as received, but recrystallization [4] is sometimes required. Both electrolytes are hygroscopic and must be dried in vacuo for 24–48 h at 90–100°C. The dried lots should be kept over P_2O_5 or $Mg(ClO_4)_2$ in a desiccator or in a dry box (e.g., Vacuum Atmosphere Corp.).

Table 1 Solvent-Supporting Electrolyte Systems Used in ECL[a]

Solvent	Supporting electrolyte[b]	Potential range[c] (V vs. SCE)	Remarks[d]
Acetonitrile TBAP, TEAP, TBAPF$_6$	TBABF$_4$,	−2.8 to +2.8	Good stability for both radical anions and cations; potential range strongly depends upon purification
Benzonitrile	TBABF$_4$	−1.8 to +2.5	Similar to acetonitrile in terms of ion stabilities; commercial spectrograde solvent can be used without purification
N,N-Dimethyl-formamide	TBAP	−2.8 to +1.5	Good stability for radical anions; poor for cations; difficult to purify and tends to decompose or hydrolyze on standing
Dimethyl-sulfoxide	TEAP	−1.8 to +0.9	Purified by vacuum distillation, collecting the middle 60%; can be stored on molecular sieves; limited positive potential range
Methylene chloride	TBAP	−1.7 to +1.8	Excellent stability of cations; limited neg. potential range; easily purified; quite resistive
Propylene carbonate	TEAP	−2.5 to +2.0	Purified by reduced pressure distillation at 120–130°C, collecting the middle 60% fraction; potential range depends greatly on purity; good stability for radical cation
Tetrahyd-rofuran	TBAP	−3.0 to +1.4	Excellent stability of anions; easily purified and dried with alkali metals; limited positive potential range; quite resistive

Table 1   Continued

Solvent	Supporting electrolyte[b]	Potential range[c] (V vs. SCE)	Remarks[d]
Acetonitrile–benzene mixed solvent[d]	TBABF$_4$, TBAP, TBAPF$_6$	−2.3 to +2.0	Better solubility for some aromatic compounds such as BPQ-PTZ; quite resistive, depending on the ratio of benzene to acetonitrile

Source: From Refs. 1 and 6, See also RN Adams. Electrochemistry at Solid Electrodes. New York: Marcel Dekker, 1969, pp. 19–37; DT Sawyer, A Sobkowiak, JL Roberts, Jr. Electrochemistry for Chemists. 2nd ed., New York:Wiley, 1995; AJ Fry. Solvents and Supporting Electrolytes. In: PT Kissinger, WR Heineman, eds. Laboratory Techniques in Electroanalytical Chemistry. 2nd ed., New York: Marcel Dekker, 1996, pp. 469–485.
[a]Other possible solvents include 1,2-dimethoxyethane and hexamethylphosphoramide.
[b]Abbreviations: TBABF$_4$, tetra-n-butylammonium fluoborate; TBAP, tetra-n-butylammonium perchlorate; TBAPF$_6$, tetra-n-butylammonium hexafluorophosphate; TEAP, tetraethylammonium perchlorate; BPQ-PTZ, 3,7-[bis[4-phenyl–2-quinolyl]]–10-methylphenolthiazine.
[c]At a Pt working electrode.
[d]A benzene/acetonitrile ratio of 1.5:1 to 4:1 has been used. See, e.g., Ref. 13.

Excellent purification procedures are described in the literature for dimethylformamide (DMF) [5–7] and acetonitrile (ACN) [6,8]. Purified ACN is frequently kept in a flask that can be attached to a vacuum line (with a pressure of $<10^{-5}$ torr), and from this apparatus it is distilled through the line as needed. Usually, it is first collected in an auxiliary vessel containing fresh P$_2$O$_5$. The solvent is further dried and degassed by two vacuum distillations over P$_2$O$_5$ and finally stored over Super I Woelm alumina N (ICN Biomedicals, Eschwege, Germany). After purification under argon or helium DMF is kept in a special flask having an outlet line that extends from the bottom of the flask through a stopcock located well above the solvent level. When needed, it can be dispensed through the spigot by slightly pressurizing the flask with the inert gas on a vacuum line. If possible, greaseless fittings should be used in all facilities handling solvents. Tetrahydrofuran (THF) can be treated as described by Mann [6], or it can be stored in CaH$_2$ and distilled as needed through a vacuum line to a secondary vessel containing a fresh sodium surface. It is left in contact with the alkali metal for several hours before being redistilled into the cell. In handling THF, care must always be taken to prevent the formation of explosive peroxides.

Mann [6] provides a general reference for more information on these and other electrochemical solvents.

Anhydrous high-purity (spectrophotometric grade) solvents listed in Table 1 are currently commercially available (e.g., Burdick and Jackson; Aldrich;

Matheson, Coleman, and Bell) and are satisfactorily used as received. These solvents should be isolated from the atmosphere during storage, and all solutions should be prepared inside an oxygen-free dry box containing helium and sealed in an appropriate airtight cell for measurements completed outside the dry box. Because of the long-term, continuous removal of water and oxygen in such a system, it is possible that the dry box yields superior materials from the standpoint of dryness. Frank and Park [9] provide general information on the experimental procedures about electrochemistry in a dry box.

However, for vigorous ECL studies, vacuum systems probably offer the best compromise between convenience, flexibility, and effectiveness in handling materials. Deoxygenation is rigorously accomplished by three or four freeze–pump–thaw cycles, and the cell can be protectively filled with argon or helium afterward. This procedure results in negligible loss of solvents for nonvolatile DMF and benzonitrile, but care is needed for volatile solvents such as acetonitrile and tetrahydrofuran. If the cell is not thermally fragile, deaeration is best carried out in situ; otherwise, auxiliary vessels designed particularly to accomplish this operation are required. One common design for this purpose is shown in Figure 1 [10]. This particular device allowed for the solvent in a storage flask to be transferred to a tube where the solution could be prepared and deaerated via several freeze–pump–thaw cycles. In the last pumping period, with the solution frozen, the stopcock to the cell was opened so that it was evacuated to about 10^{-5} torr. The system was then filled with about 0.5 atm of inert gas and the cell stopcock was closed. When the sample solution had melted, the auxiliary vessel was pressurized and inverted. The cell then filled with solution when its stopcock was reopened. A cell incorporated with this device was the one designed by Tachikawa (see Ref. 10) for use in studying magnetic field effects on ECL (Figure. 2). This cell was very fragile and required the auxiliary vessel for vacuum degassing. Light from the working electrode was observed through the window in the large Pt counter electrode (1.5×4.5 cm). Experimentation in the magnetic effect has not encountered special problems with respect to luminescence generation, and most studies have involved transient techniques in which ECL intensity has been measured by the heights of single pulses. The cell was designed to fit into the sample compartment of a special phosphorimeter that was expressly intended for the study of samples immersed in a magnetic field [10]. Hence it specifically dealt with the primary experimental pitfalls, the important effects of stray magnetic fields on PMT operation. Careful shielding of the PMT with foil of high magnetic permeability and removal of the tube to about 30 in. from the field are requisites. Therefore attention must be given to the optical coupling between sample and detector, so that maximum collection efficiency of ECL intensity can be obtained. Concerning the application of vacuum line techniques in electrochemistry, Katovic et al. [11] supply an excellent chapter.

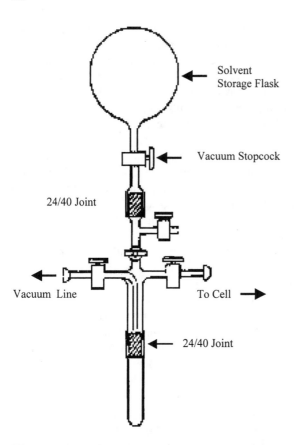

Solvent
Storage Flask

Vacuum Stopcock

24/40 Joint

Vacuum Line

To Cell

24/40 Joint

Figure 1 Auxiliary vessel for sample preparation outside the electrochemical cell. (Adapted from Ref. 10.)

Bubbling of the sample solutions with an inert gas is a long-established technique for deaeration and works well for aqueous solutions. Oxygen's high solubility in organic solvents renders it more difficult to remove from them than from aqueous media. Long bubbling times are usually required, and one must worry about the cumulative extraction of impurities from the bubbling stream. It is doubtful that gas bubbling ever removes oxygen as completely as vacuum deaeration or operation in a dry box. It has the advantage of simplicity, but careful scrubbing and deoxygenation of the gas is mandatory.

Several factors besides the constraints imposed by photometric apparatus must be considered in the design of electrochemical cells for ECL studies. An important one is the proposed method of deoxygenation, because this choice affects the selection of materials. Metal-to-glass seals that are

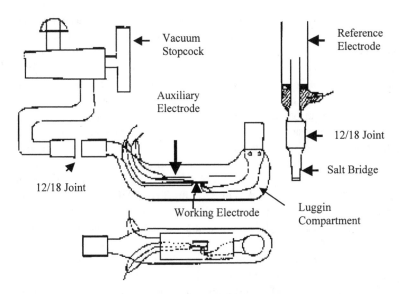

Figure 2 Cell design by Tachikawa for studies of magnetic field effects on ECL. (Adapted from Ref. 10.)

usually used in vacuum lines, for example, are difficult to cool routinely to liquid nitrogen temperatures without breakage in freeze–pump–thaw cycles. If parts of the cell must have such seals or other thermally fragile mountings, the solution must be vacuum deaerated outside the cell and transferred to it afterward, or the fragile parts must be added to the cell when degassing is complete.

Several materials, such as Pt, Au, glassy carbon, indium tin oxide (ITO), and heavily boron-doped diamond thin film, have been used successfully as the working electrode for ECL studies. The size and configuration of the working electrode are usually determined by the purpose of the experiment. If decay curve analysis is to be performed, a disk electrode is required for compatibility with the theoretical assumption, and a small electrode (usually <0.1 cm^2) helps to limit waveform distortion by uncompensated resistance effects. In decay curve analysis or total quantum output measurements, it is vital that the current density be uniform across the face of the working electrode. Nonuniform current distributions are especially common in nonaqueous media, where low electrolyte conductivity causes high electric field gradient in solution. To counter such effects, one often seeks coaxial configurations between working and counter electrodes and separates them by an appreciable distance. For multicycle experiments used in spectral recording, configurations yielding large effective areas are favored to get a high light level for detection.

Luminescence from the counter electrode is another nuisance to overcome. Because the emitting processes occurring there are usually not well defined, such emission can cause interference to spectral studies or pulse-shape analysis. When a large cylindrical counter electrode (e.g., a Pt foil of 1.5 × 4.5 cm) is used, double-layer charging satisfies virtually all its current requirements and negligibly weak luminescence originates there.

The reference electrode used in early ECL studies was an aqueous SCE with an integral aqueous agar salt bridge, which itself contacted the solution through a Luggin tip filled with the nonaqueous electrolyte [10]. The aqueous reference electrode, therefore, did not contaminate the working solution. A reference electrode in the solvent of interest [e.g., Ag/AgNO$_3$ (0.01 M), ACN] contained in a tube closed with a plug of porous Vycor (Corning Type 7930 "thirsty glass") can also be used. In a relatively short-term experiment, one can use a silver wire as a quasi-reference electrode (Ag QRE); its potential needs to be calibrated by adding ferrocene as an internal reference after a series of experiments. Because of the possibility of drift in its potential, a QRE is not suitable for long-term experiments.

One experimental constraint imposed on cell design is to study the temperature dependence of the ECL efficiency. The electrochemical cell employed by Wallace and Bard [12] for the variable-temperature ECL is shown in Figure 3. The cell consists of three sections, the first of which allows the insertion and parallel positioning of the working (a Pt disk sealed in uranium glass; area 0.064 cm^2), auxiliary (a Pt foil, 2.9 × 1.0 cm; total area 5.8 cm^2), and reference (Ag QRE) electrodes in the sample compartment. The second section consists of a 10 mL graduated degassing arm and high-vacuum stopcock with associated joints, allowing complete cell assembly, connection to a vacuum line, and vacuum-tight isolation from the atmosphere. The third section consists of a 5 mL sample compartment integrally sealed into a double-walled evacuated Pyrex Dewar flask (100 mL capacity) that is silvered to minimize light loss through scattering. The Pt disk electrode can be spectroscopically observed directly through the two flat sections of Pyrex glass at the bottom of the Dewar. Within the sample compartment the auxiliary foil electrode is folded in a cylindrical arrangement around the working electrode, with the Ag QRE being positioned between the Pt foil and disk as close to the disk as possible. The resulting geometry promotes even current distribution over the surface of the Pt disk and minimizes the uncompensated solution resistance between the working and reference electrodes. The use of a large-area counter ensured that most of the counter electrode current was non-Faradaic, i.e., was involved in double-layer charging, so no light was generated at the counter electrode. Solutions were prepared by transferring 5 mL of the solvent under vacuum at −30°C into the degassing arm containing the supporting electrolyte and the compound, which had been predried for 12 h at <10^{-5} torr. After two freeze–pump–thaw cycles the solution was transferred directly to the sample compartment. In constructing the cell, sufficient care was

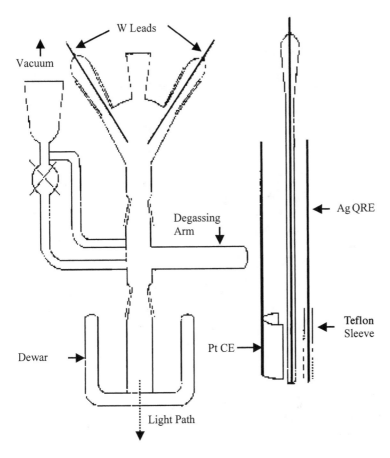

Figure 3   Variable temperature ECL and electrochemical cell. (Adapted from Ref. 12.)

taken that direct contact of the ECL solution with grease (Dow Corning high-vacuum silicone lubricant) was avoided. Temperature variation and stabilization were accomplished either by using a methanol–dry ice mixture or by using a temperature control apparatus equipped with a constant-temperature bath fluid pump that circulated methanol or water through a copper coil heat exchanger in the Dewar of the ECL cell.

The cell used for solution preparation in a dry box has fewer constraints; hence one can be quite flexible in choosing the materials and component design. The group at the University of Texas nowadays employs a conventional three-electrode configuration, airtight vessel. The cell was designed to fit in front of the

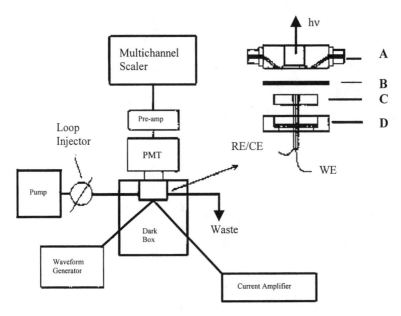

Figure 4 Simplified block diagram of ECL flow cell and equipment (A) Stainless steel cell body housing an optical window; (B) polyethylene spacer; (C) epoxy-encased microelectrode; (D) stainless steel electrode retainer. (Adapted from Ref. 14.)

entrance slit to a CCD spectrometer and has a total volume of 4–5 mL [13]. A metal through Teflon or glass airtight cap (ACE Glass) is sealed with a *Torr-seal* (Varian Associates).

One of the most recently employed ECL cells is the flow cell coupled with a flow-injection analysis (FIA) system [14–19]. For example, the flow cell used by Collinson and Wightman [14], shown in Figure 4, consists of a stainless steel body housing an epoxy-sealed glass, a 150 μm thick polyethylene spacer, and a stainless steel electrode retainer plate. The epoxy-encased microelectrode fits snugly in the electrode retainer plate, which secures to the cell body with stainless steel screws. The volume of the cell defined by the rectangular groove cut into the polyethylene spacer is about 12 μL. Solvent was continuously pumped through the cell via glass-lined, stainless steel tubing (1/16 in. o.d., Alltech) with an ISCO microflow pump at a flow rate of 300 μL/min. An SSI valve equipped with a pneumatic actuator was used to inject approximately 100 μL of the deoxygenated sample into the flow stream. Any emission from the counter electrode was prevented by masking the electrode with black electrical tape placed on the outside of the glass optical window. The system yields reproducible results with multiple injections. Because the compound is in contact with the electrode for only a short period (~ 1 min), impurities have only a very low chance to build up in solution or on the electrode and degrade the ECL response. This type of flow cell has frequently been used in commercial instruments for ECL assays of antibodies, antigen, and DNA, employing magnetic beads [20,21] (for more detailed discussions see Chapters 8 and 9).

The main experimental constraint on cell design for spectral recording is simply that it must be possible to fit the electrochemical cell into the sample compartment of the spectrometer. However, this can sometimes cause a serious obstacle because it sometimes (e.g., when the light level is low) demands the placement of several large electrodes in an uncomfortably small volume. Freed and Faulkner's design [22] (Figure 5) proved very satisfactory in this regard. The device was quite small, yet it featured a large-area working electrode made from Pt foil or coil. The concentric counter electrode was located out of the field of view of the photon detector, so its emission remained undetected. The Pt lead wires for these two electrodes were insulated with Teflon spaghetti in their rise to the wall seals. After degassing, a fiber-tipped Ag/AgCl, KCl (saturated) aqueous reference electrode with a nonaqueous salt bridge or a QRE was inserted through the working electrode support under a positive pressure of inert gas. We address the methods for spectral recording in more detail in Section II.D.2.

B. Transient Techniques

Transient methods for studying ECL ordinarily involve alternate generation of the reactants at a single electrode (the working electrode), and they rely on diffusive or natural convective transport for mixing. A single experiment cycle produces one or two light pulses whose integral or shape parameters often contain quantitative kinetic information of interest. Alternatively, one can employ multicycle reactant generation to create a train of closely spaced light pulses for spectral study. This latter procedure can be treated as an extension of the single-pulse, potential step experiment. Transient methods are better introduced in terms of this technique, because it has emerged as the most useful transient probe for mechanism and because its design and interpretation require consideration of virtually all the phenomena involved in ECL generation.

1. The Triple Potential Step Experiment

Figure 6 describes the triple potential step procedure [1]. One begins with a deaerated solution of the precursors to generate the desired reactant radicals, e.g., rubrene (RU) in benzonitrile containing 0.1 M TBAP. Figure 6a shows the cyclic voltammogram of such a solution. Prior to the start of the experiment, the working electrode, usually a Pt disk a few millimeters in diameter is held at the rest potential to allow the system to become quiescent. At zero time, the working

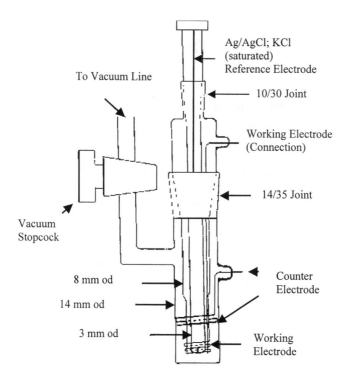

Figure 5 A cell designed by Freed for generation of ECL inside a spectrometer cuvette holder. (Adapted from Ref. 22.)

electrode potential is stepped to a potential in the mass transfer–controlled region to generate radical anions (see Fig. 6b). A progressively thickening layer of anions is created near the electrode surface, and the concentration profiles at the end of the forward step are like those of Fig. 7b. At this time, t_f, the electrode potential is stepped to a value in the diffusion-controlled region for rubrene oxidation. The second step initially destroys some $RU^{-\bullet}$ at the electrode surface by the reaction

$$RU^{-\bullet} \rightarrow RU^{+\bullet} + 2e \qquad (1)$$

So $RU^{-\bullet}$, now facing a concentration gradient toward the electrode, will begin to move in that direction. Because $RU^{+\bullet}$ diffuses only toward the solution, the two species meet and react in a very thin zone that is near the electrode surface and gradually moves away from it. The situation midway through the second step is shown in Figure 7c; the reaction produces a light pulse that decays with time as shown in Figure 6c. Time into this step is measured from its beginning by t_r. At $t_r = t_f$ the electrode potential is changed a third time to an intermediate potential

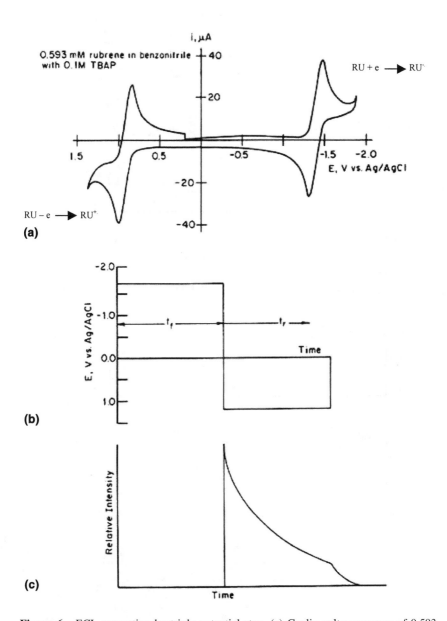

Figure 6 ECL generation by triple potential step. (a) Cyclic voltammogram of 0.593 mM rubrene in BN with 0.1 M TBAP; (b) working electrode potential program; (c) emission intensity vs. time. (Adapted from Ref. 1.)

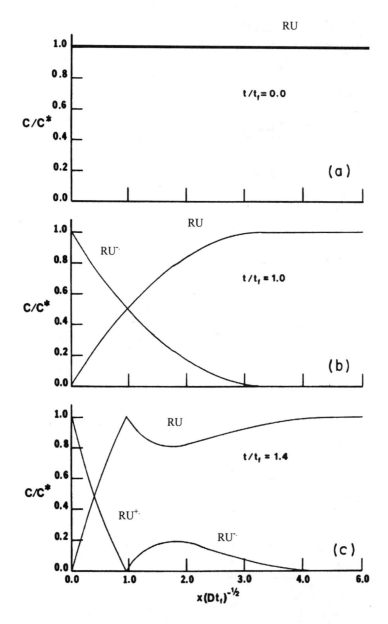

Figure 7 Concentration profiles in the triple-step generation of ECL. The initial bulk concentration of RU is represented as C^*. (Adapted from Ref. 1.)

at which both radical ions are destroyed electrolytically, thus the solution is restored quickly to its original state, Figure 7a. The opposite sequence is obviously available for an anodic forward step.

Information coming from this experiment is twofold. First, one can know the number of reactant species generated in the forward step from integrated current measurements. Comparison of this value with the total number of photons produced (integrated absolute emission–time curve) provides information about the emission efficiency, i.e., the probability of emission per charge transfer event, Φ_{ecl}. Φ_{ecl} is given by

$$\Phi_{ecl} = \frac{\int_0^\infty I \, dt_r}{\int_0^\infty N \, dt_r} \qquad (2)$$

where I is the total photonic emission rate (einsteins per second) and N is the total redox rate (moles per second). The closest experimental approximation one can make to this emission probability per charge transfer event is the coulometric efficiency Φ_c,

$$\Phi_c = \frac{F}{Q_f} \int_0^\infty I \, dt_r \qquad (3)$$

where F is the Faraday constant and Q_f is the faradaic charge passed in the first step. Φ_c differs from Φ_{ecl} because the number of charge transfer events, $\int_0^\infty N \, dt_r$, is smaller than the number of reactant species injected in the forward step, Q_f/F, by the number destroyed at the electrode in the second and third steps. The ratio $\Phi_{ecl}/\Phi_c = \theta$ can be obtained by digital simulation. It is 1.078 for stable reactants derived from the same precursor and increases with decreasing reactant stability [23]. It also depends on the ratio of reactant precursor concentrations in mixed system. For equal concentrations, it is 1.105 [24]. An interesting fact uncovered from the simulation is that virtually all reactant loss occurs during the third step. During the transition to the second step, a very small amount of the first reactant is lost to the electrode, but this loss ends as soon as the second reactant is established at the surface. Second, electron transfer kinetic information is contained in the parameters governing luminescence decay. This requires a rather sophisticated data analysis scheme to extract it (cf. Chapter 3 for details).

2. Multicycle Methods

The generation of a train of pulses, usually for spectral recording, is an obvious extension of the single-pulse, potential step technique. A series of light pulses are generated, accompanying each potential change. To attain the resolution in a spectral study, one usually seeks to achieve a pseudo-steady-state reasonably high light output; hence one desires a high alternation frequency (>10 Hz) and employs electrodes of fairly large area (>1 cm^2). Both factors imply markedly high power dissipation due to double-layer charging; hence one often encounters severe rounding of the potential waveform via uncompensated resistance effects and thermal heating of the solution resulting in convective transport. The well-defined electrode boundary conditions and diffusive transport processes no longer apply, and the pulse shape becomes unpredictable. On the other hand, one rarely cares about the pulse shape in these instances, and the effect of convection is to increase transport rates, resulting in desirably higher light output.

When possible, such experiments should be carried at controlled potential, as for the single-pulse experiment described above. Many investigators have used ac electrolysis to generate light in a cell containing only two electrodes. However, one must exercise caution under such a circumstance, for one can easily generate undesired species by applying too large a voltage to the cell. The homogeneous reaction can be safely limited to that between the most easily generated oxidant and reductant by operating the cell at a voltage just higher than the threshold value required for luminescence. In many cases, increasing the applied voltage causes the intensity to pass through a maximum as the electrode reactions reach the mass transfer–controlled region. The voltage required to generate maximum intensity usually represents an optimal applied voltage.

Although this technique has commonly been employed for spectral recording, by using an ultramicroelectrode (UME) as the generator in annihilation experiments and comparing the shape of the relative intensity vs. time curve to theoretical behavior obtained by digital simulation, it is possible to study fast annihilation reactions of the radical ions of 9,10-diphenylanthracene (DPA) and several other compounds [25]. More details about this technique are discussed in Section III.A; also see Chapter 3 of this volume for theory.

C.   Steady-State Methods

The transient techniques for ECL studies are straightforward and have been successfully applied to elucidate the reaction mechanisms, but they suffer several disadvantages. First, the current attained in transient experiments includes an appreciable amount of non-Faradaic contributions from charging of double-layer capacitance. This effect becomes more important as the switching frequency increases and step time decreases and must be taken into account in the determination of Φ_{ecl}. Second, the demands on the response time and power output of the potentiostat can be quite large in transient techniques for obtaining an accurate decay curve for analysis. Finally, steady-state ECL intensities cannot be attained at a single electrode and it is difficult to quantitatively consider ECL arising from reaction species other than those formed at the first oxidation and reduction waves. Steady-state ECL can be generated by forming the two reactants at separated electrodes and flowing these together shortly after formation. There are several steady-state methods for ECL study. These include rotating ring-disk electrode, thin-layer cell, and microband electrode arrays.

1. The Rotating Ring-Disk Electrode

The rotating ring-disk electrode (RRDE) (Figure 8) consists of a disk electrode and a concentric ring electrode electrically insulated from the disk and separated from it by a thin Teflon spacer. Upon rotation, solution at the ring-disk surface is spun out in a radial direction and is replenished by the flow of solution normal to the RRDE. In ECL experiments, one reactant (e.g., radical cation) is generated at the disk and the other (e.g., radical anion) is produced at the ring. These ions mix to form a circle of light in the ring electrode region.

Because solutions in most ECL procedures must be prepared and maintained under an inert atmosphere, the rotation of the working electrodes presents problems. The construction of an RRDE and cell for ECL studies on a vacuum line has been described in detail [1,26] and will not be described here. An alternative arrangement involves the use of an open-cell and motor system inside an inert atmosphere glove box. The electrodes themselves and bipotentiostat are now commercially available (e.g., from Pine Instrument Company, Grove City, PA; EI-400 bipotentiostat from Cypress Systems, Inc., Lawrence, KS). In this case, total light emission or emission at a particular wavelength isolated by interference filters can be measured, but spectral recording is more difficult.

Some of the experiments performed with the RRDE are of a purely electrochemical nature, such as the determination of disk current, i_d, vs. disk potential, E_d, and ring current, i_r, vs. ring potential, E_r, at constant E_d or i_r vs. E_d at constant E_r, all as functions of ω. These experiments provide standard potentials and information about the stabilities of the radical ions and other intermediates. Evidence for the stability of the electrogenerated species was obtained by determination of the collection efficiency N_c (where $N_c = |i_r/i_d|$) for generation of a species at the disk electrode and collection of this species at the ring electrode. Experimental N_c values for stable species are close to the theoretical value for total collection of the disk-generated species at an electrode of known geometry. Theoretical N_c depends only on the radius of the disk, r_1, the inner radius of the ring, r_2, and its outer radius, r_3, and is independent of ω and the bulk concentration C_R^* and diffusion coefficient D_R of the reactant. The theoretical N_c can be calculated from

$$N_c = 1 - F(\alpha/\beta) + \beta^{2/3} [1 - F(\alpha)] -$$
$$(1 + \alpha + \beta)^{2/3} \{1 - F[(\alpha/\beta)(1 + \alpha + \beta)]\} \tag{4}$$

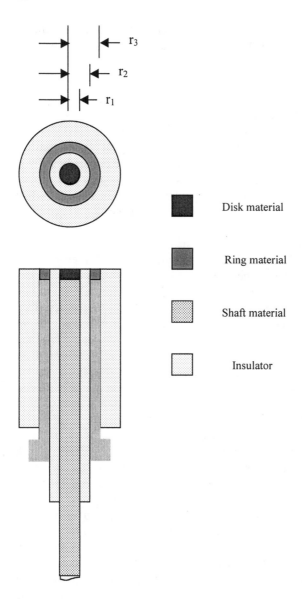

Figure 8   Ring-disk electrode.

where $\alpha = (r_2/r_1)^3 - 1$, $\beta = (r_3/r_1)^3 - (r_2/r_1)^3$, and F values are defined by

$$F(\chi) = \frac{3^{1/2}}{4\pi} \ln \left[\frac{(1 + \chi^{1/3})^3}{1 + \chi} \right] + \frac{3}{2\pi} \arctan \left[\frac{2\chi^{1/3} - 1}{3^{1/2}} \right] + \frac{1}{4}$$

(5)

The function $F(\chi)$ and values of theoretical N_c for different values of the ratios r_2/r_1 and r_3/r_2 have been tabulated [27]. One can also determine N_c experimentally for a given electrode, by measuring $|i_r/i_d|$ for a stable system. Once N_c is determined, it is a known constant for that RRDE. For example, the RRDE used in Maloy's ECL studies [26], which had a disk of platinum with $r_1 = 0.145$ cm and a platinum ring with $r_2 = 0.163$ cm and $r_3 = 0.275$ cm had an N_c value of 0.54. This implies that for a stable species generated at the disk, 54% of it reaches the ring and is collected, with the remainder escaping into the bulk solution. When the disk-generated species is unstable, smaller N_c values are observed, and N_c determined as a function of i_d or ω under these conditions can be used to measure the rate constants for the decomposition reactions [28].

In the ECL experiments, besides its use in the determination of Φ_{ecl} as described later in Section II.D.1.b, the RRDE can be employed to measure simultaneously the steady-state values of I and i_d as a function of E_d during continuous production of the appropriate reactant at the ring electrode. One representative use of the RRDE in this manner is illustrated in Figure 9, which shows the i_d and I values obtained for the $Ru(bpy)_3^{2+}$ system in acetonitrile with TBAP. The ring was maintained at a potential where $Ru(bpy)_3^{3+}$ was generated [e.g., at $+1.48$ V vs. SCE where $Ru(bpy)_3^{3+}$ is produced], and the disk was scanned from 0 to -2.0 V vs. SCE [29]. Emission is observed for production of the $+1$, 0, and -1 species at the disk in three light waves of equal height. For all waves the light is from the emitting $Ru(bpy)_3^{2+}$ state. When the disk was held at $+1.48$ V and the ring scanned to negative potentials, the ECL intensity remained constant, as expected, because with the geometry of the RRDE employed in this experiment i_r was much larger than i_d for the same electrode reaction.

The Φ_{ecl} value can be determined by adjusting E_r and E_d to appropriate values and determining the ECL intensity incident on the photon detector. The usual problems of absolute photometric measurements arise (see Section II.D.1.b), as well as others peculiar to the RRDE and its cell geometry. This is corrected for geometric factors and frequency to yield the total emission I. The disk current (corrected for any residual current) i_d in the limit of a fast redox reaction involving n electrons, represents the number of annihilations, so that

$$\Phi_{ecl} = nFI/i_d$$

(6)

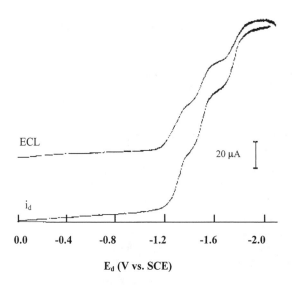

Figure 9 Steady-state current and ECL intensity for Ru(bpy)$_3^{2+}$ system. The solution contained 1.0 mM Ru(bpy)$_3$(ClO$_4$)$_2$ and 0.1 M TBABF$_4$ in ACN with E_r = +1.48 V (generation of +3 species) and E_d scanned from 0 to −2.0 V. (Adapted from Ref. 29.)

Values of Φ_{ecl} at different rotation speeds are shown in Figure 10 for the Ru(bpy)$_3^{2+}$ system [29]. The efficiency of this system (5–6%) is one of the highest and is often taken as a standard [12–14] for the calibration of Φ_{ecl} of an unknown system (see also Section II.D.1.b for the determination of Φ_{ecl}).

2. Thin-Layer Cells

Another approach to steady-state ECL uses two closely spaced electrodes in which, during dc electrolysis, the oxidized and reduced forms generated at the different electrodes diffuse and migrate together. The annihilation reaction occurs in the space between two electrodes to produce the light. The electrode spacing, ℓ, must be quite small if the rise time for emission, t_s, is to be reasonably short (i.e., $t_s \approx \ell^2/16D_R$, where D_R is the diffusion coefficient of the reactant); a 1 s rise time thus requires a 100 μm spacing. A TLC will establish a steady-state current i_{st} and emission intensity I governed by

$$\Phi_{ecl} = nFI/i_{st} \qquad (7)$$

Because the solution volume contained in a TLC is so small, a closed TLC is more sensitive to buildup of impurities and quenchers and so usually shows

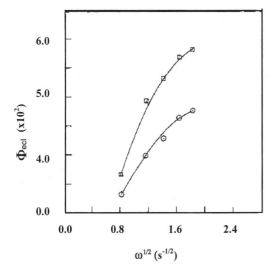

Figure 10 Φ_{ecl} vs. $\omega^{1/2}$ for 1 mM Ru(bpy)$_3$(ClO$_4$)$_2$/degassed ACN/0.1 M TBABF$_4$. ($\bigcirc$) $E_d = -1.37$ V, $E_r = +1.48$ V; ($\square$) $E_d = +1.48$ V, $E_r = -1.37$ V. (Adapted from Ref. 29.)

relatively short lifetimes. This drawback can be eliminated by combining an open TLC with a flow injection analysis (FIA) system (cf. Chapters 8 and 9 for details about flow injection ECL systems). Thin-layer cells can be used advantageously with solvent systems of much higher resistivity than those employed in conventional cells. Various aspects of thin-layer ECL cells and experimental results have previously been described [30–32].

3.   Microband Electrode Arrays

An alternative approach to the generation of steady-state ECL uses an array of microband electrodes. For demonstration, we focus on a double-band electrode configured in a generation–collection mode. In this configuration, information analogous to the classical RRDE, as discussed above, is obtained [33,34].

Because of the close proximity of the individual electrodes, arrays of band electrodes are particularly convenient tools to generate the reactants for ECL. Two members of the array are used in double-band configuration to generate both reactants at a constant potential, and an easily measured steady-state light output can be achieved. The small gap dimensions at these electrodes focus the reactant diffusion layers for the ECL reaction adjacent to one another, so a narrow, intense emission region is produced between the bands. Besides,

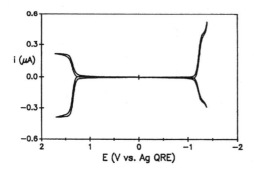

Figure 11   Collector–generator voltammograms recorded at a double band for 1.0 mM Ru(bpy)$_3$(ClO$_4$)$_2$ in ACN with 0.1 M TBAPF$_6$. The initial generator potential was 0.0 V, and scans are shown to both negative and positive potentials at 50 mV/s. The potential of the collector was 0.0 V for both scans. (Adapted from Ref. 35.)

they are easy to construct and provide relatively large current with minimal distortion from Ohmic drop. These features combined with the FIA technique make them very useful for continuous monitoring of emission efficiency. The efficiency that can most readily be determined experimentally, Φ_c, is given by Eq. (3).

The use of double-band electrodes in the collector–generator mode provides a convenient way to characterize the potentials that should be used for generation of the ECL reactants. Collector–generator voltammograms for Ru(bpy)$_3{}^{2+}$ are shown in Figure 11 [35]. The generator potential was scanned while the potential of the collector was kept constant, and the current at both electrodes was recorded. The current measured at the collector is due to diffusion of electrogenerated Ru(bpy)$_3{}^+$ or Ru(bpy)$_3{}^{3+}$ from the generator. Meanwhile the current at the generator is due to semi-infinite diffusion of Ru(bpy)$_3{}^{2+}$ from the bulk and "feedback" diffusion from the collector electrode. In addition to the potential information, the stability of the ECL reactants can be evaluated by the magnitude of the collection efficiency. The collection efficiency is defined as i_c/i_g, where i_c and i_g are the diffusion-limited currents at the collector and generator, respectively. The experimental values for the anodic and cathodic reactions are 0.58 and 0.56, in the range expected for chemically stable products on the electrochemical time scale at double-band arrays of the dimensions employed [4–5 µm in width (w), 0.3–0.5 cm in length, and 4–5 µm gap (g)] and a potential scan rate of 0.05 V/s. An empirical expression for the collection efficiency of a chemically stable product as a function of dimensionless scan rate p for $w/g \geq 1$ is given by [36]

$$i_c/i_g = 0.2415 - 0.4091(\log p) - 0.07487(\log p)^2 \qquad (8)$$

where $p = (g/2)(nFv/D_R \Re T)^{1/2}$, v is the potential scan rate, D_R is the diffusion coefficient of the redox species, $\Re$ is the gas constant, and T is the temperature. As the electrogenerated species at the generator are unstable, collection efficiency is affected by the homogeneous chemical kinetics. A working curve for the EC (a chemical reaction following the interfacial electron transfer) mechanism for double-band electrodes is approximated with very good accuracy at low collection efficiencies by [36]

$$\frac{i_c}{i_g} = 0.5 \exp\left[+\left(\frac{k}{D_R}\right)^{1/2} \left(\frac{\pi g}{2}\right) \right] \tag{9}$$

in which k is the first-order homogeneous rate constant.

In the ECL mode, one of the bands serves as the cathode and the other as the anode. The potential of the first reduction wave [formation of Ru(bpy)$_3^+$] is close to subsequent reduction waves that correspond to the formation of the neutral and -1 species. Therefore, the potential of the cathode during ECL operation is chosen near the plateau of the Ru(bpy)$_3^+$ wave. The anode potential during ECL operation is chosen to be 250 mV past the half-wave potential for the Ru(bpy)$_3^{2+}$/Ru(bpy)$_3^{3+}$ redox couple, sufficiently positive to ensure diffusion controlled formation of the reactant. The anodic current and ECL intensity during the ECL experiment at a double-band array are shown in Figure 12. Whereas the electrolysis current rapidly reaches steady state, the ECL intensity does so at a much slower rate.

Similar to thin-layer cells, microband electrodes can be used to observe ECL in solutions of low ionic strength. For Ru(bpy)$_3^{2+}$ in low ionic strength solutions, the electrochemical and ECL measurements at steady state are unchanged from those at high ionic strength. However, before steady state is achieved, the anode current is lower than the cathode current, an effect caused by migration [37]. The annihilation reaction results in a redistribution of the counter ions in the solution volume adjacent to the double band, resulting in the disappearance of migration as a contributor to the mass transport. Under all solution conditions the double-band ECL device produces light continuously with direct current application because continuous feedback occurs with reactants and products. The current flows mainly between the adjacent anode and cathode. It would be possible, therefore, to operate this ECL device with a simple voltage source between the anode and cathode, without the need for reference and auxiliary electrodes or a potentiostat. However, one must exercise caution under such a circumstance, for one can easily generate undesired species by applying too large a voltage to the cell, as discussed previously in Section II.B.2.

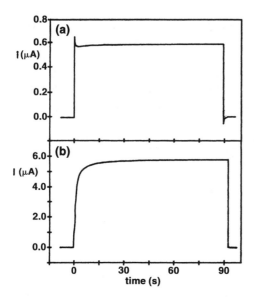

Figure 12 Currents measured during ECL. (a) Anodic current, i; (b) PMT current *I*. The solution contained 1 mM Ru(bpy)$_3$(ClO$_4$)$_2$ in ACN with 0.1 M TBAPF$_6$. (Adapted from Ref. 35.)

D. Light Detection

Important detector qualities such as spectral response, rise time, and sensitivity differ not only between different detector types but also between different detectors of the same type. These qualities are also influenced by the design of the overall measurement system, including component specification. Several parameters are used to characterize detector performance. A discussion of the many types of detectors or of the details of any one design is beyond the scope of this chapter. We summarize some of the properties of a few of the more basic devices in Table 2.

1. Intensity and Quantum Efficiency Measurements

a. Relative Measurements of ECL Intensity.   Much of the ECL work reported involved relative photometric measurements. A detector signal proportional to the absolute emission flux was followed, but the proportionality constant was unknown. In many cases, detection was made essentially by a bare photomultiplier tube (PMT). The measured signal was photocurrent converted to a voltage by either an anode resistor or an operational amplifier acting as a current follower. The group at the University of Texas presently employs a

Table 2 Basic Properties of Detector Families

Detector type	Peak sensitivity	Spectral linearity	Response time
Eye	Equivalent to PMT	Peaks near 555 nm	50 ms
Thermopile	55 V/W	Flat	1 s
Pyroelectric	2400 V/W	Flat	1 μs
PMT	10^5 A/W	Varies by type[a]	10 ns
Solar cell[b]	0.3 A/W	Peaks near 750 nm	200 μs
Photodiode[b]	0.3 A/W	Peaks near 750 nm	1 ns
Avalanche[b]	100 A/W	Peaks near 750 nm	1 ns
Vidicon tube	NA	Varies by type[a]	1 μs
CCD[c]	Varies by type	Varies by type	Varies by type

[a]The spectral response of PMT is determined by the photoemissive (cathode) material and the transmission of the window in the tube at the blue end.
[b]Refers to silicon-based detectors.
[c]See Section II.D.2 for more detailed discussion.
Source: Adapted from D O'shea. Elements of Modern Optical Design. New York John Wiley & Sons, 1985, p. 337.

PMT–electrometer (Hammamatsu R4220 or R928 PMT and Keithley 6517 electrometer) combination for general ECL intensity measurement. This simplest of detection methods has high sensitivity (≤ 20 pA photocurrent detection limit) and offers maximum flexibility in the design of other parts of the apparatus. This setup is most suitable for quasi-steady-state measurements owing to the limited response time of the electrometer (ranging from 1 ms in micro- and milliampere ranges to 2.5 s in the picoampere range). For transient measurements, the PMT signal can be amplified by a fast preamplifier (EG&G Ortec VT120A, 150 MHz bandpass, 200 gain) before going into an oscilloscope or a multichannel scaler. An alternative to the PMT could be a PIN photodiode. This kind of device has not been used extensively in transient studies but has been employed for steady-state measurement. Likewise, an alternative to determination of photocurrent is the use of photon counting. In the conventional mode, this method is used for steady-state measurement. The Hamamatsu C1230 photon counter and its related products are quite suitable for this kind of experiment. An R928p PMT was cooled at $-10°C$ with a thermoelectric device (Model TE308TSRF, Products for Research, Inc., Danvers, MA) housed in a water-jacketed refrigerated chamber connecting to a cooled-water circulator (MGW Lauda RM6, Brinkman Instruments, Inc., Westbury, NY). In the transient measurement, the photon-counting method was originally developed for studies of fluorescence decay [38]. For ECL studies, a modified technique was used by Van Duyne's group to acquire fast transients (~ 10 μs) at high precision [39]. Quite recently, by using an ultramicroelectrode (UME) as the generator in annihilation experiments and comparing the shape of the relative

Table 3   Other Liquid-Phase Chemical Actinometers[a]

Actinometer	Useful wavelengths (nm)	Remarks
Azobenzene photoisomerization[b]	230–460	No wavelength dependence of quantum yield for trans → cis reaction; no temperature dependence; reproducibility better than 2%; reusable.
Uranyl oxalate photolysis[c]	200–500	Quantum yields are accurately known; requires titrations and the results are obtained from titration˘20 differences; absorption has a minimum near 366 nm.
2,3-Dimethyl–2-butene-sensitized photooxygenation[d]	280–560	Quantum yield = 0.76; sensitized with methanolic $Ru(bpy)_3^{2+}Cl_2$ solution.
Heterocoerdianthrone[e]	400–580	Simple evaluation procedure; small error and good reproducibility.
meso-Diphenylhelianthrene self-sensitized photooxygenation[f]	475–610	Quantum yield = 0.224; independent of wavelength.
Hexakis(urea)Cr(III) chloride photoaquation[g]	452–735	Low quantum yields (0.09–0.10) and nearly independent of wavelengths.
Reinecke's salt photoaquation[h]	316–750	Quantum yields depend on wavelengths and accurately known, only slightly on temperature, and no significant dependence on light fluence; reproducibility ±2%; precision ±5%; thermal aquation becomes a problem in basic solution.

[a]For general information on chemical actinometry, see, e.g., SL Murov, I Carmichael, GL Hug. Handbook of photochemistry, New York: Marcel Dekker, 1993, Chap. 13; HJ Kuhn, SE Braslavsky, R Schmidt Chemical Actinometry. Pure & Appl. Chem. 61:187–210, 1989.

[b]See, e.g., G Zimmerman, LY Chow, UJ Paik. The Photochemical Isomerization of Azobenzene. J. Am. Chem. Soc. 80:3528–3531, 1958; G Gauglitz, S Hubig. Chemical actinometry in the UV by azobezene in concentrated solution: a convenient method. J. Photochem. 30:121–125, 1985; N Siampiringue, G Guyot, S Monti, P Bortolus. The cis → trans photoisomerization of azobenzene: an experimental re-examination. J. Photochem. 37:185–188, 1987.

[c]HA Taylor, In: JM Fitzgerald, ed. Analytical Photochemistry and Photochemical Analysis, New York: Dekker, 1971, p.91 and references cited therein.

Table 3   Other Liquid-Phase Chemical Actinometers[a]

---

[d]JN Demas, RP McBride, EW Harris. Laser Intensity Measurement by Chemical Actinometry. A photooxygeneration Actinometer. J. Phys. Chem. 80:2248–2253, 1976.

[e]HD Brauer, W Drews, R. Gauglitz, S Hubig. Heterocoerdianthrone: a new actinometer for the visible (400–580 nm) range. J. Photochem. 20:335–340, 1982.

[f]HD Brauer, R Schmidt, G Gauglitz, S Hubig. Chemical actinometry in the visible (475–610 nm) by meso-diphenylhelianthrene. Photochem. Photobiol. 37:595–598, 1983; R Schmidt, HD Brauer. Self-sensitized photo-oxidation of aromatic compounds and photocycloreversion of endoperoxides: applications in chemical actinometry. J Photochem. 25:489–499, 1984.

[g]EW Wegner, AW Adamson. Photochemistry of complex ions. III. Absolute quantum yields for the photolysis. J. Am. Chem. Soc. 88:394–404, 1966.

[h]For potassium Reineckate, $K[Cr(NH_3)_2(CNS)_4]$, see, e.g., EE Wegner, AW Adamson. Photochemistry of complex ions. III. Absolute quantum yields for the photolysis. J. Am. Chem. Soc. 88:394–404, 1966.

---

intensity vs. time curve (acquired with a single-photon counting apparatus) to theoretical behavior obtained by digital simulation, it was possible to study fast annihilation reactions of the radical ions of 9,10-diphenylanthracene (DPA) in the temporal resolution to the nanosecond regime [25]. More details about this technique are discussed in Section III.A.

b.   Absolute Measurements of Quantum Efficiency.   We noted above that measurements of Φ_{ecl} and a full analysis of transients require knowledge of the absolute photonic intensity emitted from the system under study. There are two methods for acquiring such information, namely, direct actinometry or the use of a calibrated photodetector.

The usual procedure in actinometric determinations is to expose the entire photon flux to a totally absorbing, photochemically active solution, i.e. an actinometer. The standard tool is 0.006 or 0.15 M potassium ferrioxalate in 0.1 M H_2SO_4 [40]. The latter absorbs within a 1 cm depth essentially all photons with wavelengths shorter than 450 nm, and each photon produces ferrous iron with a quantum yield that varies only slightly with wavelength and is approximately unity [41]. The ferrous species is then measured spectrophotometrically by the 1,10-phenanthroline method. The experiment must be carried out in the dark, of course, and a number of precautions must be taken, as in any actinometric measurement [40,41]. The ferrioxalate solution must be degassed and maintained under nitrogen during the experiment to prevent oxidation of any Fe(II) formed. Actinometry is very reliable but rather insensitive. One requires about 10^{16} photons/mL of solution at a minimum and several millimeters of solution is usually necessary to cover an ECL cell to a 1 cm depth. In addition, the actinometer provides only time-integrated photon output. Transient techniques yield photon fluxes of perhaps 10^{13} photons/s; thus actinometry is not suitable for single-pulse work, but it can be used for long-term, multicycle light generation. Several other liquid phase chemical actinometers covering different wavelength ranges are listed in Table 3.

It is therefore necessary to calibrate a photon detection system, but obtaining a precisely standardized low-level source for such a calibration is an obstacle. Moreover, the ECL intensity often shows a complex angular distribution for different systems, which makes it extremely difficult to define the geometry of the photometric measurement. Finally, one has to contend with the usual wavelength dependence of the detector response.

The latter two problems were overcome by using the apparatus designed by Bezman and Faulkner [42] shown in Figure 13. One essential accessory for measuring ECL and light-emitting devices is an integrating sphere. Zweig et al. [43] introduced the integrating sphere as a convenient solution to the problem of geometric definition in the ECL studies. The sphere in the Figure 13a apparatus is a modified 3 L flask painted on the inside with several layers of diffusely reflecting BaSO$_4$ white reflectance paint (Eastman Kodak), which causes a source placed anywhere inside to yield a uniform intensity over the entire inner surface. As long as the detector does not view the source directly, this surface intensity is not very dependent on either the angular distribution of source emission or the source location, so that a reproducible fraction of the emission can easily be monitored via the PMT or photodetector port. The electrolysis cell enters from a joint at the top, whose axis is directed so that the working electrode faces away from the detector. The need for a quantum flat detector response (photons per microcoulomb calibration independent of wavelength) was met by inserting a quantum counter solution (6.0 g/L Rhodamine B in ethylene glycol) [40] between the detector port and the PMT. The integrating sphere is now commercially available (e.g., from Newport Corporation, Irvine, CA or Labsphere, Inc., North Sutton, NH).

Bezman and Faulkner [42] used a complicated and tedious stepwise calibration technique based on actinometry. A simplified calibration technique for the sensitivity of the integrating sphere–based photometric apparatus was reported by Michael and Faulkner [44]. In their approach, an optical fiber light guide which was mounted in a spectrophotofluorometer so that the excitation beam at the cell position impinged on one end. If the 436 nm line from a mercury-xenon lamp was used for excitation, the monochromatic flux emerging from the end of the guide was sufficiently high that one could quantitatively measure it by immersing the tip in the ferrioxalate actinometer for about 20 min. After this exposure, the guide was then inserted into the sphere and the photocurrent due to the known flux was measured. The sensitivity for the apparatus shown in Figure 13 was 3.6×10^{13} photons/μC. Any calibration is tedious, so the apparatus was also provided with two secondary standards in the form of small tungsten lamps powered by a constant current source. The photocurrents arising from them were standardized during the initial calibration of the system; thus they were available as a routine check on the PMT sensitivity.

Because the light is reflected several times, any wavelength variation in surface reflectivity is amplified into a serious wavelength dependence of sensitivity. Such a complication will not usually arise with fresh commercial

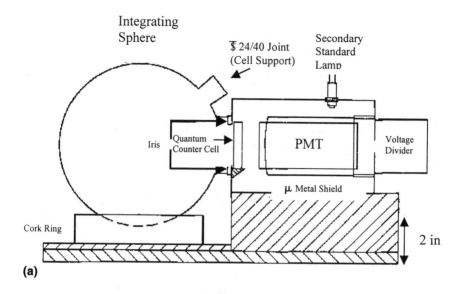

(a)

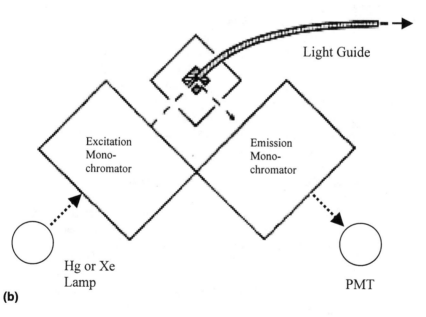

(b)

Figure 13 (a) A photometric apparatus designed by Bezman for absolute measurements of ECL output. (Adapted from Ref. 42.) (b) A simplified calibration technique for the sensitivity of the integrating sphere-based photometric apparatus reported by Michael. (Adapted from Ref. 44.)

paint unless extreme regions of the visible and near ultraviolet are approached. However, if such a complication should occur (the reflectance paint could deteriorate with time), one can easily check for this problem via the light-guide method. One should mount the guide in the spectrophotofluorometer as described above and illuminate it with light from a xenon arc lamp. Scanning the excitation monochromator then supplies a continuously variable monochromatic output at the exit tip of the guide. Its spectrum is recorded by placing the tip at the front face of the quantum counter (sphere removed) and measuring the PMT output as a function of wavelength. If the reflectance paint has flat reflectivity, the same spectral shape should be obtained upon replacement of the sphere and insertion of the guide into the position normally occupied by the ECL cell. A problem with wavelength dependence of sensitivity can also arise when an object inside the sphere adsorbs the source emission, so one should obviously avoid the use of light-absorbing materials in cell construction.

A final limitation of this particular photometric system is the operation of the fluorescent screen quantum counter. Because the screen becomes transparent abruptly at 610 nm, source emission at longer wavelengths is not scattered by the Rhodamine B but is registered directly at the PMT. This effect usually weights the longer wavelength photons much more heavily. The light-guide method allows one to evaluate the sensitivity factor as a function of wavelength, and this curve can be convoluted with the emission spectrum of interest to obtain an absolute sensitivity for the system at hand. In their absolute multicycle experiments, Pighin and Conway [45] and Schwartz et al. [46] detected light with solid-state devices (e.g., solar cells of 1 cm^2 area or silicon photocells) and used box-shaped integrating enclosures to deal with the geometry problem. Calibration was done with a standardized red-light-emitting diode. In any of these absolute measurements of ECL intensities, one may also have to account for losses caused by imperfect reflectivity of the electrode and optical absorption by solution (see the following discussions).

An alternative method for acquiring Φ_{ecl} is based on the RRDE technique. In direct actinometric measurements, the RRDE cell is surrounded by an actinometer solution, for example, a solution of 0.15 M potassium ferrioxalate in 0.05 M H_2SO_4. The ECL experiment is carried out at the RRDE while i_d is monitored, and the total number of photons absorbed by the actinometric solution is determined from the amount of Fe(II) formed. The apparatus for the ECL measurement from the thianthrene–DPA system is shown in Figure 14 [47]. The optical flat at the bottom of the cell allows the intensity to be monitored with a photodiode during the experiment; the actinometrically measured total emission is geometrically corrected for the losses through this monitoring port. A second detector monitors any light that escapes through the actinometric solution, especially the green tail of the DPA emission.

It was found that the RRDE produced emission that was uniform at all angles below the plane of the disk electrode [47]. Thus, the RRDE approach can greatly

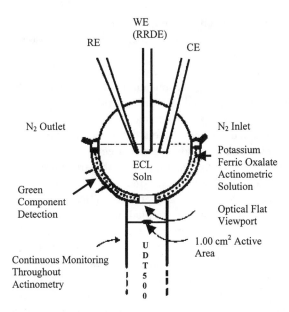

Figure 14 Apparatus used for direct actinometric determination of ECL efficiency. (Adapted from Ref. 47.)

simplify the calibration procedure for a photometric system; it is based on a calibrated photodetector of known sensitive area and geometrically determined efficiency of light collection. Especially useful as a detector is a planar diffused silicon PIN photodiode combined in a single package with a low-noise FET operational amplifier and equipped with a subtractive filter to provide a flat spectral response (in terms of incident power) over a wide spectral region (from 400 to at least 900 nm). A UDT-500 UV photodiode–amplifier combination (formerly United Detector Technology, now Graseby Optronics, Orlando, FL) with a 1.00 cm^2 active area was employed; similar suitable photodiodes are available nowadays from other manufacturers. The UDT-500 UV was calibrated using a small He-Ne laser (632.8 nm), whose power (e.g., 1.85 ± 0.05 mW) was determined with a laser power meter (e.g., at present, Model 1830-C, Detector 818-UV for the 100–1100 nm range; calibration factors for sensitivity in this wavelength range are supplied by the manufacturer, Newport Corp., Irvine, CA). In the calibration, the laser beam was passed through a 0.99% neutral density filter to prevent damage to the photodiode surface. Uniform response across the sensitive portion of the photodiode was ensured by moving the small laser beam spot to different locations on the photodiode surface. This power response can be converted into a response in terms of photons per second at any wavelength simply by multiplication by hc/λ, in which h is Planck's constant, c is the speed of light, and λ is its wavelength. The quantum response is obviously wavelength-dependent, so one must convolute the response curve with the emission spectrum in order to obtain an absolute sensitivity factor for any particular system.

Determination of Φ_{ecl} is made by locating the detector at a fixed, known distance from the electrode and measuring the quantum flux incident on the detector. The total intensity of emission equals the product of this flux and the area of the grand hemisphere ($2\pi r^2$) corresponding to the distance of the detector from the RRDE. With this calibrated system, the total intensity divided by the instantaneous disk current immediately yields the uncorrected Φ_{ecl}. Several corrections are required to determine the Φ_{ecl} that represents the true number of photons emitted per annihilation at the reaction site, and these introduce considerable uncertainty into Φ_{ecl} values. One concerns the reflectivity of the electrode. Only a certain fraction of the photons emitted in the direction of the electrode will be reflected into the lower hemisphere and be measured. The reflectivity of the platinum electrode is a function of both wavelength and potential. Moreover, reflectivity losses also occur for photons impinging on the Teflon spacer of the RRDE. An assumption that approximately 55% of the photons are lost by reflectivity effects in the blue (DPA-ECL) region has been used [47], but obviously this number has a high uncertainty with it. Another correction represents loss of photons and energy on passage through the test solution. Some photons are absorbed by solution components (e.g., radical ions and impurities) and not re-emitted. Others are absorbed by the fluorescer molecules and are re-emitted. Losses occur here because for most fluorescers the fluorescence quantum yield Φ_f is not unity, so that repeated absorption–re-emission steps cause a progressive loss in photons. Moreover, this process causes a spectral shift of the final emission to longer wavelengths and also leads to further reflectivity losses [47,48]. This absorption correction depends on the solute species, their concentrations, and the thickness of solution between electrode and cell wall.

2. Spectral Recording

Spectral studies make up a large fraction of ECL investigation, but they have usually been made with commercially available photon detection apparatus. We intend only to discuss briefly some underlying concepts and the operational principles of some of the most commonly used charge-coupled device (CCD) detectors [49] and their particular features. We then illustrate some photometric setups based on CCD cameras.

Various kinds of CCD cameras have traditionally been used to detect low light levels: CCDs (front- or back-illuminated), intensified CCDs (ICCDs), electron bombardment CCDs (EBCCDs), and one of the newest sensors, the electron-multiplying CCDs (EMCCDs). With effective sensor cooling, CCD dark-current effects are negligible. As a rule, detector noise is cut in half by each 20°C drop in temperature. The extent of cooling required depends largely on the longest integration time desired and the minimum acceptable signal-to-noise ratio. CCDs are most commonly cooled by using a liquid nitrogen Dewar or by attaching the device to a Peltier cooler. The choice depends on how low a temperature the device must be operated at to achieve the desired performance. A CCD sensor can achieve low readout noise of two to four electrons when operated at slow readout speeds. At fast readout speeds, this noise increases. In the front-illuminated mode of operation, incident photons must pass through a passivation layer as well as the gate structure in order to generate signal. Because of the absorption coefficient for short-wavelength photons in silicon, the quantum efficiency of front-illuminated CCDs is poor in the blue and UV regions. To increase short-wavelength quantum efficiency, the silicon wafer is thinned to approximately 15 μm and mounted with the gate structure against a rigid substrate. Light is incident on the exposed, thinned surface. The incident photons do not pass through the front surface electrodes and passivation layers. An enhancement layer and an antireflective coating may then be added to the back surface to increase quantum efficiency. A multiphase pinned device can also be added to reduce or eliminate the Si/SiO$_2$ interface state contribution to dark current. This is a thinned, back-illuminated, and multiphased pinned CCD.

Intensified CCDs consist of an intensifier tube over a CCD sensor. The intensifier tube contains a multichannel plate with a high accelerating voltage across it. Photoelectrons generated by incident light cascade down the plate, amplifying the incident signal. This high electron gain effectively results in an ICCD readout noise of less than one electron, enabling the detection of single photons. The intensifier tube also provides a means of ultrafast gating of the camera. However, quantum efficiencies of ICCDs are much less than those of CCDs. ICCDs have also lower resolution and are more expensive than CCDs. The operation of EBCCDs is very similar to that of ICCDs, but rather than having a vacuum they have no microchannel plates situated between the sensor and the photocathode. Thus EBCCDs have better resolution than ICCDs but have a limited lifetime because overexposure caused by direct detection can damage the CCD sensor. Like ICCDs, EBCCDs are complex and expensive and their quantum efficiency is limited by the photocathode.

An EMCCD combines the tunable signal amplification typical of an ICCD with the inherent advantage of a CCD, such as high and broad efficiency and high spatial resolution. It is reliable and robust because it is an all-solid-state device. Such advantages are very suitable for low-light imaging and spectroscopic applications. An EMCCD achieves this sensitivity by using a unique electron-multiplying structure, or gain register, built onto the silicon. The gain register is inserted between the end of the sift register of the sensor and the output amplifier of the CCD. Using impact ionization, or the avalanche effect, this structure multiplies the charge packets from each pixel before they are read off the sensor. The signal is multiplied by a suitably high gain (as high as 500) by the gain register before it reaches the output amplifier of a conventional CCD with a typical readout noise of a few electrons root mean square (rms), which produces insignificant levels of effective noise of less than one electron rms at any readout speed. However, an EMCCD has low temporal resolution, and its gain is highly dependent on the temperature of the device. A cooler sensor provides greater gain multiplication.

As a summary, a conventional CCD with a large-area format is quite useful for imaging and spectroscopic applications that require reasonably low readout noise (approximately two electrons rms), long exposure times, low dark current, and a wide field of view at high resolution. ICCDs should be used for low-light applications that require fast (less than 100 µs) gated image capture, such as time-resolved optical microscopy. In applications that require less than one electron of effective readout noise and for which nanosecond resolution gating is not necessary, such as single-molecule microscopy and chemiluminescence detection, an EMCCD camera is ideal. These cameras can be used in any low-light application in which nongated ICCDs or back-illuminated CCDs are used.

One CCD-based facility for measurements in the 230–1100 nm region, for example, consisted of a Czerny–Turner normal incidence monochromator with an f/9 aperture and a grating with a ruling frequency of 1200 lines/mm [50]. A schematic diagram of the experimental setup is shown in Figure 15. The monochromator had two exits; the light beam could be focused on one or the other of them by means of a rotating plane mirror at 45°. The CCD detector (thinned and back-illuminated, multipinned phase, with a 512 × 512 format, a 24 µm × 24 µm pixel size, and a 12.3 mm × 12.3 mm overall area, manufactured by Scientific Imaging Technologies, Inc., Tigard, OR) and the calibrated photodiodes (Graseby Optronics, Orlando, FL) were mounted on the two slits. The diodes were calibrated in the working wavelength region.

The quantum efficiency measurements were based on a comparison between the signal integrated by the CCD and the one measured on the calibrated photodiode. The latter is connected to a power meter that makes it possible to measure the incident power in the range, e.g., from 10 pW to 1 mW. The CCD total quantum efficiency η_{CCD} [50] is calculated by means of the equation

$$\eta_{CCD} = \frac{S_{CCD}\,\Im}{t_{exp}} \left(\frac{eE}{P_{ph}} \right) \check{R} \tag{10}$$

where S_{CCD} is the integrated signal on the CCD [in analog-to-digital units (ADU)], $\Im$ is the conversion factor (e/ADU), t_{exp} is the exposure time (s), e is the electron charge (1.6×10^{-19} C), P_{ph} is the incident power measured on the photodiode (W), E is the photon energy (eV), and $\check{R}$ is the reflectivity of the planar mirror.

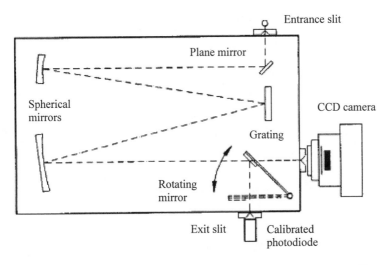

Figure 15 Experimental setup for spectral recording in the 250–1100 nm spectral region. (Adapted from Ref. 50.)

Note that here η_{CCD} is defined as the product of the effective CCD quantum efficiency, which is the fraction of incident photons producing one or more e–h pairs that are collected in the pixel wells, multiplied by the electron yield, which is the number of e–h pairs produced per interacting photon. Because visible wavelength photons generate one e–h pair and more energetic photons generate one e–h pair for each 3.65 eV of energy [51], the effective CCD quantum efficiency QE_{CCD} can be determined as

$$
\begin{aligned}
QE_{CCD} &= \eta_{CCD}, & E < 3.65 \text{ eV (or } \lambda > 340 \text{ nm)} \\
QE_{CCD} &= (3.65/E)\,\eta_{CCD}, & E > 3.65 \text{ eV}
\end{aligned}
\qquad 11
$$

The results of the quantum efficiency measurements (performed at a cooling temperature of $-30°C$ on the CCD with a multistage Peltier device) with this photometric system in the near-infrared and the visible ranges are shown in Figure 16. The statistical uncertainty of the data is from 3% to 6% rms because of the contributions of the errors to the calibrated photodiode quantum efficiency (3% in the visible range and 5% in the UV range), to the conversion factor of the CCD (2%), and to the measured incident power (1%). The maximum sensitivity is ~60% at 650 nm for this tested CCD, lower than typical values for CCDs with antireflection coating. In fact, the antireflection coating would increase the response in the visible region, but it would strongly absorb the UV radiation, causing a dramatic decrease in the response in the UV region.

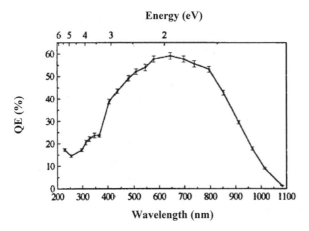

Figure 16 Effective quantum efficiency in the 230–1100 nm spectral range. (Adapted from Ref. 50.)

Because the ECL emission is sometimes quite weak and its intensity is often time-dependent, high light-gathering power, high QE (of the detector), and fast spectral recording speed are very desirable characteristics for the photometric system. In our lab, a CCD camera (e.g., Model CH260, Photometrics Ltd., Phoenix, AZ) is used for ECL imaging, intensity integration, and spectral recording. Figure 17 shows the two arrangements of the system used for light measurement [52]. The total integrated light intensities were obtained on the CCD camera by focusing it on the electrode surface using a 100 mm Pentax macro lens. The integrated intensities were then obtained by calculating the average intensities of each pixel of the focused image (Fig. 17a). ECL spectra were recorded using a Chemspec 100S (American Holographic, Littleton, MA) spectrometer (focal length 10 cm). The spectrometer was positioned such that the diffracted image was focused on the CCD detector (Fig. 17b). The camera was operated at −100°C (liquid nitrogen cooling), and all exposures were corrected for any dark or background current. For the ECL efficiency measurements, the potential was stepped to the first anodic peak potential, E_{pa}, for 0.5 s and then to the first cathodic peak potential, E_{pc}, for 0.5 s or vice versa. The camera recorded all light output from the electrode during excitation, and the cathodic current was integrated for the double pulse.

For spectral measurements, the potential of the working electrode was pulsed continuously between E_{pa} and E_{pc} with a pulse width of 0.5 s at a frequency of, e.g., 10 Hz. An exposure time of, e.g., 2 s or longer for weak ECL intensity was then used by the camera to give the resulting spectra. The

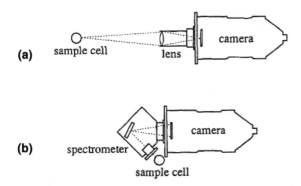

Figure 17 (a) Schematic diagram of the experimental setup for ECL imaging and light integration with the CCD camera. (b) Setup for ECL emission spectroscopy. (Adapted from Ref. 52.)

spectrometer was calibrated using an Hg-Ar or mercury test lamp (e.g., Ultra-Violet Products, San Gabriel, CA). Experiments were performed in a dark room and care was taken to eliminate stray light. Figure 18 shows the fluorescence spectrum of BPQ-PTZ in benzene–ACN (1.5:1) mixed solvent and the ECL spectrum of BPQ-PTZ in the same solvent containing 0.1 M TBAP as supporting electrolyte during repeated pulsing (pulse width = 0.5 s) between 0.72 and −2.15 V (vs. SCE) [13].

III. COMBINING ECL WITH OTHER TECHNIQUES

Carrying out ECL experiments in combination with other techniques, e.g., scanning probe microscopies (SPMs) and stopped-flow techniques, greatly increases the difficulty of the experiment. However, it supplies more independent information, e.g., topographical, spatial, and temporal resolutions of current and luminescence data, and makes it possible to observe explicitly the electron and energy transfer reactions in solutions.

A. ECL on an Ultramicroelectrode and Single-Electron Transfer Event

Electron transfer theories [53] predict that the highly exothermic production of the ground states proceeds in the inverted region. This allows the formation of the

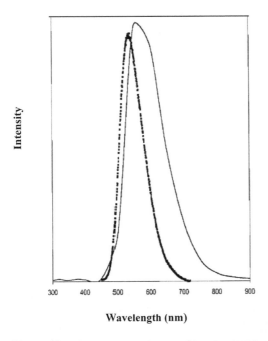

Wavelength (nm)

Figure 18   Fluorescence spectrum (20 μM) of BPQ-PTZ in benzene/ACN (squares) and ECL spectrum of 1 mM BPQ-PTZ in 0.1 M TBAP in benzene/ACN with pulsing (0.5 s) between 0.72 V and −2.15 V (vs. SCE). (Adapted from Ref. 13.)

excited state to be kinetically competitive with other nonradiative pathways that are predicted to occur near the diffusion-controlled limit. Time-resolved fluorescence quenching has commonly been used to measure the rates of formation of the separated radical ions and ground-state donor and acceptor molecules in the photoinduced electron transfer reactions (see, e.g., Ref. 54). The back electron transfer rate to re-form the emitting excited state, however, is not readily accessible from such experiments. This value can be obtained from the ECL provided the ECL efficiency and total rate of ion annihilation (k_{annih}) are known.

As carried out in Collinson and coworkers' experiments [14,25], high-speed electrochemical and photon detection techniques were used to monitor the real time rate of ECL generation of several compounds, e.g., DPA. In these experiments, a microelectrode in a flow cell (see Section II.A) is continuously stepped between the oxidation and reduction potentials of DPA to alternatively generate the radical ions. The cation and anion radicals react in a thin plane at a point where the inward and outward fluxes meet and subsequently produce light. The ECL intensity was monitored with a Hamamatsu 4632 PMT. A high-voltage power supply (Bertan Series 230) applied −800 V to the PMT. The PMT signal was amplified by a fast preamplifier (EG&G Ortec VT120A, 150 MHz bandpass, 200 gain), and the output was directed to the discriminator of a multichannel scaler (EG&G Ortec T 914). The discriminator level was set at -600 mV. A Wavetek Model 143 function generator applied a symmetric square wave to a silver counter electrode and triggered the multichannel scaler. The microelectrode was connected to the current amplifier. Particularly worthwhile to stress is that these short time steps (or high square-wave frequencies) are not suitable for larger electrodes, because the current during the step is dominated by double-layer charging and the electrode potential does not follow the applied potential step waveform.

To ensure that the electrode potentials are chosen so that the cation and anion radicals are produced at a diffusion-controlled rate at all frequencies following frequency selection of the square wave, the cathodic and anodic potentials were adjusted so that they roughly correspond to those of the redox potentials for the generation of the cation and anion radicals. Figure 19 shows two cycles of the potential waveform applied to a gold disk (radius 5 μm) and the resulting luminescence from a 0.6 mM DPA solution. Two pulses of light are observed from each cycle. When the potential is stepped positive, the cation radical reacts with the anion radical formed in the previous step in a reaction zone lying near the electrode surface. As described previously [1], the light increases sharply as the diffusion layers meet and then decays as the reactants are depleted. If the cation and anion radicals are stable during the time scale of the experiment, equal-size light pulses should be obtained on the forward and reverse steps. Increasing the frequency of the applied square wave usually results in more equivalent luminescent curves. As shown in Figure 19, at ~1 kHz a slightly smaller pulse of light is observed when the electrode is stepped from a negative to a positive potential, indicating that the DPA anion is less stable than the cation on this time scale. At ~20 kHz, both the cation and anion radicals are stable, as can be seen from the equivalent light pulses.

When the dimensionless kinetic parameter $\lambda_k = k_{annih} t_f C$ approaches infinity, the reaction layer is an infinitely thin parallel plane that moves nearly linearly away from the electrode surface with time. This plane of light broadens and becomes Gaussian-like as λ_k decreases. When λ_k drops below 1000, i.e., when t_f and C are significantly reduced, the transfer from diffusion to kinetic control begins and distinct changes in the width and the shift in the peak maximum of the ECL–time curves become apparent. The k_{annih} can be theoretically evaluated from the characteristic shapes of ECL–time curves in dilute solutions and at reduced step times (cf. Chapter 3 for detailed theoretical discussion). For illustration, we show only some comparisons of the simulated curves to the experimental data with a k_{annih} of $2 \times 10^{10} \, M^{-1} s^{-1}$ in Figure 20 [25]. Roughly agreeing with experimental observations, the simulation predicts certain features in the data such as the delay time in the initial ECL and the diminished amplitude and increased breadth of the ECL curve with decreasing step time. The delay time is due to the electrochemical

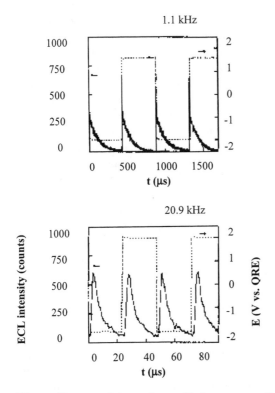

Figure 19 Potential waveform with the corresponding ECL curves from 0.6 mM DPA at a gold disk (radius 5 mm) in ACN containing 0.2 M TBAPF6 at two different frequencies. The ordinate represents the number of counts collected during 1 ms time bins. The luminescence curves were summed 100 times. (Adapted from Ref. 14.)

time constant (the product of the double-layer capacitance and the uncompensated solution resistance), whereas the diminished amplitude and increased breadth are due to the finite ECL kinetics. The simulation, however, does not predict the slow rise in emission or the substantially lower amplitude evident in Figure 20 as the step time is decreased from 48 to 5 μs. Similar results were obtained with Au or Pt disk electrodes (radius 3–10 μm), with different solvents and with half the electrolyte concentration. However, the shape of the ECL curve is dependent on the concentration of DPA. At higher concentration, e.g., 5.8 mM, distinct oscillation in the ECL intensity can be observed on the decaying emission. These features have been attributed to the direct interaction of the emission with the metal electrode due to the close proximity of the light-emitting species to the metal electrode surface [55]. These effects were least apparent with carbon fiber

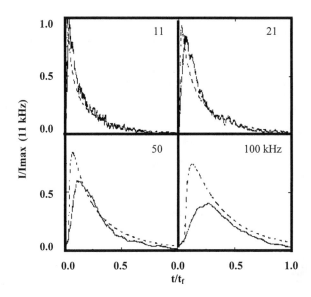

Figure 20 Normalized ECL from 0.38 mM DPA in ACN containing 0.1 M TBAPF$_6$ at a Pt disk (radius 1 µm) (solid lines) and corresponding simulated curves (dashed lines) as a function of frequency of potential pulse. Simulations for $k_{annih} = 2 \times 10^{10}$ M^{-1}s^{-1} and double-layer charging time constant, RC_{dl}, of 0.10 µs. (Adapted from Ref. 25.)

microelectrodes due to their low reflectivity and density of states. In this case, diffusion-controlled k_{annih} of 2×10^{10} M^{-1}s^{-1} was measured for DPA in ACN and 4×10^9 M^{-1}s^{-1} for DPA in propylene carbonate, a more viscous solvent.

In a later experiment [56], the DPA concentration was decreased to 15 µM and the temporal resolution to the nanosecond regime. In this case, an unsymmetrical waveform (a 500 µs anodic pulse followed by a 50 µs cathodic pulse) was used. The emission occurred predominantly during the shorter cathodic pulse (Fig. 21) as the electrogenerated radical anion diffused into the sea of DPA radical cations generated in the anodic pulse. When the photons were counted over 1 s intervals, no evidence for individual reaction events was observed. The ensemble average of the counts detected during 1000 cathodic voltage pulses also masked individual reaction events but revealed that the luminescence approached steady state (Fig. 21d). When events during a single cathodic pulse were viewed with greater temporal resolution (i.e., bin size of 5 ns), photons resulting from individual reactions were revealed (Figs 21e and 21f). Virtually no background photons were detected, so the corrections were

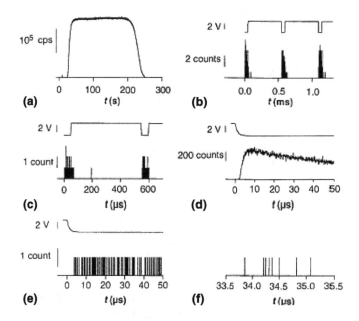

Figure 21 Temporal resolution of single-reaction events. (a) Chemiluminescence from a bolus of 15 μM DPA in ACN containing 0.1 M TBAPF$_6$. The electrode was a gold disk (radius 5.4 μm) pulsed from 1.7 to -2.1 V at 550 μs intervals. Data collected during the time interval between 50 and 200 s have been expanded through a successive decrease in the bin size from 1 s (a) to 1 μs (b), 100 ns [(c) and (d)], and 5 ns [(e) and (f)]. The double-layer capacitance (14 pf) and the solution resistance (75 kohm) were used to calculate the rise time of the voltage pulses shown in (d) and (e). The curve shown in (d) is an ensemble average of events measured during 1000 cathodic pulses. (Adapted from Ref. 56.)

unnecessary. Thus, the individual photon counts shown in Figures 21e and 21f resulted from single chemical reactions between individual DPA radical ions in solution.

The stochastic nature of these events was characterized in two ways. First, a histogram of the time between individual photons was constructed (Fig. 22a). Such a histogram for random events should be an exponential whose frequency f gives the mean rate of the events [57]. The value of f from the exponential was in excellent agreement with the mean rate of photon arrival obtained by ensemble averaging data from repetitive cathodic pulses. Second, the data followed a Poisson distribution (Fig. 22b):

$$P_m(t) = e^{-nft}(ft)^m/m! \tag{12}$$

where m is the number of cathodic pulses evaluated.

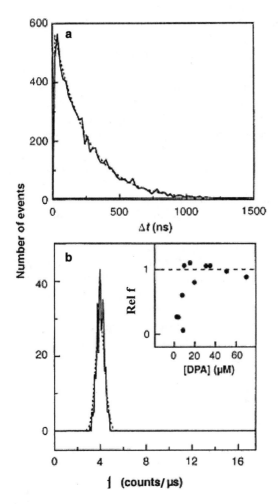

Figure 22 Poisson description of single-reaction events obtained in ACN solution containing 31 μM DPA. (a) Histogram (7000 inter-event times and a bin size of 20 ns) of interarrival times; (b) histogram (375 voltage pulses and a bin size of 0.8 counts/μs). The counts used were from the last 25 μs of the voltage pulse where the emission rate is pseudo-steady state. Dashed line: (a) exponential fit to data with rate of 205 counts/50 μs and (b) Poisson fit to data with rate of 200 counts/50 μs. Other conditions as in Figure 21. Insert: Plot of relative *f* values (*f* normalized by DPA concentration and electrode area and given as a ratio of the highest value) obtained from Poisson distribution. Data were obtained with electrodes of radii of 5.4 or 2.5 μm. (Adapted from Ref. 56).

The value of f was similar to that obtained in the exponential distribution. The measured value of f should be a function of the generation rate and diffusion of the radical ions, the rate and efficiency of the light-producing chemical reaction, and the photon collection efficiency. At concentrations greater than 20 μM, f, normalized by concentration and electrode area, was essentially constant (inset, Fig. 22b). The ratio of the mean photon rate (accounting for the detection efficiency) and the mean rate of generation of radical anions during each cathodic pulse directly yields the probability of a reaction generating a photon, i.e., Φ_{ecl}. At the high concentrations, this ratio yields a value of Φ_{ecl} of 6%, in good agreement with values reported for DPA in acetonitrile [1]. At lower concentrations, the normalized value of f decreases (inset, Fig. 22b).

In a later experiment, Wightman et al. [58] used the chemiluminescence arising from reaction of electrogenerated intermediates of DPA to generate images of microelectrodes with dimensions in the micrometer range. Lateral resolution was controlled by the use of rapid potential pulses that maintain the reaction zone in close proximity to the electrode. The solution employed, BN containing 0.1 M TBAPF$_6$, promotes high intensity because it enables dissolution of a high concentration (>25 mM) of DPA. In addition, radical anions of BN serve as a reagent reservoir to ensure efficient reaction of DPA radical cations to form the singlet excited state. Under such conditions, the measured ECL intensity could reach 3.2×10^5 photons/s per square micrometer of electrode area with a 1 kHz square-wave excitation. The images reveal that the electrode areas have quite different topography than that inferred from steady-state cyclic voltammograms.

B. Dual-Electrolysis Stopped-Flow ECL Method

If the ECL is observed in the reaction between the cation and anion radicals of an identical precursor, the reaction scheme is relatively simple as shown in Eqs. (13) and (14).

$$R^{+\bullet} + R^{-\bullet} \rightarrow {}^1R^* + R \tag{13}$$
$$^1R^* \rightarrow R + h\nu \tag{14}$$

For observation of ECL between cation and anion radicals of different precursors, however, there has been a significant restriction that arises from the energy levels of molecules when conventional methods such as the potential step method with a single electrode and the RRDE method are used. For example, it is very difficult to observe the reaction between the DPA radical anion (DPA$^{-\bullet}$) and rubrene radical cation (RU$^{+\bullet}$) by conventional techniques, because the reduction of rubrene, which is concurrent with the reduction of DPA, interferes with the observation of the reaction between DPA$^{-\bullet}$ and RU$^{+\bullet}$. Thus, for the reactions between different kinds of radical ions, ECL has been

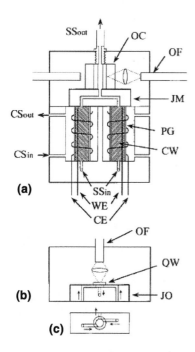

Figure 23 Schematic diagram of the cells for ECL measurements. (a) Cross-sectional view of dual electrolysis stopped-flow cell. CW, carbon wool working electrode; PG, porous glass tube; WE, Pt lead wire to carbon wool working electrode; CE, Pt wire counter electrode; JM, jet mixer; OC, optical cell; OF, optical fiber; SS, sample solution; CS, counter solution. (b) Cross-section, view of optical cell for ECL observation. OF, optical fiber; QW, quartz window; JO, jet-mixing optical cell. (c) Top view of the jet-mixing optical cell in (b). (Adapted from Ref. 60.)

unambiguously observed only for "energy-deficient" systems, e.g., the reaction between N,N,N',N'-tetramethyl-p-phenylenediamine cation radical (TMPD$^{+\bullet}$) and DPA$^{-\bullet}$ [1].

To make ECL measurements on "energy-sufficient" systems composed of different kinds of radical ions, Oyama and Okazaki [59,60] developed a novel technique using a pulse-electrolysis stopped-flow method using dual electrolysis columns. In this apparatus, after generating cation and anion radicals separately, the ECL was observed in an optical cell, and hence various energy-sufficient systems can be studied, including short-lived species. The cross-sectional view of the dual electrolysis stopped-flow cell is shown in Fig. 23a. Carbon wool (CW) with a feltlike structure was used as the working electrode. It was packed tightly in a microporous glass diaphragm tube (PG, 35 mm length $\times$ 6 mm i.d.). The CW was contacted electrically with a Pt wire

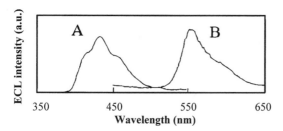

Figure 24   Emission spectra from (A) ¹DPA* and (B) ¹RU*. (Adapted from Ref. 60.)

(WE, 0.3 mm diameter) wound around the PG, which served as a counter electrode (CE). By using these column electrodes, a solution of cation radical can be generated quantitatively within a short time in one column and that of the anion radical in the other. Both constant-current and controlled potential pulse electrolyses could be performed with reference electrode attached on the back of the electrolysis columns.

For the ECL observation, a jet mixing optical cell (JO) as shown in Figure 23b was used. This cell was designed to collect the emission directly from the point where the two solutions were mixed, so that high-sensitivity measurement could be carried out. Both a photon-counting system and an image-intensified multichannel photodiode array detector (II-MCPD) were employed. The former was used to monitor the time course of the ECL intensity, and the latter for spectral recording. These two detection systems, the electrolysis cell, and the stopped-flow apparatus were constructed with the assistance of Unisoku Co. Ltd. (Hirakata, Japan). All solutions used in the ECL measurements were carefully degassed. For each solution, using a flask with two stopcocks, the oxygen was removed and replaced with nitrogen with a vacuum pump.

Oyama and Okazaki demonstrated the capability of their apparatus by studying the ECL of an energy-sufficient system, e.g., observing ECL between DPA⁻˙ and RU⁺˙, which has been difficult with conventional methods because RU is more easily reduced than DPA. In Figure 24 curve A is the emission spectrum of the ECL reaction between DPA⁺˙ and DPA⁻˙ the emission from ¹DPA* was clearly observed. For the reaction between DPA⁻˙ and RU⁺˙, the ECL was successfully observed by preparing the solutions of RU⁺˙ and DPA⁻˙ independently and mixing them directly. In this case, the observed emission spectrum was that from ¹RU* (see Figure 24, curve B), suggesting that the ECL was generated through the reaction scheme expressed by the equation

$$RU^{+\bullet} + DPA^{-\bullet} \rightarrow {}^1RU^* + DPA \tag{15}$$

Five other energy-sufficient systems—$RU^{+\bullet}$–$PY^{-\bullet}$ (pyrene radical anion), $DPA^{+\bullet}$–$RU^{-\bullet}$, $RU^{-\bullet}$–$TH^{+\bullet}$ (thianthrene radical cation), $TH^{+\bullet}$–$DPA^{-\bullet}$ and $DPA^{+\bullet}$–$PY^{-\bullet}$—were also studied by the proposed technique, and the results indicated that the ECL was emitted from the lower available singlet state after the direct electron transfer [60].

C. ECL/Scanning Electrochemical Microscopy/Scanning Optical Microscopy

We describe here the generation of visible light by the ECL technique at a scanning electrochemical microscope (SECM) [61,62] tip as the tip is moved in the vicinity of a substrate. Detection of ECL emission by this technique has potential applications to elucidation of the mechanisms of light-generating reactions and in scanning optical microscopy. ECL generation at UME tips can be accomplished either by the radical annihilation approach (in nonaqueous solvents) [14,63] or by a coreactant route (also applicable in aqueous solutions) [64,65].

1. Instrumentation and Methods

The basic principles of the experiment are illustrated in Figure 25. As in SECM and other forms of SPM [66], the UME tip is moved and positioned with high resolution by piezoelectric elements. The ECL cell (Fig. 26a) and bipotentiostat for controlling the tip and substrate potentials and generating current are similar to those employed in SECM. These electrochemical components combined with the positioning techniques used in SPM allow for manipulating a tip (or substrate) and measuring the current at high spatial resolution. The ECL intensity can be measured with a PMT (e.g., Hamamatsu R4220p or R928p) and a photon-counting system (e.g., Model C1230, Hamamatsu Corp., Middlesex, NJ) or with a PMT–electrometer (e.g., Keithley 6517 electrometer) combination. Approach curves in which the current and ECL intensity varied as functions of tip–substrate separation d are performed. Monitoring tip current and ECL intensity could be useful in approaching the tip to the substrate and supplying information on d. Constant-height current and optical images could be obtained by maintaining d and rastering the tip along the surface. In this case, sharpened UME suitable for SECM experiments, e.g., sharpened glass-encapsulated Pt disks, can be used as the tips. These tips can be prepared on the basis of the procedures described previously [61,62].

However, to bring the tip very close to the substrate surface, shear force, e.g., tuning fork–based, or another sensing method may be required to allow

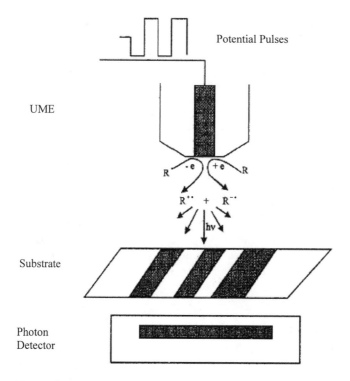

Figure 25 Schematic diagram illustrating the operating principles of ECL generation at an SECM in the alternating potential pulses mode. ECL is generated by the annihilation scheme $R^{\bullet+} + R^{\bullet-} \rightarrow 2R + h\nu$ and is detected by a PMT after attenuation by the substrate. (Adapted from Ref. 64.)

scanning at near-field distances. The block diagram of a custom-built apparatus used in our lab [65] for ECL imaging and measurement is shown in Fig. 26b. In this apparatus, a tuning fork–based shear force sensor [67] is used to monitor and control the tip-to-substrate separation. Briefly illustrated, the tuning fork is dithered laterally along the substrate surface at its resonant frequency by an attached piezoelectric element, and the alternating current (or voltage) output of the tuning fork is detected with a phase-sensitive technique. Interaction of the tip, which is attached to one prong of the tuning fork, with a surface causes a decrease in the amplitude of the oscillation. This signals when the tip touches the surface (also see Section III.D.1 for more discussion). For electrochemical and ECL measurements, a tip that is well-insulated (except at the very end) is extended beyond the end of the tuning fork by ~1 mm. A thin layer (~0.5 mm) of solution is used in the experiment, and only the tip is immersed in the

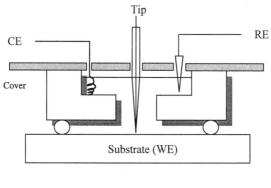

(a)

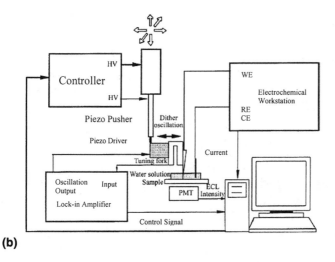

(b)

Figure 26 (a) Electrochemical cell for the four-electrode configuration. WE, working electrode (substrate); CE, counter electrode; RE, reference electrode; Tip, SECM tip. (b) Block diagram of apparatus used in the ECL imaging experiments. (Adapted from Ref. 65.)

solution, while the tuning fork remains in air to maintain its sensitivity. A LabVIEW (National Instruments, Austin, TX) program is used to control the approach and scanning of the tip over a substrate and to acquire the data.

SECM/ECL tips used in tuning fork–based shear-force sensors have at least three major restrictions: They need to be a good electrode material, i.e., have well-behaved electrochemical activity for reactants, well-insulated except at the very end and the part not immersed in the solution, and not too stiff, which would reduce the sensitivity of the tuning fork considerably. Tips satisfying these three criteria, with effective diameters from several nanometers to several micrometers and suitable for ECL experiments in aqueous solutions, have been successfully prepared based on the procedure reported previously [65,68,69]. Briefly described, a fine (25 or 250 μm diameter) Pt wire electrochemically etched or an Au-coated thin optical fiber was coated with electrophoretic paint anodically. After heating at ~160°C, an exposed conical or ring-shaped sharp apex was naturally formed.

2. Kinds of Experiments

The $Ru(bpy)_3^{2+}$ system is used in most experiments described here. It is quite possible to apply similar techniques to other systems. One experiment demonstrates the well-behaved electrochemistry of $Ru(bpy)_3^{2+}$ at the tip in SECM experiments. As shown in Fig. 27a, cyclic voltammograms of 1 mM $Ru(bpy)_3^{2+}$ in ACN containing 0.2 M TBABF$_4$ as the supporting electrolyte at a Pt UME (radius $a = 12.5$ μm) show three reduction waves at half-wave potentials of $E_{1/2} = -1.30$, -1.49, and -1.73 V vs. SSCE corresponding to the reduction of $Ru(bpy)_3^{2+}$ to the $+1$, 0, and -1 species. The oxidation of $Ru(bpy)_3^{2+}$ to $Ru(bpy)_3^{3+}$ occurs at $E_{1/2} = 1.32$ V vs. SSCE. As the tip is moved close to the ITO substrate biased at -0.30 V vs. SSCE ($d = 0.5a$, where d is the separation between tip and substrate), the steady-state tip current i_T increases to about twice the magnitude of $i_{T,\infty}$, the steady-state tip current when the tip is far from the substrate (compare curve 2 with curve 1 of Fig. 27a). This positive feedback effect is further depicted in the SECM tip approach curve (Fig. 27b), in which the tip is biased at 1.60 V vs. SSCE to oxidize $Ru(bpy)_3^{2+}$ to $Ru(bpy)_3^{3+}$ while the substrate is biased at -0.30 V vs. SSCE to reduce $Ru(bpy)_3^{3+}$ back to $Ru(bpy)_3^{2+}$, which diffuses back to the tip. As the tip is moved toward a properly biased conductive substrate surface, i_T increases with decreasing d, as expected from the theory [61]. This experiment supplies information on d and may also suggest the reaction kinetics.

a. ECL Generated by Annihilation Scheme. To generate ECL by an annihilation scheme, as discussed in Sections II.B.2 and III.A, one can continuously apply an alternating sequence of potential pulses between 1.60 and -1.40 V vs. SSCE where $Ru(bpy)_3^{3+}$ and $Ru(bpy)_3^+$ are sequentially generated at the surface of the tip. A strong and nearly steady (by using a current follower having a reasonably long time constant) ECL intensity can be achieved through the annihilation reaction

$$Ru(bpy)_3^{3+} + Ru(bpy)_3^+ \rightarrow 2\, Ru(bpy)_3^{2+} + h\nu \tag{16}$$

Figure 28 shows typical waveforms of the tip current and ECL intensity accompanied by several cycles of square-wave potential of 5 ms pulse width (τ).

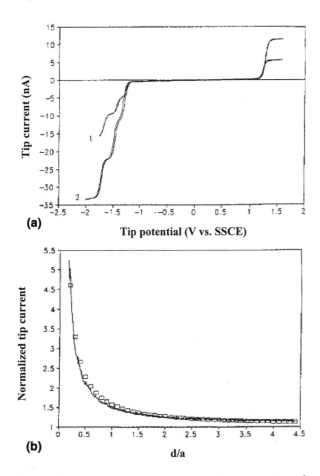

Figure 27 (a) Cyclic voltammograms of 1 mM Ru(bpy)$_3^{2+}$ in 0.2 M TBABF$_4$/ACN solution at a Pt microdisk electrode (radius 12.5 μm) with the tip far from an ITO substrate (curve 1) and 6.25 μm from the substrate (curve 2). Scan rate, 5 mV/s. (b) Dependence of tip current on relative tip displacement over an ITO substrate (substrate potential E_s = −0.30 V vs. SSCE). The solution and tip were the same as those in (a). The tip was biased at 1.60 V vs. SSCE. The tip was moved to the substrate at a speed of 0.3 μm/s. Solid curve is experimental data, and squares are simulated data for a conducting substrate. (Adapted from Ref. 64.)

An almost constant ECL intensity was obtained due to the electronic filtering of the light-measuring circuit, which had a time constant of ~0.5 ms. The shape of the tip current wave was not distorted by the current-measuring circuit and followed the current transient behavior for UME reasonably well except in the short-time-scale region where double-layer charging is important [70].

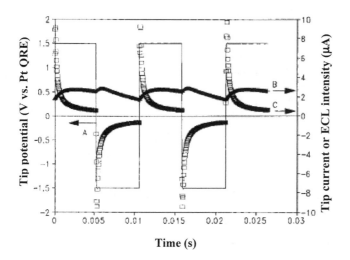

Figure 28 Typical waveforms of tip current (squares, curve C) and ECL intensity (pluses, curve B) accompanied by several cycles of square-wave potential of 5 ms pulse width (solid line, curve A). The solution contained 1 mM Ru(bpy)$_3^{2+}$ in 0.2 M TBABF$_4$/ACN. Tip and substrate ($E_s = -0.30$ V vs. SSCE) were the same as those used in Figure 27. (Adapted from Ref. 64.)

The ECL intensity increased monotonically with decreasing pulse width in the range of τ studied (50 μs to 10 ms). It was proportional to the inverse square root of τ when τ was greater than 0.5 ms. Triple-step ECL experiments have previously shown that the total photonic emission rate (einsteins/s) I can be evaluated from the Faradaic charge passed in the forward step Q_f, the coulometric efficiency Φ_c, and the pulse width τ, through Eq. (3). For a coplanar microdisk electrode, Q_f is well approximated by [70]

$$Q_f = 4nFC_R{}^*aD_R[\tau + 2a(\tau/\pi D_R)^{1/2}] \tag{17}$$

where n is the number of electrons involved in the redox reaction and $C_R{}^*$ and D_R are the bulk concentration and diffusion coefficient of the reactant. Because I is constant or a very slowly varying function of time, combining Eqs. (3) and (17) gives

$$I = 2naD_RC_R{}^* \Phi_c[1 + 2a/(\pi\tau D_R)^{1/2}] \tag{18}$$

Thus, if there are no other complications, ECL intensity is expected to increase linearly with the concentration of the luminescent reactant and quadratically with the tip radius and with the inverse square root of τ, as is reasonably confirmed experimentally when $\tau > 0.5$ ms. However, deviation from this relation occurs at very short τ, perhaps because of interference from double-layer charging.

The ECL intensity with the tip close to either an insulating or a conductive substrate will be different from that in bulk solution because of hindered diffusion and feedback effects. It was observed [64] in a multicycle pulse (τ = 10 ms) experiment that the ECL intensity was nearly steady as the tip approached a substrate (insulating or conductive) until it was a few tenths of the tip radius ($d \sim 0.3a \sim 0.4$ μm) away from the substrate surface. The ECL intensity then decreased sharply and almost linearly with decreasing d. This behavior is quite different from that of the i_T vs. d curve normally observed in SECM. For an insulating substrate, i_T at a constant tip bias decreases smoothly over a distance of a few tip radii before the tip reaches the substrate surface. For a properly biased conductive substrate, i_T increases nearly inversely with decreasing d. Although exact mathematical descriptions of these experimentally observed (ECL intensity vs. d) curves are still not established, qualitative rationalization of their shapes is possible by comparing the pulse width with the diffusional transit time t_d, expressed as d^2/D_R. In the annihilation scheme, the ECL intensity depends on the annihilation reaction, Eq. (16), occurring in a reaction zone near the tip electrode. For a given τ, this zone has a thickness of about $(2D_R\tau)^{1/2}$. As long as this thickness is small compared to d, the ECL at the tip will not sense or be appreciably perturbed by the presence of the substrate. At a distance where $d^2/2D_R$ is larger than τ, ECL intensity is dominated by the diffusion layer developed at the tip during the time of the potential transient. Over this distance range, the ECL intensity depends mainly on τ and is essentially independent of distance. Conversely, when d is so small that $d^2/2D_R$ is smaller than τ, the effect of the substrate on the response becomes important. The ECL intensity in this distance range will depend strongly on d, with the transition from one regime to the other occurring at a distance near $(2D_R\tau)^{1/2}$, which is ~5 μm for τ = 10 ms and D_R = 1.2 × 10^{-5} cm^2/s.

b. *ECL Generated by Coreactant Scheme.* As in conventional ECL experiments, an alternative method of generating ECL for an ECL/SECM experiment involves the use of a coreactant, e.g., generating Ru(bpy)$_3$$^{3+}$ in the presence of oxalate [71] or tripropylamine (TPrA) [15,72]. This allows ECL to be generated at a constant (rather than alternating) potential and permits the use of either a nonaqueous or aqueous solution [73]. This scheme is quite useful for ECL/SECM imaging because steady-state ECL can be obtained as discussed below.

In a 0.2 M TBABF$_4$ ACN solution containing only 1 mM Ru(bpy)$_3$$^{2+}$, a wave corresponding to the oxidation of Ru(bpy)$_3$$^{2+}$ occurs at $E_{1/2}$ = 1.32 V vs. SSCE (curve A, Fig. 29a). The addition of 0.5 mM TPrA to this solution produces a voltammogram (curve B, Fig. 29a) with a new additional wave at $E_{1/2}$ = 1.10 V, which is coincident with the oxidation wave observed in a 0.2 M TBABF$_4$ ACN solution containing 1 mM TPrA (curve C, Fig. 29a), suggesting

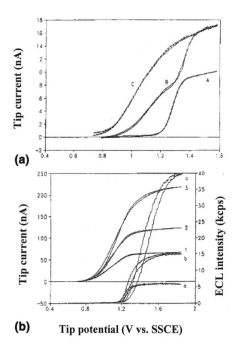

(a)

(b)   Tip potential (V vs. SSCE)

Figure 29 (a) Cyclic voltammograms of 1 mM Ru(bpy)$_3^{2+}$ in 0.2 M TBABF$_4$/ACN solution (curve A) at a Pt microdisk electrode (radius 12.5 μm), (curve B) after addition of 0.5 mM TPrA to the solution used in (a), and (curve C) in a 1 mM TPrA/0.2 M TBABF$_4$/ACN solution. Scan rate, 5 mV/s. (b) Cyclic voltammograms at the same Pt microdisk in ACN solutions containing 1 mM Ru(bpy)$_3^{2+}$, 0.2 M TBABF$_4$ as the supporting electrolyte, and 3.5 (curve 1), 6.5 (curve 2), and 12 mM (curve 3) TPrA. The corresponding ECL (intensity vs. potential) curves are shown for (a), 3.5; (b), 6.5; and (c), 12 mM TPrA. Kcps: kilocounts per second. (Adapted from Ref. 64.)

that the oxidation wave at $E_{1/2}$ = 1.10 V corresponds to the direct oxidation of TPrA. By scanning the tip potential to where both TPrA and Ru(bpy)$_3^{2+}$ are oxidized (~ 1.2 V vs. SSCE), a fairly strong and steady ECL intensity can be observed (curves a, b, and c, Fig. 29b). Note that significant ECL intensity is detected at potentials only where Ru(bpy)$_3^{2+}$ is oxidized. The reaction responsible for ECL generation in this potential region is generally believed to involve the catalytic oxidation of TPrA by the electrogenerated Ru(bpy)$_3^{3+}$ [15,72]. Refer to Chapter 5 for detailed discussions on the mechanisms of various coreactant schemes.

In Figure 30 we show the distance dependence of i_T and ECL intensity over a biased conducting substrate, such as ITO, in an ACN solution containing 1 mM Ru(bpy)$_3^{2+}$, 20 mM TPrA, and 0.2 M TBABF$_4$ as the supporting

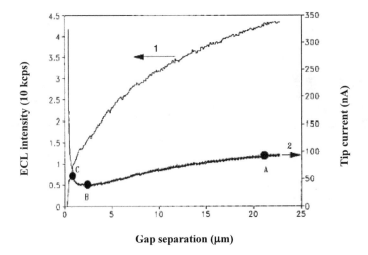

Figure 30 Distance dependence of ECL intensity (curve 1) and tip current (curve 2) over a conducting substrate (ITO) in an ACN solution containing 1 mM $Ru(bpy)_3^{2+}$, 20 mM TPrA, and 0.2 M $TBABF_4$ as the supporting electrolyte. The tip (Pt, radius 12.5 µm) was biased at 2.1 V vs. SSCE and the substrate at -0.30 V vs. SSCE. kcps: kilocounts per second. (Adapted from Ref. 64.)

electrolyte. As shown, when the tip ($a = 12.5$ µm) was biased at $+2.1$ V vs. SSCE to oxidize both TPrA and $Ru(bpy)_3^{2+}$ and the substrate was biased at -0.30 V, ECL intensity decreased monotonically as the tip approached the ITO surface because of the SECM "negative feedback" of TPrA oxidation due to its irreversibility. However, i_T decreased to a minimum and then increased as the tip came even closer to the ITO surface before it contacted the surface. The final increase in i_T could be attributed to the SECM "positive feedback" and TPrA catalytic effect of $Ru(bpy)_3^{2+}$ oxidation. The histograms of ECL intensity recorded for three different gap separations between the tip and ITO substrate (points A, B, and C in Fig. 30) showed distributions of ECL intensity spikes characteristic of discrete random events [74]. The probability density functions of all three ECL intensity distributions followed the Poisson density function, Eq. (12). As the tip approached an insulating substrate such as a quartz plate, both ECL and i_T decreased monotonically with decreasing distance, as expected for the "negative feedback" effect of SECM.

c. ECL Imaging. One of the purposes of the ECL imaging technique is to explore the possibility of using an ECL probe as a small light source for optical imaging. As a demonstration, Pt UMEs of various effective radii were used as an ECL probe to image optically an interdigitated array (IDA) consisting of Au

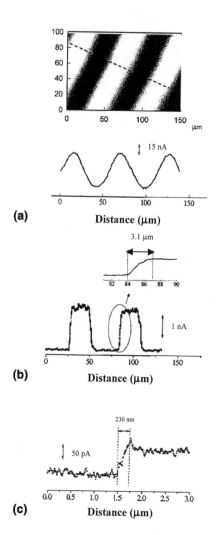

(a)

(b)

(c)

Figure 31   (a) Constant-height ECL image and its cross section of an IDA consisting of Au bands (30 μm wide) spaced 25 μm apart deposited on a glass substrate. Tip size, 25 μm diameter. Tip-to-substrate distance, ~5 μm. Data acquisition time, ~10 min. (b) Cross section of a constant-height ECL image of the same IDA. Tip size, ~2 μm diameter. Tip-to-substrate, ~1 μm. (c) Cross section of an ECL near-field image of the same IDA. A tuning fork–based shear force sensor is used to control the tip-to-substrate separation of ~24 nm. Effective tip size is 155 nm diameter. Data acquisition time for one line is ~ 25 s. (Adapted from Ref. 65.)

bands (~ 30 µm wide) spaced 25 µm apart deposited on a glass substrate [65]. Figure 31a shows the ECL image of the IDA with a 12.5 µm radius Pt disk electrode. The image was obtained with the distance between tip and substrate adjusted to a certain value using SECM feedback in 1 mM $Ru(NH_3)_6^{3+}$ solution; this was then replaced by 1 mM $Ru(bpy)_3^{2+}$/0.1 M TPrA/0.15 M phosphate buffer solution (pH 7.5) before imaging. During, lateral scan of the tip position for imaging, the tip was held at a constant height (~5 µm) above the substrate and the ECL intensity was monitored with a PMT located underneath the substrate. The edge of the sharpness of the signal obtained manifests the resolution of a microscope. The resolution obtained in Fig. 31a, although it demonstrates the capability of this technique for imaging is quite low owing to the large size of the tip. When a 1.0 µm radius Pt UME was used as the probe and the tip-to-substrate distance was held at ~1 µm, the ECL image resolution increased significantly, as shown in Fig. 31b.

To improve the resolution further to the submicrometer range and bring it closer to the near-field regime (50–150 nm), both the tip size and the tip–substrate distance need to be reduced greatly. For very small tips, force or other sensing methods may be required to allow scanning of the tip at close distances. Fig. 31c shows the cross section of the ECL image of one edge in the IDA structure using a 155 nm diameter (effective) Pt tip. A tuning fork–based shear force sensor was used to sense the substrate surface and control the tip-to-substrate separation [65]. The resolution was dramatically increased compared to that obtained using the micrometer-sized tips (Figs. 31a and 31b). The width of the ECL signal across the edge of the glass/gold bands was ~230 nm, close to about one-third of the wavelength of the ECL emission wavelength maximum (~645 nm) for the $Ru(bpy)_3^{2+}$/TPrA system.

An interesting technique developed by Wightman and coworkers [58,63] for decreasing the size of the ECL light source but still generating ECL intensity (≥1.8 pW) strong enough for high-resolution ECL imaging is the one based on high-frequency alternating pulses as described in Section III.B. Figure 32 shows a series of optical photographs of an ECL-emitting conical carbon fiber electrode at various frequencies of applied square waves in benzonitrile containing 25 mM DPA and 10 mM $TBAPF_6$. Square-wave potentials sufficient to produce the radical cation of DPA and radical anion of BN were applied to the electrode at different frequencies. At low frequencies (e.g., 200 Hz), ECL is observed to occur along the whole uninsulated region. The image appears to be diffuse. As the frequency is increased, the ECL intensity increases. At higher frequencies, the area of the cone supporting emission diminishes. At the highest frequency shown, 20 kHz, the ECL is supported only at the apex of the cone. This phenomenon allows an uninsulated conical carbon fiber electrode to form a small light source. Submicrometer resolution has been achieved with this kind of light source [63].

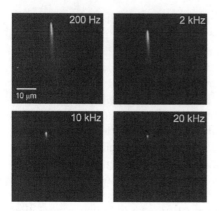

Figure 32 Optical photographs of ECL-emitting conical carbon fiber electrode at various frequencies of applied square wave. Electrode is immersed in benzonitrile containing 25 mM DPA and 10 mM TBAPF$_6$ at an angle of 45° with respect to the plane of observation. Amplitude and offset of applied potential were adjusted to maximum light intensity. (Adapted from Ref. 63.)

D. ECL/Force Microscopy on Thin Films

With interest growing in solid-state molecular electroluminescent devices (MELDs) (sometimes generically called "organic" light emitting devices or OLEDs), efforts have been made to investigate such devices based on thin films of polymers or small molecules (see, e.g., Ref. 75). The ruthenium(II) complexes of bipyridine, phenanthroline, and their derivatives have recently been used in solid-state based light-emitting devices (see, e.g., Ref. 76). High-brightness and high-efficiency emission with short delay times to reach maximum emission have been reported. An electrochemical mechanism has been suggested as operative in these solid-state devices, as originally proposed in solution phase ECL [12,29,52]. We introduce here very briefly the use of tuning fork–based scanning probe microscopy (TFSPM) in combination with ECL and other electrochemical techniques for the characterization of topographic, electrical, and electroluminescent properties of solid films used in a MELD (for mechanisms of electrogenerated luminescence of solid thin films, see Chapter 10)

1. Experimental Setup and Methods

The primary components of a transmitted-light near-field luminescence microscope for thin film MELD are shown in Figure 33. The optical elements used to collect the luminescence generated near the tip and to deliver it to the optical detectors are conventional and are described here only very briefly. A

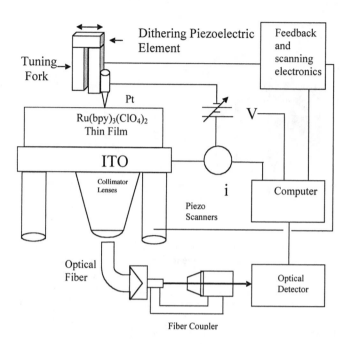

Figure 33 Schematic representation of the measurement with a tuning fork–based SPM tip contacting the surface of a $Ru(bpy)_3(ClO_4)_2$ thin film on ITO substrate. Optical components include fiber optics coupling to a microscope objective for focusing the luminescence to optical detectors through a fiber coupler. The sample is placed on a piezoelectric stage, which provides x, y, and z motion of the sample.

conventional microscope objective is used to collect near-field luminescence which is delivered to the optical detector (or a collimator–detector assembly) through a fiber coupler. The luminescence intensity is measured with either a Hamamatsu R4220 photomultiplier tube (PMT) operated at -750 V or a time resolved photon-counting system (Model T914P, EG&G). The PMT current is measured with a Keithley Model 6517 electrometer. The sample is placed on a piezoelectric stage, which provides x, y, and z motion of the sample.

For high-resolution near-field imaging the probe–sample separation must be maintained at a distance comparable to the effective size of the probe. In a typical imaging experiment, the probe–sample separation is held constant while the sample is scanned in the xy plane. The sample or tip must move up and down in response to sample topography. The electronic circuitry and piezoelectric positioners employed in this process are very similar to those used in standard SPMs [66]. However, the methods for sensing tip–sample separations are varied, for example, by use of a laser beam dither detection scheme [77], tuning fork–based shear force

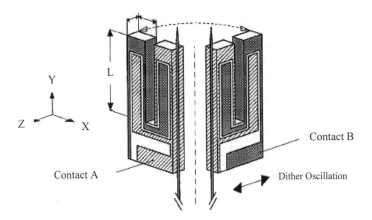

Figure 34 Front and back views of a quartz crystal tuning fork and a probe mount for nonoptical shear force feedback. The probe is cemented to one leg of the quartz crystal tuning fork. X, Y, and Z are the quartz crystal axes. Probe motion is driven at a tuning fork resonance along the X axis, which is used to sense proximity of the probe to the sample surface. The shaded and dark areas represent contact pads serving both as pickup for the piezoelectric signal and as coupling between the two prongs. L is the length of the prong. The tip protrudes ~1 mm out of the prong's end. (Based on Ref. 67.)

detection scheme [67], or tapping mode [78]. To eliminate optical interference from the laser beam, we adapted a tuning fork–based shear force feedback in this experiment.

As shown in Figure 34, a piezoelectric quartz tuning fork is used for shear force detection. Such tuning forks are commercially available for operation at 32,768 Hz or higher (Digi-Key, Thief River Falls, MN). Here, the probe (a sharply etched electrically conducting tip) is mounted directly to the tuning fork. A conventional piezoelectric tube or block is employed to drive probe dithering parallel to the sample surface; however, the resonance employed is that of the tuning fork. Therefore, probe dithering is a result of tuning fork motion rather than a probe resonance. Interaction between the probe and substrate surface alter the tuning fork resonance. The feedback signal is acquired electronically by detecting the time-dependent voltage developed across the tuning fork with a phase-sensitive technique. We usually monitor the change in the amplitude of the oscillation with a lock-in amplifier. This surface-sensing technique is highly sensitive and avoids strong interaction between the probe and the sample surface. Readers who are interested in details of the operation principles of a tuning fork–based shear force sensor can refer to Ref. 67.

Current–voltage responses of a thin film on substrate [e.g., a $Ru(bpy)_3(ClO_4)_2$ thin film on ITO] can be examined by voltage steps and sweeps across the substrate and the tip, which is positioned at the surface or inside the film. The voltage is supplied from a waveform generator (e.g., an EG&G Model 175 universal programmer or a Wavetek Model 143 function generator), controlled by an IBM PC equipped with a DT2821 interface board. The picoampere-level currents are monitored with a high sensitivity current amplifier. Specific experimental procedures and parameters are described in more detail in the following section.

2. Application on Solid-State Ru(bpy)$_3$(ClO$_4$)$_2$ Thin Film

Several experiments can be performed with this technique, including current–voltage and electroluminescence–voltage relations; current–, shear force–, and electroluminescence–distance relations; current and electroluminescence transients; and imaging.

Figure 35 shows some typical current–voltage and luminescence–voltage plots of a 100 nm thick Ru(bpy)$_3$(ClO$_4$)$_2$ layer on ITO with a Pt tip positioned inside the film (about 50 nm deep). As the tip voltage (E_a) was scanned negative with respect to the ITO substrate, significant current began to flow through the device at about -1.7 V, whereas detectable light emission was observed only at negative tip voltages of about -2.7 V. In the positive-bias region, current flow started at about $+2.3$ V, and luminescence was observed at a positive tip bias of about $+2.7$ V. These onset voltages for light emission of 2.7 V are slightly

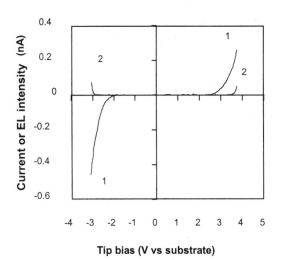

Figure 35   Typical current–voltage (curve 1) and luminescence–voltage (curve 2) plots of a 100 nm thick Ru(bpy)$_3$(ClO$_4$)$_2$ layer on ITO with a Pt tip positioned inside the film (approximately 60 nm from the ITO surface). Voltage scan rate is 0.1 V/s.

higher than those reported for solution ECL of the same compound [29] but are roughly the same as those seen in solid-state devices with different contact materials. Similar current– and luminescence–voltage behavior was observed when Au on glass, instead of ITO, was used as the substrate.

When the tip was biased at +3.0 or −3.0 V with respect to the ITO substrate and it approached the surface of a Ru(bpy)$_3$(ClO$_4$)$_2$ film, current flow was observed at the position where change in shear force was detected (Fig. 36). The current response showed only a slight delay with respect to the shear force response when the tip was biased at +3.0 V. At +3.0 V tip bias, light was emitted from the surface region of the film as soon as the current flow started through the device. However, at −3.0 V bias, light emission spatially lagged behind both current and the change in shear force. A delay of the emission with respect to the current at negative bias was also observed when Au, instead of ITO, was used as the substrate. This apparent depth profile of light emission is mainly associated with the time behavior of the carrier injection and transport as described below, because the luminescence profile is more nearly coincident with the current profile in the reverse displacement scan.

It is quite possible to estimate the local thickness of the film from the difference between the onset substrate displacement to observe current (or shear force change) and the displacement where tunneling (or abrupt change in the shear force) takes place. The estimation of a local film thickness of about 100 nm

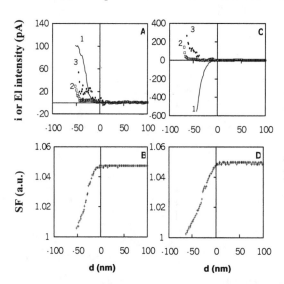

Figure 36 Current, luminescence, and shear force (SF) as functions of distance (d). In frames A and B, tip biased at +3.0 V; in frames C and D, tip was biased at −3.0 V. Curves 1 and 2 represent current i and luminescence intensity (El), respectively. Curves 3 are expanded curve 2 (5 times for frame A and 10 times for frame C).

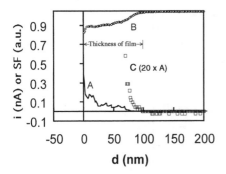

Figure 37 Film thickness measurement from the current vs. distance (A) or shear force vs. distance (B) curves. C is the expanded (20 times) curve of A. Tip was biased at -3.0 V during approach, and the tip current is inverted.

is shown in Figure 37, which agrees fairly well with that determined by making a light scratch on the surface and measuring the line profile across it with an atomic force microscope.

Another useful experiment obtains the current transient from a potential step, which supplies some information about charge carrier injection and field-driven charge transport properties through the film. Figure 38a show a series of time-dependent current flows through a $Ru(bpy)_3(ClO_4)_2$ film ~100 nm thick coated on ITO as the Pt tip was positioned approximately 60 nm away from the ITO surface and a positive voltage step was applied between tip and substrate. The voltage step produced a fast (≤ 1 ms) transient current spike (barely distinguishable from the y axis) followed by a slow rise. A smaller current spike was seen when the tip was further from the film surface. Part of the initial current spike is apparently associated with the stray capacitance of the current-measurement circuit, which has a time constant set at about 20 μs for this experiment. Part of it is associated with the ionic current of the film.

As shown in Figure 38a, at low positive tip bias (e.g., ≤ 3.0 V), the current rose slowly and approached a steady-state plateau within a few seconds. An increase of the step voltage to 4.0 V increased the initial rise speed of the current, which reached a steady-state plateau within 2 s. An increase of the step voltage to 4.5 V had no dramatic effect on the shape of the initial part of the current–time curve, but it substantially affected the current on a longer time scale. Similar but not identical current transients were obtained by stepping the tip bias from 0 V to different negative values and recording the current as a function of time. By comparing the shape of the current vs. time curves to theoretical behavior, it was possible to find the effective diffusion coefficients (or mobilities) of charge carriers. From such measurements, we found, as shown in Figure 38b, that the effective diffusion coefficients of both electrons and holes depended strongly on the electric field

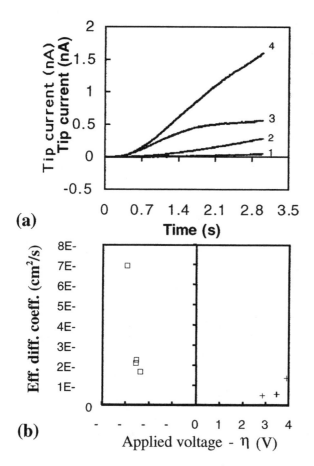

Figure 38 (a) A series of current transients through an approximately 100 nm thick $Ru(bpy)_3(ClO_4)_2$ layer on ITO as the Pt tip was positioned within the film ~60 nm away from the ITO surface and a voltage step was applied between tip and substrate. Tip voltage steps: (1) +2.8 V; (2) +3.5 V; (3) +4.0 V; (4) +4.5 V. (b) Effective diffusion coefficients of electrons (squares) and holes (pluses) as functions of applied voltage − overpotential at the tip, $(E_a - \eta)$.

strength (or the applied voltage) [79]. This information is important to understanding the slow rise in the current after a voltage step is applied to $Ru(bpy)_3^{2+}$-based MELD and why the initiation of light emission lags behind the current flow in the present experimental configuration.

Still other experiments explore the possibility of this technique for high-resolution imaging of electroluminescent $Ru(bpy)_3(ClO_4)_2$ thin film. Information about the morphology, conductance, and luminescence homogeneity of spin-cast $Ru(bpy)_3(ClO_4)_2$ films was obtained with TFSPM. Figure 39 shows the topography,

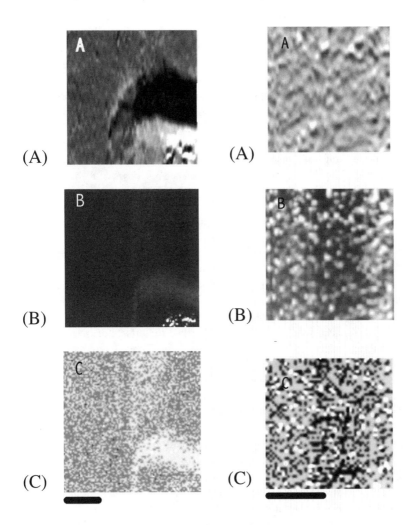

Figure 39 Topograph (A), current image at $+3.5$ V tip bias (B), and luminescence image at -3.5 V (C) of a single layer of $Ru(bpy)_3(ClO_4)_2$ on ITO, taken simultaneously with a Pt tip. Also shown in the right-hand frames of the figure are higher resolution images for the portion of surface at the upper left corner of the frames on the left. Tip raster rate is ~0.25 Hz. Marker is 1 μm.

current image at 3.5 V, and electroluminescence image at -3.5 V of a layer (~100 nm thick) of $Ru(bpy)_3(ClO_4)_2$ on ITO, taken simultaneously with a Pt tip. The topographic image was recorded at constant amplitude of shear force, which was set at 2% less than when the tip was far from the film surface. The tip bias was pulsed between $+3.5$ and -3.5 V with pulse duration of a few milliseconds, and the current at $+3.5$ V and luminescence at -3.5 V tip bias were monitored. On frame A at the left hand side of Figure 39, we show a micrometer-sized hole as a marker that was artificially made by a tip crash by approaching the tip to the substrate without turning the feedback loop on. Also shown in frame A on the right hand side of Figure 39 is a higher resolution image for the portion of surface at the upper left corner of the frame on the left. The image suggests that the film contains no particularly crystalline grains, but rather uniformly distributed nanostructures. These structures, however, produce heterogeneity in the current and luminescence responses of the film.

E. Magnetic Field Effects and Interception Techniques

Still other experiments are designed to investigate the effects of magnetic field on ECL intensities and to intercept intermediates. Studies along this line have been used for mechanistic diagnosis for certain reactions involving triplets; hence they are associated with the T route [1]. We describe here only very briefly some experimental essences of these techniques.

Since 1969, occasional reports have described the effects of magnetic field strength on the intensity of ECL from various systems. Table 4 summarizes some data by showing only the change in the relative ECL intensity at a point between 1.5 and 7.5 kG. The magnet that furnished the field in which the sample was immersed was a compact electromagnet device that was energized by a regulated but variable current supply [10,80]. Field strengths up to 7.5 kG could be applied to the samples.

In each instance for which a field effect has been seen, the enhancement increased monotonically; otherwise the change in relative ECL intensity was less than a few tenths of a percent as shown in Figure 40 [10]. As shown, enhancements in ECL intensity are seen for systems containing Wurster's Blue (WB) (radical cation of N,N,N',N'-tetramethyl-p-phenylenediamine) as an oxidant and 1,3,6,8-tetraphenylpyrene (TPP) as an emitter in DMF. The data for the mutual annihilation of TPP anion and cation radicals suggests that either no field effect on ECL intensity was observed or there was an extremely slight decrease of luminescence with increasing field strength. The results summarized in Table 4 show that all the tabulated energy-deficient systems show field enhancements, including ECL from intramolecular exciplexes of 1-amino-3-anthryl-[9]-propane derivatives [81]. For the other listed marginal case, i.e., the rubrene($+$)/rubrene($-$) annihilation system, the magnitude of the field effect is dependent on the solvent [82,83], supporting electrolyte concentration, and

Table 4 Magnetic Field Effects on ECL at Room Temperature

Reactants[a,b]	Field effect (%)[c]	Field strength (kG)	Energy classification
DPA(+)/DPA(−)	−0.3 ± 0.8	7.5	Sufficient
Thianthrene(+)/PPD(−)	−0.4 ± 0.3[e]	6.15	Sufficient[d]
TPP(+)/TPP(−)	+0.2 ± 0.2	6.15	Marginal
Rubrene(+)/Rubrene(−)	+9.5 ± 0.2	5.70	Marginal
	+6.5	1.52	
	+3.0	1.52[e]	
AAP(+)/AAP(−)	+16 − + 24	—	E route[f]
TMPD(+)/anthracene(−)	+16.3	6.25	Deficient
TMPD(+)/DPA(−)	+5.6 ± 0.3	6.00	Deficient
TMPD(+)/TPP(−)	+2.6 ± 0.2	6.15	Deficient
TMPD(+)/rubrene(−)	+28.2 ± 0.2	5.70	Deficient
TMPD(+)/tetracene	+18.0	7.50	Deficient
10-MP(+)/fluoranthene(−)	+7.0 ± 0.4	6.15	Deficient
Rubrene(+)/p-Benzoquinone(−)	+15.7 ± 0.2	5.70	Deficient

Source: Refs. 1 and 81.
[a]DPA = 9,10-diphenylanthracene; PPD = 2,5-diphenyl–1,3,4-oxadiazole; TPP = 1,3,6,8-tetra-phenylpyrene; TMPD = N,N,N',N'-tetramethyl-p-phenylenediamine; 10-MP = 10-methylphenothia-zine; AAP = 1-aminoanthryl-(9)-propane derivatives.
[b]In DMF solutions except for thianthrene(+)/PPD(-), which was in acetonitrile, and AAP(+)/AAP(−), which was in THF.
[c]Percentage change from zero-field intensity for field strength noted in adjacent column.
[d]Emission from thianthrene precursor only.
[e]From Ref. 85, at −48°C.
[f]Intramolecular exciplex formation (cf. Chapters 1 and 4 for discussions).

temperature [84,85]. The enhancement becomes smaller with decreasing solvent polarity and temperature. For the energy sufficient systems listed, no field enhancement was observed. Magnetic field effects have thus been used frequently as a diagnostic tool for the differentiation of mechanisms for ECL. However, for mixed ECL systems, e.g., the thianthrene(+)/PPD(−) system, where no magnetic field effect was observed [86] although ECL decay curve analysis showed that it was a limiting case of pure T route behavior [87], one should be more careful in interpreting magnetic field effects on ECL efficiencies (cf. Chapter 4 for insight discussions).

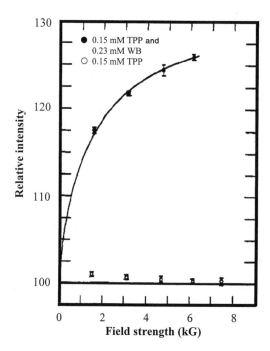

Figure 40 Magnetic field effects on luminescence from systems containing Würster's Blue as an oxidant and 1,3,6,8-tetraphenylpyrene (TPP) as an emitter in DMF containing TBAP as the supporting electrolyte. Error bars denote average deviations. Reactants: (•) TPP anion and Würster's Blue, (○) TPP anion and cation radicals. (Adapted from Ref. 10.)

A very sensitive method for probing T-route systems has involved intercepting a triplet intermediate, $^3D^*$, by triplet energy transfer to an acceptor species A, through the reaction

$$^3D^* + A \rightarrow D + {}^3A^* \tag{19}$$

The acceptor triplet can undergo triplet annihilation, and hence the addition of the acceptor may transform the ECL spectrum from emission of the donor to that of the acceptor. Alternatively, it might undergo radiationless decay, resulting in quenching of the ECL emission. Moreover, the acceptor triplet can undergo a photochemical reaction, so that an analysis of the amount of the product of the photochemical reaction created by a known number of redox events can give triplet yields. Few interception experiments have yet been carried out with ECL systems [22; see also, e.g., Ref 88], because the conditions for unambiguous transfer are quite stringent. The basic problem is the strong correlation between the energies of

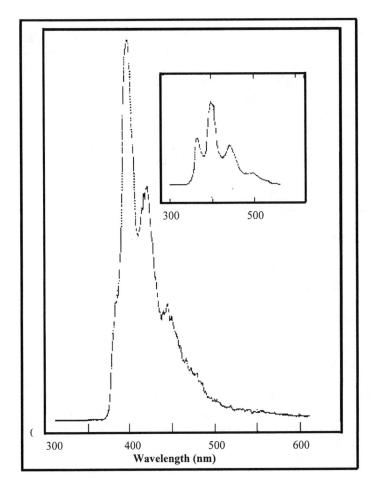

Figure 41 Chemiluminescence spectrum obtained in energy transfer experiment with fluoranthene and anthracene. Inset shows anthracene fluorescence spectrum (10^{-5} M in DMF). (Adapted from Ref. 22.)

a molecule's lowest excited states and the potentials at which it undergoes oxidation and reduction to stable radical ions. Thus, a suitable acceptor for a given system, which must have its lowest triplet at a lower energy than the donor triplet, will ordinarily reduce or oxidize more easily than one of the ion precursors. It therefore interferes severely with ECL generation, and any modification in ECL behavior upon its addition complicates the interpretation. Another important problem is the need for chemical inertness of the acceptor toward the radical ions.

One of the previously reported results demonstrating triplet energy transfer is shown in Figure 41, which arose from a study of the energy-deficient reaction between the cation radical of 10-methylphenothiazine (10-MP) and the anion radical of fluoranthene (FA) [22]. The FA molecule has small S_1–T_1 splitting, exhibits a high triplet energy (2.3 eV), and can be reduced easily at −1.70 V vs. SCE. There are several suitable acceptors for its triplet. When the FA radical anion was oxidized by the 10-MP radical cation in the presence of anthracene as an acceptor, which has a triplet energy of 1.8 eV, a strong blue-violet emission was seen. It was found that the ECL spectrum was identical to the fluorescence emission spectrum of anthracene. As shown in Figure 41, the spectrum is extremely well resolved; although the peak at 385 nm is reduced in intensity owing to self-absorption, the vibrational transitions at 400 and 420 nm due to anthracene are clearly seen, which are very different from the normal unstructured emission spectrum of FA. In a solution containing only anthracene and 10-MP (well degassed as before), when the electrode potentials were maintained at the same values as those used for generation of the FA and 10-MP radical ions, no ECL emission was observed; however, when the reducing potential was shifted to a value 0.2 V more cathodic, the characteristic anthracene emission was seen. Thus the emission of anthracene luminescence described above from the solution containing both anthracene and FA cannot be ascribed to the presence of anthracene radical anions. The photoluminescence of a solution containing 10^{-5} M in both anthracene and FA excited at 290 nm, where only FA absorbs appreciably, only FA emission was observed. This precludes the possibility of singlet–singlet energy transfer between FA and anthracene. Instead of anthracene, when *trans*-stilbene, whose triplet energy (– 2.0 eV) is lower than that of FA and thus triplet interception is facile, but which is more difficult to reduce (about −2.1 V vs. SCE), was used as the acceptor, quenching of ECL was observed, because *trans*-stilbene's triplet decays very rapidly to undergo a cis–trans isomerization. *cis*-Stilbene has higher triplet energy (– 2.5 eV) than FA, and interception in this case is inefficient, so one expects it to exert no effect, as experimentally observed. Freed and Faullener [22] carried out quantitative measurement of the triplet yield from the amount of *cis*-stilbene produced as a consequence of a redox process.

Ziebig and Pragst [89] studied triplet energy transfer from pyrazolines **I** (R = Ph, 4-biphenylyl, β-styryl) to a series of polycyclic aromatic hydrocarbons.

Kinetically or thermodynamically controlled energy transfer can occur, depending on the energy difference between the donor and the acceptor. The lifetime of the hydrocarbon triplets was estimated as approximately 3 μs on the basis of the concentration dependence of the sensitized ECL intensity. Under kinetic control conditions, ECL was observed when the concentration of the acceptor was >2 μM. Triplet energies of the donors and acceptors were determined from measurements in the thermodynamically controlled range. The effect of a magnetic field on the sensitized luminescence depends on the structure of the acceptor and on the initial excited system (**I** and the anion-forming component *N-p*-tolylphthalide).

IV. SUMMARY AND FUTURE PERSPECTIVES

The aim of this chapter has been to show the instrumentations and experimental (conventional and newly developed) techniques employed in ECL studies. The improvement of cell design and photon detection systems, e.g., by incorporating flow injection analysis (FIA) systems using UME to a single-photon counting unit, has greatly enhanced the detection limit, temporal resolution, and analytical throughput of ECL. ECL-based scanning probe methods, like other types of scanning probe microscopes, also provide unique experimental capabilities for imaging and studying charge transport either in solution or in electroluminescent thin films. Further improvements in both spatial and temporal resolutions are anticipated with the development of procedures that allow the routine fabrication of submicrometer-sized tips suitable for working in solutions. The retrieval of the interception techniques by designing new molecules and new processes and the development of new ECL-producing processes, e.g., luminescent nanostructure-based ECL systems, are encouraging for both mechanistic studies of electron and/or energy transfer reactions and their potential applications.

ACKNOWLEDGMENTS

I thank Prof. Allen J. Bard for his constant support and discussion. The support of his research by the National Science Foundation, the Robert A. Welch Foundation, and the Laboratory of Electrochemistry Organized Research Unit is also gratefully acknowledged.

REFERENCES

1. Faulkner, L.R.; Bard, A.J. Techniques of electrogenerated chemiluminescence. In *Electroanalytical Chemistry*, Bard, A.J., ed.; Marcel Dekker: New York, 1977; Vol. 10, pp 1–95.
2. Bocarsly, A.B.; Tachikawa, H.; Faulkner, L.R. Photonic electrochemistry. In *Laboratory Techniques in Electroanalytical Chemistry*, 2nd ed., Kissinger, P.T., Heineman, W.R., eds.; Marcel Dekker: New York, 1996; pp. 855–899.
3. Knight, A.W.; Greenway, G.M. Occurrence, mechanisms and analytical applications of electrogenerated chemiluminescence. Analyst **1994**, *119*, 879–890. Knight, A.W. A review of recent trends in analytical applications of electrogenerated chemiluminescence. Trends Anal. Chem. **1999**, *18*, 47–62.
4. House, H.O.; Feng, E.; Peet, N.P. A comparison of various tetraalkylammonium salts as supporting electrolytes in organic electrochemical reactions. J. Org. Chem. **1971**, *36*, 2371–2375.
5. Faulkner, L.R.; Bard, A.J. Electrogenerated chemiluminescence. I. Mechanism of anthracene chemiluminescence in N,N-dimethylformaide solutions. J. Am. Chem. Soc. **1968**, *90*, 6284–6290.
6. Mann, C.K. Nonaqueous solvents for electrochemical use. In *Electroanalytical Chemistry*, Bard, A.J., ed., Marcel Dekker: New York, 1969; Vol. 3, Chap. 2.
7. Bezman, R.; Faulkner, L.R. Mechanisms of chemiluminescent electron-transfer reactions. IV. Absolute measurements of 9,10-diphenylanthracene luminescence in N,N-dimethylformamide solution. J. Am. Chem. Soc. **1972**, *94*, 6317–6323. Mechanisms of chemiluminescent electron-transfer reactions. V. Absolute measurements of 9,10-diphenylanthracenc luminescence in benzonitrile and N,N-dimethylformamide. J. Am. Chem. Soc. **1972**, *94*, 6324–6330.
8. Osa, T.; Kuwana, T. Nonaqueous electrochemistry using optically transparent electrodes. J. Electroanal. Chem. **1969**, *22*, 389–406.
9. Frank, S.N.; Park, S.M. Electrochemistry in the dry box. In *Laboratory Techniques in Electroanalytical Chemistry*, 2nd ed., Kissinger, P.T, Heineman, W.R., eds; Marcel Dekker: New York, 1996; pp. 569–581.
10. Faulkner, L.R.; Tachikawa, H.; Bard, A.J. Electrogenerated chemiluminescence. VII. The influence of an external magnetic field on luminescence intensity. J. Am. Chem. Soc. **1972**, *94*, 691–699. Tachikawa, H., Ph.D. Dissertation, University of Texas at Austin, 1973.
11. Katovic, V.; May, M.A.; Keszthelyi, C.P. Vacuum-line techniques. In *Laboratory Techniques in Electroanalytical Chemistry*, 2nd ed., Kissinger, P.T., Heineman, W.R., eds.; Marcel Dekker: New York, 1996; pp. 543–567.
12. Wallace, W.L.; Bard, A.J. Electrogenerated chemiluminescence. 35. Temperature dependence of the ECL efficiency of $Ru(bpy)_3^{2+}$ in acetonitrile and evidence for very high excited state yields from electron transfer reactions. J. Phys. Chem. **1979**, *83*, 1350–1357.
13. Lai, R.Y.; Fabrizio, E.F.; Lu, L.; Jenekhe, S.A.; Bard, A.J. Synthesis, cyclic voltammetric studies, and electrogenerated chemiluminescence of a new donor-acceptor molecule: 3,7-[bis[4-phenyl-2-quinolyl]]-10-methylphenothiazine. J. Am. Chem. Soc. **2001**, *123*, 9112–9118.

14. Collinson, M.M.; Wightman, R.M. High-frequency generation of electro-chemiluminescence at microelectrodes. Anal. Chem. **1993**, *65*, 2576–2582.
15. Noffsinger, J.B.; Danielson, N.D. Generation of chemiluminescence upon reaction of aliphatic amine with tris(2,2′-bipyridine)ruthenium(III). Anal. Chem. **1987**, *59*, 865–868.
16. Zheng, X.; Guo, Z.; Zhang, Z. Flow-injection electrogenerated chemiluminescence determination of isoniazid using luminol. Anal. Sci. **2001**, *17*, 1095–1099.
17. Kremeskotter, J.; Wilson, R.; Schiffrin, D.J.; Luff, B.J.; Wilkinson, J.S. Detection of glucose via electrochemiluminescence in a thin-layer cell with a planar optical waveguide. Meas. Sci. Technol. **1995**, *6*, 1325–1328.
18. Zhan, W.; Alvarez, J.; Crooks, R.M. Electrochemical sensing in microfluidic systems using electrogenerated chemiluminescence as a photonic reporter of redox reactions. J. Am. Chem. Soc. **2002**, *124*, 13265–13270.
19. Arora, A.; JCT Eijkel, Morf, W.E.; Manz, A. A wireless ECL detector applied to direct and indirect detection for electrophoresis on a microfabricated glass device. Anal. Chem. **2001**, *73*, 3282–3288.
20. Bard, A.J.; Whitesides, G.M. US Patent 5221605, June 22, 1993.
21. Blackburn, G.F.; Shah, H.P.; Kenten, J.H.; Leland, J.; Kamin, R.A.; Link, J.; Peterman, J.; Powell, M.J.; Shah, A.; Talley, D.B.; Tyagi, S.K.; Wilkins, E.; Wu, T.G.; Massey, R.J. Electrochemiluminescence detection for development of immunoassays and DNA probe assays for clinical diagnostics. Clin. Chem. **1991**, *37*, 1534–1539.
22. Freed, D.J.; Faulkner, L.R. Mechanisms of chemiluminescent electron-transfer reactions. I. The role of the triplet state in energy-deficient systems. J. Am. Chem. Soc. **1971**, *93*, 2097–2102.
23. Bezman, R.; Faulkner, L.R. Theoretical and practical considerations for measurements of the efficiencies of chemiluminescent electron-transfer reactions. J. Am. Chem. Soc. **1972**, *94*, 3699–3707.
24. Bezman, R.; Faulkner, L.R. Mechanisms of chemiluminescent electron-transfer reactions. VI. Absolute measurements of luminescence from the fluoranthene-10-methylphenothiazine system in N,N-dimethylformamide. J. Am. Chem. Soc. **1972**, *94*, 6331–6337. Corrections for mechanisms of chemiluminescent electron-transfer reactions. IV, V, and VI. J. Am. Chem. Soc. **1973**, *95*, 3083.
25. Collinson, M.M.; Wightman, R.M.; Pastore, P. Evaluation of ion-annihilation reaction kinetics using high-frequency generation of electrochemiluminescence. J. Phys. Chem. **1994**, *98*, 11942–11947.
26. Maloy, J.T.; Prater, K.B.; Bard, A.J. Electrogenerated chemiluminescence. V. The rotating-ring-disk electrode. Digital simulation and experimental evaluation. J. Am. Chem. Soc. **1971**, *93*, 5959–5968. Electrogenerated chemiluminescence. VI. Studies of the efficiency and mechanisms of 9,10-diphenylanthracene, rubrene, and pyrene systems at a rotating-ring-disk electrode. J. Am. Chem. Soc. **1973**, *93*, 5968–5981. Maloy, J.T., Ph.D. Thesis, University of Texas at Austin, 1970.
27. Albery, W.J.; Hitchman, M. Ring-disc electrodes. Oxford: Clarendon, 1971, Chap. 3.
28. Bard, A.J.; Faulkner, L.R. Electrochemical methods, fundamentals and applications. New York: Wiley, 2001, Chap. 9.
29. Tokel-Takvoryan, N.E.; Hemingway, R.E.; Bard, A.J. Electrogenerated chemiluminescence. XIII. Electrochemical and electrogenerated chemiluminescence studies of ruthenium chelates. J. Am. Chem. Soc. **1973**, *95*, 6582–6589.

30. Laser, D.; Bard, A.J. Electrogenerated chemiluminescence. XXIII. On the operation and lifetime of ECL devices. J. Electrochem. Soc. **1975**, *122*, 632–640.

31. Brilmyer, G.H.; Bard, A.J. Electrogenerated chemiluminescence. XXXVI. The production of steady state direct current ECL in thin layer and flow cells. J. Electrochem. Soc. **1980**, *127*, 104–110.

32. Schaper, H.; Kostlin, H.; Schnedler, E. New aspects of D-C electro-chemiluminescence. J. Electrochem. Soc. **1982**, *129*, 1289–1294.

33. Bard, A.J.; Crayston, J.A.; Kittleson, G.P.; Shea, T.V.; Wrighton, M.S. Digital simulation of the measured electrochemical response of reversible redox couples at microelectrode arrays: Consequences arising from closely spaced ultramicroelectrodes. Anal. Chem. **1986**, *58*, 2321–2331.

34. Chidsey, C.E.; Feldman, B.J.; Lundgren, C.; Murray, R. Micrometer-spaced platinum interdigitated array electrode: Fabrication, theory, and initial uses. Anal. Chem. **1986**, *58*, 601–607.

35. Bartelt, J.E.; Drew, S.M.; Wightman, R.M. Electrochemiluminescence at band array electrodes. J. Electrochem. Soc. **1992**, *139*, 70–74.

36. Fosset, B.; Amatore, C.A.; Bartelt, J.E.; Michael, A.C.; Wightman, R.M. Use of conformal maps to model the voltammetric response of collector-generator double-band electrodes. Anal Chem. **1991**, *63*, 306–314.

37. Amatore, C.; Fosset, B.; J Bartelt, Deakin, M.R.; Wightman, R.M. Electrochemical kinetics at microelectrodes. Part V. Migrational effects on steady or quasi-steady-state voltammograms. J. Electroanal. Chem. **1988**, *256*, 255–268.

38. Ware, W.R. Transient luminescence measurements. In *Creation and Detection of the Excited State*, Lamola, A.A., ed.; Marcel Dekker: New York, 1971; Vol. 1A, Chap. 5.

39. Van Duyne, R.P.; Fischer, S.F. A nonadiabatic description of electron transfer reactions involving large free energy changes. Chem. Phys. **1974**, *5*, 183–197.

40. Parker, C.A. Photoluminescence of solution. Amsterdam: Elsevier, 1968, and references contained therein.

41. Hatehard, C.G.; Parker, C.A. A New sensitive chemical actinometer. II. Potassium ferrioxalate as a standard chemical actinometer. Proc. Roy. Soc. (Lond.) **1956**, *235A*, 518–536.

42. Bezman, R.; Faulkner, L.R. Construction and calibration of an apparatus for absolute measurement of total luminescence. Anal. Chem. **1971**, *43*, 1749–1753.

43. Zweig, A.; Hoffmann, A.K.; Maricle, D.L.; Maurer, A.H. An investigation of the mechanism of some eletrochemiluminescence processes. J. Am. Chem. Soc. **1968**, *90*, 261–268.

44. Michael, P.R.; Faulkner, L.R. Comparison between the luminol light standards and a new method for absolute calibrations of light detectors. Anal. Chem. **1976**, *48*, 1188–1192.

45. Pighin, A.; Conway, B.E. A correlation between the quantum efficiency of ECL and the redox potentials of rubrene in various solvents. J. Electrochem. Soc. **1975**, *122*, 619–624.

46. Schwartz, P.M.; Blakeley, R.A.; Robinson, B.B. Efficiency of the electrochemiluminescence process. J. Phys. Chem. **1972**, *76*, 1868–1871.

47. Keszthelyi, C.P.; Tokel-Takvoryan, N.E.; Bard, A.J. Electrogenerated chemi-luminescence. Determination of absolute luminescence efficiency in ECL: 9,10-diphenylanthracence–thianthrene and other systems. Anal. Chem. **1975**, *47*, 249–256.

48. Keszthelyi, C.P.; Ph.D. Dissertation, University of Texas at Austin, 1973.

49. Devlin, S. Low-light detectors; Sensitive CCDs see the light. Laser Focus World **2002**, *38*, 61–64.

50. Poletto, L.; Boscolo, A.; Tondello, G. Characterization of a charge-coupled-device detector in the 1100–0.14-nm (1-eV to 9-keV) spectral region. Appl. Opt. **1999**, *38*, 29–36.

51. Klein, C.A. Bandgap dependence and related features of radiation ionization energies in semiconductors. J. Appl. Phys. **1968**, *39*, 2029–2038.

52. McCord, P.; Bard, A.J. Electrogenerated chemiluminescence. Part 54. Electro-generated chemiluminescence of ruthenium(II)-4,4′-diphenyl-2,2′-bipyridine and ruthenium(II) 4,7-diphenyl-1,10-phenanthroline systems in aqueous and acetonitrile solutions. J. Electroanal. Chem. **1991**, *318*, 91–99.

53. Marcus, R.A. On the theory of chemiluminescent electron-transfer reactions. J. Chem. Phys. **1965**, *43*, 2654–2657.

54. Kavarnos, G.J. Fundamental concepts of photoinduced electron transfer. In *Topics in Current Chemistry*, Marray, J., ed.; Springer-Verlag: New York, 1990; Vol. 156, pp 21–58.

55. Collinson, M.M.; Pastore, P.; Maness, K.M.; Wightman, R.M. Electrochemi-luminescence interferometry at microelectrodes. J. Am. Chem. Soc. **1994**, *116*, 4095–4096.

56. Collinson, M.M.; Wightman, R.M. Observation of individual chemical reaction in solution. Science **1995**, *268*, 1883–1885.

57. van Kampen, N.G. Stochastic processes in physics and chemistry. Amsterdam: North-Holland, 1981.

58. Wightman, R.M.; Curtis, C.L.; Flowers, P.A.; Maus, R.G.; McDonald, E.M. Imaging microelectrodes with high-frequency electrogenerated chemi-luminescence. J. Phys. Chem. B**1998**, *102*, 9991–9996.

59. Oyama, M.; Okazaki, S. Pulse-electrolysis stopped-flow method for electrogenerated chemiluminescence in energy sufficient systems. J. Electrochem. Soc. **1997**, *144*, L326–L328.

60. Oyama, M.; Okazaki, S. Development of a dual-electrolysis stopped-flow method for the observation of electrogenerated chemiluminescence in energy-sufficient systems. Anal. Chem. **1998**, *70*, 5079–5084.

61. Bard, A.J.; Fan, F.R.F.; Mirkin, M.V. Scanning electrochemical microscopy. In *Electroanalytical Chemistry*, Bard, A.J., ed.; Marcel Dekker: New York, 1994; Vol. 18, pp 243–373.

62. Scanning Electrochemical Microscopy; Bard, A.J., Mirkin, M.V., eds.; New York: Marcel Dekker, 2001.

63. Maus, R.G.; McDonald, E.M.; Wightman, R.M. Imaging of nonuniform current density at microelectrodes by electrogenerated chemiluminescence. Anal. Chem. **1999**, *71*, 4944–4950. Maus, R.G., Wightman, R.M. Microscopic imaging with electrogenerated chemiluminescence. Anal. Chem. **2001**, *73*, 3993–3998.

64. Fan, F.R.F.; Cliffel, D.; Bard, A.J. Scanning electrochemical microscopy. 37. Light emission by electrogenerated chemiluminescence at SECM tips and their application to scanning optical microscopy. Anal. Chem. **1998**, *70*, 2941–2948.

65. Zu, Y.; Ding, Z.; Zhou, J.; Lee, Y.; Bard, A.J. Scanning optical microscopy with an electrogenerated chemiluminescent light source at a nanometer tip. Anal. Chem. **2001**, *73*, 2153–2156.

66. Scanning Probe Microscopy and Spectroscopy; *Theory, Techniques, and Applications*, Bonnell, D., ed.; New York: Wiley-VCH, 2001.
67. Karrai, K.; Grober, R.D. Piezoelectric tip-sample distance control from near field optical microscopes. Appl. Phys. Lett. **1995**, *66*, 1842–1844.
68. Slevin, C.J.; Gray, N.J.; Macpherson, J.V.; Webb, M.A.; Unwin, P.R. Fabrication and characterization of nanometer-sized platinum electrodes for voltammetric analysis and imaging. Electrochem. Commun. **1999**, *1*, 282–288. Conyers, J.L., White, H.S. Electrochemical characterization of electrodes with submicrometer dimensions. Anal. Chem. **2000**, *72*, 4441–4446.
69. Lee, Y.; Amemiya, S.; Bard, A.J. Scanning electrochemical microscopy. 41. Theory and characterization of ring electrodes. Anal Chem. **2001**, *73*, 2261–2267.
70. Bard, A.J.; Faulkner, L.R. Electrochemical methods, fundamentals and applications. New York: Wiley, 2001, Chap. 5.
71. Chang, M.M.; Saji, T.; Bard, A.J. Electrogenerated chemiluminescence. 30. Electrochemical oxidation of oxalate ion in the presence of luminescers in acetonitrile solutions. J. Am. Chem. Soc. **1977**, *99*, 5399–5403.
72. Leland, J.K.; Powell, M.J. Electrogenerated chemiluminescence: An oxidative-reduction type ECL reaction using tripropyl amine. J. Electrochem. Soc. **1990**, *137*, 3127–3131.
73. Rubinstein, I.; Bard, A.J. Electrogenerated chemiluminescence. 37. aqueous ECL systems based on $Ru(2,2'\text{-bipyridine})_3^{2+}$ and oxalate or organic acids. J. Am. Chem. Soc. **1981**, *103*, 512–516.
74. Bard, A.J.; Fan, F.R.F. Electrochemical detection of single molecules. Acc. Chem. Res. **1996**, *29*, 572–578.
75. Sheats, J.R.; Antoniadis, H.; M Hueschen, Leonard, W.; Miller, J.; Moon, R.; Roitman, D.; Stocking, A. Organic electroluminescent devices. Science **1996**, *273*, 884–888 and references cited therein, Armstrong, N.R., Wightman, R.M., Gross, E.M. Light-emitting electrochemical processes. Ann. Rev. Phys. Chem. **2001**, *52*, 391–422.
76. Gao, F.G.; Bard, A.J. Solid-state organic light-emitting diodes based on tris(2,2'-bipyridine ruthenium (II)) complexes. J. Am. Chem. Soc. **2000**, *122*, 7426–7427 and references cited therein. Rudmann, H., Shimada, S., Rubner, M.F. Solid-state light-emitting devices based on the tris-chelated ruthenium(II) complex. 4. High-efficiency light-emitting devices based on derivatives of the tris(2,2'-bipyridyl) ruthenium(II) complex. J. Am. Chem. Soc. **2002**, *124*, 4918–4921 and references cited therein. Buda, M., Kalyuzhny, G., Bard, A.J. Thin-film solid-state electroluminescent devices based on tris(2,2'-bipyridine) ruthenium(II) complexes. J. Am. Chem. Soc. **2002**, *124*, 6090–6098. Lepretre, J.C., Deronzier, A., Stephan, O. Light-emitting electrochemical cells based on ruthenium(II) using crown ether as solid electrolyte. Synth. Metals **2002**, *131*, 175–183.
77. Betzig, E.; Finn, P.L.; Weiner, J.S. Combined shear-force and near-field scanning optical microscopy. Appl. Phys. Lett. **1992**, *60*, 2484–2486.
78. Talley, C.E.; Cooksey, G.A.; Dunn, R.C. High resolution fluorescence imaging with cantilevered near-field fiber optic probes. Appl. Phys. Lett. **1996**, *69*, 3809–3811.
79. Fan, F.R.F.; Bard, A.J. Scanning probe microscopy studies of solid-state molecular electroluminescent devices based on tris(2,2'-bipyridine) ruthenium(II) complexes. J. Phys. Chem. B, 2003, 107, 1781–1787.

80. Morris, J.L.; Jr. Ph.D. Thesis, University of Illinois at Urbana-Champaign, 1978.

81. Ziebig, R.; Hamann, H.J.; Jugelt, W.; Pragst, F. Intramolecular exciplexes in electrogenerated chemiluminescence of 1-amino-3-anthryl-(9)-propanes. J. Lumin. **1980**, *21*, 353–356.

82. Tachikawa, H.; Bard, A.J. Electrogenerated chemiluminescence. Effect of solvent and magnetic field on ECL of rubrene systems. Chem. Phys. Lett. **1974**, *26*, 246–251.

83. Periasamy, N.; KSV Santhanam. Studies of efficiencies of electrochemiluminescence of rubrene. Proc. Ind. Acad. Sci. **1973**, *80A*, 194–202.

84. Glass, R.S.; Faulkner, L.R. Chemiluminescence from electron transfer between the anion and cation radicals of rubrene. Dominant S-route character at low temperatures. J Phys. Chem. **1982**, *86*, 1652–1658.

85. Kim, J.; Faulkner, L.R. Environmental control of product states in the chemiluminescent electron transfer between rubrene radical ions. J. Am. Chem. Soc. **1988**, *110*, 112–119.

86. Keszthelyi, C.P.; Tachikawa, H.; Bard, A.J. Electrogenerated chemiluminescence. VIII. The thianthrene-2,5-diphenyl-1,2,4-oxadiazole system. A mixed energy-sufficient system. J. Am. Chem. Soc. **1972**, *94*, 1522–1527.

87. Michael, P.R.; Faulkner, L.R. Electrochemiluminescence from the thianthrene – 2,5-diphenyl-1,3,4-oxadiazole system. Evidence for light production by the T-route. J. Am. Chem. Soc **1977**, *99*, 7754–7761.

88. Faulkner, L.R.; Freed, D.J. Mechanisms of chemiluminescence electron-transfer reactions. II. Triplet yield of energy transfer in the fluoranthene-10-methylphenothiazine system. J. Am. Chem. Soc. **1971**, *93*, 3565–3568. Freed, D.J.; Faulkner, L.R. Near unit efficiency of triplet production in an electron transfer reaction. J. Am. Chem. Soc. **1972**, *94*, 4790–4792. Hemingway, R.E.; Park, S.M.; Bard, A.J. Electrogenerated chemiluminescence. XXI. Energy transfer from an exciplex to a rare earth chelate. J. Am. Chem. Soc. **1975**, *97*, 200–201. Michael, P.R.; Faulkner, L.R. Electrochemiluminescence from the thianthrene–2,5-diphenyl-1,3,4-oxadiazole system. Evidence for light production by the T-route. J. Am. Chem. Soc **1977**, *99*, 7754–7761. Ziebig, R.; Pragst, F. Electrochemical production of triplet states. VII. Triplet energy transfer in electrochemical luminescence. Z. Physik. Chem. (Leipzig), **1979**, *260*, 748–762. Ziebig, R.; Pragst, F. Electrochemical production of triplet states. VIII. Electrochemical luminescence of isobenzofurans. Z. Physik. Chem. (Leipzig), **1979**, *260*, 795–803.

89. Ziebig, R.; Pragst, F. Electrochemical production of triplet states. VII. Triplet energy transfer in electrochemical luminescence. Z. Physik. Chem. (Leipzig), **1979**, *260*, 748–762.

3

ECL Theory: Mass Transfer and Homogeneous Kinetics

Joseph T. Maloy
Seton Hall University, South Orange, New Jersey, U.S.A.

I. PROCESS FUNDAMENTALS

This chapter develops the theoretical basis for understanding the relationships between current $i(t)$ and radiant intensity $I(t)$, the analytical variables in any quantitative study of electrogenerated chemiluminescence (ECL). Because the observed CL intensity results from the reaction of species generated electrochemically at a solution/electrode interface, this development begins with an understanding of the mass transfer and mass transport of solution species to an electrochemical interface to generate the precursors to ECL. Only then is it possible to address the ECL phenomenon as these precursors undergo subsequent mass transfer and transport from the electrochemical interface to react in a spatial reaction zone within the diffusion layer. Ultimately, the observed CL intensity depends upon the rate at which these precursors react within this reaction zone to produce excited state species and the rate at which these species go on to produce photons. Each of these processes may influence the observed current as well as being the rate-determining step for the observed CL intensity.

We begin by considering the simplest version of the first ECL reaction sequence presented in Chapter 1 in which an appropriate aromatic hydrocarbon A, e.g., 9,10-diphenylanthracene (DPA), is used to produce ECL by carrying out a double potential step experiment at a single working electrode under diffusion limiting conditions:

$$A + e^- \rightarrow A^{-\bullet} \quad \text{(reduction at electrode potential } E_1) \tag{1}$$
$$A - e^- \rightarrow A^{+\bullet} \quad \text{(oxidation at electrode potential } E_2) \tag{2}$$

$$A^{-\bullet} + A^{+\bullet} \rightarrow {}^1A* + A \qquad \text{(excited state singlet formation)} \qquad (3)$$

$$^1A* \rightarrow A + h\nu \qquad \text{(CL emission from singlet state)} \qquad (4)$$

Both electrode processes take place at fixed potentials that are suitable for either the reduction (E_1) or oxidation (E_2) of A under diffusion-limiting conditions so that the concentration of A at the electrode surface is equal to zero throughout both halves of the double potential step experiment. (It should be noted that E_2 is assumed to be sufficiently positive to bring about the two-electron oxidation of $A^{-\bullet}$,

$$A^{-\bullet} - 2e^- \rightarrow A^{+\bullet} \qquad \text{(two-electron oxidation at } E_2) \qquad (5)$$

also under diffusion-limiting conditions.

II. THEORETICAL FUNDAMENTALS

The theoretical treatment leading to the expression for the current flow during reaction (1) is fundamental to the study of electrochemical methods. The problem is recognized as a boundary value problem in which the material and charge flux are governed, in the absence of migration and convection as modes of mass transport, by Fick's laws of diffusion [1]. Electrode and semi-infinite solution boundary conditions are specified for the partial differential equations in time and space representing Fick's laws, and these are then transformed into ordinary differential equations via the Laplace transformation. Solution of these ordinary differential equations for the current expression yields the Cottrell equation, also given in Chapter 1:

$$i(t) = FAC_A^* D_A^{1/2}/(\pi t)^{1/2} \qquad (6)$$

This expression defines $i(t)$ during the first half of the double potential step in terms of the electrode area and the bulk concentration and diffusion coefficient of A, C_A^* and D_A, respectively [2]. This treatment also results in equations for $C_A(x,t)$ and $C_{A^{-\bullet}}(x,t)$, the concentration profiles of A and $A^{-\bullet}$. Because $i(t)$ is the variable of primary experimental interest, little further attention is directed toward these concentration profiles in the development of electroanalytical methods. Theoretical investigations of ECL, however, rely heavily upon an understanding of their behavior because $I(t)$ results from a reaction of these solution species within the diffusion layer, and the CL emission is controlled to a great extent by the transport phenomena that govern these spatial molecular distributions.

    A similar treatment, i.e., solution of the partial differential equations representing the given boundary value problem using Laplace transforms, results in a closed-form solution [3] for $i(t)$ in a double potential step experiment that begins at and returns to a potential E_0 where A is electrochemically inactive; that is, at E_0, $C_A-(0,t) = 0$:

$$A + e^- \rightarrow A^{-\bullet} \qquad \text{(reduction at electrode potential } E_1) \qquad (1)$$

$$A^{-\bullet} - e^- \rightarrow A \qquad \text{(oxidation at electrode potential } E_0) \qquad (7)$$

The expression of this closed form solution for the current–time characteristic is not important for this work, because the double potential step described above does not result in ECL emission. Although E_0 is sufficiently positive to oxidize $A^{-\bullet}$ to A, as shown in reaction (7), it is not sufficiently positive to produce $A^{+\bullet}$, as required in reaction (2).

What is noteworthy is the level of operational mathematics that is necessary to obtain this closed-form expression for the current–time characteristic (and the corresponding concentration profiles) in a double potential step that reverses only one of the reactions necessary to produce the two radical ion precursors of the ECL phenomenon. This difficulty is compounded when one attempts to take into account the rate of subsequent chemical reactions, such as one of the solution-phase reactions in the ECL mechanism:

$$A^{-\bullet} + A^{+\bullet} \rightarrow {}^1A^* + A \qquad \text{(excited state singlet formation)} \qquad (3)$$

In this simplest case, the local rate of excited state singlet formation would be given by

$$\text{Local rate} = \frac{d[C_A^*(x,t)]}{dt} = k_2[C_{A^-}(x,t)][C_{A^+}(x,t)] \qquad (8)$$

where k_2 is the second-order rate constant for reaction (3), which is presumed to be rate-determining, and the concentrations of all species are distributed over time and space. The production of excited state species via reaction (3) will alter the concentration of all three species (A, $A^{-\bullet}$, and $A^{+\bullet}$) within the diffusion layer and may thereby influence the material flux at the electrode during the second half of the double potential step. Thus, the differential equations describing the ECL rate law must be included in the system of ordinary differential equations to be solved following the transformation of Fick's laws to ordinary form by using Laplace transforms. Even though the resulting system of ordinary differential equations can be solved in transform space for the simplest ECL system described above, it not always possible to find an appropriate inverse transform to express both halves of the current–time characteristic in closed-form. Thus, even though the Cottrell equation may describe $i(t)$ during the first half of the double potential step ECL sequence, there is no guarantee that closed form expressions for the current during the latter (ECL) half-step will be available, because the transport phenomena that control the current are complicated by the chemical reactions that produce (or quench) the ECL.

An additional problem arises in the double potential step ECL experiment. As indicated above, the reaction of $A^{-\bullet}$ and $A^{+\bullet}$ results in a distribution of excited state species within the diffusion layer. The width of this distribution depends upon the magnitude of the various diffusion coefficients, the bimolecular rate constant k_2, and the subsequent rate of photon production via Eq. (4). Whatever its width, however, this band of ECL must be integrated over its spatial coordinates in order to obtain $I(t)$, the overall ECL intensity. This required integration of the solution to a system of differential equations adds an additional degree of complexity to the quest for a closed-form solution for $I(t)$ describing ECL.

Mathematical complexity, of course, is no stranger to the development of the theoretical basis for electrochemical methods. At a very minimum, the boundary value problem governed by the system of partial differential equations given by Fick's laws must be solved because diffusion is ever-present as a mode of mass transfer. Fick's first law,

$$-J_O\ (x,t) = D_O\ \frac{\partial C_O(x,t)}{\partial x} \tag{9}$$

is actually a part of a more complicated differential equation that is used to describe the material flux in an electrochemical system, the Nernst–Planck equation [4], which is given in one dimension as

$$-J_j(x,t) = D_j\ \frac{\partial C_j(x,t)}{\partial x} - \left(\frac{Z_j\ F}{RT}\right) D_j\ C_j\ \frac{\partial \phi(x,t)}{\partial x} - C_j V(x,t) \tag{10}$$

where D_j is the diffusion coefficient of the jth species (cm²/s), $\varphi(x,t)$ is the electrostatic potential, and $v(x,t)$ is the fluid velocity in centimeters per second. This equation not only accounts for diffusion as a mode of mass transfer but also includes terms for the longer range modes of mass transport, migration and convection. If closed-form expressions for the current transient exist for systems in which the complete Nernst–Planck equation governs the flux, they are not widely recognized parts of the corporate electrochemical memory. Because these additional modes of mass transport can sometimes lead to long-time steady state behavior by producing a material flux that is greater than the charge flux at the electrode, a steady-state current may result. Sometimes a closed-form approximation may be found for this steady-state current. For example, combination of diffusion and convection as modes of mass transport in the rotating disk electrode (RDE) and the rotating ring-disk electrode (RRDE) results in a steady-state expression for the current known as the Levich equation [5],

$$i_{Levich} = 0.62nFAD_O^{2/3}\omega^{1/2}v^{-1/6}C_O^* \tag{11}$$

where A is the electrode area (cm²), ω is the angular velocity (s⁻¹), v is the kinematic viscosity (cm²/s), and C_O^* is the bulk concentration (in mol/cm³) of the species electrolyzed at the disk. Indeed, combination of orthogonal modes of diffusion produces a material flux that is sufficient to counterbalance the charge flux at an ultamicroelectrode (UME) [6] to produce a steady-state current in the long-time limit. The magnitude of this steady-state current depends somewhat upon the geometry of the UME, but the long-time numerical approximation of the series solution to the boundary value problem obtained by using Laplace transforms over three dimensions yields the steady-state current expression

$$i_{ss} = 4nFD_OC_O*r_0 \tag{12}$$

for an ultramicro disk electrode of radius r_0. Each of these steady-state equations results from a series approximation for the current transient taken to a long- time limit. The numerical approximation is required, generally, because a closed-form solution does not exist.

One of the earliest and most familiar electrochemical applications of numerical methods can be attributed to Nicholson and Shain [7,8] and their development of the theory for linear sweep voltammetry (LSV) and cyclic voltammetry (CV). These numerical current transients, expressed parametrically as $i(E)$, result from a solution of the partial differential equations of Fick's laws under the influence of a potential-dependent boundary condition where the extent of reaction (1) and its reverse, reaction (7), depends upon the electrode potential, which varies between E_0 and E_1 as a linear function of time,

$$E = vt \tag{13}$$

where v is the potential sweep rate (in volts per second). Because a closed-form solution does not result from this theoretical analysis, the tabulated numerical results [9] must be examined to develop diagnostic criteria for the interpretation of the experimental results. Examples of diagnostic criteria resulting from an interpretation of the numerical solutions of the partial differential equations for LSV and CV include the equations for $E°$, the standard electrode potential, and i_p presented in Chapter 1,

$$E° = E_p \pm 1.109(RT/F) = E_p \pm 28.5 \text{ mV at } 25° \text{ C} \tag{14}$$

$$i_p = (2.69 \times 10^5) AD_A C_A^* v^{1/2} \tag{15}$$

where E_p is the peak potential and i_p is the peak current. Although these diagnostic criteria are quite useful in the interpretation of electrode processes using LSV and CV, they cannot be used directly in the interpretation of ECL produced during cyclic voltammetry. The boundary value problem leading to the numerical LSV–CV results does not include the differential equations that describe the processes related to the production of ECL, e.g., Eq. (8).

This, then, is the situation confronting those who wish to develop a theoretical basis for the interpretation of ECL: The closed-form and approximate solutions to the problems that are classical components of the theoretical repertoire cannot be applied directly to the ECL phenomenon because ECL redefines the problem. And, although it might be possible to obtain numerical solutions to the ECL boundary value problem based upon the Laplace transform, it certainly would not be convenient to do so. Each new set of boundary conditions adds an additional layer of complexity to a problem that is already unwieldy on account of the chemical reactions (and their associated rate expressions) that are necessary to produce the ECL. When subjected to analysis by operational mathematics, each new problem employs a different set of mathematical techniques to obtain solutions that are generally numerical, rather than functional, in form. Aware that this additional mathematical complexity generally leads to a numerical outcome, early ECL theoreticians turned to finite difference methods—digital simulations—in order to obtain the approximate numerical results while reducing the mathematical complexity of the work. It is the intent of this chapter to serve as a guide to this endeavor.

III. THE ORIGINS OF ELECTROCHEMICAL DIGITAL SIMULATIONS

The use of finite difference methods in the development of electrochemical theory is rooted in Feldberg and Auerbach's pioneering work employing a "computer approach" to model second-order kinetic effects in chronopotentiometry [10]. Soon thereafter, Feldberg began to use these digital simulations to predict $I(t)$ during the double potential step ECL experiment described above [11]. By using this new numerical method Feldberg was able to show graphically both $i(t)$ and $I(t)$ in the double potential step ECL experiment described above. These graphical results for $I(t)$ are displayed in Figure 1.

Displayed in semilogarithmic form, these dimensionless plots of $I(t)$, where $\omega \propto I(t)$, predict that the ECL decay approaches exponential behavior when $k_2 t_f C_A^*$ is greater than 1000 and the light producing reaction is diffusion-controlled:

$$\log \omega = -1.45 \, (t/t_f)^{1/2} + 0.71 \tag{16}$$

where the duration of the initial potential step is given by t_f and t is measured from the onset of the second step. Thus, the slope of a plot of $\log I(t)$ vs. $(t/t_f)^{1/2}$, known universally as a Feldberg plot, can be used as a diagnostic criterion for the diffusion-controlled ECL emission. Unfortunately, however, this "working curve" criterion is not unambiguous, and the slope is subject to other mechanistic influences. Nevertheless, its use exhibits fundamental methodology in employing the results of finite difference simulations in the interpretation of experimental data.

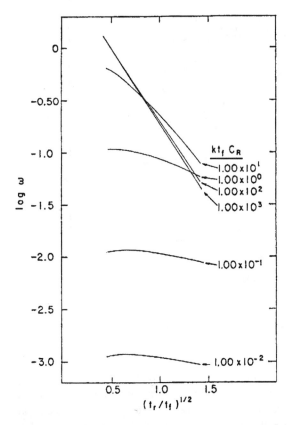

Figure 1 Dimensionless $I(t)$, identified by Feldberg as ω, as a function of dimensionless time. In our context $I(t)$ is shown for different values of $k_2 t_f C_A{}^*$, the dimensionless rate constants for the ECL-producing reaction (3). (From Ref. 11.)

Perhaps more important, Feldberg's initial ECL paper displayed, for the first time, concentration profiles showing the distribution of all electroactive species (A, $A^{-\bullet}$, and $A^{+\bullet}$) within the diffusion layer during the ECL half-cycle. These concentration profiles are reproduced in Figure 2. In addition, Figure 2 shows the spatial distribution of the ECL at the designated instant. Integrated over space, this then represents $I(t)$.

These two figures and the accompanying methodology were sufficient to convince workers in Allen Bard's research group of the practicality of applying digital simulation to develop a theoretical foundation for the study of ECL. Directly influenced by Feldberg's papers, several workers within the Bard research group began to investigate these finite difference methods for ECL studies in particular and the study of homogeneous electrochemical kinetics in general.

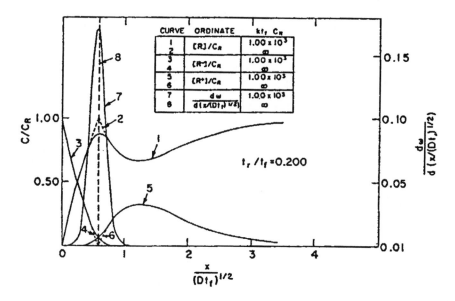

Figure 2 Normalized concentration profiles and ECL emission profile. These snapshots were obtained for $k_2 t_f C_A^* = 1000$ when $t/t_f = 0.2$. (From Ref. 11.)

At that time, however, computer-based concepts were not easily transmitted between groups working in different locations. The Internet as we now know it did not exist, and computer programs were written in dialects that were more or less specific for the platform at hand (and submitted on stacked punched cards for batch processing). Collaboration beyond the level of concept discussion was difficult, and the Bard group went on independently to develop its own school of digital simulation methodology and nomenclature. This work provided the theoretical basis for numerous publications in electrochemical kinetics [12–15] and ECL [16–18]. In addition, it developed a group of practitioners [19–25] who went on to employ these finite difference methods in their own research. The development of separate schools has proven to be beneficial to the field because it has permitted the independent determination of numerical results for the purposes of comparison when analytical solutions of limiting cases were not available.

Work that was initiated during this early period forms the basis of published tutorials on digital simulation [26–29]; the reader is referred to these works for more details. Our discussion of the use of digital simulation in the development of ECL theory begins with a brief review of the concepts contained therein.

IV. A REVIEW OF FINITE DIFFERENCE CONCEPTS

Many advances in digital simulation have taken place since the beginning of this work more than three decades ago. Some of these advances have occurred in hardware through the development of the personal computer. Others have taken place by the development of commercial software that will perform specific kinds of simulations, e.g., DigiSim™ for cyclic voltammetry [30]. The advent of common platforms has permitted the development of a computer environment (e.g., a spreadsheet) that will allow one to do a demonstration simulation without even having to write a computer program; to go beyond the demo stage using Microsoft Excel, however, now requires some knowledge of Microsoft's Visual Basic. Finally, there have been theoretical advances where newer methods [31–33] and implicit algorithms [34–36] are used to perform the simulations described below. However, most of the ECL simulations described in this chapter employ the more intuitive explicit methods that were originally used in this endeavor.

This review does not attempt to take many of these recent advances into account but is meant to provide a rigorous foundation for writing and understanding programs that will perform explicit finite difference simulations. It does so in the hope that the reader will develop an appreciation of the method and, more important, its limitations. Not the least of these limitations results from the absence of an analytical solution to almost any problem of consequence except in the limits of the boundary value problem. The appreciation of this limitation and how one can minimize its impact is essential to the utilization of digital simulations in any application.

Because diffusion is the ever-present mode of mass transport, any discussion of the digital simulation of problems involving transport phenomena begins with consideration of Fick's second law [37],

$$\frac{\partial C_A(x,t)}{\partial t} = D_A \frac{\partial^2 C_A(x,t)}{\partial x^2} \tag{17}$$

for the concentration $C_A(x,t)$, a function of position and time, where D_A is the diffusion coefficient. This law can be expressed in finite difference form and rearranged to obtain an expression for $\Delta C_A(x,t)$, the change in concentration at any time and place due to diffusion:

$$\Delta C_A(x,t) = C_A(x,t+\Delta t) - C_A(x,t)$$

$$= \lim_{\substack{\Delta x \to 0 \\ \Delta t \to 0}} \left[D_A \frac{\Delta t}{(\Delta x)^2} \right] [C_A(x+\Delta x,t) - 2C_A(x,t) \tag{18}$$

$$+ C_A(x-\Delta x,t)]$$

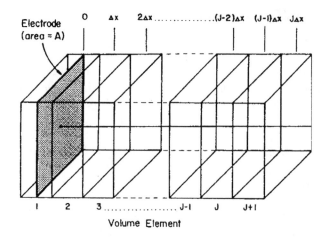

Figure 3 Model volume element array used in digital simulations of potential-dependent electrochemical problems. Note that the planar electrode has been placed in the middle of the first volume element. (From Ref. 40.)

This equation can be simplified further by dividing both sides by C_A^*, the bulk concentration, in order to express all concentrations in fractional form, and by defining a model diffusion coefficient D_M,

$$D_M = \lim_{\substack{\Delta x \to 0 \\ \Delta t \to 0}} \left[D_A \frac{\Delta t}{(\Delta x)^2} \right] \tag{19}$$

This allows one to write the finite difference diffusion algorithm in fractional concentration form:

$$f_A(x,t) = D_M[f_A(x + \Delta x, t) - 2f_A(x,t) + f_A(x - \Delta x, t)] \tag{20}$$

To employ this equation to calculate the fractional concentration change resulting from a diffusion process, one partitions the medium (the solution in the vicinity of the electrode) into an as yet unspecified number of volume elements of thickness Δx and requires that $f(x,t)$ be uniform within each element. In the simplest case, these elements would form a linear array originating at a planar surface (the electrode) of area A located at $x = 0$. Therefore, the volume of each element will be $A \, \Delta x$, as represented in Figure 3. Each element can be assigned a serial number J, as indicated in this figure, so that any distance x is given as

$$x = (J - 1) \Delta x \tag{21}$$

provided that the electrode is envisioned in the center of the first (J = 0) volume element. This placement is particularly convenient for modeling potential-dependent phenomena.

This having been done, the fundamental diffusion algorithm for explicit finite difference simulations can be written

$$\Delta f_A(J) = D_M[f_A(J + 1) - 2f_A(J) + f_A(J - 1)] \tag{22}$$

to describe the change in fractional concentration occurring in the Jth element at any time due to diffusion. In this form, D_M is seen as the proportionality constant governing flux between adjacent elements containing different fractional concentrations. Applied successively to each element in the array, this algorithm predicts the new fractional concentration, $f_A'(J)$

$$f_A'(J) = f_A(J) + \Delta f_A(J) \tag{23}$$

that will occupy each element after diffusion has occurred during the interval Δt. The use of this algorithm first requires the assignment of an appropriate value to D_M to ensure that Δx and Δt approach 0. One then establishes initial ($t = 0$) conditions and boundary conditions so that concentration gradients can be obtained. With time, distance, and all input/output variables expressed in dimensionless parameters, one merely writes the computer code that will apply Eq. (23) repeatedly in order to compute the new array values from the old values. New is made old, the time index I is incremented by 1, and the process is repeated until I reaches some predetermined limit. Actual examples of the required computer code appear in Fortran [26] and Basic [27]. Before these programs can be used, some additional comments are necessary regarding the terminology used therein.

Although the assignment of a value to D_M is arbitrary, it is by no means unrestricted. It is clear from its definition, Eq. (19), that D_M must be positive. To find the upper limit for D_M, one need only substitute the Einstein definition of the diffusion coefficient [38] into Eq. (19):

$$D_M = \lim_{\substack{\Delta x \to 0 \\ \Delta y \to 0}} \left[\frac{(\Delta y)^2}{2(\Delta x)^2} \right] \tag{24}$$

where Δy is the infinitesimally small distance a diffusing molecule travels, according to Einstein, during the infinitesimally brief time Δt, and Δx is the finite length of the volume element containing that molecule. Although Δy can approach zero, it can never exceed Δx because Δx is finite. Thus, for explicit finite difference simulations of diffusion phenomena,

$$1/2 \geq D_M \geq 0 \tag{25}$$

Because the assignment of its value is arbitrary, it would appear that setting D_M equal to 1/2 (its maximum value) would tend to minimize Δx with respect to Δt;

see Eq. (19). This is true, but setting D_M equal to 0.50 in the simulation of Cottrell behavior leads to oscillations in the computed current transient. Because the extent of oscillation decreases with decreasing D_M, Feldberg has recommended using $D_M = 0.45$ to obtain reliable simulations of current–time behavior [39], but other values up to $D_M = 0.49$ have been employed with equal success [40].

As indicated above, Eq. (22) is applied successively to each element in the volume element array to simulate material transfer during the interval Δt. To define a Δt unit, one must select some known time (t_k) in the physical experiment and partition that known time into a specified number of equal intervals, L. Thus,

$$\Delta t = t_k/L \tag{26}$$

and increasing L, the number of iterations representing a known time in the physical experiment, causes Δt to approach zero. The known time parameter t_k may change as different ECL generation schemes are considered. For example, in the simulation of a double potential step experiment, t_k is usually taken to be t_f, the duration of the forward step. In the simulation of cyclic voltammetry $\Delta E/v$ [see Eq. (13)] can be used as t_k. In the simulation of RDE and RRDE behavior [see Eq. (11)] t_k might be taken as the reciprocal of the angular velocity, $\omega\,(s^{-1})$, although a better suggestion is offered below.

Real times t can be related to t_k through Eq. (26). Thus, if Eq. (22) has been applied to all the volume elements in the spatial array I times, diffusion has occurred for a time $t = (I - \frac{1}{2})\Delta t$; $\frac{1}{2}$ is subtracted from I because time is measured midway through each time iteration. The real time t can be related to the known time t_k by eliminating Δt from Eq. (26):

$$t/t_k = (I - \tfrac{1}{2})/L \tag{27}$$

This provides a dimensionless representation of time throughout the simulation. Because Δx, Δt, and D_M are related by Eq. (24), the specification of D_M and L and the identification of tk result in the definition of Δx,

$$\Delta x = [D_A t_k/D_M L]^{1/2} \tag{28}$$

Thus, increasing L causes both Δt and Δx to approach zero. It is important to note, however, that Δx decreases with $L^{1/2}$; therefore, once reasonable accuracy has been obtained by increasing L to a large value (approximately 1000), little additional diminishment of the spatial grid is to be gained by increasing it any further.

With mass transfer variables all assigned, the simulation of the electrode process

$$A \pm e \rightarrow B \tag{29}$$

then proceeds by establishing the initial and boundary conditions for the method under consideration. Initial conditions usually require that the relative concentration of the first electroactive species, A, be set equal to 1.0 in each theoretical volume element and that of B be set equal to 0.

$$f_A(J) = 1.0, \qquad f_B(J) = 0.0, \qquad \text{for all } J \tag{30}$$

In the simulation of double potential step ECL generation, the first electrode boundary condition sets the concentration of species A equal to zero in the first volume element (representing the electrode). In this case, the electrode surface concentration of electroactive species A will be maintained at zero throughout the initial potential step:

$$f'_A(1) = f_A(1) = 0 \tag{31}$$

The electrode boundary condition for product B is less straightforward and must be written

$$f'_B(1) = f_B(1) + D_{MA}f_A(2) - D_{MB}[f_B(1) - f_B(2)] \tag{32}$$

where the second term on the right-hand side accounts for the A converted to B by the electrode reaction, and the third term accounts for the diffusion of B out of the electrode volume element. In this form Eq. (32) allows for different model diffusion coefficients for species A and B.

At the other (semi-infinite) boundary, bulk concentration of A must be maintained at some finite distance from the electrode, and the concentration of B will be zero at the same point. This distance may be regarded as the diffusion layer thickness ($J_{max} \Delta x$). In terms of the simulation, the establishment of the semi-infinite boundary condition requires the determination of the maximum number of volume elements (J_{max}) making up the diffusion layer. J_{max} may arbitrarily be set equal to I, the time iteration number; this allows the time and space grid to expand equally. A great deal of computation time can be saved by estimating the maximum number of spatial volume elements that are required during the Ith iteration for a typical electrochemical diffusion layer thickness of $x = 6(Dt)^{1/2}$. Appropriate substitution yields [41]

$$J_{max} = 3(2I_{max})^{1/2} \tag{33}$$

where I_{max} is the maximum time iteration number.

As noted in our discussion of Feldberg's original ECL simulations [11], the experimental variables used for input/output must be expressed in dimensionless form, and this transformation of variables is a necessary requirement of the method. The concept of a dimensionless representation of time has already been introduced in Eq. (27). The dimensionless distance parameter may be found by substituting Eq. (28), the definition of Δx, into Eq. (21), the definition of x, and rearranging:

$$\frac{x}{(Dt_k)^{1/2}} = \frac{(J-1)}{(D_M L)^{1/2}} \tag{34}$$

It is in this manner that J, the ambiguous "box number," is transformed into a meaningful dimensionless distance. Note that the process of formulating the proper dimensionless parameters always entails the segregation of experimental variables on one side of an equation and computer variables on the other.

Dimensionless current is obtained by first writing the expression for the current in finite difference form. For the system illustrated in Figure 3 with a planar electrode placed at $x = 0$ and A electroactive, the fractional material flux at the electrode is given by $D_{MA}[f_A(2) - f_A(1)]/\Delta t$. Multiplication by C_A^*, the bulk concentration of A; $A\Delta x$, the element volume; and nF, the Faraday conversion, yields an expression for the current,

$$i(t) = nFA\,\Delta x\; C_A^*\; D_{MA} \left[\frac{f_A(2) - f_A(1)}{\Delta t} \right] \tag{35}$$

Substitution of the definitions of Δx and Δt into this equation followed by segregation of experimental and computer variables yields

$$\frac{i(t)t_k^{1/2}}{nFAC_A^*\,D_A^{1/2}} = \frac{L}{D_{MA}} - D_{MA}\,[f_A(2) - f_A(1)] \tag{36}$$

This gives $i(t)t_k^{1/2}/nFAC_A^*D_A^{1/2}$, the dimensionless current that passes during each time iteration I. This dimensionless current is identified in the Bard group work as Z(I), so that

$$Z(I) = L/D_{MA}^{1/2}\,D_{MA}[f_A(2) - 0] \tag{37}$$

for diffusion-limited current. Whatever the correct expression for the dimensionless current, a plot of Z(I) vs. $(I - 1/2)/L$ yields the dimensionless expression of the current transient for that system. Of course, the correct formulation of Z(I) will change as the electrode boundary conditions change. In the case of the simple reversal of A and B under consideration, Eq. (29), the change in electrode potential brings about an analogous change in the expression for Z(I):

$$Z(I) = -[L/D_{MA}]^{1/2}D_{MB}[f_B(2) - 0] \tag{38}$$

The simple, double potential step ECL system under consideration, Eqs. (1)–(4), requires only minor variations in setup and interpretation. Letting $A^{-\bullet} = B$ and $A^{+\bullet} = C$ in computational nomenclature, the ECL sequence becomes

$$A \pm e \rightarrow B \qquad \text{(reduction or oxidation at } E_1) \tag{39}$$

$$A \mp e \rightarrow C \qquad \text{(oxidation or reduction } E_2) \tag{40}$$

$$B + C \rightarrow A^* + A \qquad \text{(excited state formation)} \tag{41}$$

$$A^* \rightarrow A + h\nu \qquad \text{(ECL emission)} \tag{42}$$

This requires separate volume element arrays for A, B, and C; a fourth array for A* must be included if any lifetime is to be attributed to the excited state. With only A present initially, the initial half-cycle is identical to that described above for A $\rightarrow$ B. Potential switching brings about new electrode conditions, however, because both A and B become electroactive:

$$B \pm 2e \rightarrow C \tag{43}$$

$$A \pm e \rightarrow C \tag{44}$$

and the C that is produced during this half-cycle reacts with the B that was generated during the first half-cycle to cause the ECL. This requires only a minor variation in the expression for the dimensionless current after potential switching [27,42]. The dimensionless currents obtained in this manner agree with those reported by Feldberg [11].

As stated in our introductory discussion, digital simulations are ideally suited to the development of ECL theory because of their ability to model the homogeneous kinetics of the reactions responsible for the ECL, as distributed over time and space. Thus, the real power of finite difference method techniques lies in their ability to predict $i(t)$ and $I(t)$ for any set of electrochemical transport boundary conditions as under the influence of a kinetically controlled reaction sequence leading to the production of ECL. Mastery of these kinetic complications requires the use of another type of dimensionless parameter.

For the rubrene system considered above, the rate-determining step in the production of ECL is given by Eq. (3), which has a rate law given by Eq. (8). Given the computational symbolism for Eq. (3) that is expressed in Eq. (41), the rate law can be expressed in finite difference form after both sides are divided by C_A^{*2} to obtain fractional concentrations:

$$\text{Local rate} = \frac{\Delta f_{A^*}(J)}{C_A^* \Delta t} = k_2 f_B(J) f_C(J) \tag{45}$$

Substitution of the definition of Δt given in Eq. (26) followed by rearrangement yields

$$\Delta f_{A^*}(J) = (k_2 t_f C_A^*) f_B(J) f_C(J)/L \tag{46}$$

where the experimental variables have been grouped together into a single dimensionless factor $(k_2 t_f C_A^*)$ and t_f has been recognized the appropriate known time (t_k) for potential step experimentation. It is this quantity that is used, along with L and D_M as input parameters, at the outset of execution. Because the dimensionless rate constant $(k_2 t_f C_A^*)$ is made up of three variables, it must take all three into account, and an increase in the arbitrary value of this input parameter could be taken as an increase in any one of the three with the other two being held constant. Because kinetic effects are of most interest, variations in $k_2 t_f C_A^*$ are usually considered as variations in k_2, with t_f and C_A^* being held conceptually constant.

Inclusion of the rate expression given by Eq. 46 allows one to compute the change in the fractional concentration of excited states, $\Delta f_{A*}(J)$, that takes place in the Jth volume element as a result of the reaction of B and C. Note that $f_B(J)$ and $f_C(J)$ are fractional quantities, and one of them must represent the limiting reagent in the production of A*. Thus, L is a reasonable upper limit for the magnitude of $k_{2t_f}C_A^*$ that can be safely selected.

Once $\Delta f_{A*}(J)$ is computed for each volume element, the actual photon flux from that element can be computed by assigning a rate of radiative decay to Eq. (42). Additional subtleties can be ascertained by assigning a (concentration-dependent) quantum efficiency (Φ) to the emission process. The total emission is then computed by integrating (summing) the photon flux emanating from all volume elements. By assigning an infinitely fast radiative decay to A* (so that all A* formed by the reaction of B and C was assumed to immediately produce a photon), Feldberg was able to generate the data shown in Figure 2. The indicated values of $k_{2t_f}C_A^*$ range from kinetic control to diffusion control of the ECL emission process, and in the limit of diffusion control, Eq. (16), the oft-cited Feldberg plot is obtained. It is interesting to note that Feldberg approached diffusion control when $k_{2t_f}C_A^* = 1000$. This probably indicates that, in this nomenclature, L = 1000 was employed in that work.

Although sufficient background in finite difference methodology has been presented to provide an understanding of the following discussion on transient ECL phenomena observed at a single, stationary electrode, one additional topic is necessary to the understanding of steady-state ECL generation at a rotating ring-disk electrode (RRDE). Our introduction to steady-state simulations begins, however, with the central component of the RRDE, the rotating disk electrode (RDE).

The study of RDE behavior provides a unique opportunity to develop a finite difference model that predicts electrochemical behavior under the influence of mass transfer (diffusion) and mass transport (convection). The effect of diffusion is modeled as described above; convection is treated by integrating an approximate velocity equation to determine the extent of the convective flow occurring during a given Δt interval. Mass transport, then, is allowed to take place by simply transferring fluid (and its contents) from one volume element to another in accord with the convection algorithm. This process is repeated once during each time iteration; it results in a steady-state concentration profile and Eq. (11), the Levich steady-state representation of the current.

In rotating-disk hydrodynamics, an approximate equation describing the velocity of the fluid in the vicinity of the electrode is given by Levich [5]:

$$v_x = \frac{dx}{dt} = -0.51x^2 \left[\frac{\omega^3}{\nu} \right]^{1/2} \tag{47}$$

where ω is the angular velocity of the rotating electrode (s^{-1}), v is the kinetic viscosity of the solution (cm^2/s), and x is measured on an axis perpendicular to the rotating disk. Integration of Eq. (47) yields

$$\int_{x_2}^{x_1} \frac{dx}{x^2} = \frac{1}{x_2} - \frac{1}{x_1} = -0.51 \left[\frac{\omega^3}{v} \right]^{1/2} \int_{t_1}^{t_2} dt = -0.51 \left[\frac{\omega^3}{v} \right]^{1/2} \Delta t \tag{48}$$

where x_2 represents the position of a given volume element of fluid *initially* (at t_1) and x_1 represents the position of that same fluid after the interval Δt. (At t_1 the fluid under consideration is situated at x_2, a greater distance from the electrode than x_1.) Equation (48) can be solved for x_2 to obtain an expression for the initial position of the fluid volume element that resides at x_1 at the conclusion of the Δt interval:

$$x_2 = \frac{x_1}{1 - 0.51[\omega^3/v]^{1/2} \Delta t \, x_1} \tag{49}$$

Substitution of Eqs. (26) and (34) into Eq. (49) eliminates x_1, x_2, and Δt and results in

$$J_2 - 1 = \frac{J_1 - 1}{1 - 0.51(\omega^3 D_A t_k^3/v)^{1/2} (J_1 - 1)/[D_M L^3]^{1/2}} \tag{50}$$

where J_1 and J_2 are volume element numbers corresponding to x_1 and x_2. Examination of Eq. (50) reveals a nice simplification if

$$t_k \equiv (v/D_A)^{1/3} \omega^{-1} \tag{51}$$

in this simulation; Eq. (50) then becomes

$$J_2 - 1 \frac{J_1 - 1}{1 - 0.51(J_1 - 1)/[D_M L^3]^{1/2}} \tag{52}$$

Thus, given the volume element J_1, one may calculate the serial number of that volume element J_2 that supplies material convectively to element J_1 during that iteration. In theory, one then simply transfers the contents of element J_2 to element J_1 in order to simulate the convective effect. Of course, it is not quite that simple. Given an integer value for J_1, it is quite unlikely that an integer will be obtained for J_2. Thus, some sort of interpolation scheme must be devised to obtain the relative concentrations of the species of interest at noninteger values of J_2; these computed concentrations are then transferred to volume element J_1. In a typical simulation, this process would take place once during each iteration after diffusion effects have been computed throughout the spatial array but before any kinetically controlled processes are allowed to take place. During this process, some care must be exercised to carry out convective transfer to interior elements first so that one does not write over newly calculated values in the spatial array.

Finally, it should be noted that the choice of t_k indicated in Eq. (51) provides the proper hydrodynamic expression for the dimensionless current Z(I). It has been reported (Ref. 40, p. 614) that the steady-state value of Z(I) obtained by this method (with L = 1000) is 0.61.

$$\lim_{I \to \infty} Z(I) = \frac{i_{Levich} \, v^{1/6}}{nFAC * D_A^{5/6} \, \omega^{1/2}} = 0.61 \tag{53}$$

Within the reliability of the simulation, this expression is identical to the Levich equation.

V. FELDBERG PLOTS AND FUNDAMENTAL ECL ENERGETICS

As was shown in Chapter 1, the free energy of the ion annihilation reaction has been compared to the radiant energy of the emitting singlet state from the early days of these investigations. If the annihilation reaction has sufficient energy to produce the excited singlet state directly, Eq. (3) proceeds as written, and the process is deemed energy-sufficient. One such system that appears to have sufficient electrochemical energy to produce the excited state singlet species directly is DPA, which was used in an early example.

On the other hand, if the annihilation reaction lacks sufficient energy to produce the excited state singlet species that is responsible for the ECL emission, some other mechanism must be invoked. One such energy-deficient mechanism is triplet–triplet annihilation (TTA) of a triplet species formed during the reaction of the precursor ions:

$$A^{-\bullet} + A^{+\bullet} \to {}^3A* + A \qquad \text{(triplet formation)} \tag{54}$$

$$^3A* + {}^3A* \to {}^1A* + A \qquad \text{(excited state singlet formation)} \tag{55}$$

Whereas reaction (54) is assumed to be quite rapid (diffusion-controlled), reaction (55) is subject to kinetic control, with a bimolecular rate constant k_T. ECL emission occurs via either route from the excited state singlet as shown in Eq. (4). Because its enthalpy calculations are, at best, ambiguous from an energetic perspective, a likely candidate for luminescence via this ECL-TTA mechanism is rubrene (5,6,11,12-tetraphenylnaphthacene).

Conjecture regarding the viability of the TTA mechanism arose out of early investigations of ECL phenomena employing the methodology developed by Feldberg using the digital simulations described above [43–45]. Feldberg plots—graphs of log $I(t)t_f^{1/2}$ vs. $(t/t_f)^{1/2}$, which, according to these simulations, should have exhibited a slope of -1.45 for prompt emission under diffusion-controlled conditions—sometimes exhibited more negative slopes. Whereas the model used in these simulations could account for a less negative slope by invoking kinetic control (see Fig. 1), it could not account for this experimental observation. Speculation then arose that this discrepancy might be due to a different light-producing pathway such as the TTA mechanism, and Feldberg published a second "digital simulation" paper, which justified more negative slopes when the TTA mechanism was operative [46].

In this paper Feldberg invoked the TTA mechanism and, at the suggestion of Marcus, investigated the role of triplet quenching in the production of energy-deficient ECL under diffusion-limiting conditions. In this latter work Feldberg abandoned the notion of radical annihilation kinetic control entirely and assumed that reaction (3) proceeds under diffusion control. Given the charge and electronic state of the reacting radical ions and the magnitude of the free energy that is liberated, this assumption appears to be justified.

From the perspective of the simulation this assumption requires the assignment of a reaction volume within which the annihilation reaction will proceed to completion, i.e., until the limiting species is exhausted. This is most easily accomplished by allowing the diffusion of both species to occur via Eqs. (22) and (23) and then computing the extent of chemical reaction in each of the two volume elements in which $A^{-\bullet}$ and $A^{+\bullet}$ coexist during the Ith iteration. For example, if B represents $A^{-\bullet}$ and C represents $A^{+\bullet}$, one would search the B and C volume element arrays after applying the diffusion algorithm and find the two adjacent volume elements X and Y in which both new concentrations $f_B(X)$ and $f_C(X)$ were nonzero and $f_B(Y)$ and $f_C(Y)$ were nonzero. Annihilation would then be simulated by subtracting the minor fractional concentration from the major one in each case. For example, if $f_B(X) > f_C(X)$ and $f_C(Y) > f_B(Y)$, then (56)

$$f_B'(X) = f_B(X) - f_C(X) \quad \text{and} \quad f_C'(X) = 0$$
$$f_C'(Y) = f_C(Y) - f_B(Y) \quad \text{and} \quad f_B'(Y) = 0$$

where the primed quantities refer to postannihilation conditions, and $\delta(I)$, the fractional production of excited state species during the Ith iteration, is given by the sum of the minor components in the adjacent volume elements:

$$\delta(I) = f_C(X) + f_B(Y) \tag{57}$$

This is, in effect, an integration over two volume elements. (Note that this integration can be carried out over all volume elements without searching for X and Y because no ECL can be generated at mid-iteration in any volume element where either ion is absent under diffusion-controlled conditions.) The rate of production of excited state species is obtained by multiplying $\delta(I)$ by the element volume–bulk concentration product followed by division by Δt. The ECL intensity $I(t)$ is given by multiplying this quantity by Φ, the ECL efficiency,

$$I(t) = \Phi\delta(I)C_A^* A \frac{\Delta x}{\Delta t} \tag{58}$$

The dimensionless expression for $I(t)$ depends upon the definitions of Δx and Δt, but in the case of the double potential step having forward duration t_f, these are defined by Eqs. (28) and (26), respectively, so that

$$\frac{I(t)t_f^{1/2}}{\Phi C_A^* AD_A^{1/2}} = \delta(I)(L/D_M)^{1/2} \tag{59}$$

The dimensionless expression for $I(t)$ given in the left hand member is identical to that defined by Feldberg [11] except that he employed ϕ, the fluorescence quantum efficiency, in place of Φ, the overall ECL efficiency, in his definition of ω for use in his diagnostic plots,

$$\omega = \frac{I(t)t_f^{1/2}}{\phi C_A^* AD_A^{1/2}} \tag{60}$$

In addition to generating the plots shown in Figure 1 using various levels of kinetic control, Feldberg employed a similar diffusion control algorithm to confirm the validity of Eq. (16) in this limit and to describe a narrow ion annihilation reaction zone (two volume elements) when diffusion control was operative (see Fig. 2).

Equation (16), rather than a new simulation, became the basis of this first treatment of the TTA mechanism. Feldberg came to realize that the simulation of the TTA problem was what he later identified as a "stiff" problem, one involving two time-dependent processes occurring on vastly different time scales. The diffusional processes treated by the simulation took place on the millisecond scale, whereas triplet lifetimes might be on the microsecond scale. Thus, although 1000 iterations might be sufficient to simulate a diffusion process explicitly, a million iterations would be necessary to superimpose upon this simulation an exact representation of the luminescent processes. Because this was beyond computational feasibility at the time, Feldberg elected to use the simulation only to model the electron transfer rate within an even narrower reaction zone in which all luminescent processes took place. Within this reaction zone it was assumed that steady-state triplet concentrations were maintained throughout each iteration. These steady-state triplet concentrations were used to control the ECL emission produced via reactions (54) and (55).

The mechanism employed in this work accounts for ECL produced by reactions (3) and (4), emission resulting from the direct formation of excited state singlet species (hereafter called the S route), and ECL produced via the TTA mechanism described by reactions (54), (55), and (4) (hereafter identified as the T route). Feldberg used the rate constants for the ion annihilation reactions (3) and (54) to define the quantity γ, the fraction of radical cation–anion reactions producing triplets, and $1 - \gamma$ as the fraction producing singlets. Thus, when $\gamma = 0$, only S-route ECL may be produced, and when $\gamma = 1$, only T-route ECL may be produced. When $1 > \gamma > 0$, ECL may be produced via a mixed mechanism known as the ST route. In this case, it might be assumed that γ is a probabilistic factor determined, in part, by the energetics of the system.

Feldberg then assumed that the overall rate of production of excited state species, both singlets and triplets, is given by $I(t)/\phi$, which can also be expressed in terms of ω_n, the dimensionless S-route ECL parameter given in Eq. (60):

$$I(t)/\phi = (\omega_n C_A^* A D_A^{1/2})/t_f^{1/2} \tag{61}$$

and ω_n is approximated by a modified form of Eq. (16),

$$\log \omega_n = -1.45(t/t_f)^{1/2} + 0.71 \tag{62}$$

in recognition of the fact that this expression merely measures the rate of the diffusion-controlled ion annihilation reaction, regardless of the energy of the product formed. The quantity $I(t)/\phi$ is then partitioned between S-route and T-route ECL by using the fraction γ, so that the rate of triplet formation is given by $\gamma I(t)/\phi\Delta$, where Δ is an arbitrary reaction volume defined as

$$\Delta = fA(D_A t_f)^{1/2} \tag{63}$$

In making this assignment, Feldberg was attempting to restrict the width of the ECL reaction zone, Δ/A, to a dimension less than Δx in order to account for the rate disparity between ion annihilation and luminescent phenomena. Comparison of Eq. (63) with the definition for Δx given in Eq. (28) reveals that

$$\frac{\Delta/A}{\Delta x} = \frac{f}{(D_M L)^{1/2}} \tag{64}$$

Using a simulation of about 1000 iterations, Feldberg was able to convince himself that f was equal to 0.525, thereby implying that the presumed constant width of the ECL–TTA reaction zone was approximately 2.5% of the width of a single volume element. Bezman and Faulkner [19] would subsequently reconsider some of the assumptions leading to this conclusion and would redefine f as a variable function of time. Using known triplet state lifetimes, they would suggest that $f(t) = 0.1$ was a more reasonable upper limit for this dimensionless reaction zone width, but they would point out that both f and $f(t)$ are inseparable from a dimensionless parameter used to account for triplet quenching.

Feldberg took this triplet quenching into account by assuming the presence of a quencher Q of known bulk concentration C_Q that reacts with triplet species according to

$$^3A* + Q \rightarrow A + Q \tag{65}$$

having a known bimolecular rate constant k_Q. Taken together with Eq. (55), Eq. (65) provides a second route for the depletion of $^3A*$ that can be juxtaposed against the rate of $^3A*$ formation, $\gamma I(t)/\phi\Delta$, to obtain an expression for $d[^3A*]/dt$. By invoking the steady-state assumption one may set $d[^3A*]/dt = 0$ and rearrange the resulting expression to obtain

$$\gamma I(t)/\phi\Delta = k_T[^3A*]_{ss}^2 + k_Q C_Q[^3A*]_{ss} \tag{66}$$

a quadratic equation that can readily be solved for the steady-state value of $[^3A*]_{ss}$ that is responsible for the T-route emission during that iteration. Consideration of the kinetics of Eq. (55) reveals that the intensity of ECL emission resulting from the TTA process occurring within the reaction volume is given by

$$\gamma I(t) = \phi\Delta k_T[^3A*]_{ss}^2/2 \tag{67}$$

Solution of Eq. (66) for $[^3A*]_{ss}$ yields

$$[^3A*]_{ss}^2 = \frac{\gamma I(t)}{k_T\,\phi\Delta} + \left[\frac{1}{4}\left(\frac{C_Q k_Q}{k_T}\right)^2\left\{1-\left(1+\frac{4\gamma I(t)k_T}{\phi\Delta k_Q^2\,C_Q^2}\right)^{1/2}\right\}\right] \tag{68}$$

Recognizing that γ partitions $I(t)$ into T- and S-route components, one may write

$$I(t) = \gamma I(t) + (1-\gamma)I(t) \tag{69}$$

and combine this with Eqs. (67) and (68) to obtain

$$I(t) = \frac{\gamma I(t)}{2} + \frac{1}{4}\frac{(C_Q k_Q)^2}{k_T}\,\phi\Delta\left[1-\left(1+\frac{4\gamma I(t)k_T}{\phi\Delta k_Q^2\,C_Q^2}\right)^{1/2}\right] \tag{70}$$

$$+(1-\gamma)I(t)$$

Equation (70) can be simplified by defining β, a parameter that compares the extent of triplet quenching, Eq. (65), with the extent of triplet–triplet annihilation, Eq. (55),

$$\beta \equiv \frac{C_Q^2 \, k_Q^2 \, t_f}{k_T \, C^*} \tag{71}$$

Combination of Eq. (71) with Eq. (60), the definition of ω, and Eq. (63), the definition of f, reveals that

$$\frac{\beta f}{\omega} = \frac{C_Q^2 \, k_Q^2 \, \phi \Delta}{k_T \, I(t)} \tag{72}$$

and division of both members of Eq. (70) by $I(t)$ followed by substitution of Eq. (72) yields

$$1 = \frac{\beta f}{4\omega}\left[1 - \left(1 + \frac{4\gamma\omega}{\beta f}\right)^{1/2}\right] + \left(\frac{1-\gamma}{2}\right) \tag{73}$$

Multiplication of both sides of this equation by ω yields the *only exact* relationship derived by Feldberg [46]:

$$\omega_i = \frac{\beta f}{4}\left[1 - \left(1 + \frac{4\gamma\omega_n}{\beta f}\right)^{1/2}\right] + \left(\frac{1-\gamma}{2}\right)\omega_n \tag{74}$$

which corresponds to his equation A-8, written in the recursion form notation subsequently employed by Bezman and Faulkner [19].

Equation (74) might be used to simulate S-route, T-route, and ST-route ECL observed under different conditions of triplet quenching simply by varying γ and βf—note that these latter two parameters cannot be separated—and using ω_n recursively to obtain ω_i during each iteration of the simulation. Neither Feldberg nor Bezman and Faulkner did this, however, choosing instead to use Eq. (16) to estimate ω_n as a function of time and then using the results of this computation parametrically with Eq. (74) to obtain ω_i, expressed in the linear form of the original plots. Details of Feldberg's linearization appear in the Appendix of his paper. By assuming $\beta f \gg 4\gamma\omega_n$ (extensive triplet quenching), he was able to show that

$$\omega_i = (1-\gamma)\omega_n + \gamma^2 \omega_n^2 / 2\beta f \tag{75}$$

In the event that ECL is produced exclusively via the T route, $\gamma = 1$ and Eq. (75) reduces to

$$\log \omega_i = 2 \log \omega_n - \log 2 - \log \beta f$$
$$= -2.90 \, (t/t_f)^{1/2} + 1.12 - \log \beta f \qquad (76)$$

which Feldberg also derived in his Appendix. Because it does not attempt to separate β and f, Eq. (76) can be regarded as more accurate than the result appearing in the body of the text. Regardless of this shortcoming, the equation reported by Feldberg,

$$\log \omega_i = -2.90(t/t_f)^{1/2} + 1.42 - \log \beta \qquad (77)$$

displayed for the first time a theoretical basis for obtaining plots that had more negative slopes than -1.45, as had been observed experimentally. In addition, this methodology yielded the conventional slope when $\gamma = 0$, indicative of S-route ECL production. Thus it was that Feldberg proposed that the slope of a plot of $\log \omega_i$ vs. $(t/t_f)^{1/2}$ might be used to distinguish between the two mechanisms.

The method leaves much to be desired, however. Because it was necessary to assume that βf was large in order to obtain the desired linear form, the diagnostic requires that the experimental system exhibit strong triplet quenching. In the event that triplet–triplet annihilation is favored over triplet quenching, the analysis is no longer valid. In addition, the proposed linear form does not allow for a transition between -1.45 and -2.90 in plot slopes with variations in γ as required by the consideration of ST-route ECL (and observed experimentally). And, perhaps most important, in his zeal to attach meaning to the slopes of his plots, Feldberg did not address the significance of their intercepts. This development would come six years later in the theoretical investigations of ECL efficiency by Bezman and Faulkner [19].

VI. FELDBERG PLOTS AND RADICAL ION DECOMPOSITION KINETICS

Because of Feldberg's work, early characterizations of potential step ECL phenomena usually employed plots of $\log I(t)$ vs. $(t/t_f)^{1/2}$ to display experimental results, and a great deal of early significance was given to the slopes of these plots. Accordingly, theoretical treatments using finite difference methods were geared to produce results in the same format. It was clearly demonstrated that the kinetics of the ion annihilation reaction, Eq. (3), could affect the slope and intercept of these plots; however, the direct production of excited state singlet species was shown to yield slopes that were less negative than those observed experimentally [11]. It was also shown that in the special case of diffusion-controlled ion annihilation leading to a triplet product, triplet–triplet annihilation in the presence of a triplet quencher could lead to slopes greater than those observed experimentally [46]. Neither of these early theoretical treatments provided a plausible explanation for the variations in the slopes and intercepts that were being observed at the time [44,45]. Cruser and Bard [16] first provided this insight.

Cruser and Bard set out to rationalize ECL variations due to homogeneous side reactions of electrogenerated radical ions that might take place within the diffusion layer in competition with the ion annihilation reaction. Although side reactions of either (or both) of the ion precursors to ECL could be treated using digital simulation, it was widely known from conventional cyclic voltammetric investigations that the radical cation was most subject to chemical attack, and so the reaction

$$A^{+\bullet} (+Z) \to X \qquad \text{(pseudo-first-order decay reaction)} \qquad (78)$$

was added to the theoretical mix. It is generally agreed that this cation decay to the electroinactive species X takes place via reaction with solvent or supporting electrolyte species Z that is present in such abundance that reaction (78) can be regarded as a first-order process with a rate constant k_1. This contention is supported by the variety of apparent first-order results obtained for reactions following the oxidation of aromatic hydrocarbons in solvent systems of varying degrees of purity.

Cruser and Bard took this kinetic process into account by writing the rate law for Eq. (78),

$$\frac{d[A^{+\bullet}]}{dt} = -k_1 [A^{+\bullet}] \qquad (79)$$

in finite difference form, allowing C to correspond to $A^{+\bullet}$ as was done previously:

$$\Delta f_C(J)/\Delta t = -k_1 f_C(J) \qquad (80)$$

Rearrangement and substitution of the definition of Δt [Eq. (26)] for a potential step experiment yields

$$\Delta f_C(J) = -\frac{k_1 t_f}{L f_C(J)} \qquad (81)$$

where $k_1 t_f$ is a dimensionless parameter that defines the magnitude of the pseudo-first-order rate constant. Because $\Delta f_C(J)$ cannot exceed $f_C(J)$, $k_1 t_f$ cannot exceed L. Applied repeatedly to the C array between successive iterative calculations of diffusion, Eqs. (22) and (23), Eq. (81) results in a very effective model for the effect of cation decay on ECL phenomena.

One of the first simulations attempted by Cruser and Bard was that of multiple potential step diffusion-controlled ECL generation under the influence

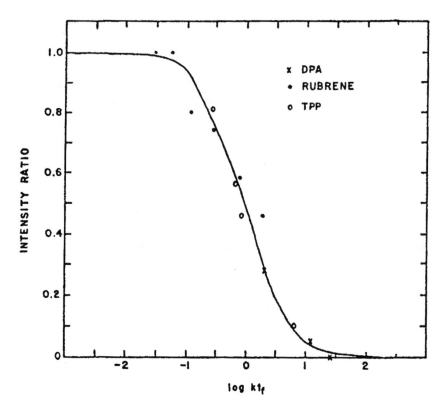

Figure 4 Theoretical plot of the integrated intensity ECL ratio for successive peaks vs, log $(k_1 t_f)$. This working curve can be used for the evaluation of k_1 from experimental data. Experimental fits for DPA (225 s^{-1}), rubrene (0.56 s^{-1}), and TPP (13 s^{-1}), are shown; parenthetic results show the values of k_1 used to fit the experimental data to the working curve. (From Ref. 16.)

of cation decay kinetics. Plots of their dimensionless parameter ZLUM, identical to Feldberg's ω (within the exact interpretation of Φ, "the quantum efficiency"), are presented in their paper over 50 cycles, each of duration $2t_f$ for three different values of $k_1 t_f$. In the absence of cation decay, $k_1 t_f = 0$, the simulation predicts identical integrated ECL intensity over successive odd and even half-cycles, so that the integrated intensity during the $2j$th peak is equal to that of the $(2j-1)$th, and the intensity ratio of successive even and odd peaks is equal to 1. When $k_1 t_f$ is greater than zero, the simulation predicts that these integrated peak intensities will not be identical, and decay in ECL intensity commensurate with the extent of cation reactivity is predicted. Cruser and Bard noted that although the long-term effect of cation decay kinetics was a gradual diminishment in ECL intensity—which had been observed experimentally—the ratio of the integrated intensity of the $2j$th peak to that of the $(2j-1)$th peak was the same for all successive peak pairs for a given value of $k_1 t_f$. This observation led Cruser and Bard to propose an ECL method for determining the magnitude of k_1 using a "working curve" that plotted the simulated ECL peak intensity ratio against log $k_1 t_f$. This plot appears in Figure 4. This plot is noteworthy because it illustrates, perhaps for the first time, a semilogarithmic method for comparing results obtained experimentally with those obtained via digital simulation. Bezman and Faulkner would subsequently [47] employ this working curve concept extensively in their work on the absolute measurements of DPA-ECL.

In addition to developing this working curve for the experimental determination of k_f, Cruser and Bard investigated the effect of cation decay, Eq. (79), upon the ECL transient as manifested in the Feldberg plot. These plots are shown for several different values of $k_1 t_f$ in Figure 5a. Note that both the slopes and the intercepts of these plots decrease with increasing $k_1 t_f$ according to these authors, with the slope decreasing from a maximum value of -1.45, as predicted for S-route ECL by Feldberg. Cruser and Bard failed to note, however, that the slope of the simulated Feldberg plot is highly dependent upon whether the unstable ion was generated during the forward or reverse potential step. (Bezman and Faulkner [19] would subsequently confirm that behavior like that illustrated in Figure 5 is predicted only when the unstable species is generated during the reverse step.) In addition, they did not note the significance of the Feldberg plot slope variations with increasing values of $k_1 t_f$.

In retrospect, this significance is quite apparent. While Feldberg was focusing on luminescence phenomena to account for slope variations in his plots, Cruser and Bard were demonstrating that slope variations could also be caused by the electron transfer processes. It would remain for Bezman and Faulkner to note that the real virtue of the digital simulation was to provide an ongoing measurement of the rate of the electron transfer processes taking place within the diffusion layer and to use the results of simulations like those carried out by Cruser and Bard to provide this rate as input for the luminescent interpretation that had been developed by Feldberg. By combining these two ideas, Bezman and Faulkner were able to rationalize a wide variety of experimental Feldberg plots and were able to obtain absolute ECL measurements for several systems.

VII. ABSOLUTE ECL MEASUREMENTS AND TOTAL TRANSIENT ANALYSIS

At the same time (and in the same laboratory) that Cruser and Bard were employing finite difference methods to provide a theoretical basis for

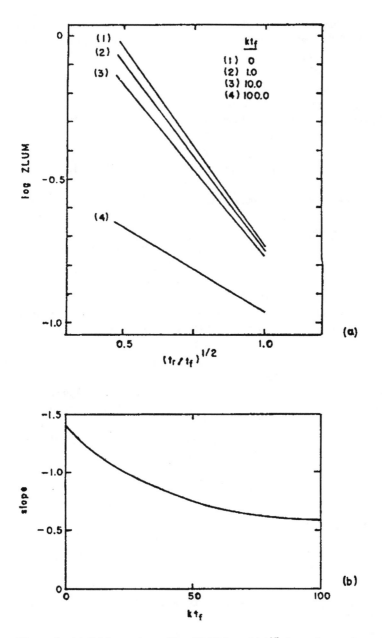

Figure 5 (a) Feldberg plots of log ZLUM vs. $(t/t_f)^{1/2}$ for various values of $(k_1 t_f)$, the dimensionless rate constant associated with pseudo-first-order cation decay. (b) A plot showing the slope of these Feldberg plots as a function of $k_1 t_f$. Note that ZLUM is the dimensionless parameter identified as ω by Feldberg. (From Ref. 16.)

understanding cation decay effects in transient ECL investigations, Faulkner and Coworkers were conducting experimental investigations of magnetic field effects upon ECL intensity [48,49]. The results of these transient experiments are reported elsewhere in this volume. What is important here is that transient ECL experiments were being conducted on a variety of systems in the presence and absence of a magnetic field, and those systems that were (or were likely to be) energy-deficient exhibited a marked increase (5–30%) in ECL intensity in the presence of an 8000 G magnetic field. Faulkner and Bard viewed this increase in intensity as evidence for triplet–triplet annihilation in these energy-deficient systems. Systems that were likely to be energy-sufficient, such as 9,10-DPA and 1,3,6,8-tetraphenylpyrene (TPP), exhibited no such intensity enhancement, and Faulkner and Bard took this as absence of triplet participation in the ECL mechanism. Thus, these magnetic field experiments offered experimental evidence indicating whether a given system produced ECL via the S route or by one of the triplet routes, the T route or the ST route. It was in search of the confirmation of this magnetic field evidence that Faulkner embarked upon a series of finite difference simulations that were to provide the first complete analysis of transient ECL phenomena employing Feldberg plots [19].

Significance was attached to both the slope and the intercept as a result of this simulation, and criteria were provided to distinguish T-route production of ECL from that produced by the direct formation of singlets. Taken in conjunction with the magnetic field results, this work provided profound insight into the ECL mechanism.

The results reported by Bezman and Faulkner relied heavily upon the luminescent triplet quenching analysis employed by Feldberg [46]. However, they considered several processes that were not included in that work. In addition to the fundamental S- route sequence described in Eqs (1)–(4) and the T-route sequence of Eqs. (54), (55), and (65) proposed by Feldberg, Bezman and Faulkner explicitly took the following additional reactions into account:

$$A^{-\bullet} + A^{+\bullet} \rightarrow 2A \qquad \text{(ion annihilation to ground state)} \qquad (82)$$

$$^3A* + {}^3A* \rightarrow {}^3A* + A \qquad \text{(triplet deactivation)} \qquad (83)$$

$$^3A* + {}^3A* \rightarrow 2A \qquad \text{(TTA to ground state)} \qquad (84)$$

Reaction (82) was regarded as diffusion-controlled, but the luminescent processes represented by Eqs. (82) and (83) were assigned binary rate constants (k_t and k_g, triplet and ground state channels, respectively). The binary rate constant for Eq. (55) that Feldberg identified as k_T was specified as k_s, i.e., the singlet-state channel. Bezman and Faulkner also pointed out that the triplet quenching reaction described in Eq. (65) was just one component of a total triplet quenching sequence,

$$^3A* (+Q) \rightarrow A \qquad \text{(triplet quenching—all sources)} \qquad (85)$$

The authors assigned a pseudo-first order rate constant τ^{-1} to Eq. (85), where τ is the triplet lifetime in the absence of TTA.

Given this luminescent framework, Bezman and Faulkner proceeded to develop their equivalents of Feldberg plots, which they expressed in the form

$$\log \omega_y = a_y + b_y(t/t_f)^{1/2} \tag{86}$$

In their work, the subscript $y = i$ was used to refer to experimental data, while the subscript $y = n$ was used to refer to the results of a simulation. Thus, the recursion relationships noted previously, e.g., in Eq. (74), would be viewed by these authors as an expression of an experimental consequence, ω_i, to a simulated result, ω_n. In making this distinction they were able to view ω_n, as Feldberg had done but in less lucid fashion, as a theoretical redox reaction rate parameter obtained via digital simulation. This formalism readily displays what might not be obvious. For example, if Φ is the ECL efficiency, then

$$\omega_i = \Phi\omega_n \tag{87}$$

and the combination of Eq. (87) with the experimental and theoretical versions of Eq. (86) at $t = 0$ reveals that

$$\log \Phi = a_i - a_n \tag{88}$$

Thus, this comparison of the experimental intercept of the Feldberg plot, a_i, with the theoretical intercept a_n determined via digital simulation can be used to determine the ECL quantum efficiency. This relationship was subsequently confirmed experimentally by the authors. It must be emphasized, of course, that the value of a_n can depend upon mechanistic parameters, as illustrated in Figure 5.

Employing this nomenclature, Bezman and Faulkner set out to obtain finite difference results for T-route generation of ECL that would exhibit mechanistic variations in the slope and intercepts of the simulated Feldberg plots. Their methodology was quite similar to that employed by Feldberg, with the already noted exceptions of the additional included reactions and the fact that the dimensionless width of the luminescent reaction zone, $f(t)$, was taken to be a function of time. Rather than attempting to take S-route or ST-route ECL generation into account by employing the fraction γ as Feldberg did, e.g., in Eq. (66), the authors simply declared Eq. (3) to be inoperative so that only T-route ECL was subjected to luminescent analysis. That bridge having been crossed, they then proceeded to write an expression for $d[^3A^*]/dt$ that they set equal to zero as they invoked the steady-state assumption. The resulting quadratic was solved for $[^3A^*]_{ss}$, and this quantity was used in an expression for $I([^3A^*]_{ss})$ to obtain an expression for $I(t)$ or ω_i in terms of the roots of the quadratic expression for $[^3A^*]_{ss}$. This procedure is directly analogous to that employed by Feldberg in the development of Eq. (74), but the result is somewhat more complicated due to the fact that the additional reactions were taken into account:

$$\omega_i = \beta[1 - (1 + \alpha\omega_n/\beta)^{1/2}] + {}^{1/2}\alpha\omega_n \qquad (89)$$

where α is a T-route efficiency parameter

$$\alpha \equiv \phi_t \phi_{tt} \phi_f/(1-g) \qquad (90)$$

and β is a T-route quenching parameter

$$\beta \equiv f(t)t_f \phi_{tt} \phi_f/8k_a \tau^2 C_A^*(1-g)^2 \qquad (91)$$

[Note that the quantity defined as β in Eq. (91) is not the same quantity defined as β by Feldberg in Eq. (71), even though both refer to the triplet quenching process.]

The symbolism employed in these definitions of α and β has, for the most part, been defined above, with the additions

$$k_a = k_g + k_t + k_s \qquad (92)$$

and

$$g = k_t/2k_a \qquad (93)$$

while ϕ_f is the fluorescence efficiency, ϕ_t is the probability of triplet formation per electron transfer event, and ϕ_{tt} is the probability of excited state singlet formation in triplet–triplet annihilation.

Having derived Eq. (89) as the general expression for T-route ECL, Bezman and Faulkner then sought to identify those values of α and β that would produce linear renditions of log ω_i vs. $(t/t_f)^{1/2}$. Using a minor modification of the finite difference methodology employed by Cruser and Bard, they were able to refine Feldberg's estimate for ω_n, the dimensionless representation of the rate of solution-phase charge transfer,

$$\log \omega_n = -1.483 (t/t_f)^{1/2} + 0.724 \qquad (94)$$

which they used, in the absence of cation decay, as input to Eq. (89). Using this technique they were able to obtain predictions of linear experimental Feldberg plots for $(\alpha/\beta) > 0.1$,

$$\log \omega_i = a_i + b_i(t/t_f)^{1/2} \qquad (95)$$

where a_i and b_i as output depend upon α and β as input to the computation. Hence, it is possible to construct working curves or relationships that will allow one to determine α and β from a_i and b_i. Working graphically, the authors found that the appropriate solutions were of the form

$$\log \alpha = r(b_i) + a_i \qquad (96)$$

and

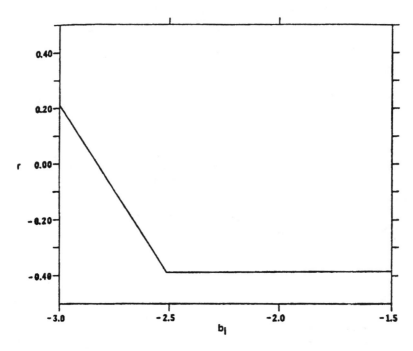

Figure 6 Working curve for $r(b_i)$, which may be used in conjunction with Eq. (96) to evaluate α, the T-route efficiency parameter, from a_i and b_i, the intercept and slope of the experimental Feldberg plot. (From Ref. 19.)

$$\log \beta = s(b_i) + a_i \tag{97}$$

where $r(b_i)$ and $s(b_i)$ are displayed in Figures 6 and 7. Thus, if it is possible to establish that ECL is produced via the T-route process, one can readily determine α and β from the intercept and slope of the experimental Feldberg plot, a_i and b_i, respectively. Bezman and Faulkner were quick to point out that while α was of greater interest because it alone is proportional to φ_t, the probability of triplet formation per electron transfer event, β is of more use diagnostically in T-route processes because of its association with triplet quenching. This is of little surprise, of course, because it was upon the suggestion of Marcus regarding triplet-quenching effects that Feldberg launched this investigation in the first place. What is noteworthy, however, is that by taking all triplet processes into account, Bezman and Faulkner were able to demonstrate that Feldberg plots might exhibit a range of slopes under the influence of T-route ECL generation. From Figure 7 it is clear that most of this variability is due to β. They have, in fact, reported results [50,51] where $10^{-4} < \beta < 10^{-2}$, which corresponds to experimental slopes in the range -1.9 to -2.5, so that Feldberg plots with slopes

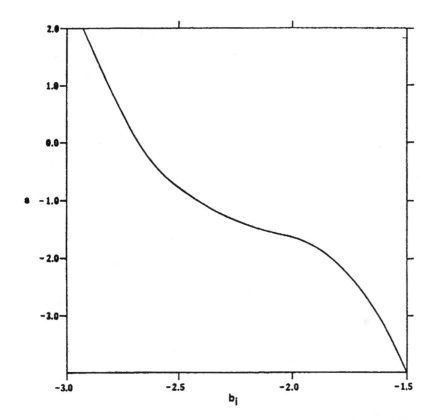

Figure 7 Working curve for $s(b_i)$, which may be used in conjunction with Eq. (97) to evaluate β, the T-route quenching parameter, from a_i and b_i, the intercept and slope of the experimental Feldberg plot. (From Ref. 19.)

in this range might be taken as a preliminary indication of T-route generation. It must be observed, however, that cation decay can produce the same effect upon the slope.

It has already been noted in Eq. (88) that Φ, the ECL efficiency, can be determined by comparing a_i and a_n, the experimental and theoretical intercepts, respectively. Bezman and Faulkner sought to compare this value of Φ with a value for Φ obtained from Φ_{coul}, which they defined as the ratio of the total quantum output to the number of reactant ions produced in the system during the forward potential step. In order to measure Φ_{coul} experimentally it was necessary to set up and calibrate an integrating sphere that could measure the absolute luminescence produced during the double (or in Faulkner's nomenclature, triple) potential step experiment and to compare this with the charge that passed during the forward step. Chronocoulometry was employed to determine Q_f, the total charge given to ion production during the forward step, and Bezman and Faulkner used this quantity along with the integrated intensity to determine Φ_{coul}.

$$\Phi_{coul} = \frac{F}{Q_f} \int\limits_{t_f}^{\infty} I(t) \, dt \tag{98}$$

A problem arises, however; $\Phi \neq \Phi_{coul}$ because Q_f is not an exact measure of the number of ion annihilation reactions that take place in the diffusion layer. Some of the ions produced during the forward step diffuse away from the electrode and are not ever available to react with the ions produced during the reverse step. This fraction, θ^{-1} in the authors' notation, can easily be computed using digital simulation, and they determined that in the absence of any homogeneous kinetic complications $\theta = 1.078$, thereby indicating that roughly 93% of the ions generated during the forward step are able to undergo ion annihilation thereafter, so that

$$\Phi = \theta \, \Phi_{coul} = 1.078 \, \Phi_{coul} \tag{99}$$

in the absence of any homogeneous reactions such as cation decay that, in effect, remove reactive ions from the diffusion layer.

In the event that homogeneous reactions alter the fundamental ion annihilation reaction rate, ω_n, as given by Eq. (94), the effect of this instability must be taken into account. Bezman and Faulkner did this by conducting additional simulations in order to derive expressions for ω_i under the influence of a pseudo-first-order reaction of an electrogenerated ion leading to an inactive product, as in Eq. (78). Although this work mimicked that of Cruser and Bard, it resulted in a more thorough presentation of the outcome in the form of simulated Feldberg plots:

$$\log \omega_n = a_n(k_1 t_f) + b_n(k_1 t_f) \, (t/t_f)^{1/2} \tag{100}$$

that expressed the coefficients a_n and b_n as functions of the dimensionless simulation input parameter $k_1 t_f$. These coefficients are shown as a function of $k_1 t_f$ in Figures 8 and 9 for the cases of unstable ion generation during the forward and reverse step, respectively.

These working curves are applicable any time a pseudo-first-order following reaction causes a depletion of one of the ionic ECL precursors to take place within the diffusion layer. Feldberg plots constructed from them can be used to interpret S-route systems directly. In addition, the resulting equations may be taken as the redox rate input equation for T-route ECL, Eq. (89). Feldberg plots constructed from the output of this process could be used, in theory, to model T-route ECL under the influence of ion precursor instability, but in practice this is both difficult (because it involves a multiparameter fit) and unnecessary (because energy-deficient T-route systems generally have

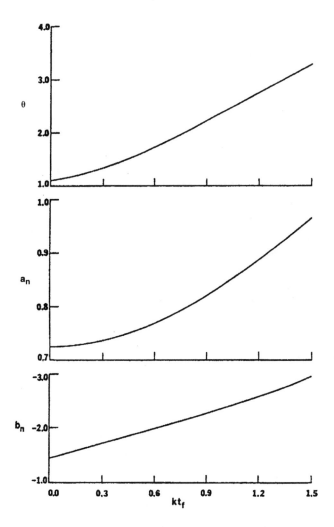

Figure 8 Working curves depicting the effects of the generation of unstable reactant during the forward step. The Feldberg plot coefficients a_n and b_n are shown as functions of the dimensionless rate constant $k_1 t_f$. The reciprocal of the ion annihilation collection efficiency is given by θ. (From Ref. 19.)

unreactive precursors). Energy-sufficient systems frequently have reactive redox products, however, and the working curves shown in Figures 8 and 9 are applicable in these cases.

Particularly interesting is the fact that the order of unstable ion generation produces markedly different results. Figure 8 shows the well-behaved results of

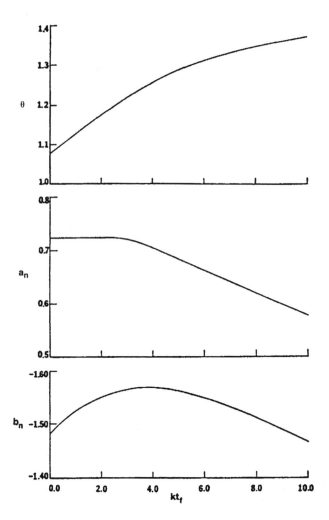

Figure 9 Working curves depicting the effects of unstable reactant generation during the reversal step. The Feldberg plot coefficients a_n and b_n are shown as functions of the dimensionless rate constant $(k_1 t_f)$. The reciprocal of the ion annihilation collection efficiency is given by θ. (From Ref. 19.)

generating the unstable ion first. Both the (negative) slope and the intercept of the theoretical Feldberg plot increase with increasing $k_1 t_f$. Note that the Feldberg slope can range anywhere between its maximum, -1.483, and -3.00 depending upon the magnitude of $k_1 t_f$; this suggests that ion decay under these conditions can be far more responsible for Feldberg slope variation than any effects due to luminescence kinetics. The reciprocal of the ion annihilation collection efficiency also increases with k_1t_f, indicating that only about 25% of the unstable ions generated during the forward step are collected by ion annihilation if k_1t_f exceeds 1.5. The behavior shown in Figure 9 for unstable ion generation during the reverse step is, like that reported in Figure 5, less well behaved. At low values of k_1t_f the coefficient a_n remains constant as b_n becomes more negative. At higher values of k_1t_f the coefficient a_n decreases and b_n becomes more positive. This behavior was subsequently observed experimentally by Bezman and Faulkner in studies of DPA ECL [47], and it required them to limit their investigation of the effects of DPA cation instability on anodic forward steps where the working curves of Figure 8 apply.

Armed with the results of this monumental paper, Bezman and Faulkner set out to experimentally investigate systems that were believed, on the basis of magnetic field experiments, to proceed to ECL via the different routes. In three successive papers they considered the ECL of DPA [47], rubrene [50], and the mixed fluoranthene–10-methylphenothiazine system (10-MP) [51]. The reader is directed to these papers for details, i.e., absolute measurement of ECL using potential steps at several different concentrations, etc., but a synopsis of the applications of this theory to these experimental systems follows:

1. DPA.   This system was found to exhibit Feldberg slopes in the -1.8 to -2.2 range. Because DPA is energy-sufficient and never reported to exhibit any magnetic field effect, slopes much closer to the minimum value of -1.48 were anticipated for this system but were not observed. Bezman and Faulkner were able to attribute the disparity between fact and anticipation entirely to the instability of the DPA cation (using the working curves of Figure 8 with values for k_1 obtained from cyclic voltammetry) and to reconcile these slopes with the S-route character of DPA. Favorable comparisons were obtained at all concentrations for the ECL efficiency Φ determined from the intercept of the Feldberg plots using Eq. (88) and from $\theta\ \Phi_{coul}$ using Eq. (99). The ECL efficiency increased with increasing concentration and approached a value of 2.5 $\times\ 10^{-3}$ in the asymptotic limit.

2. Rubrene.   Suspicions that rubrene ECL proceeded via the T route on the basis of magnetic field effects were confirmed by applying the results from the finite difference simulations described above to these experimental observations. Feldberg slopes in the -1.9 to -2.8 range were observed with no indication of instability for either ionic precursor to ECL. This required Bezman and Faulkner to fit these data using the T-route methodology described above. At each rubrene concentration in DMF and benzonitrile solutions the slope and the intercept of the Feldberg plot were recorded and the absolute value of Φ_{coul} was measured. From these data α and β were computed using Eqs. (96) and (97) along with the working curves shown in Figures 6 and 7. The efficiency Φ was obtained using Eq. (99) and found to be 10^{-3} for rubrene in benzonitrile and 5

$\times 10^{-4}$ for rubrene in DMF. Analysis of the results reported for α at each concentration yielded an estimate for ϕ_t, the probability for triplet formation per electron transfer event. Bezman and Faulkner estimated this in the 10–30% range, with the remainder of collisions producing ground-state molecules. Whereas Φ exhibited little dependence on the direction of the initial potential step, the efficiency parameter α and the quenching parameter β both exhibited considerable dependence upon reaction sequence in benzonitrile. This was attributed to the oxidation of water impurities in these solutions.

3. Fluoranthene–10-MP.   This study is noteworthy because it considers a mixed system that was believed to be energy-deficient on the basis of magnetic field effects [49] and triplet quenching experiments [52]. In this work, Bezman and Faulkner carried out a series of potential step experiments on solutions containing concentrations of fluoranthene and 10-MP in the vicinity of 1 mM in order to observe ECL corresponding to fluoranthene luminescence. Potentials were switched between -1.70 and $+0.77$ V (vs. QRE) in order to produce the fluoranthene anion and the 10-MP cation as the ionic precursors to ECL. Slopes of the resulting Feldberg plots were in the -2.0 to -2.9 range, with most observations approaching the upper limit -2.96 that can be expected for the T-route mechanism. On the basis of these observations Bezman and Faulkner tentatively made the T-route assignment and went on to carry out a series of digital simulations that resulted in the computation of a_n, b_n, and θ for the mixed system. Rather than having to resort to working curves as they did in their original T-route simulations, Bezman and Faulkner found that these computed quantities could all be expressed as simple functions of R, the ratio of the bulk concentration of the precursor of the ion formed during the first potential step to that of the precursor of the ion formed upon reversal. Using these minor computational adjustments, they were able to linearize Eq. (89) and employ the result along with Eq. (99) to obtain a complete ECL characterization of this mixed system. Values of α, β, and Φ_{coul} are reported for each fluoranthene–10-MP concentration pair along with the slopes and intercepts of each Feldberg plot. From these data, Bezman and Faulkner determined that Φ for this system is approximately 8×10^{-5} and estimated that ϕ_t, the triplet yield of the ion annihilation reaction, is approximately 30%.

VIII.   ANALYSIS OF STEADY STATE ECL AT THE RRDE

Most of the early ECL investigations were carried out using the transient current methods described above; then workers in the Bard group began to use the rotating ring-disk electrode (RRDE) in an attempt to observe and interpret steady-state ECL emission. As early as 1968, Maloy et al. [53] reported the initial observation of ECL at the RRDE. This work was noteworthy because it demonstrated that the ECL phenomenon occurs as a result of the reaction of solution-phase species rather than resulting from electrode surface processes, a possibility that could not easily be excluded on the basis of single-electrode transient experiments. Having demonstrated that steady-state ECL resulted from homogeneous reactions that yielded products that could be observed at the RRDE, these investigators went on to develop a quantitative model for this electrode so that it could be used to investigate the kinetics of this process.

Since its introduction by Frumkin and Nekrasov [54], the rotating ring-disk electrode has become widely used in the study of the homogeneous kinetics of electrogenerated species. Shown in Figure 10, the RRDE consists of concentric coplanar ring and disk electrodes mounted in a common inert matrix, coaxial with a shaft about which the entire assembly can be rotated, and separated from each other by an inert insulating material. Electrical contact is provided for each electrode so that the ring and disk can be independently connected to an external source with brush contacts.

When the RRDE is used conventionally, a species is electrogenerated at the disk electrode while the ring electrode is maintained at a potential at which this species will be electroactive; thus, the reverse of the disk reaction occurs at the ring. If A is the species present in the bulk of the solution and B is the species generated from A at the disk, the reactions occurring in conventional RRDE studies are

$$A \rightarrow B \quad \text{(disk reaction)}$$
$$B \rightarrow A \quad \text{(ring reaction).} \tag{101}$$

[Equation (101) is identical to Eq. (29), with the exception that two distinct electrodes—the ring and the disk—are specified.]

The ratio of the steady-state current at the ring to that at the disk under these conditions is called the collection efficiency (N) because it measures the fraction of the material generated at the disk that ultimately reaches the ring. If the species that is generated at the disk is stable, then N is determined only by the geometry of the electrode and is less than unity and, for the most part, calculable [55]. If the disk-generated species is unstable, yielding an electroinactive product X [see Eq. (78)], then N will be an increasing function of the angular velocity ω of the electrode which, at high ω, approaches that value of N which would be obtained in the absence of reaction (78). This variation of N with ω has been treated theoretically by exact [56] and approximate [13] methods so that the lifetime of the unstable species can be determined experimentally.

A case in point is the lifetime of the cationic ECL precursor DPA+ in DMF. It has already been shown that in DPA this cation is unstable, yielding an unknown product X, which is electroinactive at potentials somewhat more negative than the oxidation of DPA [16]. Although the cation probably reacts

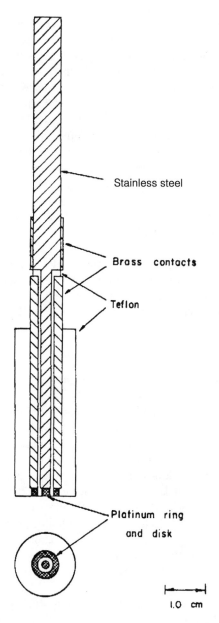

Figure 10   The rotating ring-disk electrode (RRDE). (From Ref. 29.)

with the solvent system, the rate law for the process is usually written in the "pseudo-first-order" form of Eq. (78). At the RRDE the cation can be generated at either the ring or the disk so that Eq. (78) could be taken as either

$$B \xrightarrow{k_1} X \qquad \text{(disk-generated species unstable)} \qquad (102)$$

or

$$B \xrightarrow{k_1} X \qquad \text{(ring-generated species unstable)} \qquad (103)$$

The effect of the first-order decay of the disk-generated species [Eq. (102)] on the collection efficiency is shown in Figure 11. Here the simulated collection efficiency is plotted as a function of a dimensionless rotation rate parameter; the filled circles are results obtained from the simulation. As expected, the collection efficiency is predicted to rise with increasing ω, ultimately approaching the

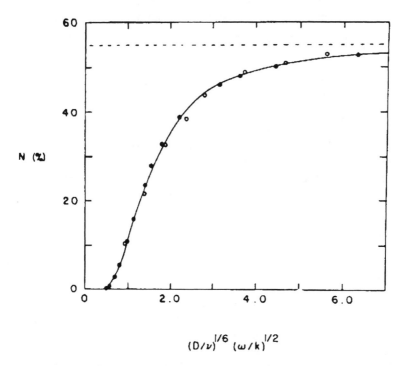

$$(D/\nu)^{1/6} (\omega/k)^{1/2}$$

Figure 11 RRDE collection efficiency results for DPA+ ●. The dimensionless plot was obtained for the given electrode using digital simulation. Filled circles show simulated points; open circles represent experimental data fitted to the simulated curve. Experiment was fit to theory by setting $k_1 = 4.1 \text{ s}^{-1}$. (From Ref. 29.)

theoretical collection efficiency of the electrode in the absence of kinetic complications. The open circles are points obtained from collection efficiency measurements made on the disk-generated DPA cation; the experimental curve was fit to the simulated curve by setting

$$k_1 = 4.1 \text{ s}^{-1} \tag{104}$$

in the dimensionless simulation parameter shown as the abscissa of this plot. Hence, the lifetime of the DPA cation in DMF was found to be 0.25 s by this study. (This lifetime is considerably longer than that reported previously [16]; the two measurements were made using different techniques, and the solutions were prepared from solvents purified in different ways. This increase in cation lifetime probably reflects improvements made in the solvent purification process rather than an error made in either determination. Indeed, further improvements in solvent purification techniques resulted in even slower decay kinetics for the same process reported subsequently by Bezman and Faulkner [47] and described above.) This illustrates the conventional use of the RRDE in kinetics studies.

In the vast majority of conventional ECL studies, the anion and cation radicals of the precursor are alternately generated at a single electrode, as described previously. These methods are fraught with difficulties due to the complex nature of the resulting current–time–intensity behavior. In order to obtain meaningful quantitative results, both the current and the ECL intensity must be integrated, as in the work of Bezman and Faulkner [47], but even when this is done one must rely upon the results of a finite difference simulation to connect observation with interpretation.

These transient ECL problems can be overcome through the unconventional use of the RRDE; instead of the ring electrode being used to collect the species made at the disk, it is used to generate a species other than that being formed at the disk. Thus, the least complicated set of electrode reactions in the RRDE-ECL experiment can be written

$$\begin{aligned} &A \rightarrow B \quad \text{(disk reaction)} \\ &A \rightarrow C \\ &B \rightarrow C \quad \text{(ring reaction)} \end{aligned} \tag{105}$$

where the latter ring reaction is thermodynamically unavoidable. These simultaneous reactions are identical to the serial processes described for a single electrode in Eqs. (39)–(43). Generation of B and C, the radical ion ECL precursors, is followed by their reaction according to Eq. (41) to produce A*, just as in the conventional case. Thus, in RRDE-ECL the disk-generated species is collected by a solution species generated at the ring rather than being collected by the ring itself. The efficiency of this collection depends upon the rate of the ECL process relative to the rotation rate of the electrode.

There are some advantages in using the RRDE in this manner for ECL studies. As noted above, the central RDE is particularly convenient for quantitative work because the solutions to the hydrodynamic equations that describe mass transfer to it are known and produce a closed-form steady-state current function given earlier in Eq. (11), the Levich equation. The same hydrodynamics govern the ring electrode, and the steady state can be attained quickly at both electrodes at even moderate rotation rates. Therefore, the ring and the disk may be set at independent potentials that can cause the simultaneous generation of the radical ion precursors to produce steady-state ECL under steady-state current conditions. Because there is no alternation of potential at either electrode, charging currents are negligible once the steady state is attained, and the current at either electrode can be easily measured. As an added advantage, the typically larger electrode area and greater material flux of the rotating ring-disk results in greater quantities of ECL emission.

Even though the RRDE has many experimental advantages in the study of ECL phenomena, its use also requires the results of a finite difference simulation in order to relate observation with results. As a consequence of these simulations and some pertinent experimental observations, it has been proposed that RRDE-ECL studies are an ideal way to determine Φ_{coul}, the efficiency of the current at producing luminescence [29]. What follows, then, is a description of this theoretical development.

The simulations used in the interpretation of RRDE-ECL are based upon those developed by Prater [28] for use in the RRDE investigations of homogeneous kinetics described in Eqs. (102) and (103). These RRDE simulations are based, in turn, upon the RDE simulations described previously in Eqs. (47)–(53).

In extending these RDE results to RRDE and RRDE-ECL with digital simulation, one models the hydrodynamic layer of the solution by dividing it into a two-dimensional grid of volume elements as shown in Figure 12. The concentration of any species in solution is considered to be uniform within a given element. Because of the cylindrical symmetry of the RRDE, the solution is divided into several layers of thickness Δx parallel to the electrode surface, with each layer containing a cylindrical element of diameter Δr that is centered on the axis of rotation. A series of concentric annular elements, each of width Δr, make up the remainder of each layer. Each element is numbered with its layer number J and its radial number K; hence, the relative concentration of any species (its real concentration divided by the bulk concentration of a species present initially) can be specified as $f_i(J,K)$. Thus, in the nomenclature developed by Prater [28], $f_B(J, K)$ is "the relative concentration of the Bth species in the Jth layer and the Kth ring."

Diffusion, convection, and chemical reactions (kinetics) of the species that occupy these volume elements are all functions of time. These temporal changes

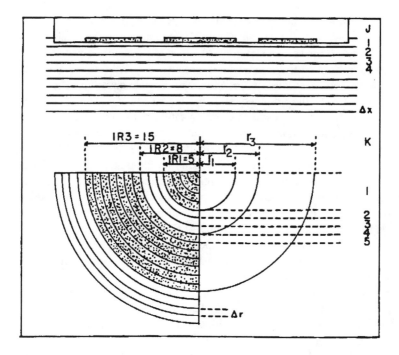

Figure 12 Two-dimensional volume element grid employed in the simulation of RRDE and RRDE-ECL phenomena. Fractional concentrations are specified as f(J,K), where J corresponds to the direction that is normal to the electrode and K corresponds to the radial direction. (From Ref. 28.)

are modeled iteratively, each iteration of duration Δt generating a new two-dimensional concentration array for each species present. The finite difference expressions for modeling linear diffusion, convection normal to the electrode, and chemical kinetics have been given in Eq. (22), Eq. (52), and, e.g., Eq. (46), respectively. Modeling the RRDE via digital simulation requires that one additional mode of mass transport—radial convection—be taken into account. (Actually, radial diffusion is also present as an additional mode of mass transfer, but its effects are assumed to be negligible with respect to radial convection.)

Radial convection is simulated in the same way that normal convection is simulated, as described in Eqs. (47)–(52). The laws of hydrodynamics are used to calculate the distance the solution in a given volume element travels radially during the interval Δt, and a new array is generated by effecting these translations for each volume element. Because the hydrodynamic equation for radial convection differs from Eq. (47) given above for normal convection, the radial convection algorithm begins with a differential equation for radial hydrodynamic transport:

$$v_r = \frac{dr}{dt} = -0.51\omega^{3/2}v^{-1/2}xr \tag{106}$$

where r is the radial distance from the axis of rotation and all other factors are as described previously. This expression can be rearranged and integrated between t_2 and t_1 (holding x constant) to obtain

$$\ln(r_2) - \ln(r_1) = -0.51\ \omega^{3/2}v^{-1/2}x(t_2 - t_1) \tag{107}$$

where r_2 and r_1 are the radial positions of solution volumes at times t_2 and t_1, respectively. This expression can be further rearranged by combining it with Eq. (21) and setting $t_2 - t_1 = \Delta t$:

$$\ln(r_1/r_2) = 0.51\ \omega^{3/2}v^{-1/2}\ (J-1)\ \Delta x\ \Delta t \tag{108}$$

Upon setting $r_1 = K_1\Delta r$ and $r_2 = K_2\Delta r$ and making appropriate substitutions for Δx and Δt, Eq. (108) becomes

$$\ln\left(\frac{K_1}{K_2}\right) = \frac{0.51(J-1)}{D_M^{1/2}\ L^{3/2}} \tag{109}$$

This equation can be rearranged to obtain an expression for K_2, the serial number of the radial volume element containing the solution that will be found at radial volume element K_1 at the end of the Δt interval:

$$K_2 = K_1\left\{\exp\left[\frac{0.51(J-1)}{D_M^{1/2}L^{3/2}}\right]\right\}^{-1} \tag{110}$$

This equation is employed repeatedly to compute radial convection within the Jth layer the same way that Eq. (52) is used to compute normal convection: The contents of the computed volume element K_2 are transferred to each volume element K_1 to simulate radial convection:

$$f_i(J, K_1) = f_i(J, K_2) \tag{111}$$

The algorithm is then carried out over the elements within the next layer until radial convection has been modeled in all the normal layers. Because K_2 need not be an integer, some interpolation may be necessary, and some care must be taken to avoid overwriting recently computed data. Moreover, because the first radial volume element is a cylinder with a base equal in area to the area of the disk electrode, values of K must be assigned to the annular ring elements that are proportional to the radial displacement of that element. Details are given by Prater [28,57].

Variables of electrochemical interest are specified by the electrode boundary conditions that give rise to the concentration gradients in solution.

Current is measured as the flux of electroactive material into the electrode under a given set of potential-dependent electrode conditions. Although it is easy to simulate any reversible electrode potential or any magnitude of constant current, the simulations reported herein all assume a potential such that the concentration of the electroactive species goes to zero at the electrode and the limiting current is achieved.

As noted previously, the validity of this kind of simulation depends upon the extent to which the dimensions of Δx, Δr, and Δt approach zero, in order that the problem may approach differential form. This is achieved by judiciously defining Δt as in Eq. (26) and Δx as in Eq. (28) in terms of D_M, the model diffusion coefficient, and t_k, a known time that is appropriate for the experiment under consideration. In these simulations the model diffusion coefficient of all species was set at $D_M = 0.45$. As noted in Eq. (51), it is appropriate to express t_k in terms of the reciprocal of the angular velocity $\omega-1$, D_A, and v (the kinematic viscosity of the solvent) in RDE and RRDE simulations. Therefore, Δt is defined as in Eq. (26) and becomes

$$\Delta t = \frac{t_k}{L} = \left(\frac{v}{D_A} \right)^{1/3} \left(\frac{\omega^{-1}}{L} \right) \tag{112}$$

where L is an assigned number of time iterations. In the course of this work it was found that by setting $L = 50$, it was possible to attain steady-state conditions after 3L iterations. Note that the dimensionless coefficient $(v/D_A)^{1/3}$ appears as a result of the fact that hydrodynamic behavior depends upon solution viscosity.

Finally, Δr is defined by I_{R1}, the number of elements representing the disk electrode,

$$\Delta r = r_1/(I_{R1} - 0.5) \tag{113}$$

where r_1 is the actual radius of the disk electrode and I_{R1} is an assigned variable. For the electrode used in the experimental work that is modeled below,

$$r_1 = 0.145 \text{ cm}, \qquad r_2 = 0.163 \text{ cm}, \qquad r_3 = 0.275 \text{ cm}$$

and the simulation variables I_{R2} and I_{R3}, which respectively represent the inner and outer radius of the ring electrode, were in the same proportion to I_{R1} as r_2 and r_3 were to r_1. Thus, specifying $I_{R1} = 81$ sets $I_{R2} = 91$ and $I_{R3} = 154$. These values were used in all the simulations reported herein. Strictly speaking, the conclusions reached as a result of these simulations are valid only for an electrode of this geometry.

Because the basic RRDE program written by Prater calculates collection efficiencies, some modification of boundary conditions is required to treat RRDE-ECL according to Eq. (105). For the disk electrode, these are unchanged by the introduction of C, a ring-generated species. They are

$$f_A'(1, 1) = f_A(1, 1) = 0 \tag{114a}$$

$$f_B'(1, 1) = f_B(1, 1) + D_{MA} [f_A(2, 1)] - D_{MB}[f_B(1, 1) - f_B(2, 1)] \tag{114b}$$

$$f_C'(1, 1) = f_C(1, 1) = 0 \tag{114c}$$

where primed quantities exist at the start of the next iteration. Note that only one element is required to simulate the I_{R1} elements making up the disk electrode, and $K = 1$ in these equations; this is possible because the disk is a uniformly accessible surface so that the same conditions would obtain for each element. The quantity $D_{MA} [f_A(2, 1)]$ represents the amount of species A that is converted to B at the disk; this is proportional to the disk current.

Because the ring is not uniformly accessible, each ring element must be treated separately. The boundary conditions for each ring element become

$$f_A'(1, K) = f_A(1, K) = 0 \tag{115a}$$

$$f_B'(1, K) = f_B(1, K) = 0 \tag{115b}$$

$$f_C'(1, K) = f_C(1, K) + D_{MA} [f_A(2, K)] + D_{MB} [f_B(2, K)]$$
$$- D_{MC} [f_C(1, K) - f_C(2, K)] \tag{115c}$$

where K is a radial element number of each element within the ring. The current at the ring is proportional to the area-weighted sum of the currents of each ring element:

$$\sum_{K_R} \left\{ D_{MA} \left[f_A (2,K) \right] + 2 D_{MB} \left[f_B (2,K) \right] \right\} \frac{A(K)}{A_D}$$

This summation is carried out over all K_R elements representing the ring; $A(K)$ is the area of the Kth ring and A_D is the area of the disk. Note that the total current is the sum of current contributions from both species A and species B.

Diffusion and convection are treated just as in previous RRDE simulations; therefore, the only remaining additional complication of the RRDE-ECL simulation is the treatment of Eq. (41), the kinetically controlled reaction of B and C to produce A*. Species B and C, of course, are the radical ions of A, the hydrocarbon present initially, and the reaction in question is the solution-phase electron transfer reaction given in Eq. (3), which has been the subject of much of the prior transient analysis. This analysis results in a two-dimensional representation of Eq. (46), the finite difference rate expression for the production of A*:

$$\Delta f_{A^*} (J,K) = (k_2 C_A^* t_k) f_B (J, K) \frac{f_C (J, K)}{L} \tag{116}$$

which becomes the appropriate rate law for RRDE-ECL by substituting the appropriate expression for t_k given in Eq. (51):

$$\Delta f_{A*} (J, K) = \left[\left(\frac{v}{D_A} \right)^{1/3} k_2 C_{A*} \ \omega^{-1} \right] f_B (J, K) \frac{f_C (J, K)}{L}$$

(117)

The bracketed quantity $(v/D_A)^{1/3} k_2 C_A* \omega^{-1}$ is the resulting dimensionless rate constant. This is the final variable input parameter used in the simulation of RRDE-ECL; variations in $(v/D_A)^{1/3} k_2 C_A* \omega^{-1}$ determine the extent to which rotation competes with kinetics.

Equation (117) is used to determine the chemical changes due to radical ion reaction within each element and to quantify the chemiluminescence occurring as a result of these changes. Because the conversion of the excited singlet to ground state singlet is quite rapid, the net material effect of each fruitful encounter of B and C is the production of 2 A. The relative concentration changes in each element are computed accordingly:

$$f_A'(J, K) = f_A(J, K) + 2\Delta f_{A*}(J, K)$$

(118a)

$$f_B'(J, K) = f_B(J, K) - \Delta f_{A*}(J, K)$$

(118b)

$$f_C'(J K) = f_C(J, K) - \Delta f_{A*} (J, K)$$

(118c)

The number of photons emitted from any element during Δt can be written

$$\Delta f_{A*} (J, K) \ \Phi_{ECL} \ C_A* \ [\Delta x \ A(K)]$$

(119)

so that $I(t)$, the total ECL intensity (quanta per second) is the sum of the emission from all the elements divided by Δt:

$$I(t) = \Sigma \{\Delta f_{A*}(J, K) \ \Phi_{ECL} \ C_A* \ [\Delta x \ A(K)]\}/\Delta t$$

(120)

Substitution of expressions for Δx and Δt and division of both members by A_D yields, after rearrangement,

$$\frac{I(t)}{\Phi_{ECL} \ A_D \ C_{A*} \ D_A^{1/3} \ v^{1/6} \ \omega^{1/2}} = \left(\frac{L}{D_M} \right)^{1/2} \sum_{J, K} \Delta f_{A*} (J, K) \frac{A(K)}{A_D}$$

(121)

where all those quantities on the right-hand side are simulation variables, and the left-hand side is a valid dimensionless representation of ECL intensity. In this representation $I(t)$ is rendered dimensionless by ω, an experimental variable; hence, any use of this parameter in a comparison of real and simulated results requires that each experimentally observed intensity be normalized by dividing it by $\omega^{1/2}$ prior to comparison.

This requirement can be eliminated by dividing both sides of Eq. (121) by $[(v/D_A)^{1/3} k_2 C_A* \omega^{-1}]^{1/2}$ and rearranging:

$$\frac{I(t)}{\Phi_{ECL}\, A_D\, [D_A\, (C_A^*)^3\, k_2\,]^{1/2}}$$

$$= \left[\frac{L}{(v/D_A)^{1/3}\, k_2 C_{A*}\, \omega^{-1} D_M} \right]^{1/2} \sum_{J,K} \Delta f_{A*}\, (J,K)\frac{A(K)}{A_D}$$

(122)

where the left hand side is now a dimensionless RRDE-ECL intensity parameter that does not depend on ω.

This representation of $I(t)$ is used in the rotation rate simulations that are reported below. In these simulations the rate parameter $(v/D_A)^{1/3}\, k_2 C_A^*\omega^{-1}$ was varied in order to obtain the dimensionless representation of ω described in Eq. (122). The results of this simulation are displayed in Figure 13, which shows the variation of $I(t)/[\Phi_{ECL}A_D[D_A(C_A^*)^3\, k_2]^{1/2}]$, the dimensionless ECL parameter defined in Eq. (122), with the square root of $(v/D_A)^{1/3}k_2C_A^*\omega^{-1}$.

Figure 13 actually consists of two graphs, that have the same ordinate and lie adjacent at a common point; the abscissa of the graph on the right, however, is the reciprocal of that on the left plotted in reverse. Hence, the total abscissa is drawn with ω increasing in a nonlinear fashion from left to right, and all possible values of ω are shown. Also note that the denominator of the ordinate variable contains only factors that are constants, given a set of experimental conditions; similarly, the abscissa variables can change only when experimental values of ω change. Thus, Figure 13 is the dimensionless representation of the $I(t)$ vs. ω behavior of RRDE-ECL using an electrode of the geometry specified.

The simulation predicts that RRDE-ECL intensity will increase with increasing rotation rate, go through a maximum, and decrease thereafter. The maximum (point b on the graph) is predicted when

$$(D_A/v)^{1/6}\, (\omega/k_2C_A^*)^{1/2} = 0.4 \tag{123}$$

or, since $v/D_A \cong 10^3$ for systems of electrochemical interest, when

$$\omega = 1.6k_2C_A^* \tag{124}$$

Because the maximum rotation rate possible with typical instrumentation is $\sim 10^3$ s^{-1} and the minimum practical concentration of electroactive species is $\sim 10^{-3}$ M, the ECL maximum can be reached only if k_2 is less than 10^6 M^{-1} s^{-1}. This is far below a bimolecular rate constant approaching the diffusion control limit suggested by the transient experiments described above.

In addition to point b, two additional points are marked in Figure 13, and vertical lines passing through these points divide the graph into three regions. Interior to point a the simulation predicts linear behavior, with the straight line

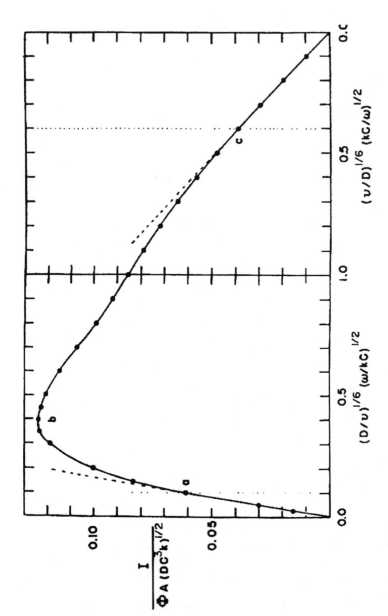

Figure 13 Variation of steady-state ECL intensity with rotation rate in RRDE-ECL. Note that this figure consists of two graphs sharing a common ordinate but having reciprocal abscissas that decrease in opposite directions and are joined together where both are equal to 1. (From Ref. 29.)

passing through the left origin; exterior to point c a straight line passing through the right-hand origin is predicted. This indicates that as $I(t)$ is increasing, it increases with $\omega^{1/2}$, but as it decreases, it decreases with $\omega^{-1/2}$. More important, it may be noted that, in theory, three distinctly different types of ECL behavior can be observed, depending upon whether the reaction kinetics are fast, slow, or intermediate with respect to the rotation rate.

Concentration contours at these three rotation rates (a, b, and c) for species A and C are shown in Figure 14. These graphs depict lines of equal concentration within the region bounded by a radius of the electrode and the hydrodynamic layer to a thickness of 2δ where

$$\delta = 1.8D_A^{1/3}v^{1/6}\omega^{-1/2} \tag{125}$$

Three vertical lines partition the hydrodynamic layer (which is drawn here increasing in a negative direction) into areas below the disk, gap, ring, and shield of the electrode. The lines of equal concentration are drawn every 0.10 unit of relative concentration. Figure 14(I) depicts species A so that the outermost (lowest) line is the line of 0.99 relative bulk concentration; Figure 14(II) shows species C with the outermost line of relative concentration 0.01. These differences result, of course, because A is moving toward the electrode whereas C is moving away from it. Two pertinent facts may be noted:

1. When the reaction is fast [Fig. 14(Ia)], a high concentration of A, the product of the ECL reaction, forms at the inner edge of the ring electrode.

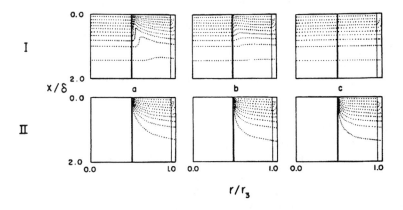

Figure 14 Concentration contours for (I) the parent hydrocarbon (A) and (II) the ring-generated species (C) in RRDE-ECL. The points a, b, and c refer to the rotation rates identified in Figure 13. The lines of equal concentration are drawn every 0.10 unit of relative concentration. (From Ref. 29.)

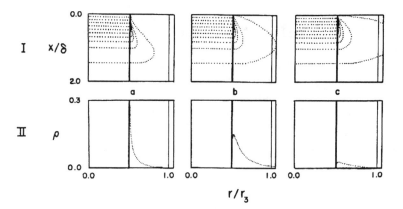

Figure 15 (I) Concentration contours for the disk-generated species (B); (II) the ECL density RRDE-ECL. The points *a*, *b*, and *c* refer to the rotation rates identified in Figure 13. In (I) the lines of equal concentration are drawn every 0.10 unit of relative concentration. (From Ref. 29.)

    2.   At any rotation rate there is very little difference in the contour for species C. The species generated at the ring always surrounds the disk with a "doughnut" of reactive material.

    Figure 15 is more informative. In this figure the concentration contours of species B are shown with the outermost line at a relative concentration of 0.01. These contours illustrate the effectiveness of the "doughnut" of C in blocking the penetration of the disk-generated species B into the bulk of the solution. In Figure 15(Ia), it is seen that in the fast kinetics regime it is quite effective. In Figure 15(Ic), (slow kinetics), much of the disk-generated species is seen to escape into the bulk of the solution.

    The effect of this on the ECL produced by the reaction of B and C is shown in Figure 15(II), where ρ, the spatial density of ECL emission (normal to the electrode), is shown as a function of electrode radius. If the ECL reaction rate is rapid with respect to the rotation rate as at point *a*, the ECL would be seen as a sharp ring of light close to the inner edge of the ring. At point *b*, the emission maximum would be displaced from the inner edge of the ring and the emission would be spread out, falling to about one-third of the maximum at midring. Finally, if the rate of the ECL reaction sequence is so slow that there is extensive penetration of disk-generated species into the bulk of the solution (point *c*), the luminescence would appear to cover the ring uniformly. From this it can be inferred that the appearance of the ECL emission can be used as an indicator of the rate of the ECL process relative to the rotation rate.

    If the process is rapid, ECL will be observed as a bright ring at the innermost edge of the ring. This region of fast kinetics, as noted above, is a region where $I(t)$ increases with $\omega^{1/2}$. In this region, one can write the equation of the straight line passing through point a to show this explicitly:

$$\frac{I(t)}{\Phi_{ECL} \, A_D \, [D_A \, (C_{A*})^3 \, k_2]^{1/2}} = 0.62 \left(\frac{D_A}{v}\right)^{1/6} \left(\frac{\omega}{k_2 C_{A*}}\right)^{1/2} \tag{126}$$

or, upon rearrangement,

$$I(t) = 0.62\Phi_{ECL}A_DC_A*D_A{}^{2/3} \, v^{-1/6}\omega^{1/2} \tag{127}$$

This is quite similar to Eq. (11), the expression for the steady state current at the disk electrode given by Levich. Because the disk current is independent of ring conditions, Eq. (11) is valid even in RRDE-ECL, and combining the simulation result of Eq. (127) with the Levich equation yields

$$I(t) = \Phi_{ECL} \, i_{Levich}/nF \tag{128}$$

which appears to be valid throughout the region of fast kinetics. Placing the upper limit of the region of fast kinetics at point a yields

$$\left(\frac{D_A}{v}\right)^{1/6} \left(\frac{\omega}{k_2 C_{A*}}\right)^{1/2} = 0.1 \tag{129}$$

which gives

$$\omega = 0.1k_2C_A* \tag{130}$$

as the rotation rate at this limit. This implies that if k_2 is greater than 10^7 M^{-1} s^{-1}, ordinary instrumentation is incapable of achieving rotation rates outside this limit, and Eq. (128) is always valid.

This consequence of a fast reaction between B and C is quite reasonable if one considers what can happen to a species B as it is swept across a ring that is generating species C:

1. It can react with C.
2. It can be converted to C at the ring electrode.
3. It can escape into the bulk of the solution.

This simulation result in the limit of fast kinetics implies that reaction with the ring-generated species is the most probable of these three events. This is due to the fact that the ECL quantum output depends not only upon Φ_{ECL} but also upon the fraction of fruitful encounters between B and C that are experienced by disk-generated species B. This contention is supported by Figure 16, which shows the computed probability of fruitful encounter, P, as a function of the same

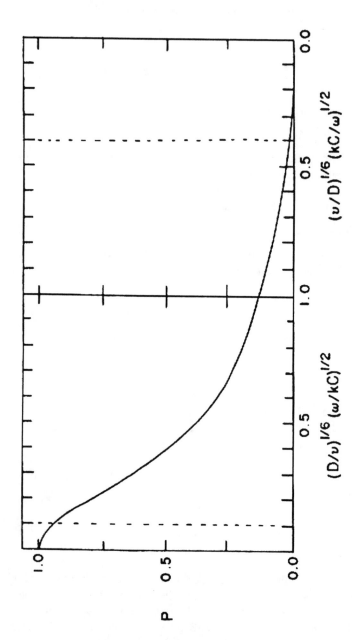

Figure 16 The probability of fruitful encounter between a disk-generated species (B) and a ring-generated species (C) to produce A* in RRDE-ECL. Note that this probability approaches unity in the region where it is predicted that $I(t)$ is linear with $\omega^{1/2}$. (From Ref. 29.)

dimensionless representation of ω used in Figure 13. Note that in the fast kinetics region where Eq. (128) is valid,

$$P \cong 1.0 \tag{131}$$

so that the disk current is an effective measure of the rate of the chemical reaction of the radical ion ECL precursors occurring within the hydrodynamic layer. Under these conditions,

$$\Phi_{ECL} = \Phi_{coul} \tag{132}$$

The results of the simulations for this system show that RRDE-ECL can be used to determine Φ_{coul}, the current efficiency of the ECL process, as long as the reactions producing the ECL are rapid. In addition, this simulation provides the diagnostic criteria by which this may be judged. Because the reaction in question is the electron transfer between the anion radical and the cation radical of a hydrocarbon, there is every reason to believe that it will be rapid. It has already been suggested that this redox reaction occurs at diffusion-controlled rates [11], and if this rate is calculated with the equation of Osborne and Porter [58],

$$k_2 = 8RT/1000\eta \tag{133}$$

with $\eta = 0.79$ cP for N, N-dimethylformamide (DMF), the solvent most commonly used in these ECL studies, a hypothetical rate constant of 1.3×10^{10} $M^{-1} s^{-1}$ is obtained. This is three orders of magnitude greater than the minimum rate constant required for Eq. (128) to be valid.

That the ECL emission is localized at the inner edge of the ring can be seen from Figure 17, a photographic reproduction of the ECL emission that results from the simultaneous oxidation and reduction of 9,10-diphenylanthracene (DPA) in dimethylformamide solution at the RRDE. Even though some clarity has been lost through the enlargement and reproduction of this print, it clearly shows that the ECL emission is concentrated at the inner edge of the ring electrode. According to the simulation, this indicates that the reaction kinetics were rapid with respect to the rotation rate ($\sim$150 s-1) when this photograph was taken. In addition, it can be observed experimentally that ECL intensity is directly proportional to disk current (and thereby proportional to $\omega^{1/2}$) just as predicted for the fast kinetics region of Figure 13.

All this indicates that the reactions producing ECL are quite rapid, so Eq. (128) is valid. Accordingly, Φ_{ECL} can be determined directly from the steady-state ratio of the total ECL intensity to the current at the disk,

$$\Phi_{coul} = nFI(t)/i_{Levich} \tag{134}$$

where both $I(t)$ and i_{Levich} result from steady-state measurements Therefore, only the precise measurements of the absolute ECL intensity and the disk current are required to determine the efficiency of the process. The results of these experimental measurements appear elsewhere[18].

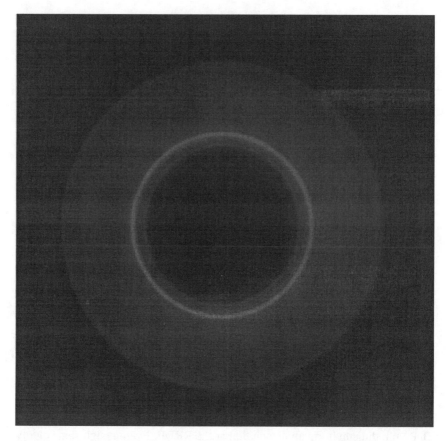

Figure 17 A photograph of the ECL from 9,10-DPA at the RRDE. Observed at a rotation rate of ~150 s^{-1}, the diameter of the inner bright ring was ~3.3 mm, corresponding to the inner ring diameter. This photograph provides evidence of very fast homogeneous kinetics in the ECL process. (Photograph contesy of T.V. Atkinson.)

The effect of cation decay upon $I(t)$, so important in the interpretation of the slope of Feldberg plots in transient studies of ECL, can also be investigated using digital simulation in RRDE-ECL. This can be accomplished by modifying the basic RRDE-ECL program to include the difference form of the differential rate law for Eq. (102) or (103). This results in an appropriate decrease in the quantity representing DPA$^{+\bullet}$ in each volume element. For example, if the generation of DPA$^{+\bullet}$ at the disk is to be simulated, one would write the rate law for Eq. (102) in relative concentration form,

$$-\Delta f_B(J, K) = (k_1\,\Delta t)\,f_B(J,K) \tag{135}$$

When combined with the appropriate definition of Δt for the RRDE, this becomes

$$-\Delta f_B (J, K) = \left(\frac{D_A}{v}\right)^{1/3} \left(\frac{k_1}{\omega}\right)\left(\frac{f_B (J, K)}{L}\right) \tag{136}$$

However, the total change in $f_B(J, K)$ is the sum of that which is depleted through cation decay, Eq. (136), and that which is used in Eq. (117), the radical redox reaction that produces ECL. The addition of these two equations yields

$$-\Delta f_{B'}(J, K) = \left(\frac{D_A}{v}\right)^{1/3} \left(\frac{f_B (J, K)}{\omega L}\right)[k_2 C_{A*} \, f_C (J, K) + k_1] \tag{137}$$

This can be simplified by setting

$$k_1/k_2 C^* = R \tag{138}$$

so that Eq. (137) becomes

$$-\Delta f_B (J, K) = \left(\frac{D_A}{v}\right)^{1/3} \left(\frac{k_2 C_{A*}}{\omega}\right)\left(\frac{f_B (J, K)}{L}\right)[\, f_C (J, K) + R] \tag{139}$$

where the initial bracketed factor is the dimensionless kinetic input parameter, which is ω-dependent, and R depends only upon the relative magnitudes of the rate constants and is ω-independent.

The results of simulations carried out using this technique appear in Figure 18. In this figure curve a shows the RRDE-ECL $I(t)$ vs. $\omega^{1/2}$ behavior predicted by Eq. (103) when an unstable species is generated at the ring. Curve b shows the effect of generating the unstable species at the disk via Eq. (102). In either case, the reaction of B and C was considered fast with respect to the first-order decay ($R = 10^{-6}$) and $I(t)$ and ω were rendered dimensionless by k_1 rather than by $k_2 C_A^*$. Note that the slope of the straight-line portion of either curve lies very close to the theoretical value of 0.62 predicted by Eq. (127) for the kinetically uncomplicated case.

Whereas curve a corresponds to the generation of DPA$^{+\bullet}$ at the ring, curve b corresponds to the collection efficiency experiment where DPA$^{+\bullet}$ is generated at the disk. In either case ECL is predicted to increase with increasing ω, but in the latter case it is quite obvious that the effect of cation decay must be overcome before $I(t)$ is apparently linear with $\omega^{1/2}$. Thus, if $I(t)$ vs. $\omega^{1/2}$ is apparently linear but with a negative ordinate intercept, this can be attributed to the generation of an unstable species at the disk. On the other hand, the generation of the unstable species at the ring is much less likely to show this variation (even though curve a is is not perfectly linear). The important result to observe here is that at rotation rates where the disk generation of the unstable species causes an apparently negative intercept in the $I(t)$ vs. $\omega^{1/2}$ curve, the ring generation of that species

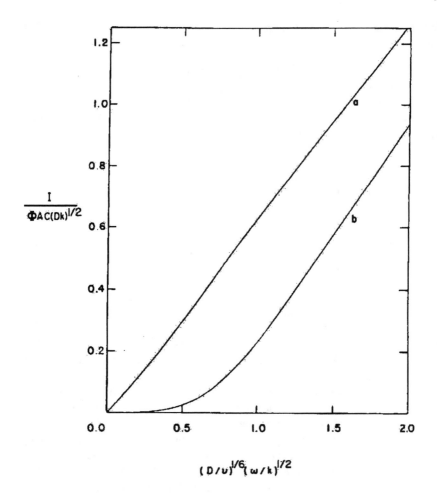

Figure 18 The effect of an unstable species on intensity–rotation rate behavior in RRDE-ECL. Curve a shows the effect of generating the unstable species at the ring, while b shows the effect of generation at the disk. In either case, k_1/k_2C^* was 10^{-6}. The slope of the straight-line portion of either curve lies very close to the theoretical value of 0.62 predicted by Eq. (127) for the kinetically uncomplicated case. (From Ref. 29.)

will produce apparent $I(t)$ vs. $\omega^{1/2}$ behavior that extrapolates through the origin. This is a necessary condition if either one (but only one) of the ionic precursors of ECL undergoes reaction to form a product that is inactive with respect to the other.

Armed with these theoretical predictions, Maloy and Bard set out to determine the experimental ECL efficiencies of a number of systems using the

RRDE. This work required the development of a rather elaborate airtight RRDE cell [17] that would allow the ECL produced at the electrode/solution interface to be observed and measured. Because the absolute intensity of the ECL emission was to be measured, the photometric detector used in this work required actinometric calibration. Initially, it was hoped that the increased ECL output observed at the RRDE would be sufficient to use it as the light source in chemical actinometry; thus, the cell was immersed directly in the actinometric bath. It was soon discovered, however, that the ECL emission from DPA, the strongest photon source, decayed to a very low intensity in about 100 min, well before a detectable amount of product developed. Therefore, it was necessary to employ a surrogate light source with a geometry similar to that of the RRDE in this calibration. Recognizing this limitation, Maloy and Bard were still able to use the RRDE to make the first reliable estimates [17,18] of the efficiency of the ECL process for a variety of systems:

1. DPA. The DPA system was found to exhibit the greatest ECL efficiency, and efficiencies in the range of 0.2–0.9×10^{-3} were observed [18] at various rotation rates. These values for Φ were quite similar to those that were subsequently reported by Bezman and Faulkner [47] (0.6–2.5×10^{-3}) using transient methods and more reliable measurements of absolute luminescence. The ideal behavior predicted by the simulation, however, was not observed for either disk-generated radical ion. Plots of $I(t)$ vs. $\omega^{1/2}$ for the ECL observed by generating $DPA^{+\bullet}$ and $DPA^{-\bullet}$ at the disk exhibited positive $\omega^{1/2}$ intercepts not unlike that predicted in curve b of Figure 18 for an unstable disk-generated species. This necessitated the estimation of Φ from the slope of the linear portions of these plots, and it was found to vary with ω. These investigators attributed this deviation from the behavior predicted by the simulation to a kinetic effect that was not taken into account in the theoretical development.

2. Rubrene. This compound was found to exhibit $I(t)$ vs. $\omega^{1/2}$ behavior similar to that of DPA, but the observed ECL efficiency was significantly less, falling in the 0.01–0.07×10^{-3} range in DMF, and varyied with ω [18]. This measurement was made directly from the slopes of the linear portions of the $I(t)$ vs. $\omega^{1/2}$ plots, and no attempt was made to take the kinetics of the TTA mechanism into account in this determination. Bezman and Faulkner would subsequently find the rubrene ECL efficiency to be significantly greater in benzonitrile (1×10^{-3}) and DMF (0.5×10^{-3}), but they specifically imposed their analysis of TTA kinetics upon the results of their transient experiments [50]. They would point out that the overall efficiency in a system that emits via the T route depends upon the volume of the reaction zone and speculate that the RRDE inherently has a larger reaction zone, thereby diminishing ECL production via TTA for the rubrene system at the RRDE.

3. DPA and TMPD (Tetramethyl-p-Phenylenediamine). Although this mixed composition T-route system was not treated theoretically using finite difference methods to model its RRDE-ECL behavior, its ECL efficiency was

estimated from the slopes of its $I(t)$ vs. i_{disk} plots obtained at a variety of rotation rates. These were found to be quite linear and yielded ECL efficiencies in the vicinity of 0.05×10^{-3} at all rotation rates [18]. The fact that the efficiency of the mixed DPA-TMPD system was found, like that of rubrene, to be 5% of the efficiency of DPA alone can be attributed to the energy-deficient nature of the mixed system. That the efficiency was found to be ω-independent may be due to the stability of both ionic precursors to ECL. Bezman and Faulkner did not report the efficiency of this system, presumably because their theoretical development for transient behavior did not include systems of mixed composition.

VII. SIMULATION OF ECL BEHAVIOR USING COMMERCIAL SOFTWARE

Since the development of these finite difference methods for the investigation of electrochemical processes and their application to the study of ECL phenomena, commercial software has been produced and marketed. Perhaps the most prominent such product that is currently available is DigiSim™, a simulation software package used to model cyclic voltammetry [30]. One of the developers of this software is S. W. Feldberg, whose early work in the simulation of ECL phenomena was reported earlier in this chapter.

Feldberg recently reported the use of DigiSim in the simulation of ECL phenomena (SW Feldberg, personal communication). Seeking to investigate the plethora of electrochemical systems that have been found to produce ECL over the past 30 years, but thwarted by the complexities of modeling systems of interest, Feldberg demonstrated the use of a "trick" that allows DigiSim to be employed for this purpose. By assigning a maximum theoretical diffusion coefficient to the photon that is produced during the ECL process, Feldberg is able to ensure that the (theoretical) photon reaches the electrode in a time that is very short with respect to the duration of other events occurring within the diffusion layer. Once the theoretical photon reaches the electrode, it can undergo a hypothetical electrochemical process so that $I(t)$ can be attributed to it as a "current." Other electrochemical processes taking place within the diffusion layer serve as the rate-determining steps in the generation of the photonic "current," $I(t)$. It is the mechanism of these kinetic processes that is to be elucidated by the interpretation of the $I(t)$ "current." The feasibility of this approach has been demonstrated through the generation of theoretical $I(E)$ plots for systems of interest (SW Feldberg, Personal communication). It is Feldberg's contention that with a few simple manipulations DigiSim can be used to simulate cyclic voltammetric responses for systems that produce ECL. There are virtually no restrictions on mechanistic complexity, and effects such as electrode capacitance and uncompensated resistance are easily examined.

REFERENCES

1. Bard, A.J.; Faulkner, L.R. *Electrochemical Methods*, 2nd ed.; Wiley: New York, 2001; p. 148.
2. p.162–163.
3. p. 207.
4. p. 138.
5. Levich, V.G. *Physicochemical Hydrodynamics*, Prentice-Hall, Englewood Cliffs, NJ, 1962
6. Bard, A.J.; Faulkner, L.R. *Electrochemical Methods*, 2nd ed.; Wiley: New York, 2001; p. 170.
7. Nicholson, R.S.; Shain, L, Anal. Chem. **1964**, *36*, 706.
8. Nicholson, R.S. Anal. Chem. **1965**, *37*, 1351.
9. Bard, A.J.; Faulkner, L.R. *Electrochemical Methods*, 2nd ed., Wiley: New York, 2001; p. 230.
10. Feldberg, S.W.; Auerbach, C. Anal. Chem. **1964**, *36*, 505.
11. Feldberg, S.W. J. Am. Chem. Soc. **1966**, *88*, 390–393.
12. Childs, W.V.; Maloy, J.T.; Keszthelyi, C.P.; Bard, A.J. J. Electrochem. Soc. **1971**, *118*, 874.
13. Prater, K.B.; Bard, A.J. J. Electrochem. Soc. **1970**, *117*, 335.
14. Puglisi, V.J.; Bard, A.J. J. Electrochem. Soc. **1972**, *119*, 833.
15. Yeh, L.S.R.; Bard, A.J. J. Electrochem. Soc. **1977**, *124*, 189.
16. Cruser, S.A.; Bard, A.J. J. Am. Chem. Soc. **1969**, *91*, 267–275.
17. Maloy, J.T.; Prater, K.B.; Bard, A.J. J. Am. Chem. Soc. **1971**, *93*, 5959–5968.
18. Maloy, J.T.; Bard, A.J. J. Am. Chem. Soc. **1971**, *93*, 5968–5982.
19. Bezman, R.; Faulkner, L.R. J. Am. Chem. Soc. **1972**, *94*, 3699–3707.
20. Lawson, R.J.; Maloy, J.T. Anal. Chem. **1974**, *46*, 559.
21. Mell, L.D.; Maloy, J.T. Anal. Chem. **1975**, *47*, 299.
22. Mell, L.D.; Maloy, J.T. Anal. Chem. **1976**, *48*, 1597.
23. Grypa, R.D; Maloy, J.T. J. Electrochem. Soc. **1975**, *122*, 377.
24. Grypa, R.D; Maloy, J.T. J. Electrochem. Soc. **1975**, *122*, 509.
25. Bezilla, Jr. B.M.; Maloy, J.T. J. Electrochem. Soc. **1979**, *126*, 579.
26. Bard, A.J.; Faulkner, L.R. *Electrochemical Methods*, 2nd ed., Wiley: New York, 2001; Appendix B, p. 785–807.
27. Maloy, J.T. In *Laboratory Techniques in Electroanalytical Chemistry*, 2nd ed., Kissinger, P.T., Heineman, W.R., Eds.; Marcel Dekker: New York, 1996; Chapter 20.
28. Prater, K.B. In *Electrochemistry: Calculations, Simulation, and Instrumentation*, Mattson, J.S., Mark, H. B. Jr., MacDonald, H.C., Eds.; Marcel Dekker: New York, 1972; Vol. 2, Chapter 8.
29. Maloy, J.T. In *Electrochemistry: Calculations, Simulation, and Instrumentation*, Mattson, J. S., Mark, H.B. Jr., MacDonald, H.C., Eds.; Marcel Dekker: New York, 1972; Vol. 2, Chapter 9.
30. DigiSim™, distributed by Bioanalytical Systems, Inc., 2701 Kent Avenue, West Lafayette, IN 47906.

31. Britz, D. *Digital Simulations in Electrochemistry*; Springer-Verlag: Berlin, 1988.
32. Joslin, T; Pletcher, D. J. Electroanal. Chem. **1974**, *49*, 171.
33. Feldberg, S.W. J. Electroanal. Chem. **1981**, *127*, 1.
34. Brumleve, T.B.; Buck, R.P. J. Electroanal. Chem. **1978**, *90*, 1.
35. Smith, G.D. *Numerical Solutions of Partial Differential Equations*; Oxford Univ. Press: Oxford, 1969.
36. Laasonen, P. Acta Math **1949**, *81*, 309.
37. Bard, A.J.; Faulkner, L.R. *Electrochemical Methods*, 2nd ed., Wiley: New York, 2001; p. 149.
38. Einstein, A. Z. Electrochem. **1908**, *14, 235.*
39. Feldberg, S.W. In *Electroanalytical Chemistry*; Bard, A.J., Ed.; Marcel Dekker: New York, 1969; Vol. 3, p. 271.
40. Maloy, J.T. In *Laboratory Techniques in Electroanalytical Chemistry*, 2nd ed., Kissinger, P.T., Heineman, W.R., Eds.; Marcel Dekker: New York, 1996; Chapter 20, p. 587.
41. p. 590.
42. Cruser, S.A. *Ph.D. Dissertation*, The University of Texas at Austin, 1968.
43. Hercules, D.M.; Lansbury, R.C.; Roe, D.K. J. Am. Chem. Soc. **1966**, *88*, 4578.
44. R.E. Visco, R.E.; E.A. Chandross, E.A. Electrochim. Acta **1968**, *13*, 1187.
45. J. Chang, J.; Hercules, D.M.; Roe, D.K. Electrochim. Acta **1968**, *13*, 1197.
46. Feldberg, S.W. J. Phys. Chem. **1966**, *70*, 3928–3930.
47. Bezman, R.; Faulkner, L.R. J. Am. Chem. Soc. **1972**, *94*, 6317–6323.
48. Faulkner, L.R.; Bard, A.J. J. Am. Chem. Soc. **1969**, *91*, 209–210.
49. Faulkner, L.R.; Tachikawa, H.; Bard, A.J. J. Am. Chem. Soc. **1972**, *94*, 691–699.
50. Bezman, R.; Faulkner, L.R. J. Am. Chem. Soc. **1972**, *94*, 6324–6330.
51. Bezman, R.; Faulkner, L.R. J. Am. Chem. Soc. **1972**, *94*, 6331–6337.
52. Freed, D.J.; Faulkner, L.R. J. Am. Chem. Soc. **1971**, *93*, 3565–3568.
53. Maloy, J.T.; Prater K.B.; Bard, A.J. J. Phys. Chem. **1968**, *72*, 4348.
54. Frumkin, A.N.; Nekrasov, L.N. Dokl. Akad. Nauk SSSR, **1959**, *126*, 115.
55. Albery, W.J.; Bruckenstein, S. Trans. Faraday Soc. **1966**, *62*, 1920.
56. Albery, W.J.; Bruckenstein, S. Trans. Faraday Soc. **1966**, *62*, 1946.
57. Prater, K.B. *Ph.D. Dissertation*, The University of Texas at Austin, 1969.
58. Osborne, A.D.; Porter, G. Proc. Roy. Soc. (Lond), Ser.A **1965**, 284.
59. Feldberg, S.W. personal communication.

4

Electron Transfer and Spin Up-Conversion Processes

Andrzej Kapturkiewicz
Institute of Physical Chemistry, Polish Academy of Sciences, Warsaw, Poland

I. INTRODUCTION

Exploration and explanation of the relationships between molecular structure and chemical properties and reactivity are the main activities of physical chemistry. This is also true in the case of electrogenerated chemiluminescence (ECL) processes, where interpretation and prediction of the spectral characteristics and the efficiency of emitted light in a given ECL system seem to be the most important tasks from the theoretical as well as practical points of view. Current understanding of the ECL mechanisms allows for prediction of light emission during a given chemical reaction in solution but quantitative description remains rather difficult because of the many factors that affect or interfere in chemiluminescence processes.

The most important step leading to light emission can be formulated as the intramolecular electron transfer between chemiluminescence precursors, strong reductant A^- and strong oxidant D^+ obtained by means of common chemical (e.g., Weller and Zachariasse [1,2]) or electrochemical (e.g., Bard and coworkers [3,4]) ways. Both methods for the preparation of the reactants can be used, but the electrochemical one (an ECL experiment) seems to be more useful in the cases of relatively unstable intermediates and when very selective oxidation and reduction are necessary. Subsequent annihilation of the A^- and D^+ species (by means of homogeneous electron transfer) can lead to light emission arising from the energy released in an elementary step. The $A^- + D^+$ pattern corresponds to so-called mixed ECL systems with A^- and D^+ reactants obtained from noncharged, usually organic, A and D precursors. In some cases (so-called single ECL systems) both A^- and D^+

species are generated from the same precursor Q. In such cases A/A^- and D/D^+ pairs should be respectively replaced (in all reaction schemes discussed below) by Q/Q^- and Q/Q^+. The same is true for charged, ionic ECL precursors where one-electron reduction of $A^{\pm m}$ or one-electron oxidation of $D^{\pm n}$ leads to $A^{\pm m-1}$ or $D^{\pm n+1}$, respectively.

Electron transfer processes occur over a very short time scale, thus requiring rapid dissipation of a large amount of energy into the vibrational modes of the molecular frameworks. This is rather difficult for the reacting system, leading to kinetic inhibition of direct formation of the stable ground state products A and D. The formation of the excited states (A^* or D^*) is less exergonic and correspondingly less thermodynamically favored, but the process may be kinetically preferred because a relatively small amount of energy needs to be vibrationally dissipated. Consequently the ECL phenomenon can be understood in terms of competition of (at least) two electron transfer reactions.

The above approach, first proposed by Marcus [5], allows for a qualitative as well as quantitative discussion if the appropriate electron transfer rates are available from other experimental data or from theoretical considerations. It should be noted, however, that the simple Marcus formalism is somewhat oversimplified from the mechanistic point of view. In real cases an electron transfer excitation is preceded and followed by diffusion of reactants from and products into the bulk solution. Moreover, ECL reactants and products are species with distinctly different spin multiplicities, which may lead to additional kinetic complications because of spin conservation rules. Correspondingly, the spin up-conversion processes (e.g., triplet–triplet annihilation to produce an excited singlet) cannot be a priori excluded from the kinetic considerations. Consequently a kinetic description of an ECL system (similarly to other bimolecular reactions, e.g., electron transfer quenching reactions in solution) may be a very complex task, especially if the distance dependence of all processes occurring is taken into account. Solution of the problem is in principle possible within the framework of the integral encounter theory (reviewed recently by Burshtein [6]), but even in the simplest ECL cases the equations obtained are quite complex [7,8] and until now, to our best knowledge, have not been applied to the quantitative description of ECL results. It should be emphasized that, despite mathematical complexity, the integral encounter theory has some weaknesses. Description of the reaction medium as a continuous space seems to be somewhat factitious, especially at short distances between reacting species where the discrete structure of real liquids may play a very important role. One can expect that some of the reaction distances are preferred, as was experimentally confirmed by Gould et al. [9,10] for electron transfer quenching processes where reactants appear to be in the form of a contact ion pair (cip) or solvent-separated ion pair (ssip) with dynamical equilibrium between them.

Annihilation and generation of the excited states in the electron transfer reactions should be (as pointed out by Meyer and coworkers [11]) essentially the

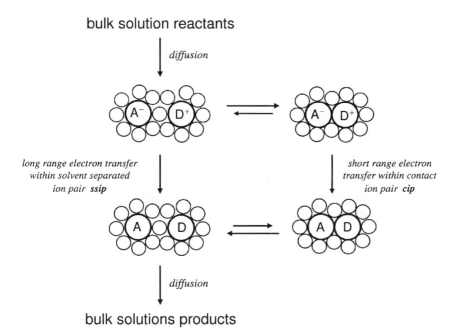

bulk solution reactants

diffusion

long range electron transfer within solvent separated ion pair **ssip**

short range electron transfer within contact ion pair **cip**

diffusion

bulk solutions products

Figure 1 Schematic representation of two forms of activated complexes in an electron transfer reaction: a contact ion pair (cip) and a solvent-separated ion pair (ssip).

reverse processes that preclude the possibility of the same method in their kinetic description. Thus, a less advanced, primitive, but more intuitive phenomenological "contact approximation" can be used in the qualitative as well as in the quantitative description of the ECL processes. This approach assumes the occurrence of the electron transfer and spin up-conversion reactions within an activated complex $(A^-\cdots D^+)$ (presumably cip) formed from the ECL precursors A^- and D^+ in the diffusion-controlled process (cf. Fig. 1). The model may be more advanced, taking into account equilibrium between cip and ssip configurations, i.e., the presence of two forms of an activated complex $(A^-\cdots D^+)$ and $(A^-\cdots solv\cdots D^+)$, respectively. The cip configuration $(A^-\cdots D^+)$ seems to be preferred in the case of organic ECL systems because of the Coulombic attraction between oppositely charged A^- and D^+ ions. In the ECL systems involving uniformly charged reactants (e.g., $A^{\pm m-1}$ and $D^{\pm n+1}$ species generated from transition metal complexes $A^{\pm m}$ and $D^{\pm n}$, respectively), Coulombic repulsion may lead to an increase in the electron transfer distance with a preference for the ssip configuration $(A^{\pm m-1}\cdots solv\cdots D^{\pm n+1})$ of the activated complex. In the cases of cip–ssip equilibrium, the kinetic description is obviously more complex than in the cip or ssip configurations alone. The approach applied by Birks [12] for photochemical processes can be adopted in such cases, but the number of kinetic parameters going into the final equations usually does not permit a quantitative analysis without additional assumptions [13]. In a first order approximation, however, the presence of one dominant conformation of activated complex in the explanation and interpretation of the ECL mechanism can be assumed. This leads to the simplest possible (from the kinetic point of view) model, which, however, should be treated as the basis for understanding the role of any kinetic parameters involved in the description of ECL mechanisms.

II. ROUTES OF LIGHT EMISSION IN ELECTROCHEMILUMINESCENCE—KINETIC DESCRIPTION

A. Energetics of Ion Annihilation

The Gibbs energy of A^- and D^+ ion annihilation ΔG° can be straightforwardly calculated from the redox potentials of reduction E_{red} and oxidation E_{ox} of the parent species A and D. The magnitude of $F(E_{red} - E_{ox})$ corresponds to the difference in energies of the isolated ECL reactants and products (bulk value). This value should be corrected for work terms because in the electron transfer reaction the reactants are required to bring on an activated complex

$$\Delta G^O = F(E_{ox} - E_{ox}) - w_R + w_P \tag{1}$$

where F is the Faraday constant. The energy w_R corresponds to the interaction between reactants and is the energy required to bring the reactants together to the most probable separation distance d at which the electron transfer takes place. Analogously, w_P is the energy required to bring the products into the activated complex. Quantities w_R and w_P correspond mainly to the electrostatic interactions and may be calculated using the ordinary Coulomb equation

$$w = \frac{N_A z_A z_D e_0^2}{4 \pi \varepsilon \varepsilon_0 d} \tag{2}$$

where z_A and z_D are the charges of the reactant ($w = w_R$ for $\pm m-1$ and $\pm n+1$) or product ($w = w_P$ for $\pm m$ and $\pm n$) species. N_A, e_0, ε, and ε_0 are the Avogadro constant, the elementary charge, the dielectric constant of the given solvent, and the permittivity of vacuum, respectively. The familiar Debye expression, however, seems to be more appropriate in solutions containing supporting electrolyte with ionic strength μ (cf. Ref. 14):

$$w = \frac{N_A z_A z_D e_0^2}{4 \pi \varepsilon \varepsilon_0 d \left(1 + d \beta \mu^{1/2}\right)} \tag{3}$$

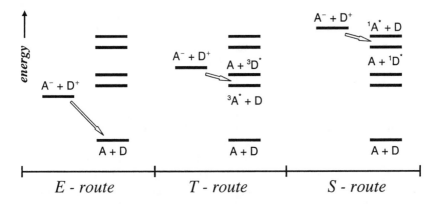

Figure 2 Schematic representation of the energy terms in ECL processes.

with parameter β (inverse radius of the ionic atmosphere) given correspondingly by

$$\beta = (2N_A^2 \, z_A \, z_B \, e_o^2/1000\varepsilon\varepsilon_o \, RT)^{1/2} \tag{4}$$

where R and T are the gas constant and absolute temperature.

Without a doubt the energy of ion annihilation is one of the most important parameters governing the mechanism of an ECL reaction, because the molecular energy levels are quantized and reactions yielding electronic excitation must deliver, in one step, the appropriate amount of energy. Efficient formation of an electronically excited ECL product (with energy E_i) requires $-\Delta G^o$ to be comparable with E_i. One can expect that the number of accessible excited states and the number of reaction channels to be considered in an ECL reaction depend strongly on the amount of released energy (cf. Fig. 2). Usually for moderately exergonic processes only the excited triplets $^3A^*$ and/or $^3D^*$ are effectively produced (T route). However, for strongly exergonic processes, the excited singlet states $^1A^*$ or/and $^1D^*$ may also be formed (S route). Obviously, if the electron transfer reaction is energetically sufficient to populate the excited singlet directly, the formation of the lower energetically lying excited triplets must also take place (mixed ST route). If the exothermicity of the given ion annihilation is still smaller than the energy of the excited triplet states ($\Delta G^o + E_i > 0$), the reaction is generally not of interest, although it has been shown (cf. Chapter 6) that such systems may still produce light. This last case corresponds formally to the direct radiative electron transfer from A^- to D^+ (E route). It can be described in terms of the competition between radiative and radiationless transition in the Marcus inverted region.

Light emission in an ECL process can be observed directly via radiative deactivation of the excited species or by any up-conversion of the nonemissive species, depending on the nature of the populated excited states. The first is the more important case, because the experimental values of the ECL efficiencies ϕ_{ecl} (expressed in number of photons emitted per electron transferred) are directly connected with the efficiencies of the excited state formation ϕ_{es}:

$$\phi_{ecl} = \phi_{es} \times \phi_o \tag{5}$$

where ϕ_o is the emission quantum yield, an intrinsic property of the emitting state. Emission of photons may be observed from the excited singlets (usual case of organic emitters, cf. Chapter 6) as well as from the excited triplets (usual case of inorganic emitters with phosphorescence allowed by the spin–orbit coupling induced by the heavy metal ions, cf. Chapter 7). Sometimes, however, organic triplets can also be active emitters, e.g., benzophenone triplet, which has a high triplet phosphorescence quantum yield due to the spin–orbit coupling by its carbonyl group [15]. These cases are of considerable interest from a practical standpoint and also because they can provide direct insight into mechanistic features of highly exergonic electron transfer reactions, providing information on the relative rates for bimolecular electron transfer in the normal and inverted Marcus regions (see below).

Emission of photons can be also observed in an ECL system where nonemissive $^3A^*$ or $^3D^*$ takes part in any up-conversion reactions. Typical examples involve bimolecular triplet–triplet annihilation (see Chapter 6), excitation energy transfer to an emissive but nonelectroactive species presented in the solutions studied (e.g., europium chelate [16]), or delayed fluorescence (e.g., from the organic intramolecular charge transfer triplets [17]). All subsequent processes lead obviously to additional complications of the ECL reaction mechanism with a much more difficult kinetic description.

B.   E Route—Direct Radiative Electron Transfer

In the simplest case A^- and D^+ ion annihilation occurs according to the scheme presented in Figure 3, where k_{as} and k_{dis} are the diffusion-controlled forward and reverse rate constants respectively, for the formation of an activated complex $(A^- \cdots D^+)$ prior to electron transfer. The activated complex is formed in two different spin states, singlet $^1(A^- \cdots D^+)$ and triplet $^3(A^- \cdots D^+)$, with (according to the spin statistics rule [18,19]) the branching ratio 3:1. Spin-allowed electron transfer within the activated complex in the singlet state $^1(A^- \cdots D^+)$ occurs at the rate k_{fg} and leads directly to generation of the ground-state product $^1(A \cdots D)$ followed by separation of A and D species into bulk solutions. Back electron transfer from $^1(A \cdots D)$ to $^1(A^- \cdots D^+)$ can be neglected because a large energy gap ($-\Delta G°$ in the range of 2–4 eV is necessary for the emissive electron transfer in ECL systems. The scheme presented in Figure 3 also neglects very slow, spin-forbidden deactivation of the activated complex in the triplet state $^3(A^- \cdots D^+)$,

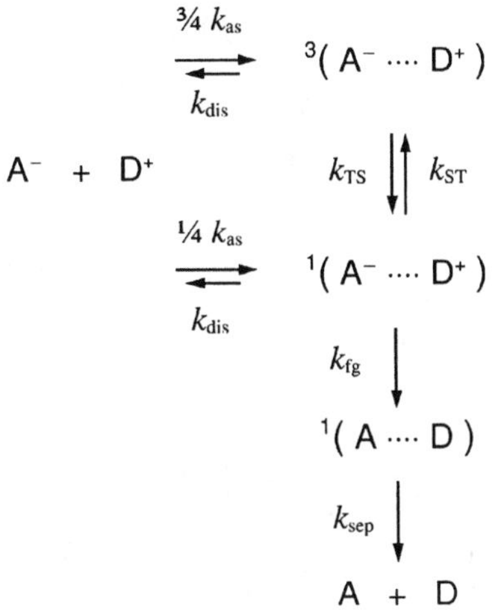

$$\xrightarrow{\frac{3}{4}\,k_{as}} \ \ {}^3(\,A^- \cdots D^+\,)$$

$$\underset{k_{dis}}{\rightleftharpoons}$$

$$A^- + D^+ \qquad\qquad k_{TS} \updownarrow k_{ST}$$

$$\xrightarrow{\frac{1}{4}\,k_{as}} \ \ {}^1(\,A^- \cdots D^+\,)$$

$$\underset{k_{dis}}{\rightleftharpoons}$$

$$k_{fg} \downarrow$$

$${}^1(\,A \cdots D\,)$$

$$k_{sep} \downarrow$$

$$A + D$$

Figure 3 Reaction mechanism of electrochemiluminescence processes occurring according to the E route.

because of their long expected lifetimes [20,21]. Triplet–singlet up-conversion, a necessary step before electron transfer, can occur directly between both forms of the activated complex $(A^- \cdots D^+)$ (at the rate k_{TS}) or by dissociation of ${}^3(A^- \cdots D^+)$ into the bulk solution (at the rate k_{dis}). Both processes occur on the nanosecond time scale and are many orders of magnitude faster than the direct ${}^3(A^- \cdots D^+) \to {}^1(A \cdots D)$ transition. Thus the spin up-conversion process seems to be a necessary step that makes back electron transfer from ${}^3(A^- \cdots D^+)$ to the ground state product ${}^1(A \cdots D)$ possible. Taking into account presumably very small energy splitting between two spin forms of the activated complex $[{}^3(A^- \cdots D^+)$ and ${}^1(A^- \cdots D^+)$, respectively] one can also conclude that $3k_{TS} \cong k_{ST}$.

The diffusion-controlled forward k_{as} and reverse k_{dis} rate constants can be straightforwardly estimated from the familiar Einstein–Smoluchowski equations [22],

$$k_{as} = \frac{8\,RT}{3\,\eta}\left(\frac{w_R\,/\,RT}{\exp\,(w_R\,/\,RT)-1}\right) \tag{6}$$

$$k_{dis} = \frac{2\,RT}{\pi\,\eta\,N_A\,d^3}\left(\frac{w_R\,/\,RT}{1-\exp\,(-w_R\,/\,RT)}\right) \tag{7}$$

where η is the solvent viscosity. Equation (7) can be correspondingly applied for estimation of the k_{sep} rates with the w_R term being replaced by w_P. Combination of Eqs. (6) and (7) simply leads to expressions for the equilibrium constant

$(K_{as} = k_{as}/k_{dis})$ for the formation of the activated complex (the Eigen–Fuoss model [23,24]):

$$K_{as} = \frac{4 \pi N_A d^3}{3} \exp\left[\frac{-w_R}{RT}\right] \tag{8}$$

Although the above formalism is widely used in the discussion of diffusion-controlled reactions, the assumptions of the model upon which it is based are subject to criticism, and another formulation of K_{as}, the "encounter pre-equilibrium" model [25], has been proposed:

$$K_{as} = 4\pi N_A d^2 \, \delta r \exp\left[-w_R/RT\right] \tag{9}$$

Equation (9) allows for electron transfer over a range of separation distances between d and $(d + \delta r)$, with δr typically equal to 0.06–0.10 nm. It should be noted, however, that for many systems the typical separation distance d between A^- and D^+ reactants is 0.4–0.6 nm and values of K_{as} as calculated from Eqs. (8) and (9) do not differ appreciably (because $\delta r \cong d/3$). Einstein–Smoluchowski formalism seems to be applicable in most cases (at least to the first order of approximation) because quantitative predictions of the rate constants for other processes involved in an overall ECL mechanism are subject of still greater uncertainties.

As mentioned above, the triplet form of activated complex $^3(A^- \cdots D^+)$ undergoes reaction to the ground-state products after relatively fast equilibration to the singlet states, and the k_{fg} rate governs electron transfer from $^1(A^- \cdots D^+)$ to $^1(A \cdots D)$. This process can in principle occur on two parallel radiative (at rate k_{em}) and nonradiative (at rate k_{nr}) ways ($k_{fg} = k_{em} + k_{nr}$). Consequently the observed ECL efficiency ϕ_{ecl} is simply given by

$$\phi_{ecl} = \frac{k_{em}}{k_{nr} + k_{em}} \tag{10}$$

Equation (10) is valid only when all other electron transfer channels (e.g., population of the excited states A* and/or D*) are excluded. If this is not the case (cf. Figs. 4 and 6), the E-route efficiency is correspondingly lowered:

$$\phi_{ecl} = \frac{k_{em}}{k_{nr} + k_{em}} \left[1 - \sum_i \phi_{es}(E_i) \right] \tag{11}$$

where $\phi_{es}(E_i)$ is the ith excited state efficiency with the summation performed over the whole range i of all accessible reaction channels. Note that use of Eqs. (10) and (11) requires that a given emission be really an electron transfer (or charge transfer). It can be confirmed by a correlation of emission maxima $\tilde{v}_{em}^{max}$ with the difference in the redox potential (e.g., Ref. 26),

$$hc\tilde{v}_{em}^{max} = F(E_{ox} - E_{red}) - \Delta E_{FC} \tag{12}$$

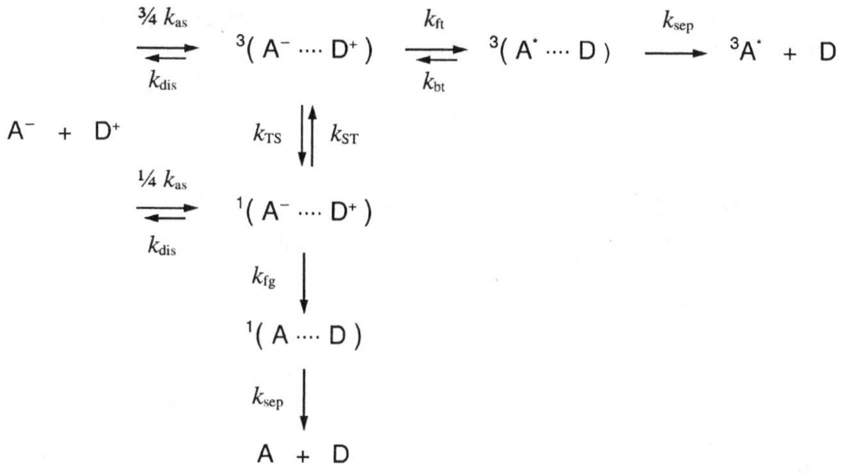

Figure 4 Reaction mechanism of electrochemiluminescence processes occurring according to the T route.

where ΔE_{FC} is the energy of the Franck–Condon state (see below) reached after the emission process. Usually this can be done by ECL studies for a homologous series (required to keep the ΔE_{FC} term nearly constant) of acceptor A molecules with the same donor D (or vice versa). The solvent polarity and temperature effects on the charge transfer emission energy and intensity (e.g., Ref. 27) can be used as well.

C. T Route—Formation of Excited Triplets

An increase in the ion annihilation exothermicity to values comparable with the excited triplet-state energies opens an additional electron transfer channel. In the simplest case only one excited triplet $^3A^*$ or $^3D^*$ becomes accessible, as presented in Figure 4. The mechanism discussed includes additional forward electron transfer within the triplet form of the activated complex $^3(A^- \cdots D^+)$ and a competitive step where redox product $^3(A^* \cdots D)$ separates into the bulk solution (at rate k_{sep}). The $^3(A^* \cdots D)$ product appears (at the forward electron transfer rate k_{ft}) also in the triplet state (because of the spin conservation rule). Only electron transfer within the activated complex in the triplet state $^3(A^- \cdots D^+)$ leads directly to generation of the excited $^3A^*$. Triplet–singlet up-conversion is a necessary step before the electron transfer from $^3(A^- \cdots D^+)$ to the ground state product occurs. The activated complex in the singlet state $^1(A^- \cdots D^+)$ exhibits exactly the opposite behavior. Electron transfer leads directly to the ground-state products with excited $^3A^*$ formation preceded by the singlet–triplet up-conversion. If the separation of $^3(A^*\cdots D)$ into $^3A^*$ and D is sufficiently slow, back electron transfer (at rate k_{bt}) can also take place and cannot be a priori excluded from kinetic considerations. The reaction scheme presented in Figure 4 assumes population of $^3A^*$, but generation of $^3D^*$ can be discussed by the same approach (also applicable if more than one electron transfer channel leading to triplet excitation is introduced in the overall reaction mechanism).

The kinetic scheme discussed can be solved simply by using a steady-state approximation applied for the concentrations of the $^1(A^-\cdots D^+)$, $^3(A^-\cdots D^+)$, and $^3(A^*\cdots D)$ species. Then the excited triplet efficiency $\phi_{es}(T_1)$ is given by

$$\frac{\phi_{es}(T_1)}{1-\phi_{es}(T_1)} = \frac{k_{sep}\,[^3(A^*\cdots D)]}{k_{fg}\,[^1(A^-\cdots D^+)]} = \frac{k_{sep}k_{ft}/(k_{sep}+k_{bt})\,[^3(A^-\cdots D^+)]}{k_{fg}\,[^1(A^-\cdots D^+)]} \tag{13}$$

where $[^3(A^*\cdots D)]$, $[^1(A^-\cdots D^+)]$, and $[^3(A^-\cdots D^+)]$ denote stationary concentrations of the corresponding species. The right-hand side of Eq. (13) arises from the relationship $[^3(A^*\cdots D)] = k_{sep}k_{ft}/(k_{sep}+k_{bt})\,[^3(A^-\cdots D^+)]$ as obtained simply from the steady-state approximation applied to $^3(A^*\cdots D)$ alone by solving the equation

$$\frac{d}{dt}[^3(A^*\cdots D)] = -(k_{sep}+k_{bt})\,[^3(A^*\cdots D)] + k_{ft}\,[^3(A^-\cdots D^+)] \cong 0 \tag{14}$$

In a similar way, the stationary value of the $[^3(A^-\cdots D^+)] / [\,^1(A^-\cdots D^+)]$ ratio can be found by solving the following set of equations:

$$\frac{d}{dt}[^3(A^-\cdots D^+)] = -(k_{dis}+k_{ft}+k_{TS})\,[^3(A^-\cdots\ldots D^+)]$$
$$+\frac{3}{4}k_{as}\,[A^-][D^+]+k_{ST}\,[^1(A^-\cdots D^+)]+k_{bt}\,[^3(A^*\cdots D)] \cong 0 \tag{15}$$

$$\frac{d}{dt}[^1(A^-\cdots D^+)] = -(k_{dis}+k_{ft}+k_{ST})\,[^1(A^-\cdots D^+)]$$
$$+\frac{1}{4}k_{as}\,[A^-][D^+]+k_{TS}[^3(A^-\cdots D^+)] \cong 0 \tag{16}$$

Taking into account that $3k_{TS} \cong k_{ST}$ and that $[^3(A^*\cdots D)] = k_{sep}\,k_{ft}/(k_{sep}+k_{bt})[^3(A^-\cdots D^+)]$, the stationary value of the $[^3(A^-\cdots D^+)] / [^1(A^-\cdots D^+)]$ ratio is given by

$$\frac{[^3(A^-\cdots D^+)]}{[^1(A^-\cdots D^+)]} = \frac{3(k_{dis}+4k_{TS}+k_{fg})}{k_{dis}+4k_{TS}+k_{sep}k_{ft}/(k_{sep}+k_{bt})} \tag{17}$$

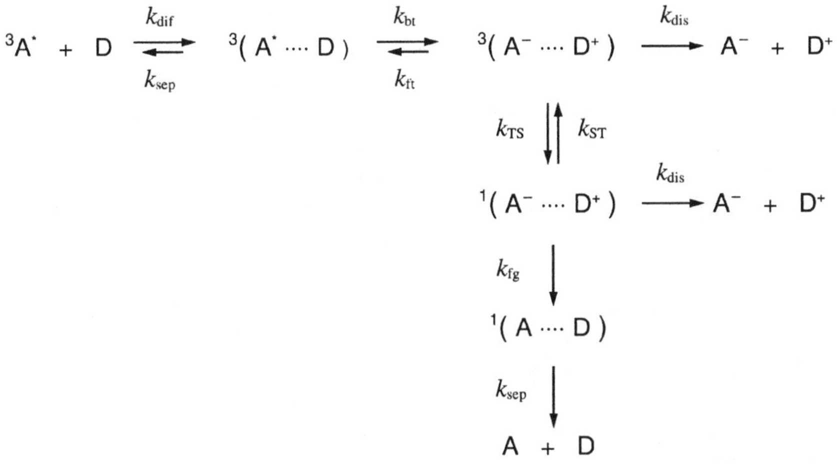

$$^3A^* + D \underset{k_{sep}}{\overset{k_{dif}}{\rightleftarrows}} \, ^3(A^* \cdots D) \underset{k_{ft}}{\overset{k_{bt}}{\rightleftarrows}} \, ^3(A^- \cdots D^+) \overset{k_{dis}}{\longrightarrow} A^- + D^+$$

$$k_{TS} \updownarrow k_{ST}$$

$$^1(A^- \cdots D^+) \overset{k_{dis}}{\longrightarrow} A^- + D^+$$

$$k_{fg} \downarrow$$

$$^1(A \cdots D)$$

$$k_{sep} \downarrow$$

$$A + D$$

Figure 5 Reaction mechanism of electron transfer quenching of the excited $^3A^*$ state.

leading correspondingly to the final equation for the excited triplet state efficiency:

$$\frac{\phi_{es}(T_1)}{1 - \phi_{es}(T_1)} = \frac{k_{sep}k_{ft}/(k_{sep}+k_{bt})}{k_{fg}} \left[\frac{3(k_{dis}+4k_{TS}+k_{fg})}{k_{dis}+4k_{TS}+k_{sep}k_{ft}/(k_{sep}+k_{bt})} \right] \quad (18)$$

As expected, the final relationship is quite complex and rather difficult to use in interpretation of the experimental values of $\phi_{es}(T_1)$ without additional information (e.g., from the expected relationship between electron transfer generation and annihilation of the excited $^3A^*$) about possible values of the elementary rates involved in the overall reaction scheme. Electron transfer quenching of the excited $^3A^*$ should be essentially the reverse process to that proposed for the ECL reactions and can be quantitatively discussed according to the reaction mechanism scheme presented in Figure 5, where k_{dif} and k_{sep} are the diffusion controlled forward and reverse rate constants for the formation of an activated complex between the excited $^3A^*$ and the quencher D. The activated complex is formed from the excited triplet state so that (because of the spin conservation rule) the electron transfer product also appears (at the forward electron transfer rate k_{bt}) in the triplet state. As usual, the redox products A^- and D^+ can be separated in solutions, at the rate k_{dis} and efficiency ϕ_{dis}, if the recombination to the singlet-ground-state product is sufficiently slow. The latter is allowed from the singlet precursor $^1(A^- \cdots D^+)$ but forbidden from the triplet one $^3(A^- \cdots D^+)$. Therefore, the triplet–singlet conversion, occurring at rate k_{TS}, is the necessary step that makes back electron transfer to the ground-state product (at rate k_{fg}) possible. The scheme also includes the reverse electron transfer corresponding to the back electron transfer to the excited state (at rate k_{ft}) and the singlet–triplet conversion (at rate k_{ST}). The kinetic scheme as discussed above can be also solved in the steady-state approximation [28] by relating of the observed bimolecular quenching rate constant k_q to the rate constants of all the reaction steps. For a given quencher D, all the rate constants in the ET quenching scheme can be determined by means of fluorescence quenching and transient absorption data [29]. Taking into account the intuitive relationship between two classes of electron transfer processes, one can expect that the same set of kinetic parameters can be applied to a quantitative description of the ECL efficiencies. The above approach was recently applied for ECL systems involving ruthenium(II) complexes and organic coreactants [30–32].

In many cases, however, the above approach cannot be applied because of the lack of necessary quenching data, and some additional (somewhat arbitrary) assumptions are necessary for the use of Eq. (18). The simplest one takes into account two limiting cases of the spin up-conversion between the triplet and singlet forms of the activated complex (with $k_{TS} \to \infty$ and $k_{TS} \to 0$, respectively). If the spin up-conversion processes are very slow (i.e., $k_{TS} \to 0$), Eq. (18) can be simplified to

$$\frac{\phi_{es}(T_1)}{1-\phi_{es}(T_1)} = \frac{k_{sep}k_{ft}/(k_{sep}+k_{bt})}{k_{fg}}\left(\frac{3(k_{dis}+k_{fg})}{k_{dis}+k_{sep}k_{fg}/(k_{sep}+k_{bt})}\right) \quad (19)$$

or [for $\phi_{es}(T_1) \ll 1$] to

$$\frac{\phi_{es}(T_1)}{1-\phi_{es}(T_1)} = \frac{3k_{sep}k_{ft}/(k_{sep}+k_{bt})}{k_{fg}}\left(\frac{k_{dis}+k_{fg}}{k_{dis}}\right) = \frac{3k_{sep}k_{ft}/(k_{sep}+k_{bt})}{k_{dis}k_{fg}/(k_{dis}+k_{fg})} \quad (20)$$

In this particular case diffusional limitations may play an important role in both intrinsic electron transfer channels, i.e., in excited- and ground-state formation, as applied in the interpretation of the ECL systems involving molybdenum halide clusters and organic coreactants [33] (see also Refs. 34 and 35). In the case of very fast spin up-conversion (i.e., $k_{ST} \to \infty$), however, only the formation of the excited triplet state becomes a diffusion-limited step:

$$\frac{\phi_{es}(T_1)}{1-\phi_{es}(T_1)} = \frac{3k_{sep}k_{ft}/(k_{sep}+k_{bt})}{k_{fg}} \quad (21)$$

If electron transfer processes are slow compared to diffusion, Eqs. (19) and (21) can be further simplified to the expression

$$\frac{\phi_{es}(T_1)}{1-\phi_{es}(T_1)} = \frac{3k_{ft}}{k_{fg}} \quad (22)$$

corresponding to the classic Marcus case, where only electron transfer rates determine the efficiency of excited state formation.

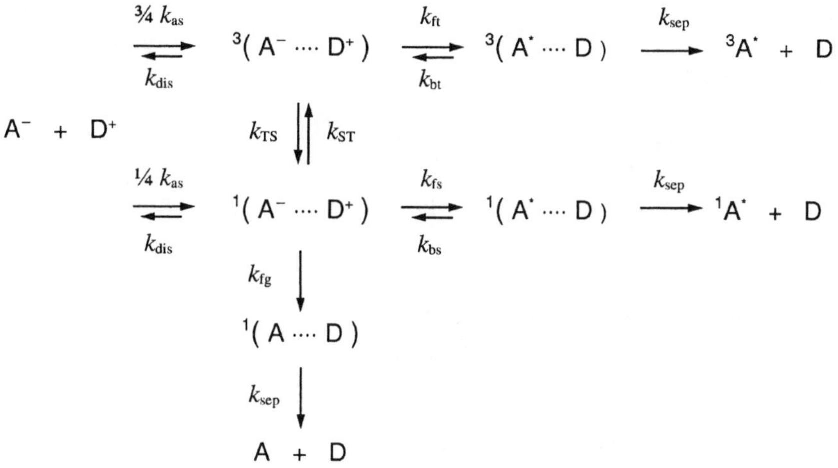

Figure 6 Reaction mechanism of electrochemiluminescence processes occurring according to the S route.

D. S Route—Formation of Excited Singlets

Further increase of the ion annihilation exothermicity to values comparable with the excited singlet-state energies opens an additional electron transfer channel, e.g., formation of ^{1}A* as presented in Figure 6. The scheme includes additional forward electron transfer (at rate k_{fs}) within the singlet form of the activated complex ^{1}(A$^-$···D$^+$) followed by separation of the redox product ^{1}(A*···D) in the bulk solution (at rate k_{sep}). Back electron transfer (at rate k_{bs}) can take also place if the separation of ^{1}(A*···D) into ^{1}A* and D is sufficiently slow. Because of the spin conservation rule, only electron transfer within the activated complex in the singlet state ^{1}(A$^-$···D$^+$) leads directly to the generation of excited ^{1}A*. Opening of the excited singlet channel leads to additional complications in the kinetic description, but as with the T-route case the problem can be solved using steady-state approximations. This can be done assuming a stationary condition for ^{1}(A*···D):

$$\frac{d}{dt}[^1(A*\cdots D)] = -(k_{sep}+k_{bs})[^1(A*\cdots D)] + k_{fs}[^1(A^-\cdots D^+)] \cong 0 \qquad (23)$$

and correspondingly for the ^{1}(A$^-$···D$^+$) species:

$$\frac{d}{dt}[^1(A^-\cdots D^+)] = -(k_{dis}+k_{fg}+k_{ST}+k_{fs})[^1(A^-\cdots D^+)]$$

$$+\frac{1}{4}k_{as}[A^-][D^+]+k_{TS}[^3(A^-\cdots D^+)]+k_{bs}[^1(A*\cdots D)] \cong 0 \qquad (24)$$

In a similar way, as was done for the T route, the total efficiency of excited and ground products in the singlet state [sum of $\phi_{es}(S_1)$ and $\phi_{gs}(S_0)$, respectively] can be expressed as

$$\frac{\phi_{gs}(S_0) + \phi_{es}(S_1)}{1 - \phi_{gs}(S_0) - \phi_{es}(S_1)} = \frac{k_{fg} + k_{sep}k_{fs} / (k_{sep} + k_{bs})}{k_{sep}k_{ft} / (k_{sep} + k_{bt})}$$

$$\times \frac{k_{dis} + 4k_{TS} + k_{sep}k_{ft} / (k_{sep} + k_{bt})}{3\left[k_{dis} + 4k_{TS} + k_{fg} + k_{sep}k_{fs} / (k_{sep} + k_{bs})\right]} \quad (25)$$

with the branching ratio between ϕ_{es} (S_1) and ϕ_{gs} (S_0) given by

$$\frac{\phi_{es}(S_1)}{\phi_{gs}(S_0)} = \frac{k_{sep}k_{fs} / (k_{sep} + k_{bs})}{k_{fg}} \quad (26)$$

The expression obtained is still more complicated than that derived for the T-route case; additional simplifications seem to be necessary before applying Eq. (25) in a kinetic discussion of the S route. Of course, introduction of additional electron transfer processes, (e.g., formation of the excited T_2, T_3, ... and/or S_2, S_3, ... products) is possible, but the expressions obtained will be still more complex. One can expect that, similarly by to the T-route case, data from electron transfer quenching of the excited $^1A^*$ (reaction scheme presented in Figure 7) may be very useful. To our best knowledge, however, such an approach has not been applied in ECL studies.

Less complicated formulation for the $\phi_{es}(S_1)$ efficiency can be obtained for $k_{ST} \to \infty$ and $k_{TS} \to 0$. If the spin up-conversion between two forms of the activated complex is infinitely fast, the second term in Eq. (25) can be neglected, giving

$$\frac{\phi_{gs}(S_0) + \phi_{es}(S_1)}{1 - \phi_{gs}(S_0) - \phi_{es}(S_1)} = \frac{k_{fg} + k_{sep}k_{fs} / (k_{sep} + k_{bs})}{3k_{sep}k_{ft} / (k_{sep} + k_{bt})} \quad (27)$$

Taking into account relationship (26), one can also obtain

$$\frac{\phi_{es}(S_1)}{1 - \phi_{es}(S_1)} = \frac{k_{sep}k_{fs} /(k_{sep} + k_{bs})}{k_{fg} + 3k_{sep}k_{ft} /(k_{sep} + k_{bt})} \quad (28)$$

Similar to the T-route case, the effective rates of formation of both excited states ($^1A^*$ and $^3A^*$) may correspond to the diffusion-limited case. Correspondingly, for $k_{TS} \to 0$ all three accessible channels are diffusion-limited processes if the intrinsic electron transfer steps are fast enough:

$$\frac{\phi_{es}(S_1)}{1 - \phi_{es}(S_1)} = \frac{k_{sep}k_{fs} /(k_{sep} + k_{bs})}{k_{dis}k_{fg} /(k_{dis} + k_{fg}) + 3k_{sep}k_{ft} /(k_{sep} + k_{bt})} \quad (29)$$

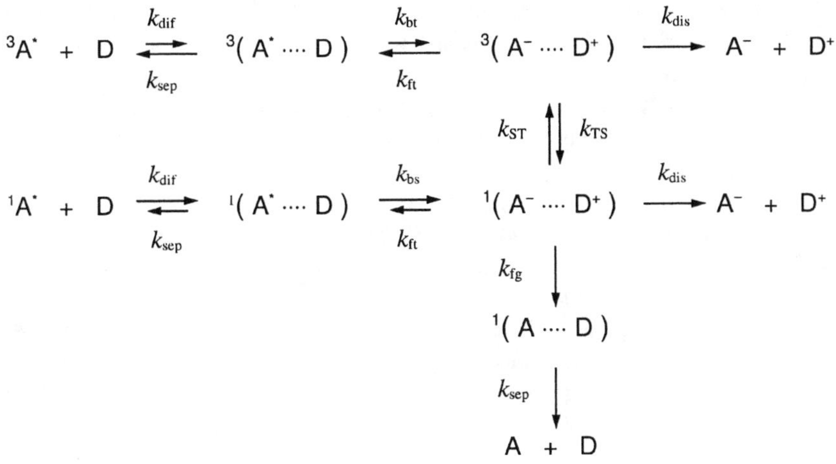

$$^3A^* + D \overset{k_{dif}}{\underset{k_{sep}}{\rightleftharpoons}} {}^3(A^* \cdots D) \overset{k_{bt}}{\underset{k_{ft}}{\rightleftharpoons}} {}^3(A^- \cdots D^+) \overset{k_{dis}}{\longrightarrow} A^- + D^+$$

$$k_{ST} \updownarrow k_{TS}$$

$$^1A^* + D \overset{k_{dif}}{\underset{k_{sep}}{\rightleftharpoons}} {}^1(A^* \cdots D) \overset{k_{bs}}{\underset{k_{ft}}{\rightleftharpoons}} {}^1(A^- \cdots D^+) \overset{k_{dis}}{\longrightarrow} A^- + D^+$$

$$k_{fg} \downarrow$$

$$^1(A \cdots D)$$

$$k_{sep} \downarrow$$

$$A + D$$

Figure 7 Reaction mechanism of electron transfer quenching of the excited $^1A^*$ state.

If the electron transfer processes are slow compared to diffusion, both Eqs. (28) and (29) can be further simplified to

$$\frac{\phi_{es}(S_1)}{1-\phi_{es}(S_1)} = \frac{k_{fs}}{k_{fg}+3k_{ft}} \tag{30}$$

Equation (30) was applied in the quantitative discussion of the $\phi_{es}(S_1)$ efficiencies in ECL systems involving rubrene [36] or intramolecular donor–acceptor organic compounds [37]. In both cases this was done, after appropriate modification, by taking into account more than one accessible reaction channel leading to triplet excitation.

III. ELECTRON TRANSFER PROCESSES

A. Main Concepts—Normal and Inverted Marcus Regions

In view of the above considerations, it becomes clear that a quantitative description of the electron transfer processes is essential for understanding ECL reactions. The role of different factors determining electron transfer rates has been the subject of very extensive theoretical and experimental studies. Electron transfer reactions in solution have been summarized and reviewed by many authors in works concerning different aspects of the theory as well as experimental results (e.g., Refs. 38–49). Thus, in this chapter the relevant ideas and equations are only briefly summarized, to serve as a basis for understanding the ECL results.

The crux of the electron transfer problem is the fact that the equilibrium nuclear configuration of the reacting species changes (in the intramolecular bond lengths and angles as well as in the vibrations and rotations of the surrounding solvent dipoles) when the reaction partners gain or lose an electron. The equilibrium configurations of the reduced and oxidized forms of a redox couple, like the ground and excited states of a molecule, are generally different. As a consequence, the rates of thermally activated electron transfer reactions, radiative transitions, and nonradiative deactivation processes can be discussed with a common formalism in which the rate is a product of an electronic factor and a nuclear factor. The former is a function of the electronic interaction in the reacting system, whereas the latter depends on the nuclear configuration changes between the reactants and products. The role of both factors in determining the rate of electron transfer can be described quantitatively. The formalism describing these processes provides a unified description of homogeneous (intra- and intermolecular) and heterogeneous electron transfer reactions, with both radiative and radiationless natures.

The theoretical description of the kinetics of electron transfer reactions starts from the pioneering work of Marcus in which a convenient expression for the free energy of activation $\Delta G^{\ddagger}$ was defined. However, the preexponential factor in the expression for the reaction rate constant was left undetermined in the framework of that classical macroscopic theory. More sophisticated, semiclassical or quantum-mechanical, approaches (e.g., Refs. 50–54), avoid this inadequacy. Typically, they are based on the Franck–Condon principle, i.e., assuming separation of the electronic and nuclear motions. The Franck–Condon principle states that the interatomic distances and nuclear moments are identical in the final and initial states at the time of electron transfer. Electron transfer between two states is a relatively instantaneous event compared to the slower nuclear motions that must take place to accommodate the new electronic configuration. Before the transfer of an electron, the nuclear geometry of the initial state, including the surrounding solvent molecules, must be converted into a high-energy "nonequilibrium" or distorted configuration. The transition state consists of two high-energy species that have the same nuclear conformation but different electronic configurations. Taking into account the assumptions presented above, the probability per unit time (first-order rate constant k_{et}) that an initial state will pass to a final product may be given by time-dependent perturbation theory as

$$k_{et} = \frac{4\pi^2}{h} V_{if}^2 \, (\text{FCF}) \tag{31}$$

where h is the Planck constant and V_{if} is the electronic matrix element describing the electronic coupling between the initial and final states, defined as

$$V_{if} = \langle \Psi_i | \mathbf{H} | \Psi_f \rangle \tag{32}$$

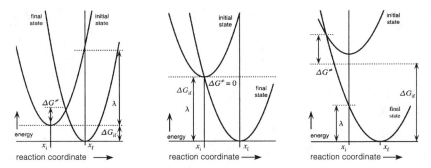

Figure 8 Electron transfer in the normal (left) and inverted (right) Marcus regions. Barrierless case with $\Delta G^{\ddagger} = 0$ (middle) corresponds to $\lambda + \Delta G_{if} = 0$. Plots of the potential energies of the reactant and product as a function of the nuclear (reaction) coordinate in the zeroth-order approximation.

where Ψ_i and Ψ_f are the electronic wave functions of the reacting system in the initial and final states and $\mathbf{H}$ is the energy operator. The potential energies of the system in its initial (V_i) and final (V_f) states are given correspondingly by

$$V_i = \langle \Psi_i | \mathbf{H} | \Psi_i \rangle \quad \text{and} \quad V_f = \langle \Psi_f | \mathbf{H} | \Psi_f \rangle \tag{33}$$

The Franck–Condon factor (FCF) is a sum of the products of overlap integrals of the vibrational and solvation wave functions of the reactants with those of the products, suitably weighted by Boltzmann factors. The value of the Franck–Condon factor can be expressed analytically by considering the effective potential energy curves of both initial and final states as a function of their nuclear configurations. Relatively simple relationships can be derived if the appropriate curves are harmonic with identical force constants f. Under these conditions,

$$V_i = f(x_i - x)^2 \quad \text{and} \quad V_f = f(x_f - x)^2 + \Delta G_{if} \tag{34}$$

where x is the dimensionless nuclear (reaction) coordinate (with x_i and x_f values corresponding to the thermodynamically equilibrated initial and final configurations) and ΔG_{if} is the Gibbs energy change between the initial and final states.

Using the above formalism, the potential energy curves can be constructed in the zeroth-order approximation as presented in Figure 8. The electron transfer can then be described in terms of the crossing of the system from one potential energy curve to the other. For a given λ value, $\Delta G^{\ddagger}$ depends on the electron transfer exothermicity. For a moderately exergonic reaction (with $\lambda - \Delta G_{if} > 0$), the $\Delta G^{\ddagger}$ barrier increases with increasing reaction exothermicity and when $\lambda - \Delta G_{if} = 0$ the reaction rate is maximized and the process is barrierless with $\Delta G^{\ddagger} = 0$ (the free energy region or normal Marcus region). A further increase in the reaction exothermicity again causes $\Delta G^{\ddagger}$ values greater than zero (if $\lambda -$

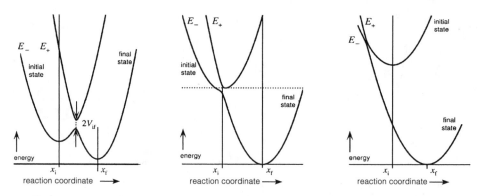

Figure 9 Electron transfer in the normal (left) and inverted (right) Marcus regions. Barrierless case with $\Delta G^{\ne} = 0$ (middle) corresponds to $\lambda + \Delta G_{if} = 0$. Plots of the potential energies of the reactant and product as a function of the nuclear (reaction) coordinate in the first-order approximation.

$\Delta G_{if} < 0$), which leads to a lowering of the reaction rate (the abnormal free energy region or Marcus inverted region). Such behavior is a simple consequence of the parabolic shapes of the potential energy curves.

In the zeroth-order approximation ($V_{if}^2 = 0$), the electron is required to remain localized on the individual (initial or final) system. No electron transfer is possible as long as this condition is imposed; electronic coupling of the reactants (i.e., $V_{if}^2 > 0$) is necessary for the system to pass from the initial state to the final state. Electronic coupling removes the degeneracy at the intersection of the V_i and V_f curves (see Fig. 9) and leads to the formation of two new states, the adiabatic states of the system, obtained by solving the secular equation

$$\begin{vmatrix} V_i - E & V_{if} \\ V_{if} & V_f - E \end{vmatrix} = 0 \tag{35}$$

The roots of this equation are

$$E_{\pm} = \frac{1}{2}\left(V_i + V_f\right) \pm \frac{1}{2}\left[\left(V_i - V_f\right)^2 + 4V_{if}^2\right]^{1/2} \tag{36}$$

where E_- and E_+ describe the lower and upper surface, respectively. The above considerations are in principle valid if $V_{if} \ll V_i$ and $V_{if} \ll V_i$. These requirements are fulfilled in most intermolecular (outer-sphere) reactions where the coupling of the initial and final states of the system is relatively weak. The shapes of the potential energy curves E_+ and E_- depend distinctly on the ΔG_{if} and λ values. For moderately exergonic processes ($\lambda + \Delta G_{if} > 0$) the system passes from the initial to final state by remaining on the same curve E_-. For strongly exergonic processes ($\lambda + \Delta G_{if} < 0$), inherently nonadiabatic electron transfer involves a transition from the E_+ to the E_- curve. In the normal Marcus region ($\lambda + \Delta G_{if} > 0$), nonadiabacity of the electron transfer is manifested (if V_{if} is sufficiently small) by the reacting system jumping between the lower E_- and upper E_+ curves.

The formalism presented above is applied for processes that lead in principle to one product only. If the electron transfer is exergonic enough (with a $-\Delta G_{if}$ value comparable with the excitation energies of the A or D species), additional reaction channels become available. This leads to a somewhat more complicated picture with more than two interacting states. The problem can be discussed in the zeroth-order as well as first-order approximations, leading to construction of potential energy curves as presented in Figure 10. One can expect that for a strongly exergonic reaction the formation of the ground state lies far in the Marcus inverted region, and this particular channel may be relatively slow compared with the less exergonic formation of the excited states. Such behavior is responsible for production of the excited states in an ECL process.

B. Electron Transfer Activation Energies and Preexponential Factors

The λ and V_{if} parameters are crucial, because their values are directly responsible for values of the activation energies and the preexponential factors of the electron transfer processes. Simple mathematics leads to the familiar Marcus expression for the activation energy $\Delta G^{\ddagger}$ as a function of ΔG_{if} and λ terms:

$$\Delta G^{\ddagger} = \frac{\left(\lambda + \Delta G_{if}\right)^2}{4\lambda} \tag{37}$$

where λ is equal to $f(x_f - x_i)^2$. Note that Eq. (37) is exactly valid if the zeroth-order approximation is used to construct the potential energy curves for the electron transfer process. If the first-order approximation is applied, the splitting (equal to $2V_{if}$) that occurs at the intersection of the two potential energy curves lowers the effective activation energy:

$$\Delta G^{\ddagger} = \frac{\left(\lambda + \Delta G_{if}\right)^2}{4\lambda} - V_{if} \tag{38}$$

Usually the V_{if} term in Eq. (38) is assumed to be negligible compared to $(\lambda + \Delta G_{if})^2/4\lambda$ and Eq. (38) reduces to Eq. (37). Note that both expressions correspond to the classical treatment of the molecular (solvent as well as solute) motions that are responsible for readjustment of nuclear coordinates associated with electron transfer.

Parameter λ (usually called the reorganization energy) describes the energy required to bring a system from its initial equilibrium conformation to the nonequilibrium one characteristic of the final state (cf. Fig. 8). λ is the sum of two contributions: the inner (λ_i), required for bond length and angle changes, and the outer (λ_o), necessary for the solvent coordination shells reorganization. The

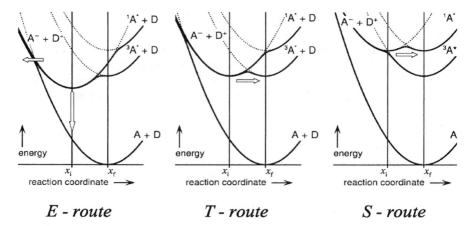

E - route T - route S - route

Figure 10 Reaction coordinate diagrams for ECL processes occurring according to the E route (left), T route (middle), and S route (right). Potential energy curves are presented in the zeroth-order (dotted curves) and first-order (solid E_- curves) approximations with the degeneracy at the potential energy curve crossing points removed in the latter.

energy required to reorganize the solvent, λ_o, can be obtained by treating the medium outside the reactant (or reactants) as a dielectric continuum with the polarization made up of two parts, a relatively rapid electronic part and a slower vibrational–orientational one. The latter has to adjust to a nonequilibrium value appropriate to the final state, whereas the former does not. On the basis of the Born solvation theory, λ_o is given by (if one electron is transferred)

$$\lambda_o = \frac{N_A e_o^2}{4\pi\varepsilon_o}\left(\frac{1}{n^2} - \frac{1}{\varepsilon}\right)\left(\frac{1}{2r_A} + \frac{1}{2r_D} - \frac{1}{d}\right) \tag{39}$$

where n is the refractive index of the reaction medium. r_A and r_D are the effective radii of the redox centers involved in the electron transfer reaction, with the center-to-center separation distance d. Usually radii r_A and r_D are calculated from the molar volumes of A and D species and the electron transfer distance $d = r_A + r_D$ is assumed. Note that Eq. (39) is exactly valid only for spherical molecules with uniform charge distribution. In cases in which the charge is nonuniformly redistributed and/or for nonspherical molecules, λ_o may be estimated on the basis of appropriate, more sophisticated extensions of the simple Born model [55–57].

The energy λ_i required to reorganize the intramolecular bonds can be calculated by using a harmonic oscillator approximation (if no bonds are broken and no new bonds are formed in the electron transfer reaction). The potential energy needed to change the atomic distances from their equilibrium values in the initial state to those appropriate to the final state can be calculated by taking into account the force constants in the reactant and product (f_i and f_f, respectively) and the changes in equilibrium values of all affected bonds (Δq_{if}):

$$\lambda_i = \sum \frac{f_i f_f}{f_i + f_f} (Dq_{if})^2 \tag{40}$$

The needed quantities can be obtained from the infrared spectroscopic (f_i and f_f) and crystallographic (Δq_{if}) parameters if appropriate data are known for both redox forms involved in the electron transfer process [58]. An alternative method for λ_i estimation is based on semiempirical quantum-chemical calculations [59]. The difference between the computed values of the heat of formation of the donor D in its equilibrium nuclear geometry and in the conformation corresponding to the equilibrium geometry of its oxidized form (D^+) is added to the respective value calculated for the acceptor A and its reduced form (A^-) [60].

The splitting (equal to $2V_{if}$) that occurs at the intersection of the two potential energy curves is essential for the electron transfer. Within the Landau–Zener framework their value is responsible for the electron hopping frequency ν_{et} describing the probability of the intrinsic electron transfer act:

$$\nu_{et} = \frac{4\pi^2}{h} V_{if}^2 \left(\frac{1}{4\pi\lambda RT} \right)^{1/2} \tag{41}$$

Expression (41) has been derived by using classical treatment of the solvent and solute, intramolecular motions. Their applicability is limited by the framework of the perturbation theory of the interaction causing transition. In this theory the dynamics of both intramolecular and solvent motions have no influence on the rate constant of the reaction. The formation of some favorable geometry of the reactant(s) that allows reaction is fast compared to the elementary electron transfer step. In conjunction with this "frozen-reactant" approximation is the assumption that the orientation of the reactant(s) can be regarded as fixed. This is justified if the electron hopping frequency is relatively small and the intramolecular and solvent shell changes occur on a much faster time scale. In most cases this is fulfilled for bond length and angle adjustment ($\nu_i \approx 10^{13} - 10^{14}$ s^{-1}). The solvent motions, however, are distinctly slower ($\nu_o \approx 10^{11} - 10^{12}$ s^{-1}). Therefore, for relatively small values of the electronic coupling element $V_{if} \approx$ 0.01–0.02 eV, the number of reacting subsystems (at the intersection points of the potential energy curves) may be smaller than expected from the equipartition theorem, and the electron transfer reaction may be slower than expected from Eq. (41). According to somewhat oversimplified interpretation, this corresponds to the lowering of the velocity with which the reacting system is moving on its potential energy curve.

For a quantitative description of solvent effects on the preexponential factor, it is necessary to go beyond of the framework of perturbation theory. The stochastic approach to description of the solvent dynamical effect in the rate constant was developed about 20 years ago [61–65]. On the basis of this approach it was found that the rate of electron transfer is proportional to the frequency of longitudinal reorientational relaxation (expressed by the longitudinal relaxation time τ_L) of the solvent [66].

The theory developed starts with a description of the dielectric loss spectra, frequency-dependent permittivity of the solvent $\varepsilon(\omega)$, in the framework of the Debye model [67], in which the reorientation of the solvent dipoles gives the main contribution to the relaxation of the solvent polarization:

$$\varepsilon(\omega) = \varepsilon_\infty + \frac{\varepsilon - \varepsilon_\infty}{1 + i\omega\tau_D} \tag{42}$$

where ω is the angular frequency. The longitudinal relaxation time $\tau_L = \tau_D \varepsilon_\infty / \varepsilon$ is related to the Debye relaxation time τ_D and the dielectric permittivity of the given solvent (in the near-infrared region ε_∞ and at the static electric field ε).

Generally the possible influence of the solvent dynamics can be expressed by the modified Eq. (41) [68]:

$$\nu_{et} = \frac{4\pi^2}{h} V_{if}^2 \left(\frac{1}{4\pi\lambda_o RT} \right)^{1/2} \left(\frac{1}{1 + 8\pi^2 V_{if}^2 \tau_L / \lambda_o h} \right) \tag{43}$$

Equation (43) exhibits the transition from the solvent-controlled limit (for sufficiently large values of V_{if}) when ν_{et} is reciprocally proportional to τ_L,

$$\nu_{et} = \frac{1}{\tau_L} \left(\frac{\lambda_o}{16\pi RT} \right)^{1/2} \tag{44}$$

to the nonadiabatic limit (for small values of V_{if}) when k_{et} is essentially controlled by the frequency of electron hopping, Eq. (41).

The electronic coupling element V_{if} is also connected with the structure of the activated complex ($A^- \cdots D^+$), and its value should in principle be a function of the distance between the A^- and D^+ reactants and their mutual orientation. Moreover, the V_{if} values depend on the nature of the molecular orbitals involved in the particular electron transfer process. The role of these factors can be quantitatively discussed using the quantum-mechanical approach based on the Dogonadze model [39]:

$$V_{if} = \sum_A \sum_D c_A c_D \beta_{AD} \tag{45}$$

where c_A and c_D are LCAO coefficients of the molecular orbitals involved in the electron transfer process. β_{AD} is the resonance integral for the given atomic pair. Appropriate calculation of the V_{if} values are possible with a known (e.g., for the intramolecular electron transfer processes [69,70]) or assumed structure of the activated complex (cf. Ref. 58), because the β_{AD} values depend strongly on the distance and orientation of the A^- and D^+ reactants. Thus, practical applications of Eq. (45) are strongly limited, especially if the electron transfers do not correspond to simple LUMO/HOMO orbital transitions.

Simple combination of Eq. (37) with Eq. (41) or (44) leads to the final expressions for the electron transfer rate k_{et}. For relatively small (nonadiabatic limit) or large (adiabatic limit) values of V_{if} one can obtain

$$k_{et} = \frac{4\pi^2}{h} V_{if}^2 \left(\frac{1}{4\pi\lambda RT} \right)^{1/2} \exp\left(\frac{-(\lambda + \Delta G_{if})^2}{4\lambda RT} \right) \tag{46}$$

and correspondingly,

$$k_{et} = \frac{1}{\tau_L} \left(\frac{\lambda_o}{16\pi RT} \right)^{1/2} \exp\left(\frac{-(\lambda + \Delta G_{if})^2}{4\lambda RT} \right) \tag{47}$$

Of course, for intermediate values of V_{if}, expression (43) instead of (41) or (44) will describe the value of the preexponential factor.

C. Radiationless Electron Transfer Processes in the Marcus Inverted Region

Electronic excitation is possible only for very exergonic electron transfer reactions with the energy excess greater than the energy of the given excited state. The vibrational intramolecular excitation is, however, already possible for a moderately exergonic process. Consequently, the reaction rate in the Marcus inverted region (with $-\Delta G_{if} > \lambda_o$) may be much faster than that expected from the simplified theory neglecting vibrational excitation of the high-frequency modes accompanying the electron transfer.

In the theoretical description of the resulting reaction rate, different Franck–Condon factors $(FCF)_j$ should be applied for the all the accessible channels. Usually this is done with the assumption that the electronic coupling element V_{if} does not depend on the quantum number j of the excited vibrational mode:

$$k_{et} = \frac{4\pi^2}{h} V_{if}^2 \sum_{j=0} (FCF)_j \tag{48}$$

Each summand in Eq. (48) represents the rate for a single contribution to the total rate from a $0 \rightarrow j$ nonradiative vibronic transition (cf. Fig. 11). The Franck–Condon principle holds for each of the single contributions. Similarly, as in the one-mode approximation, the corresponding Franck–Condon factors $(FCF)_j$ are a sum of the products of overlap integrals of the vibrational wavefunctions of the reactants with those of products, weighted by the appropriate Boltzmann factors. It is assumed (for simplicity) that only one (averaged over all of the changing bonds) high-frequency internal vibrational

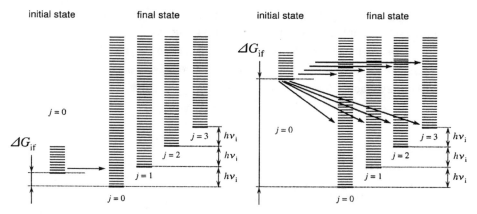

Figure 11 Schematic illustration of nonradiative electron transfer (horizontal arrows) in the normal (left) and inverted (right) Marcus regions. Associated with each vibronic state is a stack of sublevels representing low-frequency (mainly) solvent modes. In the initial state only one vibrational mode, with $j = 0$, is occupied, whereas in the final state various vibrational modes, with $j = 0, 1, 2, \ldots$ may be accessible. Diagonal arrows illustrate the radiative electron transfer (charge transfer emission in the inverted Marcus region).

mode, of frequency ν_i, undergoes reorganization. (At normal temperatures, $h\nu_i > RT$ and vibrational excitation of the reactants may be neglected.) Low-frequency modes, mainly associated with the solvent shell ($h\nu_o < RT$) are treated classically. Under the above assumptions the resulting expression for the total reaction rate in the nonadiabatic limit is [71,72]

$$k_{et} = \frac{4\pi^2}{h} V_{if}^2 \left(\frac{1}{4\pi\lambda_o RT} \right)^{1/2} \sum_{j=0} \frac{e^{-S} S^j}{j!} \exp\left(\frac{-(\lambda_o + \Delta G_{if} + jh\nu_i)^2}{4\lambda_o RT} \right) \qquad (49)$$

where S (the so-called electron-vibration coupling constant) is equal to the inner reorganization energy ν_i expressed in units of vibrational quanta $h\lambda_i$:

$$S = \frac{\lambda_i}{h\nu_i} \qquad (50)$$

where ν_i denotes mean vibration frequency of the reactant and product bonds.

In low exergonic electron transfer processes the reaction product is mainly formed in the vibrational ground state ($j = 0$) only. Thus, in the normal Marcus region, only the first $0 \rightarrow 0$ nonradiative vibronic transition contributes to the overall reaction rate:

$$k_{et} = \frac{4\pi^2}{h} V_{if}^2 \left(\frac{1}{4\pi\lambda_o RT} \right)^{1/2} e^{-S} \exp\left(\frac{-(\lambda_o + \Delta G_{if})^2}{4\lambda_o RT} \right) \qquad (51)$$

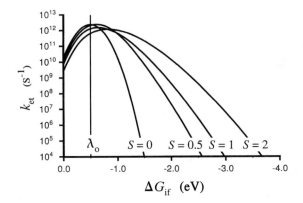

Figure 12 Plots of the electron transfer rate as a function of the reaction exothermicity ΔG_{if}. Case $S = 0$ corresponds to neglect of the vibrational excitation. If $S > 0$, vibrational excitations are taken into account. k_{et} values calculated for $V_{if} = 0.01$ eV, $\lambda_o = 0.5$ eV, and $h\nu_i = 0.2$ eV with $\lambda_i = 0.1$ ($S = 0.5$), 0.2 ($S = 1$), and 0.4 eV ($S = 2$), respectively.

However, when the process is more exergonic, vibrationally excited sublevels with $[j < (\lambda_o + \Delta G_{if})/h\nu_i]$ may be accessible (if $S > 0$). This results in strongly pronounced enhancement of the reaction rate (cf. Fig. 12). Sometimes the enhancement may be so large that the Marcus inverted region cannot be distinctly observed. On the other hand, unequivocal evidence of the Marcus inverted region has been presented in many works [9,10,48,73–79]. The quantum effects (in the Marcus inverted region) are expected to not only modify the free energy relationship but also affect the temperature dependence of the electron transfer rate [80,81], with an essential difference for small and large λ_i values. In both cases the strong temperature dependence predicted by the classical Marcus model becomes considerably weaker.

Depending on the values of S and V_{if}, some of the accessible reaction channels may be affected by the solvent molecular dynamics. This problem has been discussed by Jortner and Bixon [82] with the main conclusion that the overall reaction rate can be expressed as

$$k_{et} = \frac{4\pi^2}{h} V_{if}^2 \left(\frac{1}{4\pi\lambda_o RT}\right)^{1/2} \sum_{j=0} \frac{e^{-S} S^j}{j!} \left(1 + \frac{8\pi^2 \tau_L}{h\lambda_o} V_{if}^2 \frac{e^{-S} S^j}{j!}\right)^{-1}$$

$$\times \exp\left(\frac{-(\lambda_o + \Delta G_{if} + jh\nu_i)^2}{4\lambda_o RT}\right) \qquad (52)$$

Equation (52) exhibits the transition from the solvent-controlled limit (for sufficiently large values of the $V_{if}^2 e^{-S} S^j/j!$ term) when k_{et} is reciprocally

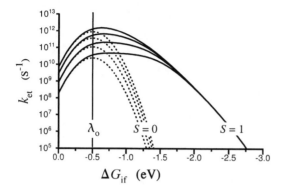

Figure 13   Plots of the electron transfer rate as a function of the reaction exothermicity with and without limitations caused by the solvent molecular dynamics. Cases $S = 0$ correspond to neglect of the vibrational excitation; if $S = 1$ the vibrational excitation is taken into account. k_{et} values calculated for $V_{if} = 0.01$ eV, $\lambda_o = 0.5$ eV, $h\nu_i = 0.2$ eV, and $\lambda_i = 0.2$ eV with $\tau_L = 0$, 1, 5, and 25 ps, respectively.

proportional to τ_L to the nonadiabatic limit (when $V_{if}^2 e^{-s}S^j/j!$ becomes sufficiently small). The overall reaction rate constitutes a superposition of solvent-controlled and nonadiabatic contributions. In the normal Marcus region the influence of the solvent molecular dynamics leads to an absolute reduction of the electron transfer rate (cf. Fig. 13). In contrast to that, in the Marcus inverted region, relative enhancement of the electron transfer rates is more pronounced than in the "pure" nonadiabatic limit.

    Similarly, as in the case of the classical approach, Eq. (52) reduces (for sufficiently large V_{if} values) to

$$k_{et} = \frac{1}{\tau_L}\left(\frac{\lambda_o}{16\pi RT}\right)^{1/2} \exp\left(\frac{-(\lambda_o + \Delta G_{if})^2}{4\lambda RT}\right) \tag{53}$$

when only the $0 \rightarrow 0$ contribution is taken into account (in the normal Marcus region).

    Calculation of the electron transfer rates for all accessible reaction channels is essential for the quantitative description of an ECL process. This is a rather difficult and usually somewhat speculative task because of the number of parameters going into the theory. However, some, at least qualitative or semiquantitative, conclusions, can be obtained from simple calculations performed for typical values of the necessary parameters. Typical results are presented in Figure 14, where the same set of $\tau_L = 0.2$ ps, $V_{if} = 0.01$ eV, $\lambda_o = 0.5$ eV, $\lambda_i = 0.2$ eV, and $h\nu_i = 0.2$ eV (values typical for acetonitrile solutions) has been applied for all accessible channels within a hypothetical ECL system

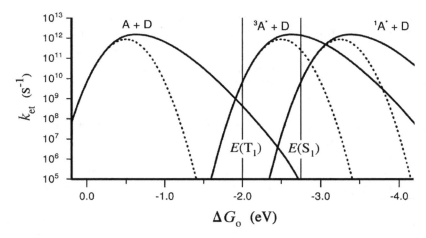

Figure 14 Plots of the electron transfer rate of the population of the ground state, the excited triplet state [with energy $E(T_1) = 2.0$ eV], and the excited singlet state [with energy $E(S_1) = 2.75$ eV] as a function of reaction exothermicity ΔG_{if}. k_{et} values calculated with $V_{if} = 0.01$ eV, $\lambda_o = 0.5$ eV, $h\nu_i = 0.2$ eV, and $\lambda_i = 0.2$ eV according to Eq. (47) (solid curves with $S = 1$) and Eq. (49) (dotted curves with $S = 0$), respectively.

with a population of excited $^3A^*$ and $^1A^*$ states with energies $E_T = 2.0$ and $E_S = 2.75$ eV. As expected, for $-\Delta G_{if} < E_T < E_S$, the population of the ground-state products prevails and eventual light emission can occur only by the E route. Efficient population of the $^3A^*$ state is possible only if $-\Delta G_{if} > E_T$. A T-route exergenicity larger than approximately -0.5 eV may be necessary for very high $\phi_{es}(T_1)$ efficiencies. This is probably the case of the well-known and widely studied ECL system involving $Ru(bipy)_3^{3+}/Ru(bipy)_3^+$ ion annihilation [83]. Similar behavior is expected for other ECL systems involving transition metal complexes, but somewhat surprisingly this does not seem to be a general rule (cf. Chapter 7). Population of the $^1A^*$ state (S route) becomes possible for still more negative ΔG_{if} (i.e., for $-\Delta G_{if} \approx E_S$). In such a case, however, production of the $^3A^*$ state remains the dominant reaction channel: $\phi_{es}(S_1) < \phi_{es}(T_1) \approx 1$ (cf. Refs. 36,37,84). Taking into account typical values of the energy gap between $^3A^*$ and $^1A^*$ states [85], one can expect that the formation of excited triplets should prevail in most organic ECL systems, especially because this process is also favored by spin statistics. Population of the excited triplets will be considerably inhibited with $\phi_{es}(S_1) \to 1$ if $-\Delta G_{if} + E_S$ is more negative than approximately -0.5 eV. Simple considerations from a molecular orbital description of a one-electron reduction/oxidation indicate that this is a rather unrealistic case (cf. Ref. 13). On the other hand, large values of the $\phi_{es}(S_1)$ efficiencies are expected for ECL systems involving organic compounds with a small energy splitting between the excited triplet and singlet states [17]. Note that all of the above conclusions arise from calculations performed according to Eq. (49) with the vibronic excitation of the electron transfer products taken into account. Application of a simplified model, e.g., Eq. (51), leads to results completely incompatible with the ones experimentally obtained.

D. Radiative Electron Transfer Processes in the Marcus Inverted Region

Electronic coupling of the initial and final states of the system also allows radiative electron transfer between redox centers. Such a process, important for the E route in ECL systems, can also be discussed in terms of Marcus theory. There are two principal differences between thermal (nonradiative) and optical (radiative) electron transfer processes [86–88]. The first is the amount of dissipated energy. Both radiative and nonradiative electron transfer are spontaneous transitions in which the entire electronic energy is dissipated into intramolecular and solvent motions. The electronic energy dissipated in the charge transfer emission at a given photon energy $hc\nu_{em}$ is $-\Delta G_{if} - hc\tilde{\nu}_{em}$ (where c is the light velocity), i.e., smaller than in the corresponding nonradiative electron transfer. In charge transfer absorption, the energy of the absorbed photon $hc\tilde{\nu}_{abs}$, however, is greater than $-\Delta G_{if}$; with the excess electronic energy equal to $-\Delta G_{if} + hc\tilde{\nu}_{abs}$ (cf. Fig. 15). The second important difference is connected with the operator coupling initial and final states. In optical transitions the dipole moment operator $\mathbf{M}$ instead of the energy operator $\mathbf{H}$ is suitable:

$$\mu = \langle \Psi_i | \mathbf{M} | \Psi_f \rangle \tag{54}$$

where μ is the electronic transition moment. If the contribution of the excited states can be neglected, the quantity μ (μ_{abs} and μ_{em} for absorption and emission, respectively) is connected with the electronic matrix element V_{if} [89,90]:

$$\mu_{abs} = \frac{V_{if}}{hc\nu_{abs}} \mu_{CT} \quad \text{and} \quad \mu_{em} = \frac{V_{if}}{hc\nu_{em}} \mu_{CT} \tag{55}$$

where μ_{CT} corresponds to the change of the dipole moment value between the initial and final states ($\mu_{CT} = e_o d$). In a more detailed analysis, however, contribution of the excited states ($^1A^*$ and/or $^1D^*$) should also be taken into account [91,92].

Pursuing the analogy between CT optical spectroscopy and thermal electron transfer processes, the rate $I(\tilde{\nu}_{em})$ of the emission of a given photon with energy $hc\tilde{\nu}_{em}$ (in photons per molecule per unit time per unit spectral energy) is given by the equation

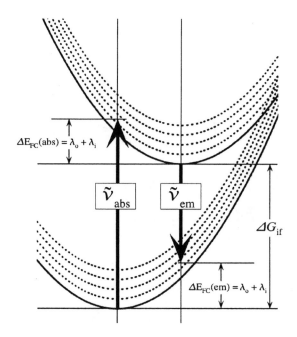

Figure 15 Radiative transition in the inverted Marcus region: charge transfer absorption (arrow up) and charge transfer emission (arrow down).

$$\frac{I(\tilde{\nu}_{em})}{n^3\tilde{\nu}_{em}^{\;3}} = \frac{64\pi^4}{3hc^3}\left(\frac{\mu_{em}^2}{4\pi\varepsilon_o}\right)\left(\frac{1}{4\pi\lambda_o RT}\right)^{1/2}\sum_{j=0}^{\infty}\frac{e^{-S}S^j}{j!}$$
$$\times\exp\left(\frac{-(\lambda_o+\Delta G_{if}+jh\nu_i+hc\tilde{\nu}_{em})^2}{4\lambda_o RT}\right) \tag{56}$$

In a similar way, the molar absorbance $\varepsilon(\tilde{\nu}_{abs})$ of the charge transfer absorption of a given photon with energy $hc\tilde{\nu}_{abs}$ may be expressed as

$$\frac{\varepsilon(\tilde{\nu}_{abs})}{n\tilde{\nu}_{abs}} = \frac{8\pi^3 N_A}{3\ln(10)}\left(\frac{\mu_{abs}^2}{4\pi\varepsilon_o c}\right)\left(\frac{1}{4\pi\lambda_o RT}\right)^{1/2}\sum_{j=0}^{\infty}\frac{e^{-S}S^j}{j!}$$
$$\times\exp\left(\frac{-(\lambda+\Delta G_{if}+jh\nu_i-hc\tilde{\nu}_{abs})^2}{4\lambda_o RT}\right) \tag{57}$$

where $\varepsilon(\tilde{\nu}_{abs})$ is defined by $\log(I/I_o) = \varepsilon(\tilde{\nu}_{abs})CL$, where C is the molar concentration of the solute, I and I_o are the incident and transmitted light intensity, and L is the optical path length.

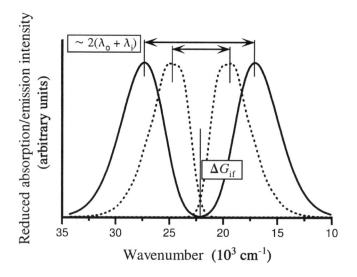

Figure 16 Reduced intensities $\varepsilon(\tilde{\nu}_{abs})/\tilde{\nu}_{abs}$ and $\bar{N}_{em})/\tilde{\nu}^3_{em}$ of the charge transfer absorption and emission spectra. Profiles calculated according to Eqs. (55) and (54), respectively, with $\Delta G_{if} = 2.75$ eV, $h\nu_i = 0.2$ eV, $\lambda_i = 0.2$ eV, and $\lambda_o = 0.2$ eV (dotted curves) or $\lambda_o = 0.5$ eV (solid curves).

Equations (56) and (57) describe also the shape of the CT emission and absorption bands, respectively (cf. Fig. 16). Thus, complementary to the kinetic data for nonradiative electron transfer, analysis of the charge transfer emission and/or absorption band shapes allows the determination of the λ_o, λ_i, $h\nu_i$, and ΔG_{if} quantities, treating their values as free fit parameters. Correspondingly, V_{if} quantities can be evaluated according to Eq. (55), using experimental values of the electronic transition moments μ_{abs} and/or μ_{em}. Thus, the absolute rate constants for nonradiative electron transfer can be predicted from information obtained from analysis of the corresponding radiative processes, as was reported for inter- (e.g., Ref. 93) and intramolecular (e.g., Ref. 94) systems.

The formalism presented above may be very useful in a quantitative interpretation of ECL processes following the E route, where the ϕ_{ecl} efficiencies are determined by the ratio of k_{em} and k_{nr} constants, according to Eq. (10). Integration of Eq. (56) over the entire range of $\tilde{\nu}_{em}$ leads to the familiar expression for the radiative rate constant k_{em} [95],

$$k_{em} = \frac{64\pi^4 n^3 \mu_{em}^2}{12\pi\varepsilon_o hc^3}\left(\tilde{\nu}_{em}^{max}\right)^3 \tag{58}$$

with the emission maximum $hc\tilde{\nu}_{em}^{max}$ equal approximately to $-\Delta G_{if} - \lambda_o - \lambda_i$.

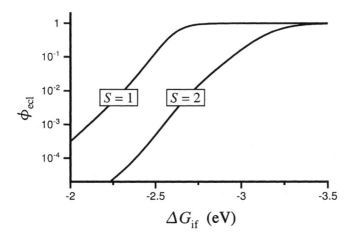

Figure 17 Plots of the calculated ϕ_{ecl} efficiencies (E route) as a function of the reaction exothermicity ΔG_{if} for a hypothetical ECL system with $\lambda_o = 0.5$ eV, $h\nu_i = 0.2$ eV, $e_o d = 25$ D, and $\lambda_i = 0.2$ eV ($S = 1$) or 0.4 eV ($S = 2$). $\phi_{ecl} = k_{em}/(k_{nr} + k_{em})$ with the k_{em} and k_{nr} values as calculated according to Eqs. (57) and (47), respectively.

Taking into account Eq. (55), i.e., the relationship between μ_{em} and V_{if}, one can straightforwardly obtain

$$k_{em} = \frac{64\pi^3 n^3 \mu_{CT}^2}{12\pi\varepsilon_o hc^5}\tilde{\nu}_{em}^{max} \approx \frac{64\pi^3 n^3 e_o^2 d^2}{12\pi\varepsilon_o hc^5}\left(-\Delta G_{if} - \lambda_o - \lambda_i\right) \qquad (59)$$

With a set of the ΔG_{if}, λ_o, λ_i parameters one can simply calculate the k_{em}/k_{nr} ratio (the value of V_{if} cancels) and correspondingly the expected ϕ_{ecl} efficiencies. Results of such calculations, performed for a hypothetical ECL system, are presented in Figure 17 with the data obtained for $\lambda_o = 0.5$ eV and $h\nu_i = 0.2$ eV with $\lambda_i = 0.2$ ($S = 1$) or $\lambda_i = 0.4$ eV ($S = 2$). As expected, the ϕ_{ecl} efficiency increases with the ion annihilation exergonicity, approaching values close to unity at sufficiently negative ΔG_{if} values. The case of $\phi_{ecl} \sim 1$, however, seems to be hardly reliable, because the ΔG_{if} values required for that are quite negative, negative enough for the efficient population of the $^3A^*$ and/or $^3D^*$ states. Opening of the T route forces an increase in the k_{nr} rate with a decrease in the ϕ_{ecl} efficiency, cf. Eq. (11). A strong dependence of the ϕ_{ecl} efficiency on λ_i as well as on λ_o (mostly through the k_{nr} rate) can also be predicted. In a similar way, the experimental data obtained from the ECL emission (using the band shape analysis) may reproduce the measured ϕ_{ecl} efficiencies of the "pure" E route. To our best knowledge, however, such approach in the description of ECL phenomena has not been applied.

IV. SPIN UP-CONVERSION PROCESSES

A. Outline of Spin Chemistry

Processes occurring in the electron transfer generation of ECL emission, like all chemical reactions, are spin-selective and are allowed (because of the fundamental principle of spin conservation) only for those spin states of reactants whose total spin is identical to that of the products. The unique character of the spin chemistry arises from magnetic interactions with their almost negligible contributions to the energetics of the reacting system. The magnetic interactions, however, are the only ones that are able to change the electron spin of reactants and switch the nature of the reaction from a spin-forbidden to a spin-allowed process. Moreover, one can expect that any perturbation of electromagnetic nature, such as an external magnetic field, microwave radiation, or the presence of paramagnetic species, can influence the kinetics of the spin up-conversion processes.

Two types of spin up-conversion processes, namely intra- and intermolecular processes, are in principle involved in an ECL reaction. The former corresponds to intersystem crossing between the singlet and triplet manifolds within the excited $^1A^*$ or $^3A^*$ (or $^1D^*$ or $^3D^*$) formed in the elementary electron transfer step. Intersystem crossing processes, responsible for the photophysical behavior of the generated excited states, are the same for states generated by photo- or chemical reactions. It should be noted, however, that the intersystem crossing rates may be important in the description of the ECL processes with light emission via delayed fluorescence (the T route). On the other hand, the intersystem crossing processes are partly responsible for the excited state lifetime, and therefore they may play an indirect role in the spin up-conversion occurring after electron transfer excitation. This is because any bimolecular reactions involving excited states are noticeably efficient only if the excited state lifetime is distinctly longer than any competitive process. Taking into account a typical concentration of a reacting species ($\sim 10^{-3}$ M in an ECL experiment) and the maximal possible values of the bimolecular reaction rates (with the diffusional limit of $\sim 10^9$–10^{10} $M^{-1}s^{-1}$), one can conclude that only species with lifetimes longer that microseconds can be considered. Thus, one can expect that excited singlets ($^1A^*$ or $^1D^*$) with lifetimes in the range of nanoseconds cannot take part in any bimolecular process. However, the excited triplet states ($^3A^*$ or $^3D^*$) with a lifetime in the range of seconds (for organic compounds) can effectively take part in any bimolecular spin up-conversion processes. The simultaneous presence of excited triplets and a high concentration of radical ions (doublets) is the typical situation in organic ECL systems following a T or mixed ST route. Thus, if the electron transfer excitation produces $^3A^*$ or $^3D^*$ states, the obtained triplets can subsequently participate in triplet–triplet annihilation and/or in triplet quenching processes [96]. Both of the above-mentioned processes, however, seem to play a much less important role in ECL systems in which generation of the short-lived triplet species (e.g., the excited states of transition metal complexes with a lifetime in the microsecond range) take place.

Unlike to processes following electrochemical excitation, the spin up-conversion preceding the electron transfer step (as included in the ECL reaction schemes presented in Figures 3, 4, and 6) occurring within the radical pair should be considered for all ECL systems. It should be noted, however, that any mechanisms causing a $^1(A^-\cdots D^+) \leftrightarrow {}^3(A^-\cdots D^+)$ conversion may have a significant effect only if exchange of the radical pair multiplicity occurs within the lifetime τ_{ac}, of the activated complex (with typical values close to $1/k_{dis}$ and a range between 10^{-9} and 10^{-10} s [97]). Thus, if spin multiplicity is much slower or much faster than 10^9–10^{10} s^{-1}, a spin-less approach may be appropriate for a description of the observed ECL behavior (cf. Sections II.C and II.D).

Generally spin processes seem to be most important for organic ECL systems following the T route and more or less negligible in other cases. On the other hand, knowledge of the nature of the spin up-conversion and, at least, its qualitative description may be crucial for a better understanding of what really occurs during electron transfer excitation. Rates of the spin up-conversion processes, similar to electron transfer reactions, can be discussed using the Fermi golden rule, taking into account a magnetic contribution in the energy operator responsible for interaction of the total spin electronic eigenstates in the reactant pair [98–103].

B. Spin Up-Conversion Before the Electron Transfer Step

The presence of four radical pair states (cf. Fig. 18) in the doublet pair is responsible for 1 : 3 ratio of two forms [the singlet $^1(A^-\cdots D^+)$ and the triplet $^3(A^-\cdots D^+)$] of the activated complex [97,102]. Thus, spin up-conversion occurring between $^1(A^-\cdots D^+)$ and $^3(A^-\cdots D^+)$ is a general phenomenon that occurs before the intrinsic electron transfer step. Kinetic considerations (cf. Sections II.C and II.D) suggest, however, that the observed ϕ_{es} efficiencies of excited state formation are not affected by the spin up-conversion rate k_{TS} in either of the limiting, spin-less cases with $k_{TS} \rightarrow 0$ or $k_{TS} \rightarrow \infty$. Any effect will be observable only if the k_{TS} rate is comparable with that characteristic for the activated complex dissociation k_{dis}. It is a necessary, but not sufficient, requirement; at least one of the electron transfer steps (responsible for generation of the ground or excited state, respectively) must be distinctly faster than the spin up-conversion process. The effect should be more pronounced for ground-state formation (occurring at the rate k_{fg}) because the electron transfer processes resulting in excited state formation (occurring at rates k_{ft} or k_{fs}, correspondingly) become a diffusion limited step if the k_{ft} or k_{fs} rates are sufficiently large. Assuming that $k_{fg} \gg k_{TS}$ and that $4k_{TS} \gg k_{dis}$, one can (for T-route cases) simplify Eq. (18) to

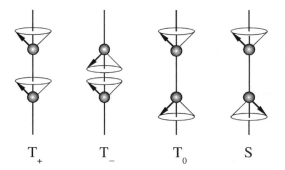

Figure 18 Vector representation of the T_+, T_B, T_0, and S radical pair states.

$$\frac{\phi_{es}(T_1)}{1-\phi_{es}(T_1)} = \frac{3k_{sep}k_{ft}/(k_{sep}+k_{bt})}{4k_{TS}+k_{sep}k_{ft}/(k_{sep}+k_{bt})} \tag{60}$$

or to

$$\frac{\phi_{es}(T_1)}{1-\phi_{es}(T_1)} = \frac{3k_{ft}}{4k_{TS}+k_{ft}} \approx \frac{3k_{ft}}{4k_{TS}} \tag{61}$$

if formation of the excited triplet state becomes sufficiently slow [31,32]. One can also conclude that generation of the excited triplet becomes negligible if both processes, ground-state formation and spin up-conversion (with rates k_{fg} and k_{TS}, respectively), are extremely fast. The same is also true for ECL processes following the S route, where a similar approach can be used to describe the $\phi_{es}(S_1)$ efficiencies. This may be simply done by taking into account the fraction of the electron transfer events populating both the excited and ground-states singlets:

$$\phi_{gs}(S_0)+\phi_{es}(S_1)=1-\phi_{es}(T_1) \approx \frac{4k_{TS}}{4k_{TS}+3k_{sep}k_{ft}/(k_{sep}+k_{bt})} \tag{62}$$

and the ratio of $\phi_{es}(S_0)$ and $\phi_{es}(S_1)$ efficiencies, i.e., Eq. (26),

$$\phi_{es}(S_1)=\left[\frac{4k_{TS}}{4K_{TS}+3k_{sep}k_{ft}/(k_{sep}+k_{bt})}\right]\left[\frac{k_{sep}k_{fs}/(k_{sep}+k_{bs})}{k_{fg}+k_{sep}k_{ft}/(k_{sep}+k_{bt})}\right] \tag{63}$$

Equation (63) may be subject to further simplification (as for the T route) assuming the lack of or presence of diffusional limitations in the electron transfer channels leading to excited state generation.

For the spin up-conversion between $^1(A^-\cdots D^+)$ and $^3(A^-\cdots D^+)$, a spin flip of one spin vector of the unpaired electrons is required. A semiclassical description of this motion [102,104] is based on the classical precession of the electron spin about different axes. As mentioned previously, magnetic

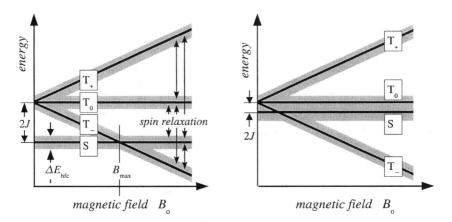

Figure 19 Energy diagram of electronic spin states of a radical pair in a magnetic field. Systems with exchange integral $J > \Delta E_{hfc}$ (left) and $J < \Delta E_{hfc}$ (right). ΔE_{hfc} is an average bandwith due to hyperfine coupling.

interaction between unpaired electrons in the A^- and D^+ radicals is responsible for the spin up-conversion process. Usually three types of magnetic interaction are considered in the radical pair mechanism: (1) the interradical exchange interaction J is responsible for coupling of the precessional motion of the two-electron spin and for the difference in the energies of the singlet and triplet forms of the activated complex $(A^-\cdots D^+)$ (cf. Fig. 19); (2) the hyperfine interactions of the individual electrons with nearby magnetic nuclei are responsible for a certain width of each T_-, T_+, T_0, and S sublevel corresponding to the average of the isotropic hyperfine coupling strength ΔE_{hfc}; and (3) the isotropic electronic Zeeman interaction caused by the presence of an external magnetic field B_0. $^1(A^-\cdots D^+) \leftrightarrow {}^3(A^-\cdots D^+)$ transitions become possible when the unpaired electrons process at different rates (ω_1 and ω_2) and when the singlet and triplet sublevels have similar energies.

The spin up-conversion rates are determined by magnetic interactions, with the mechanism of intersystem crossing depending on the methods by which $\Delta\omega = \omega_1 - \omega_2$ is caused to change from zero. In cases of organic radical pairs, this is mainly caused by the Zeeman and hyperfine interaction present in the system. The precession frequency (the so-called Larmor frequency) for the electron spin is given by [104–106]

$$\omega = \frac{2\pi g \mu_B B}{h} \tag{64}$$

where g and μ_B are the gyromagnetic factor and the Bohr magneton, respectively. In the semiclassical picture, an active magnetic field B is a vector sum of two components: the external magnetic field B_o and effective magnetic field B_{hfc}, resulting from the sum of the hyperfine coupling of the various nuclear spins in the A^- and D^+ radicals. Depending on the relative values of J and ΔE_{hfc}, the effective singlet–triplet mixing necessary for the spin up-conversion occurs at different values of the external magnetic field B_o: at $B_o \to 0$ for $J << \Delta E_{hfc}$ or at $B_o = B_{max} \cong 2J/g\mu_B$ for $J >> \Delta E_{hfc}$. The time scale on which the singlet–triplet mixing occurs can be estimated by using the semiclassical effective hyperfine field $B_{1/2}$ defined as [104,105]

$$B_{1/2} = \frac{2(B_1^2 + B_2^2)}{B_1 + B_2} \tag{65}$$

where the individual B_i values can be calculated with the equation

$$B_i = \left[\sum A_N^2 I_N (I_N + 1) \right]^{1/2} \tag{66}$$

on the basis of interaction between nuclear spin I_N and the unpaired electron in each radical with the isotropic hyperfine coupling constant A_N (usually estimated from the electron spin resonance data). Typical $B_{1/2}$ values are 0.5–5 mT, yielding a Larmor frequency in the range of 10^8–10^9 s^{-1}. This allows one to conclude that organic ECL systems correspond mostly to a limiting case with $k_{TS} \to 0$ and that Eq. (21) or (22) is adequate for a kinetic description of the T route [or Eq. (29) or (30) for the S route]. This is because spin relaxation is too slow to meet the condition $k_{TS} > 1/\tau_{ac}$ (also at $B_o \cong 0$).

The situation can be distinctly different in the case of inorganic ECL systems involving transition metal complexes, where, because of strong spin–orbit coupling, typical spin relaxation times are in the range of 10^{-9}–10^{-13} s^{-1} [106]. Thus, during the lifetime of the activated complex the spin up-conversion may occur very efficiently. In such cases, Eq. (60) or (61) seems to be appropriate for a kinetic description of the ECL efficiencies. Moreover, strong spin–orbit coupling causes considerable deviation of the g factor of inorganic paramagnetic species from the values characteristic for organic radicals (~2.0023 and close to that for a free electron). This could result in considerable differences in the spin precession frequencies in the A^- and D^+ species, because of different g factors (a Δg mechanism), affording intersystem crossing processes, with the rate of the $^1(A^- \cdots D^+) \leftrightarrow {}^3(A^- \cdots D^+)$ transition increasing linearly with the magnetic field. This effect, however, becomes important at very high B_o values (a few tesla) [107–109].

C. Spin Up-Conversion After the Electron Transfer Step

As mentioned above, spin up-conversion processes following electron transfer excitation may play an important role in the case of organic ECL systems, where

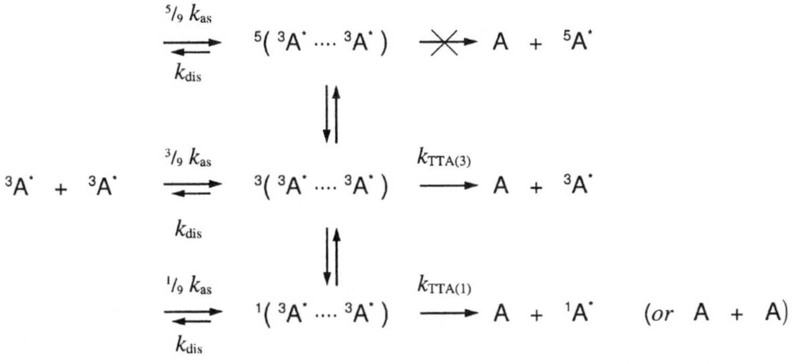

$$\xrightarrow[k_{dis}]{^{5/9}\,k_{as}} \quad ^5(\,^3A^* \cdots ^3A^*\,) \quad \xslashedrightarrow{} \quad A \;+\; ^5A^*$$

$$\Updownarrow$$

$$^3A^* \;+\; ^3A^* \quad \xrightarrow[k_{dis}]{^{3/9}\,k_{as}} \quad ^3(\,^3A^* \cdots ^3A^*\,) \quad \xrightarrow{k_{TTA(3)}} \quad A \;+\; ^3A^*$$

$$\Updownarrow$$

$$\xrightarrow[k_{dis}]{^{1/9}\,k_{as}} \quad ^1(\,^3A^* \cdots ^3A^*\,) \quad \xrightarrow{k_{TTA(1)}} \quad A \;+\; ^1A^* \quad (or \; A \;+\; A)$$

Figure 20 Reaction mechanism of the triplet–triplet annihilation processes.

triplet–triplet annihilation results in the observed emission from the excited singlet states. The excited triplet states of organic species ($^3A^*$ or $^3D^*$) with a lifetime in the range of seconds can effectively participate in triplet–triplet annihilation:

$$^3A^* + {}^3A^* \to {}^1A^* + A \qquad \text{or} \qquad ^3D^* + {}^3D^* \to {}^1D^* + D \qquad (67)$$

followed by fluorescence from the excited singlets $^1A^*$ or $^1D^*$. A more detailed scheme of $^3A^*$–$^3A^*$ annihilation is presented in Figure 20. In diffusion-controlled processes the triplet pair forms an activated complex in three different spin states (with nine spin sublevels with a branching ratio of 1 : 3 : 5). The excitation energy fusion within $^5(^3A^* \cdots {}^3A^*)$ is not very probable because of the usually very high energies of the excited quintet $^5A^*$. Correspondingly, the excited singlet $^1A^*$ or triplet $^3A^*$ can be produced or recovered from the $^1(^3A^* \cdots {}^3A^*)$ or $^3(^3A^* \cdots {}^3A^*)$ spin forms, respectively. Overall, the efficiency of excited singlet $^1A^*$ generation in the triplet–triplet annihilation process is governed by the rates of all processes occurring in parallel. One can argue that in the simplest case (i.e., for $k_{TTA(3)} \gg k_{dis}$ and $k_{TTA(1)} \gg k_{dis}$) the maximal value of the triplet–triplet annihilation efficiency ϕ_{TTA} is 0.25.

On the other hand, interaction between a triplet and a doublet (A^- or D^+) leads to triplet deactivation with dissipation of electronic excitation energy,

$$^3A^* + A^- \to A + A^- \qquad \text{and} \qquad ^3A^* + D^+ \to A + D^+ \qquad (68)$$

according to the reaction scheme presented in Figure 21. Similarly, as in triplet–triplet annihilation processes, the triplet–doublet pair forms an activated complex but in two different spin states (with six spin sublevels with a branching ratio of 2 : 4). Excitation energy fusion within the $^4(^3A^* \cdots A^-)$ spin form of the activated complex is usually an inefficient channel because of the relatively high

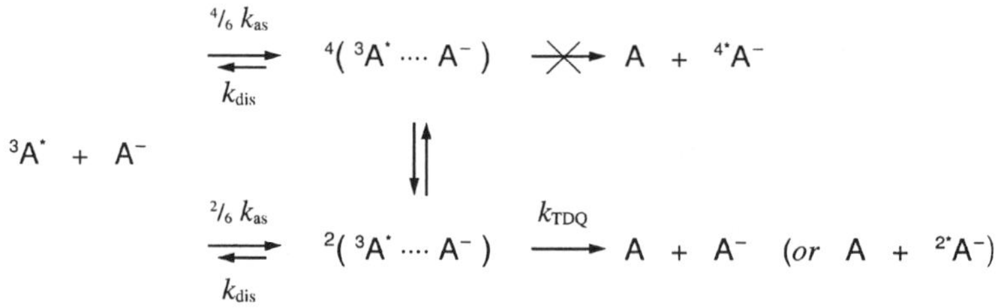

$$\xrightleftharpoons[k_{dis}]{^4/_6\, k_{as}} \quad {}^4(\,{}^3A^{\cdot} \cdots A^-\,) \quad \xcancel{\longrightarrow} \quad A \ + \ {}^{4\cdot}A^-$$

$${}^3A^{\cdot} \ + \ A^- \qquad\qquad \updownarrow$$

$$\xrightleftharpoons[k_{dis}]{^2/_6\, k_{as}} \quad {}^2(\,{}^3A^{\cdot} \cdots A^-\,) \quad \xrightarrow{k_{TDQ}} \quad A \ + \ A^- \quad (or \ A \ + \ {}^{2\cdot}A^-)$$

Figure 21 Reaction mechanism of the triplet–doublet quenching processes.

energies of the excited quartet ${}^{4*}A^-$. Energy dissipation within ${}^2({}^3A^* \cdots A^-)$ leads to effective quenching of the excited ${}^3A^*$, recovering A and A^- species (with a maximal value of 1 for the triplet–doublet quenching efficiency ϕ_{TDQ}. The two processes (triplet annihilation and triplet quenching) occur in parallel and usually lead to rather low ϕ_{ecl} efficiencies in organic ECL systems following the T route:

$$\phi_{ecl} = \phi_{es}\phi_o\phi_{TTA}(1 - \phi_{DDQ}) \tag{69}$$

where the term $\phi_{TTA}(1 - \phi_{DDQ})$ describes the effective triplet–triplet annihilation efficiency in an ECL process. Taking into account the fact that in ECL systems following the S route, $\phi_{ecl} = \phi_o\phi_{es}$, one can also expect that the directly produced excited singlets become the dominant light source, even if $\phi_{es}(T_1)$ is much higher than $\phi_{es}(S_1)$ [84].

Similarly, as in the case of the doublet–doublet pair, magnetic interactions between unpaired electrons in the ${}^3A^* \cdots {}^3A^*$ or ${}^3A^* \cdots A^-$ species are responsible for the spin up-conversion processes. Intramolecular spin–spin and spin–orbit interactions (zero field splitting within the triplet molecule) play a general role in triplet–triplet annihilation as well as in triplet–doublet quenching. In the case of a triplet–doublet pair, however, one of the zero field splitting terms is replaced by the isotropic hyperfine interaction (within the radical ion). Moreover, exchange interactions and the Zeeman interaction (in the presence of an external magnetic field) also take place. Examples of a particular ordering of energy levels are illustrated schematically in Figures 22 and 23 for the triplet–triplet and triplet–doublet pairs, respectively. Both energy level diagrams allow straightforward qualitative discussion of the external magnetic field influence on particular spin up-conversion processes.

D. Magnetic Field Effects

The spin up-conversion processes usually involve nearly degenerate sets of spin levels, which is the reason that Zeeman splittings much smaller than thermal

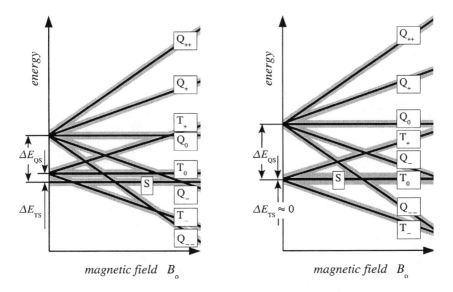

Figure 22 Energy diagram of electronic spin states of a triplet–triplet pair in a magnetic field. Systems with large (left) and small (right) values of the triplet–singlet splitting energies ΔE_{TS}. Eigenstate energies are not sharply defined because both ΔE_{TS} and ΔE_{QS} quantities depend on the random orientation (in liquid solutions) of the $^3A^*$ species within the activated complex $^3A^* \cdots ^3A^*$.

energies may have appreciable effects. Thus application of the external magnetic field can be exploited as a diagnostic tool not only for the spin up-conversion processes themselves but also for the entire reaction mechanism of which they are a part. As mentioned above, the observed effects can be understood on the basis of the energy diagrams presented in Figures 19, 22, and 23. In all of the cases presented, application of an external magnetic field may lead to decoupling or coupling of the energy levels characteristic of a particular spin up-conversion process. The first effect is responsible for a lowering of the spin up-conversion rate, because intersystem transitions may occur only if the involved levels approach each other within their energetic width. The second effect is connected with energy level crossings, at some higher magnetic fields, allowing for intersystem transitions. This may lead to an increase in the spin up-conversion rates, which at still higher fields are suppressed again (resonance in the magnetic field dependence). Within this picture (Merrifield's treatment [98]), intersystem transitions are expected to be suppressed at a sufficiently high magnetic field. This is because for high magnetic fields the energy differences between some of the spin states are too large for any significant mixing to occur. At low magnetic fields more states are nearly degenerate, which may result in enhancement of the spin

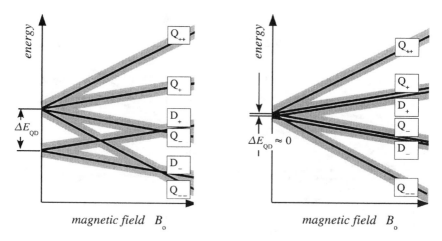

Figure 23 Energy diagram of electronic spin states of a triplet–doublet pair in a magnetic field. Systems with large (left) and small (right) values of the quartet–doublet splitting energies ΔE_{QD}. Eigenstate energies are not sharply defined because ΔE_{QD} depends on the random orientation (in liquid solutions) of the $^3A^*$ and A^- species within the activated complex $^3A^*\cdots A^-$.

up-conversion rate. It should be noted, however, that a simple monotonic decrease of the spin-up conversion rate is expected only for systems with very small energy splitting at zero field. Generally the state mixing is maximal when the Zeeman and other magnetic interactions (e.g., hyperfine interactions) present in the given system are of a similar magnitudes (cf. appropriate energy level schemes).

Application of the external magnetic field may in principle influence all classes of the spin up-conversion processes involved in an ECL system. Thus the observed effects (changes in the ECL intensity as a function of the applied external magnetic field) are a superposition of all changes occurring in the spin up-conversion kinetics. In the simplest cases, when only processes occurring before an intrinsic electron transfer step are taken into account, one can expect a lack of any intrinsic effects, as was briefly discussed in Section IV.B. Kinetic considerations (Sections II.C and II.D) indicate that the spin up-conversion between two forms of the activated complex, $^1(A^-\cdots D^+)$ and $^3(A^-\cdots D^+)$, is governed by the sum of the k_{dis} and k_{TS} rates. Thus, the eventual decrease (or increase) in the k_{TS} values leads to much less pronounced changes in the $k_{dis} + 4k_{TS}$ sum. Lack of a magnetic field effect reported [110] for the ECL system involving 9,10-diphenylanthracene agrees nicely with the above conclusion.

However (see Section IV.C), magnetic field effects should be much more pronounced for organic ECL systems following the T route, where both triplet–triplet annihilation and triplet–doublet quenching reactions are expected to be influenced by an external magnetic field. Effects of this kind have been reported by Bard and coworkers [111–114] for the organic ECL systems with the ion annihilation energies insufficient to populate the excited singlet states directly. A monotonic increase in the ECL intensity (up to 30% at a field strength of ~8000 G) was found, although independent measurements [115] have indicated that the triplet–triplet annihilation efficiency is inhibited at all magnetic field strengths up to 10^4 G. Evidently the magnetic field induced a decrease in the rate of the triplet–doublet quenching prevailing in the ECL systems studied. This conclusion has been further supported in the studies of the magnetic field effect on the triplet–doublet quenching processes [116].

The fact that organic ECL systems exhibit different types of behavior suggests that the influence of an external magnetic field on the ECL efficiency can be used as an additional classification criterion that allows one to distinguish between S and T routes. Thus, in ECL systems that are not too strongly energy-deficient (rubrene cation + rubrene anion), changes in the solvent or supporting electrolyte can cause a shift from the T route to the S route, which is borne out in changes in the observed magnetic field effect [114,117,118]. It should be noted, however, that there are some exceptions to the general rule, e.g., the mixed ECL system involving thianthrene and 2,5-diphenyl-1,3,4-oxadiazole precursors [119], where no magnetic effect was observed. Thus one should be careful in interpreting magnetic field effects on ECL efficiencies [120].

V. CONCLUDING REMARKS

Electrochemiluminescence can be used as a tool for studying the kinetics and mechanism of electron transfer reactions, which, unfortunately, is rather difficult task. The combined requirements of reductant and oxidant chemical stability (in the presence of electrodes, electrolyte, and solvent) and a lack of chemical complications following the initial electron transfer to and from the electrode still pose a problem. Thus the chemistry occurring in solution after electrolysis must be examined very carefully. The photochemical stability and desired high fluorescence efficiencies of the generated excited states are an additional problem. All these requirements drastically limit the types of compounds suitable for use in the quantitative studies of the ECL phenomenon. The mentioned complications may lead to misinterpretation of the experimental results. On the other hand, if all of the interferences are removed by the appropriate conditions of the experiment and if only simple ion annihilation takes place during the ECL process without any competitive reactions, the obtained data (ECL efficiencies) allow for a quantitative discussion of the electron transfer excitation.

As discussed above, the Marcus model for ECL processes can be used for qualitative as well as quantitative descriptions of such kinds of electron transfer reactions. However, a more sophisticated approach, taking into account the vibronic excitation in the reaction products (important in the Marcus inverted region), solvent molecular dynamics (important in the case of large values of the electronic coupling elements), and eventual changes in the electron transfer distance, seems to be necessary. Moreover, additional kinetic complications caused by diffusional limitations as well as by the presence of spin up-conversion processes should also be taken into account in a more detailed discussion. It seems to be clear that, because of the large number of parameters going into the theoretical description, a quantitative description of the ECL phenomenon remains an open question. This is also true for the relatively well understood ECL systems involving ruthenium(II) complexes (see Refs. 30–32).

The Marcus model may also be applied in principle for predicting ECL behavior (i.e., spectral characteristics and ECL efficiencies) using information from pertinent electrochemical, spectroscopic, and photophysical data of the A and D molecules. This particular aspect of the theoretical descriptions of ECL offers the possibility of designing new ECL systems with extremely high efficiencies. At the present stage, however, it is only a very promising possibility. One can trust that it will really be possible when the roles of all factors affecting electron transfer excitation are much better recognized. Further quantitative investigations (e.g., of temperature effects on ECL efficiencies) are necessary to give a decisive answer to remaining questions and doubts, especially concerning the adequacy of a simple ion-pair approximation for the ECL activated complex.

ACKNOWLEDGMENTS

This work was a part of 7T09A11620 Research Project sponsored by the Committee of Scientific Research.

REFERENCES

1. Weller, A.; Zachariasse, K.A. Chemiluminescence from radical ion recombination. In *Molecular Luminescence*; Lim, E.C., Ed.; Benjamin: New York, 1969; pp 859–905.
2. Zachariasse, K.A. Exciplexes in chemiluminescent radical ion recombination. In *The Exciplex*; Gordon, M., Ware, W.R., Eds.; Academic Press: New York, 1975; pp 275–303.
3. Faulkner, L.R.; Bard, A.J. Techniques of electrogenerated chemiluminescence. Electroanal. Chem. **1977**, *10*, 1–95.

4. Bard, A.J.; Santhanam, K.S.V.; Cruser, S.A.; Faulkner, L.R. Electrogenerated chemiluminescence. In *Fluorescence*; Guibault, G.G., Ed.; Marcell Dekker: New York, 1967; pp 627–651.

5. Marcus, R.A. On the theory of chemiluminescent electron-transfer reactions. J. Chem. Phys. **1965**, *43*, 2654–2657.

6. Burshtein, A.I. Unified theory of photochemical charge separation. Adv. Chem. Phys. **2000**, *114*, 419–587.

7. Burshtein, A.I.; Neufeld, A.A.; Ivanov, K.L. Reversible electron tranfer in photochemistry and electrochemistry. J. Chem. Phys. **2001**, *115*, 2652–2663.

8. Burshtein, A.I.; Neufeld, A.A.; Ivanov, K.L. Fluorescence and phosphorescence resulting from electrochemical generation of triplet excitation. J. Chem. Phys. **2001**, *115*, 10464–10471.

9. Gould, I.R.; Young, R.H.; Moody, R.E.; Farid, S. Contact and solvent-separated geminate radical ion pairs in electron-transfer photochemistry. J. Phys. Chem. **1991**, *95*, 2068–2080.

10. Gould, I.R.; Ege, D.; UMoser, J.F.; Farid, S. Efficiences of photoinduced electron-transfer reactions: role of the inverted Marcus region in return electron transfer within geminate radical-ion pairs. J. Am. Chem. Soc. **1990**, *112*, 4290–4301.

11. Bock, C.R.; Connor, A.R.; Gutierrez, A.R.; Meyer, T.J.; Whitten, D.G.; Sullivan, B.P.; Nagle, J.K. Estimation of the excited state redox potentials by electron-transfer quenching. Application of electron-transfer theory to excited-state redox processes. J. Am. Chem. Soc. **1979**, *101*, 4815–4823.

12. Birks, J.B. Reversible photophysical processes. Nouv. J. Chim. **1977**, *1*, 453–459.

13. Kapturkiewicz, A. Marcus theory in the qualitative and quantitative description of electrochemiluminescence phenomena. Adv. Electrochem. Sci. Eng. **1997**, *5*, 1–60.

14. Jensen, T.J.; Gray, H.B.; Winkler, J.R.; Kuznetsov, A.M.; Ulstrup, J. Dynamic ionic strength effects in fast bimolecular electron transfer between a redox metalloprotein of high electrostatic charge and an inorganic reaction partner. J. Phys. Chem. B **2000**, *104*, 11556–11562.

15. Park, S.-M.; Bard, A.J. Electrogenerated chemiluminescence. The production of benzophenone phosphorescence in fluid solution by radical ion reaction. Chem. Phys. Lett. **1976**, *38*, 257–262.

16. Hemigway, R.E.; Park, S.-M.; Bard, A.J. Electrogenerated chemiluminescence. XXI. Energy transfer from an exciplex to a rare earth chelate. J. Am. Chem. Soc., **1975**, *97*, 200–201.

17. Kapturkiewicz, A.; Herbich, J.; Nowacki, J. Highly efficient electrochemical generation of the fluorescent intramolecular charge-transfer states. Chem. Phys. Lett. **1997**, *275*, 355–362.

18. Saltiel, J.; Atwater, B.W. Spin-statistical factors in diffusion-controlled reactions. Adv. Photochem. **1988**, *14*, 1–90.

19. Hoytink, G.J. Electrochemiluminescence of aromatic hydrocarbons. Disc. Faraday Soc. **1968**, *45*, 14–22.

20. Hayashi, H.; Nagakura, S.; The e.s.r. and phosphorescence spectra of some dicyanobenzene complexes with methyl-substituted benzenes. Mol. Phys. **1970**, *19*, 45–53.

21. Herbich, J.; Kapturkiewicz, A.; Nowacki, J. Phosphorescent intramolecular charge-transfer states. Chem. Phys. Lett. **1996**, *262*, 633–643.

22. Rice, S.A. Diffusion-controlled reactions. In *Comprehensive Chemical Kinetcs*; Bauford, C.H., Tipper, C.F.H., Compton, R.G., Eds.; Elsevier: Amsterdam, 1985; Vol. 25. pp. 1–400.

23. Eigen, M. Über die Kinetik sehr schnell verlafender Ionenreaktionen in wässeriger Lösung. Z. Phys. Chem. NF **1954**, *1*, 176–200.

24. Fuoss, R.M. Ionic association. III. The equilibrium between ion pairs and free ions. J. Am. Chem. Soc. **1958**, *80*, 5059–5061.

25. Rosseinsky, D.R. Reactant approach: the spherical shell model for molecular juxtaposition. Comments Inorg. Chem. **1984**, *3*, 53–70.

26. Bard, A.J.; Park, S.-M. Exciplexes in electrogenerated chemiluminescence. In *The Exciplex*; Gordon, M., Ware, W.R., Eds.; New York: Academic Press, 1975; pp. 305–326.

27. Park, S.-M.; Bard, A.J. Electrogenerated chemiluminescence. XXII. On the generation of exciplexes in the radical ion reactions. J. Am. Chem. Soc. **1975**, *97*, 2978–2985.

28. Rehm, D.; Weller, A. Kinetic und Mechanismus der Elektronübertragung bei der Fluoreszenzlöschung in Acetonitril. Ber. Bunsenges. Phys. Chem. **1969**, *73*, 834–839.

29. Hofmann, M.Z.; Boletta, F.; Moggi, L.; Hug, G.L. Rate constants for the quenching of the excited states of metal complexes in fluid solution. J. Phys. Chem. Ref. Data. **1989**, *16*, 219–544.

30. Szrebowaty, P.; Kapturkiewicz, A. Free energy dependence on tris(2,2′-bipyridine)ruthenium(II) electrochemiluminescence efficiency. Chem. Phys. Lett. **2000**, *328*, 160–168.

31. Kapturkiewicz, A.; Szrebowaty, P.; Angulo, G.; Grampp, G. Electron transfer quenching and electrochemiluminescence comparative studies of the system containing N-methylpyridinium cations and Ru(2,2′-bipyridine)$_3$$^{2+}$ and Ru(1,10′-phenanthroline)$_3$$^{2+}$ complexes. J. Phys. Chem. A **2002**, *105*, 1579–1685.

32. Kapturkiewicz, A.; Szrebowaty, P. Electrochemically generated chemiluminescence of the tris(2,2′-bipyridine)ruthenium(II), tris(1,10-phenanthroline)ruthenium(II) and tris(4,7-diphenyl-1,10-phenanthroline)ruthenium(II) complexes. J. Chem. Soc. Dalton Trans. **2002**, 3219–3225.

33. Mussell, R.D.; Nocera, D.G. Effect of long-distance electron transfer on chemiluminescence efficiencies. J. Am. Chem. Soc. **1988**, *110*, 2764–2772.

34. Mussell, R.D.; Nocera, D.G. Partitioning of the electrochemical excitation energy in the electrogenerated chemiluminescence of hexanuclear molybdenum and tungsten clusters. Inorg. Chem. **1990**, *29*, 3711–3717.

35. Mussell, R.D.; Nocera, D.G. Role of solvation in the electrogenerated chemiluminescence of hexanuclear molybdenum cluster ion. J. Phys. Chem. **1991**, *95*, 6919–6934.

36. Kapturkiewicz, A. Solvent and temperature control of the reaction mechanism and efficiency in the electrogenerated chemiluminescence of rubrene. J. Electroanal. Chem. **1994**, *372*, 101–116.

37. Kapturkiewicz, A. Electrochemical generation of excited TICT states. Part V. Evidence of inverted Marcus region. Chem. Phys. **1992**, *166*, 259–273.

38. Bockris, J.O.M.; Khan, S.U.M. *Quantum Electrochemistry*, Plenum: New York, 1979.

39. Dogonadze, R.R.; Kuznetsov, A.M.; Marsagishvili, T.A. The present state of the theory of charge transfer processes in condensed phase. Electrochim. Acta **1980**, *25*, 1–28.

40. Cannon, R.D. *Electron Transfer Reactions*, Butterworths: London, 1980.

41. Sutin, N. Nuclear, electronic and frequency factors in electron-transfer reactions. Acc. Chem. Res. **1982**, *15*, 275–282.

42. Eberson, L. Electron-transfer reactions in organic chemistry. Adv. Phys. Org. Chem. **1982**, *18*, 79–185.

43. Sutin, N. Theory of electron transfer reactions: insights and hindsights. Prog. Inorg. Chem. **1983**, *30*, 441–498.

44. Marcus, R.A.; Sutin, N. Electron transfer in chemistry and biology. Biochim. Biophys. Acta **1985**, *811*, 265–322.

45. Newton, M.D.; Sutin, N.; Electron transfer reactions in condensed phases. Ann. Rev. Phys. Chem. **1984**, *35*, 437–480.

46. Newton, M.D. Quantum chemical probes of electron transfer kinetics: the nature of donor-acceptor interactions. Chem. Rev. **1991**, *91*, 767–792.

47. Fawcett, R.W.; Opallo, M. The kinetics of heterogeneous electron transfer in polar solvents. Angew. Chem. Int. Ed. **1994**, *33*, 2131–2143.

48. *Photoinduced Electron Transfer Parts A–D*, Fox, M.A., Chanon, M., Eds.; Elsevier: Amsterdam, 1998.

49. Barzykin, A.V.; Frantsuzov, P.A.; Saki, K.; Tachiya, A. Solvent effects in nonadiabatic electron-transfer reactions. Theoretical aspects. Adv. Chem. Phys. **2002**, *123*, 511–616.

50. Fischer, S.F.; Van Duyen, R.I. On the thory of electron transfer reactions. The naphthalene$^-$/TCNQ system. Chem. Phys. **1977**, *26*, 9–16.

51. Ulstrup, J.; Jortner, J. The effect of intramolecular quantum modes on free energy relationship for electron transfer reactions. J. Chem. Phys. **1975**, *63*, 4358–4368.

52. Kestner, N.R.; Logan, J.; Jortner, J. Thermal electron transfer reactions in polar solvents. J. Phys. Chem. **1974**, *78*, 2148–2166.

53. Webman, I.; Kestner, N.R. Low temperature free energy relations for electron transfer reactions. J. Chem. Phys. **1982**, *77*, 2387–2398.

54. Siders, P.; Cave, R.J.; Marcus, R.A. A model for orientation effects in electron-transfer reactions. J. Chem. Phys. **1984**, *81*, 5613–5624.

55. Peover, M.E.; Powell, J.S. Dependence of electrode kinetics on molecular structure. Nitrocompounds in dimethylformamide. J Electroanal Chem. **1969**, *20*, 427–433.

56. German, E.D.; Kuznetsov, A.M. Outer sphere energy of reorganization in charge transfer processes. Electrochim. Acta **1981**, *26*, 1595–1608.

57. Fawcett, W.R.; Kharkats, Y.I. Estimation of the free energy of activation for electron transfer reactions involving dipolar reactants and products. J. Electroanal. Chem. **1973**, *47*, 413–418.

58. Grampp, G.; Jaenicke, W. Kinetics of diabatic and adiabatic electron exchange in organic systems. Comparison of theory and experiment. Ber. Bunsenges. Phys. Chem. **1991**, *95*, 904–927.

59. Grampp, G.; Cebe, M.; Cebe, E. An improved bond-order/bond-length relation for the calculation of inner sphere reorganization energies of organic electron transfer reactions. Z. Phys. Chem. N.F. **1990**, *166*, 93–101.

60. Cosa, G.; Chesta, C.A. Estimation of the bimolecular rate constant for exciplex formation from the analysis of its emission spectrum. J. Phys. Chem. A. **1997**, *101*, 4922–4928.

61. Zusman, L.D. Outer-sphere electron transfer in polar solvents. Chem. Phys. **1980**, *49*, 295–304.

62. Zusman, L.D. The theory of transition between electronic states. Application to radiationless transition in polar solvent. Chem. Phys. **1983**, *80*, 29–43.

63. Alexandrov, I.V. Physical aspects of charge transfer theory. Chem. Phys. **1980**, *51*, 449–457.

64. Calef, D.F.; Wolynes, P.G. Classical solvent dynamics and electron transfer. I. Continuum theory. J. Phys. Chem. **1983**, *87*, 3387–3400.

65. Calef, D.F.; Wolynes, P.G. Classical solvent dynamics and electron transfer. II. Molecular aspects. J. Chem. Phys. **1983**, *78*, 470–482.

66. Bagchi, B. Dynamics of solvation and charge transfer reactions in dipolar liquids. Ann. Rev. Phys. Chem. **1989**, *40*, 115–1141.

67. Fröhlich, H. Theory of Dielectrics. Oxford Univ. Press: Oxford, 1958.

68. Rips, I.; Jortner, J. Dynamic solvent effects on outer-sphere electron transfer. J. Chem. Phys. **1987**, *87*, 2090–2104.

69. Kapturkiewicz, A.; Herbich, J.; Karpiuk, J.; Nowacki, J. Intramolecular radiative and radiationless charge recombinations in donor-acceptor carbazole derivatives. J. Phys. Chem. A. **1997**, *101*, 2323–2344.

70. Herbich, J.; Kapturkiewicz, A. Electronic structure and molecular conformation in the excited charge transfer singlet states of p-acridyl and other aryl derivatives of aromatic amines. J. Am. Chem. Soc. **1998**, *120*, 1014–1029.

71. Efrima, S.; Bixon, M. On the role of vibrational excitation in electron transfer reactions with large negative free energies. Chem. Phys. Lett. **1974**, *25*, 34–37.

72. Efrima, S.; Bixon, M. Vibrational effects in outer-sphere electron transfer reactions in polar media. Chem. Phys. **1976**, *13*, 447–460.

73. Miller, J.R.; Beitz, J.V.; Huddleston, R.H. Effect of free energy on rates of electron transfer between molecules. J. Am. Chem. Soc. **1984**, *106*, 5057–5068.

74. Miller, J.R.; Calcaterra, L.T.; Closs, G.L. Intramolecular long-distance electron transfer in radical anions. The effect of free energy and solvent on the reaction rates. J. Am. Chem. Soc. **1984**, *106*, 3047–3049.

75. Miller, J.R.; Calcaterra, L.T.; Closs, G.L. Distance, stereoelectronic effects and the Marcus inverted region in intramolecular electron transfer in organic radical anions. J. Phys. Chem. **1986**, *90*, 3673–3683.

76. Wasielewski, M.P.; Niemczyk, M.P.; Swec, W.A.; Pewitt, E.B. Dependence of rate constants for photoinduced charge separation and dark charge recombination on the free energy of reaction in restricted-distance porphyrin-quinone molecules. J. Am. Chem. Soc. **1985**, *107*, 1080–1082.

77. Johnson, M.D.; Miller, J.R.; Green, N.J.; Closs, G.L. Distance dependence of intramolecular hole and electron transfer in organic radical ions. J. Phys. Chem. **1989**, *93*, 1173–1178.

78. Irvine, M.P.; Harrison, R.J.; Beddard, G.S.; Leighton, P.; Sanders, J.K.M. Detection of the inverted region in the photo-induced intramolecular electron transfer of capped porphyrines. Chem. Phys. **1986**, *104*, 315–324.

79. Ohno, R.; Yoshimura, A.; Shioyama, H.; Mataga, N. Energy gap dependence of spin-inverted electron transfer within geminate radical pairs formed by the quenching of phosphorescent states in polar solvents. J. Phys. Chem. **1987**, *91*, 4365–4370.

80. Bixon, M.; Jortner, J.; Non-Arrhenius temperature dependence of electron-transfer rates. J. Phys. Chem. **1991**, *95*, 1941–1944.

81. Jortner, J. Temperature dependent activation energy for electron transfer between biological molecules. J. Chem. Phys. **1878**, *64*, 4860–4867.

82. Jortner, J.; Bixon, M. Intramolecular vibrational excitations accompanying solvent-controlled electron transfer reactions. J. Chem. Phys. **1988**, *88*, 167–170.

83. Wallace, W.L.; Bard, A.J. Electrogenerated chemiluminescence. 35. Temperature dependence of the ECL efficiency of $Ru(bpy)_3^{2+}$ in acetonitrile and evidence for very high excited state yields from electron transfer reactions. J. Phys. Chem. **1979**, *83*, 1350–1357.

84. Ritchie, E.L.; Pastore, P.; Wightman, R.M. Free energy control of reaction pathways in electrogenerated chemiluminescence. J. Am. Chem. Soc. **1997**, *119*, 11920–11925.

85. Murov, S.L.; Carmichael, I.; Hug, G.L. *Handbook of Photochemistry*, Marcel Dekker: New York, 1993; pp. 1–98.

86. Hush, N.S. Intervalence-transfer absorption. Part 2. Theoretical considerations and spectroscopic data. Prog. Inorg. Chem. **1967**, *8*, 391–444.

87. Hush, N.S. Homogeneous and heterogeneous optical and thermal electron transfer. Electrochim. Acta **1968**, *13*, 1005–1023.

88. Marcus, R.A. Relation between charge-transfer absorption and fluorescence spectra and the inverted region. J. Phys. Chem. **1989**, *93*, 3078–3086.

89. Mulliken, R.S. Molecular compounds and their spectra. II. J. Am. Chem. Soc. **1952**, *74*, 811–824.

90. Murrell, J.N. Molecular complexes and their spectra. IX. The relationship between the stability of a complex and the intensity of its charge-transfer bands. J. Am. Chem. Soc. **1959**, *81*, 5037–5043.

91. Beens, H.; Weller, A. Excited molecular π-complexes in solution. In *Organic Molecular Photophysics*; Birks, J.B., Ed.; Wiley: New York, 1975; Vol. 2, pp.159–215.

92. Mataga, N. Electronic structures and dynamical behaviour of some exciplex systems. In *The Exciplex*; M Gordon and W R Ware, Eds.; Academic Press: New York, 1975; pp.113–144.

93. Gould, I.R.; Noukakis, D.; Gomez-Jahn, L.; Young, R.H.; Goodman, J.L.; Farid, S. Radiative and non-radiative electron transfer in contact radical-ion pairs. Chem. Phys. **1993**, *176*, 439–456.

94. Borowicz, P.; Herbich, J.; Kapturkiewicz, A.; Opallo, M.; Nowacki, J. Radiative and nonradiative electron transfer in donor-acceptor phenoxazine and phenothiazine derivatives. Chem. Phys. **1999**, *249*, 49–62.

95. Förster, T. Fluorescenz organischer Verbindungen. Vandenhoeck & Ruprecht: Göttingen, 1951; pp. 67–83.

96. Faulkner, L.R. Chemiluminescence in the liquid phase electron transfer. Int. Rev. Sci. Phys. Chem. Ser. Two **1976**, *9*, 213–263.

97. Salikhov, K.M.; Mollin, Y.N.; Sagdeev, R.Z.; Buchachenko, A.L. Spin Polarization and Magnetic Effects in Radical Reactions. Elsevier: Amsterdam, 1984.

98. Merrifield, R.E. Magnetic effects on triplet exciton interactions. Pure. Appl. Chem. **1971**, *27*, 481–498.

99. Swenberg, C.E.; Geacintov, N.E. Exciton interactions in organic solids. In *Organic Molecular Photophysics*, Birks, J.B., Ed.; Wiley: New York, 1973; Vol. 1, pp. 489–564.

100. Avakian, P. Influence of magnetic field on luminescence involving triplet excitons. Pure. Appl. Chem. **1974**, *37*, 1–19.

101. Geacintov, N.E.; Swenberg, C.E. Magnetic field effects in organic spectroscopy. In *Luminescence Spectroscopy*; Lumb, M.D., Ed.; Academic: London, **1978**, pp. 239–298.

102. Gould, I.R.; Turro, N.J.; Zimmt, N.J. Magnetic field and magnetic isotope effects on the product of organic reactions. Adv. Phys. Org. Chem. **1984**, *20*, 1–53.

103. Steiner, U.E.; Ulrich, T. Magnetic field effects in chemical kinetics and related phenomena. Chem. Rev. **1989**, *89*, 51–147.

104. Schulten, K.; Wolynes, P.G. Semiclassical description of electron spin motion in radicals including the effect of electron hopping. J. Chem. Phys. **1978**, *68*, 3292–3297.

105. Weller, A.; Nolting, F.; Staerk, H. A quantitative interpretation on the magnetic effect on hyperfine-coupling-induced triplet formation from radical ion pairs. Chem. Phys. Lett. **1983**, *96*, 24–27.

106. Molin, Y.N.; Shalikov, K.M.; Zamaraev, K.I. Spin Exchange. Springer: Berlin, 1982.

107. Bürssner, D.; Wolff, H.J.; Steiner, U.E. Magnetic spin effects on photooxidation quantum yield of RuII-tris(bipyridine) type complexes in magnetic fields up to 17.5 tesla. Angew. Chem. Int. Ed. Engl. **1994**, *33*, 1772–1775.

108. Wolff, H.J.; D Bürssner, Steiner, U.E. Spin-orbit coupling controlled spin chemistry of $Ru(bipy)_3^{2+}$ photooxidation: detection of strong viscosity dependence of in-cage backward electron transfer rate. Pure. Appl. Chem. 67:167–174, 1995.

109. Hötzer, K.A.; Klingert, A.; Klumpp, T.; Krissinel, E.; Bürssner, D.; Steiner, U.E. Temperature-dependent spin relaxation: a major factor in electron backward transfer following the quenching of $*Ru(bpy)_3^{2+}$ by methyl viologen. J. Phys. Chem. A **2002**, *106*, 2207–2217.

110. Faulkner, L.R.; Bard, A.J. Electrogenerated chemiluminescence. IV. Magnetic field effect on the electrogenerated chemiluminescence of some anthracenes. J. Am. Chem. Soc. **1969**, *91*, 209–210.

111. Faulkner, L.R.; Tachikawa, H.; Bard, A.J. Electrogenerated chemiluminescence. VII. The influence of an external magnetic field on luminescence intensity. J. Am. Chem. Soc. **1972**, *94*, 691–699.

112. Tachikawa, H.; Bard, A.J. Electrogenerated chemiluminescence. XII. Magnetic field effects on ECL in the tetracene-TMPD system; evidence for triplet-triplet annihilation of tetracene. Chem. Phys. Lett. **1973**, *19*, 287–289.

113. Keszthelyi, C.P.; Tokel-Takvoryan, N.E.; Tachikawa, H.; Bard, A.J. Electrogenerated chemiluminescence. XIV. Effect of supporting electrolyte concentration and magnetic field effects in the 9,10-dimethylanthracene–tri-p-tolylamine system in tetrahydrofuran. Chem. Phys. Lett. **1973**, *23*, 219–222.

114. Tachikawa, H.; Bard, A.J. Electrogenerated chemiluminescence. Effect of solvent and magnetic field on ECL of rubrene systems. Chem. Phys. Lett. **1974**, *26*, 246–251.

115. Faulkner, L.R.; Bard, A.J. Magnetic field effects on anthracene triplet-triplet annihilation in fluid solutions. J. Am. Chem. Soc. **1969**, *91*, 6495–6497.

116. Tachikawa, H.; Bard, A.J. Effect of concentration and magnetic field on radical (Wurster's blue cation and benzoquinone anion) quenching of anthracene triplets in fluid solution. Chem. Phys. Lett. **1974**, *26*, 10–15.

117. Periasamy, N.; Santhanam, K.S.V. Studies of efficiencies of electrochemiluminescence of rubrene. Proc. Indian Acad. Sci. Sect. A **1974**, *80*, 194–202.

118. Glass, R.S.; Faulkner, L.R. Chemiluminescence from electron transfer between the anion and cation radicals of rubrene. Dominant S-route character at low temperatures. J. Phys. Chem. **1982**, *86*, 1652–1658.

119. Keszthelyi, C.P.; Tachikawa, H.; Bard, A.J. Electrogenerated chemiluminescence. VIII. The thianthrene–2,5-diphenyl-1,2,4-oxadiazole system. A mixed energy-sufficient system. J. Am. Chem. Soc. **1977**, *94*, 1522–1527.

120. Michael, P.R.; Faulkner, L.R. Electrochemiluminescence from the thianthrene – 2,5-diphenyl-1,3,4-oxadiazole system. Evidence for light production by the T-route. J. Am. Chem. Soc. **1977**, *99*, 7754–7761.

5
Coreactants

Wujian Miao and Jai-Pil Choi
The University of Texas at Austin, Austin, Texas, U.S.A.

I. INTRODUCTION

Electrogenerated chemiluminescence (ECL) can be generated by annihilation reactions between oxidized and reduced species produced at a single electrode by using an alternating potential, or at two separate electrodes in close proximity to each other by holding one electrode at a reductive potential and the other at an oxidative potential [1]. Because annihilation reactions are very energetic (typically 2–3 eV) and potential windows for aqueous solutions are generally too narrow to allow convenient electrolytic generation of both the oxidized and reduced ECL precursors, most annihilation ECL processes have been investigated in organic solvents or partially organic solutions [2]. By adding certain species, called coreactants, into solutions containing luminophore species, ECL can also be generated with a single potential step or one directional potential scanning at an electrode; this permits ECL to be observed in aqueous solutions [1]. Depending on the polarity of the applied potential, both the luminophore and the coreactant species can first be oxidized or reduced at the electrode to form radicals, and intermediates formed from the coreactant then decompose to produce a powerful reducing or oxidizing species that reacts with the oxidized or reduced luminophore to produce the excited states that emit light. Because highly reducing intermediate species are generated after an electrochemical oxidation of a coreactant, or highly oxidizing ones are produced after an electrochemical reduction, the corresponding ECL reactions are often referred to as "oxidative reduction" ECL and "reductive oxidation" ECL, respectively [2,3]. Thus, a coreactant is a species that, upon electrochemical oxidation or reduction, immediately undergoes chemical decomposition to form a strong reducing or oxidizing intermediate that can react with an oxidized or reduced ECL luminophore to generate excited states. Common

213

coreactants for oxidative reduction ECL are oxalate (See Section II.A) and tertiary amines (See Section II.C). Upon oxidation, the oxalate ion loses CO_2, producing $CO_2^{\bullet-}$, a strongly reducing agent, whereas tertiary amines deprotonate to yield strongly reducing radical species. A typical example of coreactants used for reductive oxidation ECL would be peroxydisulfate (Section II.B), which, upon reduction, forms $SO_4^{\bullet-}$, a strong oxidant. Clearly, unlike annihilation ECL, where all starting species can be regenerated after light emission, coreactant ECL can regenerate luminophore species only while the coreactant is consumed via the ECE [1] reactions.

Coreactant ECL has been used in a wide range of analytical applications [4–12] (See also Chapters 7 and 8). Because the ECL emission intensity is usually proportional to the concentration of the coreactant and emitter, ECL can be used in the analysis of various species, in which either the coreactant, the emitter, or a species tagged with the emitter can be analyzed [4].

An understanding of the mechanism of the coreactant ECL system is important in designing and selecting new coreactants and luminophore species and in improving the sensitivity and reproducibility of the ECL system. In this chapter, commonly used coreactants and relevant ECL mechanisms are reviewed. Although there are a wide variety of molecules that exhibit ECL, the overwhelming majority of publications concerned with coreactant ECL and its analytical applications are based on chemistry involving tris(2,2′-bipyridine)ruthenium(II), bipyridine)ruthenium(II), $Ru(bpy)_3^{2+}$ (bpy = 2,2′-bipyridine), or closely related analogs as the emitting species, because of their excellent chemical, electrochemical, and photochemical properties even in aqueous media and in the presence of oxygen [13]. Consequently, much of this chapter concerns $Ru(bpy)_3^{2+}$/coreactant ECL systems.

II. GENERAL REVIEW OF COREACTANTS

A. Oxalate System

In an attempt to determine the postulated intermediate dioxetanedione from the so called peroxyoxalate chemiluminescent reaction [14], in 1977 Bard's group [15] studied the electrochemical oxidation of oxalate ion, $C_2O_4^{2-}$, in MeCN. Although no evidence of the dioxetanedione as an intermediate was found in these studies, a new chemiluminescent reaction was discovered when oxalate and a fluorescent compound were simultaneously oxidized at a platinum electrode. This was, in fact, the first account of coreactant ECL reactions. In a subsequent ECL report primarily based on $Ru(bpy)_3^{2+}$ and oxalate in aqueous solutions, Rubinstein and Bard [2] referred to the coupled reactions of oxalate, which involve first an electrochemical oxidation and then a chemical decomposition of the product to form a highly reducing intermediate ($CO_2^{\bullet-}$, $E° = -1.9$ V vs. NHE [16]) that reacts with

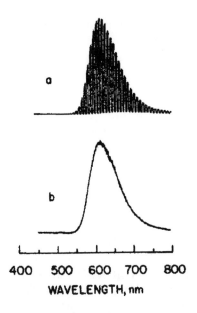

Figure 1 (a) Spectrum of ECL emission at Pt electrode in a solution of 1.0 mM Ru(bpy)$_3$(ClO$_4$)$_2$ and 50 mM Na$_2$C$_2$O$_4$ at pH 5.0, with cyclic square wave excitation, between 0.2 and 0.7 V vs mercury sulfate electrode (MSE) at 0.2 Hz. (b) Luminescence emission spectrum of 1.0 mM Ru(bpy)$_3$(ClO$_4$)$_2$ in 0.1 M H$_2$SO$_4$, with photoexcitation at 500 nm. (From Ref. 2.)

Ru(bpy)$_3^{3+}$ produced electrochemically at the electrode to give an ECL signal, as "oxidative-reductions." Since that time, this kind of ECL has been often referred to as oxidative reduction coreactant ECL.

In aqueous solutions, oxalate is irreversibly oxidized to CO$_2$ at Pt electrodes at potentials within the platinum oxide region. This oxidation is known to be completely suppressed on an oxide-covered Pt electrode surface [2,17–20]; however, on such an electrode the reversible Ru(bpy)$_3^{2+/3+}$ couple is readily observed [2]. Carbon electrodes (both pyrolytic graphite and glassy carbon) show behavior similar to that of oxide-covered Pt electrodes toward oxalate and Ru(bpy)$_3^{2+}$ oxidation; i.e., there is a very large over potential for oxalate oxidation on carbon compared to that on reduced Pt. Moreover, at an oxidized Pt electrode, in the presence of oxalate, the anodic current for Ru(bpy)$_3^{2+}$ was larger and the cathodic current on the reverse scan was smaller than in its absence, suggesting that the catalytic reaction of Ru(bpy)$_3^{3+}$ with oxalate occurred and that under these conditions oxalate is not oxidized directly at the electrode.

When a potential that is more positive than the potential for the oxidation of Ru(bpy)$_3^{2+}$ is applied to a Pt or carbon electrode in an aqueous solution containing both $Ru(bpy)_3^{2+}$ and oxalate, orange light attributed to the excited state $Ru(bpy)_3^{2+}*$ is emitted from the electrode surface (Figs. 1 and 2). The ECL mechanism of this system was proposed to be as in Scheme 1 [2].

$$Ru(bpy)_3^{2+} - e \longrightarrow Ru(bpy)_3^{3+}$$

$$Ru(bpy)_3^{3+} + C_2O_4^{2-} \longrightarrow Ru(bpy)_3^{2+} + C_2O_4^{\bullet-}$$

$$C_2O_4^{\bullet-} \longrightarrow CO_2^{\bullet-} + CO_2$$

$$Ru(bpy)_3^{3+} + CO_2^{\bullet-} \longrightarrow Ru(bpy)_3^{2+}* + CO_2$$

$$Ru(bpy)_3^{2+} + CO_2^{\bullet-} \longrightarrow Ru(bpy)_3^+ + CO_2$$

$$Ru(bpy)_3^{3+} + Ru(bpy)_3^+ \longrightarrow Ru(bpy)_3^{2+}* + Ru(bpy)_3^{2+}$$

$$Ru(bpy)_3^{2+}* \longrightarrow Ru(bpy)_3^{2+} + h\nu$$

Scheme 1

First, the $Ru(bpy)_3^{2+}$ is oxidized at the electrode to the $Ru(bpy)_3^{3+}$ cation. This species is then capable of oxidizing the oxalate ($C_2O_4^{2-}$) in the diffusion layer close to the electrode surface to form an oxalate radical anion ($C_2O_4^{\bullet-}$). This breaks down to form a highly reducing radical anion ($CO_2^{\bullet-}$) and carbon dioxide. The reducing intermediate then either reduces the $Ru(bpy)_3^{3+}$ complex back to the parent complex in an excited state or reduces $Ru(bpy)_3^{2+}$ to form $Ru(bpy)_3^+$, which reacts with $Ru(bpy)_3^{3+}$ to generate the excited state $Ru(bpy)_3^{2+}*$, which emits light with λ_{max} ~620 nm.

In MeCN, however, oxalate has been shown to be easier to oxidize than the $Ru(bpy)_3^{2+}$ complex [15]. Thus, both reactants are oxidized at the electrode during light generation. The short-lived oxalate radical anion is then transformed into the carbon dioxide radical anion close to the electrode, where most of it is oxidized. That is, the direct contribution of oxidation of oxalate at the electrode in MeCN to the overall ECL is probably small.

The annihilation reaction is probably the main route for the generation of the excited state species in MeCN. For example, in MeCN, thianthrene (TH, Fig. 3a) forms a stable radical cation ($TH^{\bullet+}$) upon oxidation at about +1.25 V vs. SCE and is not reducible up to at least −2.4 V vs. SCE, whereas 2,5-diphenyl-1,3,4-oxadiazole (PPD, Fig. 3b) forms a stable anion ($PPD^{\bullet-}$) at −2.17 V vs. SCE [21]. The lack of emission during the oxidation of TH and oxalate in the absence of PPD but the formation of excited state TH* in the presence of PPD (Fig. 4), suggests that in this case the excited state TH* is most likely generated via the annihilation reaction between $TH^{\bullet+}$ cations electrochemically oxidized at the electrode from TH and $PPD^{\bullet-}$ anions produced via homogeneous reduction of PPD with $CO_2^{\bullet-}$, and it is unlikely that it is due to the reaction between $TH^{\bullet+}$ and $CO_2^{\bullet-}$, although the energy of the latter reaction should be great enough [15]. Other examples of annihilation reactions between the oxidized fluorescer

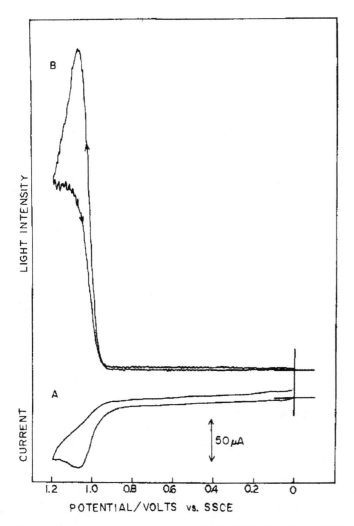

Figure 2 (A) Cyclic voltammogram for 1 mM Ru(bpy)$_3^{2+}$ in pH 5 phosphate buffer in the presence of 30 mM Na$_2$C$_2$O$_4$, at glassy carbon electrode, at a scan rate of 50 mV/s, (B) Light intensity vs. potential profile for the solution in (A). (From Ref. 22.)

D$^{\cdot+}$ at the electrode and the reduced fluorescer A$^{\cdot-}$ via CO$_2^{\cdot-}$ include D$^{\cdot+}$ [D = rubrene, DPA, TPP, bipyridyl chelates Ru(II) and Os(II)] and A$^{\cdot-}$ [A = rubrene, DPA, TPP, and chelates Ru(II) and Os(II)] [15].

In aqueous solutions, the ECL intensity of the Ru(bpy)$_3^{2+}$/oxalate system has been reported to have a maximum at ~pH 6 [2] (Fig. 5a) and also to be essentially constant from pH 4 to 8 [22,23] at macroelectrodes (Fig. 5b) and from

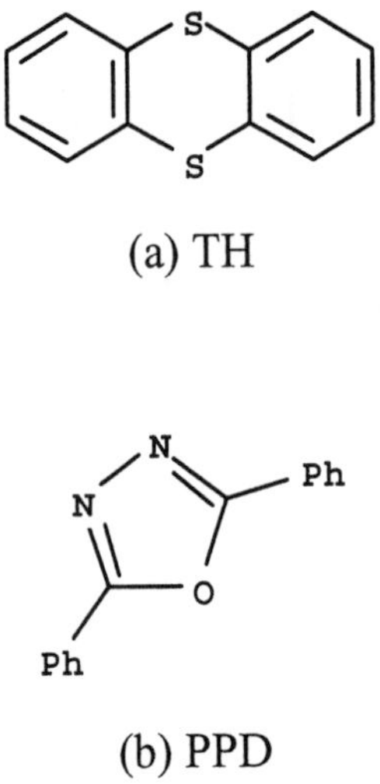

(a) TH

(b) PPD

Figure 3 Structures of (a) thianthrene (TH) and (b) 2,5-diphenyl-1,3,4-oxadiazole (PPD).

pH 5 to 8 at ultramicroelectrodes [24] (Fig. 5c). Note that the data shown in Figure 5a were measured in unbuffered solutions, so the pH at the electrode surface during oxidation was somewhat lower than in the bulk solution [2]. Addition of buffers, hence an increase in solution ionic strength, resulted in a decrease in ECL intensity [2,24]. As shown in Figure 5c, the intensity of steady emitted light correlates with the steady-state current at the ultramicroelectrode and thus to the catalytic efficiency of the reaction between $Ru(bpy)_3^{3+}$ and oxalate. That is, the ECL intensity is first governed by the first homogeneous oxalate oxidation, which acts as the rate-determining step for either the electrochemical current or the ECL measurements, as would be expected from the proposed mechanism shown in Scheme 1. Therefore, the pH and ionic strength dependence of the light intensity is approximately due to changes in the first homogeneous ET rate constant of these parameters, which can be attributed to the acid–base behavior of oxalate acid ($pK_{a1} = 1.23$; $pK_{a2} = 4.19$ at 25°C [25]) and the existence of an ion-pairing equilibrium preceding the ET process, respectively [24]. However, for Nafion- immobilized $Ru(bpy)_3^{2+}$ with oxalate in solution, ECL intensity was found to increase with increasing ionic strength; this was shown to be a phenomenon related to the Nafion film and not to the ECL reaction [23]. The decrease in the ECL intensity at high pH values was attributed to the oxygen evolution reaction at the electrode, which is shifted to less positive potentials with increasing pH [2].

a

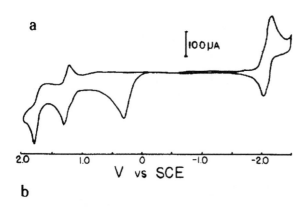

b

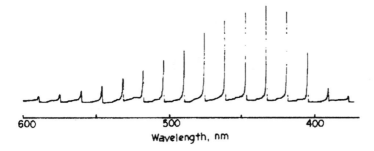

Figure 4   (a) Cyclic voltammogram of solution containing 2 mM each TH, PPD, and (TBA)$_2$Ox on 0.1 M TBAP-MeCN. (b) ECL spectrum with 5 s pulses between 0 and 1.3 V vs. SCE. No emission is observed under similar pulsing conditions in a solution lacking the PPD. (From Ref. 15.)

To examine the importance of the homogeneous ET in the overall electrochemical/ECL process, Kanoufi and Bard [24] further investigated ECL generation for different Ru(II) species $RuL'L_2^{2+}$, varying the driving force of the homogeneous ET. It was found that a large driving force results in a faster homogeneous ET and higher current and light intensities (Fig. 6). However, the luminescence emission of ECL is also related to the competition between the different pathways of the second electron transfer (oxidation of $CO_2^{\bullet-}$) [24].

The relationship between the electrochemical current and the ECL intensity has also been rationalized [24,26]. At an ultramicroelectrode, the steady emitted light I_{ls}, normalized by the initial Ru(bpy)$_3^{2+}$ concentration ($I_{ls}/[Ru^{2+}]^0$) varies linearly with the steady-state catalytic current i_s, normalized by the initial electrochemical current i_{s0} in the absence of oxalate (i_s/i_{s0}) (Fig. 7) [24]. Under ultrasonic agitation, Malins et al [26] found that the ECL intensity can be increased

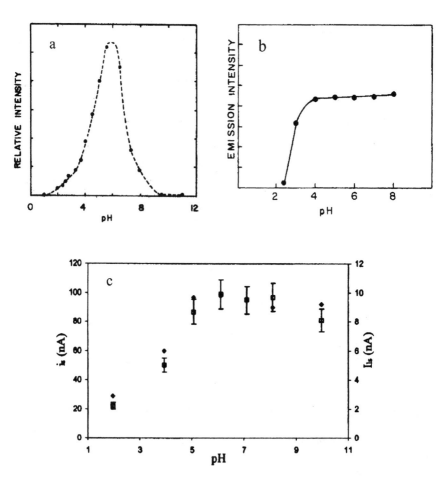

Figure 5 Dependence of ECL intensity on pH. The solution contained (a) 1.0 mM $Ru(bpy)_3^{2+}$ and 6.0 mM $H_2C_2O_4$, with the pH adjusted by addition of NaOH to 0.1 M H_2SO_4 (from Ref. 2) and (b) 0.6 mM $Ru(bpy)_3^{2+}$ and 27 mM $Na_2C_2O_4$ in phosphate buffer (from Ref. 22). (c) Variation of the plateau (□) light intensity I_{ls} and (◆) steady-state electrochemical current intensity i_s with pH obtained from an ultramicroelectrode. The solution contained 0.75 mM $Ru(bpy)_3^{2+}$ and 30 mM $C_2O_4^{2-}$ in 0.1 M NaCl + 0.1 M phosphate buffer. (From Ref. 24.)

more than 100% compared to that obtained under silent conditions and that the light intensity is linearly proportional to the square of the electrochemical current.

For the aqueous $Ru(bpy)_3^{2+}$/oxalate system, removal of oxygen from the solution is said to be unnecessary to observe ECL [2], but deaeration with

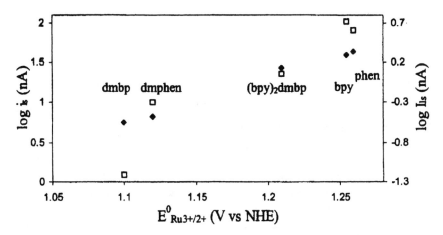

Figure 6 Variation of the logarithm of the plateau (□) light intensity I_{ls} and (◆) current intensity i_s with the standard oxidation potential of RuL L_2^{2+}. [RuL L_2^{2+}] ≈ 0.25 mM and [$C_2O_4^{2-}$] = 50 mM in 0.1 M NaCl + 0.1 M phosphate buffer, pH 6.1. (From Ref. 24.)

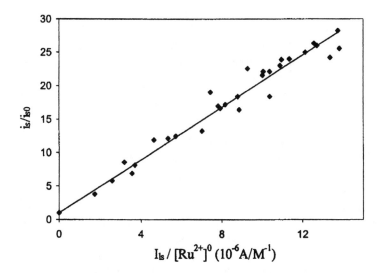

Figure 7 Variation of the normalized catalytic current i_s/i_{s0}, where i_{s0} is the current in the absence of oxalate, with the normalized plateau light intensity $I_{ls}/[Ru^{2+}]$ for 0.4 < [Ru(bpy)$_3^{2+}$] < 4 mM and 1 < [$C_2O_4^{2-}$] < 30 mM in 0.1 M NaCl + 0.1 M phosphate buffer, pH = 6.1. (From Ref. 24.)

nitrogen for several minutes increases the ECL intensity and improves reproducibility [22]. The ECL efficiency ϕ_{ECL}(photons emitted/Ru(bpy)$_3^{3+}$ generated) of this system is also affected by deaeration [2]. In aqueous deaerated solution of 1.0 mM Ru(bpy)$_3$(ClO$_4$)$_2$ and 50 mM Na$_2$C$_2$O$_4$, the estimated ϕ_{ECL} value is ~2%. In a nondeaerated solution, ϕ_{ECL} decreased to about one-third of this value.

Electrogenerated chemiluminescence from Ru(bpy)$_3^{2+}$ contained within Nafion films with oxalate dissolved in the solution has been used to study electrochemical behavior including charge and mass transport within polymer films and the ECL mechanism of the system [27,28] and to quantitate solution-phase analytes [23,29]. By combining ECL data with the usual electrochemical results, Rubinstein and Bard [28] were able to propose a model that accounts for essentially all of the experimental results obtained with the polymer film electrode. This model included oxidation of oxalate ions that penetrated into the Nafion layer mediated by Ru(bpy)$_3^{3+}$ formed upon oxidation of Ru(bpy)$_3^{2+}$ at the substrate surface, as well as direct oxalate oxidation on the substrate at high positive potentials. It also demonstrated the possibility that only a fraction of ECL is produced by direct reaction of CO$_2^{\bullet-}$ with Ru(bpy)$_3^{3+}$ and that quenching of Ru(bpy)$_3^{2+}$* by an excess of Ru(bpy)$_3^{3+}$ is also possible.

A number of Ru(bpy)$_3^{2+}$ derivatives have been immobilized covalently, entrapped in SiO$_2$ sol-gel polymer, or via the Langmuir–Blodgett method onto ITO, Au, or Pt electrode surfaces, and the ECL behavior of these Ru(II) complex films was studied in the presence of oxalate [30–34]. A red shift of the ECL spectrum [~30–60 nm compared to that found for solution-phase luminescence of the complex in organic solvents or Ru(bpy)$_3^{2+}$ in aqueous solution] was commonly observed [30–32], suggesting that there is some interaction between the immobilized molecules [31]. The ECL mechanism of this type of system with oxalate as a coreactant should be analogous to that of the solution-phase Ru(bpy)$_3^{2+}$/oxalate system (Scheme 1), although the formation of the excited state species is unlikely to occur via the annihilation reaction [32] and quenching of the excited state by electrode may take place [30,31]. Investigations of oxalate oxidation and ECL across the liquid/liquid interface has also been reported [35,36].

Other chelate complexes, such as Os(bpz)$_3^{2+}$ (bpz = 2,2′-bipyrazine) [37] and binuclear iridium(I) complexes [38], can also produce ECL in the presence of oxalate upon electrochemical oxidation in both organic and aqueous solutions.

Oxalate ions have been quantitatively determined by using the Ru(bpy)$_3^{2+}$/oxalate coreactant ECL reaction in a diverse range of biological and industrial samples, including vegetable matter [39], urine [40–42], blood plasma [40,42], and alumina process liquors [43].

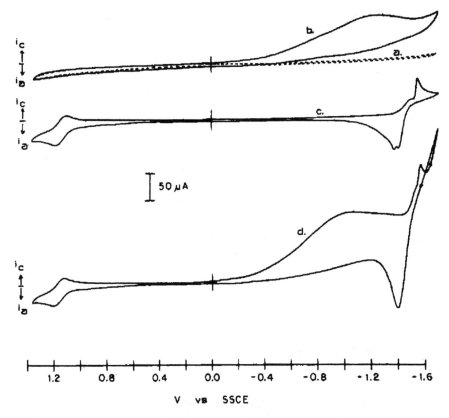

Figure 8 Cyclic voltammograms (50 mV/s) at a glassy carbon electrode in degassed MeCN–H$_2$O (1:1 by volume) solutions containing (a) supporting electrolyte (0.1 M TMAP), (b) 6.6 mM (NH$_4$)$_2$S$_2$O$_8$, (c) 2 mM Ru(bpy)$_3$(PF$_6$)$_2$, (d) 2 mM Ru(bpy)$_3$(PF$_6$)$_2$ and 6.6 mM (NH$_4$)$_2$S$_2$O$_8$. The supporting electrolyte is 0.1 M TMAP in (b)–(d). (From Ref. 3.)

B. Peroxydisulfate (Persulfate) System

The first reductive oxidation coreactant ECL reactions were introduced independently by Bolletta et al. [44] and White and Bard [3] in 1982. They described ECL production in DMF [44], MeCN [44], aqueous [44], or MeCN–H$_2$O (1:1 in v/v) [3] solution by the reaction of electrogenerated Ru(bpy)$_3^+$ [3,44], Cr(bpy)$_3^{2+}$ [44], or Os(bpy)$_3^+$ [44] with the strongly oxidizing intermediate SO$_4^{\bullet-}$ ($E^\circ \geqslant 3.15$ V vs. SCE [45]), generated during reduction of S$_2$O$_8^{2-}$. Because Ru(bpy)$_3^+$ is unstable in aqueous solutions and (NH$_4$)$_2$S$_2$O$_8$ has a low solubility in MeCN solutions, the MeCN–H$_2$O mixed solutions were chosen by the latter group to produce intense ECL emission. As shown in Figure 8, the S$_2$O$_8^{2-}$ can be reduced either at the electrode

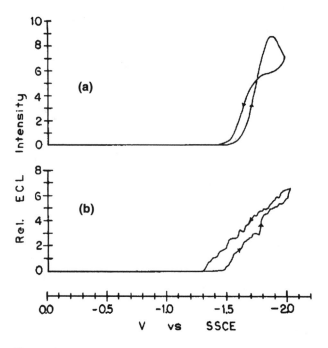

Figure 9   Relative ECL intensity–potential curves at (a) a glassy carbon electrode and (b) a Pt electrode in MeCN–H$_2$O containing 1 mM Ru(bpy)$_3^{2+}$ and 10 mM S$_2$O$_8^{2-}$. The scan rate is 50 mV/s. The noise in (b) is due to H$_2$ evolution. (From Ref. 3.)

directly or by electrogenerated Ru(bpy)$_3^+$ in MeCN–H$_2$O solution. The ECL intensity–potential profile curves (Fig. 9) indicate that no ECL is observed at potentials up to -1.4 V vs. SCE, where S$_2$O$_8^{2-}$ is reduced directly at both glassy carbon and Pt electrodes and that only when the potential is biased negative of the onset of the Ru(bpy)$_3^{2+/+}$ reduction wave (-1.4 V vs. SCE) does ECL appear [3]. Similar results were observed in DMF for Ru(bpy)$_3^{2+}$ and Os(bpy)$_3^{2+}$, in MeCN for Ru(bpy)$_3^{2+}$, and in aqueous solution for Cr(bpy)$_3^{3+}$ at a Pt electrode in the presence of S$_2$O$_8^{2-}$ [44]. These data suggest that the direct electrogeneration of one-electron reduced metal complex species, such as Ru(bpy)$_3^+$, is necessary to produce the corresponding excited state species, i.e., Ru(bpy)$_3^{2+}$*.

Scheme 2 summarizes the possible pathways for the production of Ru(bpy)$_3^{2+}$* when S$_2$O$_8^{2-}$ is used as the coreactant. This reaction sequence should be also operative for the Os(bpy)$_3^{2+}$/S$_2$O$_8^{2-}$ and Cr(bpy)$_3^{3+}$/S$_2$O$_8^{2-}$ systems, although the annihilation reaction is unlikely to occur for the Cr(bpy)$_3^{3+}$/S$_2$O$_8^{2-}$ system [44]. By using aromatic hydrocarbons (R) whose cation radicals undergo a rapid dimerization reaction for R, Fabrizio et al. [46] recently confirmed through electrochemical and ECL measurements that the sulfate radical anion $SO_4^{\bullet-}$ generated during reduction of $S_2O_8^{2-}$ can oxidize R and generate radical cations ($R^{\bullet+}$). These data indirectly support the formation of $Ru(bpy)_3^{3+}$ from $Ru(bpy)_3^{2+}$ oxidation by $SO_4^{\bullet-}$ in $Ru(bpy)_3^{2+}$–$S_2O_8^{2-}$ solution upon electrochemical reduction.

$$S_2O_8^{2-} + e \longrightarrow S_2O_8^{\bullet3-}$$

$$Ru(bpy)_3^{2+} + e \longrightarrow Ru(bpy)_3^{+}$$

$$Ru(bpy)_3^{+} + S_2O_8^{2-} \longrightarrow Ru(bpy)_3^{2+} + S_2O_8^{\bullet3-}$$

$$S_2O_8^{\bullet3-} \longrightarrow SO_4^{2-} + SO_4^{\bullet-}$$

$$Ru(bpy)_3^{+} + SO_4^{\bullet-} \longrightarrow Ru(bpy)_3^{2+*} + SO_4^{2-}$$

$$Ru(bpy)_3^{2+} + SO_4^{\bullet-} \longrightarrow Ru(bpy)_3^{3+} + SO_4^{2-}$$

$$Ru(bpy)_3^{+} + Ru(bpy)_3^{3+} \longrightarrow Ru(bpy)_3^{2+*} + Ru(bpy)_3^{2+}$$

Scheme 2

The ECL intensity of the $Ru(bpy)_3^{2+}/S_2O_8^{2-}$ system was found to be a function of $S_2O_8^{2-}$ concentration, and for 1 mM $Ru(bpy)_3^{2+}$ solution the maximum ECL intensity was obtained at 15–20 mM $S_2O_8^{2-}$ (Fig. 10) [3]. This is because the persulfate ion is a coreactant of $Ru(bpy)_3^{2+}$ ECL as well as an effective quencher of the excited state $Ru(bpy)_3^{2+*}$ [47], and Figure 10 demonstrates both the increase in ECL intensity as more excited states are formed and the quenching of the excited states by the persulfate coreactant. However, at high $S_2O_8^{2-}$ concentrations (>30 mM) not only is the excited state quenched but electrogenerated $Ru(bpy)_3^{+}$ may be removed by reaction with $S_2O_8^{2-}$ to give $Ru(bpy)_3^{2+}$, decreasing the $Ru(bpy)_3^{+}$ steady-state concentration near the electrode surface [3].

The coulometric efficiency (ϕ_{coul}, photons generated per electron injected) of this system was estimated to be ~2.5%, and this value was found to be insensitive to dissolved O_2 when $Ru(bpy)_3^{2+}$ concentrations were higher than 10 nM [3,22]. Because practically no background emission is detected in the persulfate, concentration of $Ru(bpy)_3^{2+}$ as low as 0.1 pM could be detected [22].

Sodium dodecyl sulfate (SDS) enhanced the ECL intensity by approximately sixfold when 15 mM SDS was added to a 10 μM $Ru(bpy)_3^{2+}/20$ mM $S_2O_8^{2-}$ MeCN–H_2O (1:1 by volume) solution containing 1.0 M TMAP electrolyte [48]. This behavior was attributed to the stabilization of ECL intermediates in the SDS micellar environment. In contrast, addition of cetyltrimethylammonium bromide (CTAB) and Triton X-100 can reduce the ECL intensity. The reason behind this is still unclear.

A strong ECL signal was seen at a carbon paste electrode for the $Ru(bpy)_3^{2+}/S_2O_8^{2-}$ system in purely aqueous solution [49]. A linear correlation

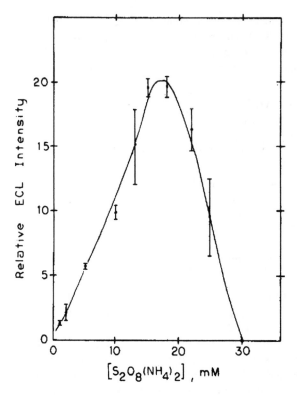

Figure 10 Relative ECL intensity as a function of $(NH_4)_2S_2O_8$ concentration in degassed MeCN–H$_2$O (1:1 by volume) solutions containing 1 mM Ru(bpy)$_3$(PF$_6$)$_2$. (From Ref. 3.)

between the light intensity and the concentration of Ru(bpy)$_3^{2+}$ over the range of 10 mM to 0.10 μM was obtained. Also, the ECL intensity was found to increase linearly with the $S_2O_8^{2-}$ concentration from 1 μM up to 0.3 mM but to drop off sharply at concentrations higher than 1 mM.

Electrogenerated chemiluminescence studies of many metal chelates other than Ru(bpy)$_3^{2+}$, including Cr(bpy)$_3^{3+}$ [44,50], Ru(bpz)$_3^{2+}$ (bpz = 2,2'-bipyrazine) [51,52], Mo$_2$Cl$_4$(PMe$_3$)$_4$ [53], Eu (III) [54], [(bpy)$_2$Ru]$_2$(bphb)$^{4+}$ (bphb = 1,4-bis(4-methyl-2,2'-bipyrindin-4-yl)benzene) [55], (bpy)$_2$Ru (bphb)$^{2+}$ [55], and Ru(phen)$_3^{2+}$ [56] complexes, using $S_2O_8^{2-}$ as a coreactant have also been reported, and some of them are reviewed in Chapter 7. For example, the ECL behavior of Ru(phen)$_3^{2+}$ studied initially in degassed MeCN/0.10 M TBABF$_4$ solution at a Pt electrode via annihilation [57] has since been investigated in aqueous phosphate buffer solution (pH 7) at a carbon paste electrode with $S_2O_8^{2-}$ coreactant [56]. The ECL of the Ru(phen)$_3^{2+}$/$S_2O_8^{2-}$ system started at ~ −1.3 V vs. Ag/AgCl, where Ru(phen)$_3^{2+}$ was directly

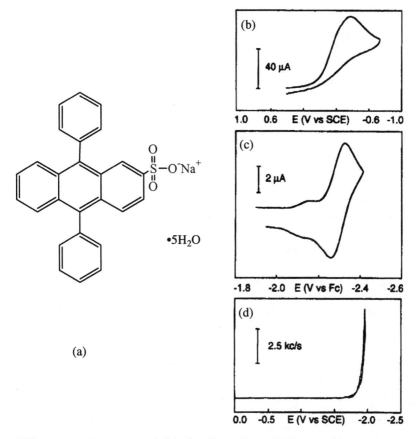

Figure 11 (a) Structure of sodium 9,10-diphenylanthracene-2-sulfate (DPAS). (b) Cyclic voltammogram of 11 mM $(NH_4)_2S_2O_8$ in an MeCN–H_2O solution (1:1 v/v) at a 1.5 mm diameter Pt electrode. Supporting electrolyte, 0.2 M TEAP; scan rate, 100 mV/s. (c) Cyclic voltammogram of 0.93 mM DPAS in MeCN at a 1.5 mm diameter Pt electrode. Supporting electrolyte, 0.10 M TBABF$_4$; scan rate, 200 mV/s. The small pre-wave was due to adsorbed DAPS. (d) The ECL emission for the reduction of 1.1 mM DPAS and 25 mM $(NH_4)_2S_2O_8$ in MeCN–H_2O solution (1:1 v/v) at a 2.0 mm diameter Pt electrode. Supporting electrolyte, 0.2 M TEAP; scan rate, 100 mV/s. (From Ref. 58.)

reduced at the working electrode, indicating that as in the case of $Ru(bpy)_3^{2+}/S_2O_8^{2-}$ the ECL was initiated by the electrochemically reduced metal chelate rather than $S_2O_8^{2-}$ or its reduction products. The light intensity versus $S_2O_8^{2-}$ concentration was linear in the range of 5 µM to 2 mM.

The cathodic reduction of sodium 9,10-diphenylanthracene-2-sulfate (DPAS) (Fig. 11a) in the presence of $S_2O_8^{2-}$ as a coreactant also produces ECL in

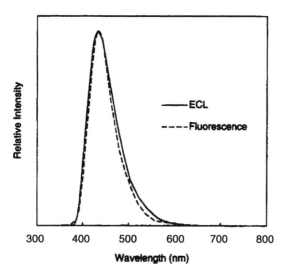

Figure 12 Fluorescence and ECL spectra for the same solution as in Figure 11c. The ECL emission was generated by pulsing a 2.0 mm diameter Pt electrode to -2.2 V (vs. SCE) and integrating the emission for 5 s. Both peak intensities are at 430 nm. (From Ref. 58.)

MeCN–H$_2$O (1:1 by volume) (Figs. 11b–d) [58]. The ECL mechanism of this system is analogous to Scheme 2, by substituting DPAS$^{+/0/-/*}$ respectively for Ru(bpy)$_3$$^{3+/2+/+/2+*}$. As shown in Figure 12, the blue ECL spectrum for the DPAS emission matches its fluorescence spectrum with the peak intensity at ~ 430 nm.

Studies of ECL of several aromatic compounds (Ar), including rubrene (RU), 9,10- diphenylanthracene (DPA), anthracene (ANT), flupranthene (FLU), 1,3,6,8- tetraphenylpyrene (TPP), and 9,10-dicyanoanthracene (DCA), in the presence of persulfate in 2:1 MeCN–benzene solutions suggest that as in the case of Ru(bpy)$_3$$^{2+}$/S$_2O_8$$^{2-}$, ECL is initiated by the electrochemical reduction of fluorophores and not S$_2$O$_8$$^{2-}$ and that formation of Ar* with the resulting ECL was primarily caused by the Ar$^{\bullet -}$–Ar$^{\bullet +}$ reaction, in which Ar$^{\bullet +}$ was generated via oxidation of Ar by SO$_4$$^{\bullet -}$ [59]. The importance of radical cation formation in the ECL mechanism is evident in the tertiary thianthrene (TH) (Fig. 3a), 2,5-diphenyl-1,3,4-oxadiazole (PPD) (Fig. 3b)/ S$_2$O$_8$$^{2-}$ system [59]. In MeCN, both TH and PPD compounds have accessible excited states, but TH can only be oxidized to a radical cation whereas PPD can only be reduced to a stable radical anion (Fig. 13) [59]. The TH fluorescence observed during the S$_2$O$_8$$^{2-}$-catalyzed reduction of PPD (Fig. 14) can be explained with the mechanism shown in Scheme 3.

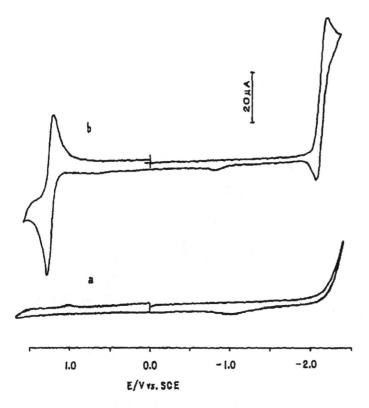

Figure 13 Cyclic voltammograms (100 mV/s) at a Pt electrode in MeCN containing (a) supporting electrolyte (0.1 M TBABF$_4$), (b) 3.0 mM PPD, 3.0 mM TH, and 0.1 M TBABF$_4$. (From Ref. 59.)

$$PPD + e \longrightarrow PPD^{\bullet -}$$
$$PPD^{\bullet -} + S_2O_8^{2-} \longrightarrow PPD + SO_4^{2-} + SO_4^{\bullet -}$$
$$TH + SO_4^{\bullet -} \longrightarrow TH^{\bullet +} + SO_4^{2-}$$
$$TH^{\bullet +} + PPD^{\bullet -} \longrightarrow TH^* + PPD$$
$$TH^* \longrightarrow TH + h\nu$$

Scheme 3

By using a micelle-forming surfactant, Brij-35 [polyoxyethylene(23) dodecanol], Ar/S$_2$O$_8^{2-}$ ECL can be also generated in aqueous solutions [60]. In this case, the emitting species was believed to be the micelle-encapsulated fluorophore rather than premicellar aggregate.

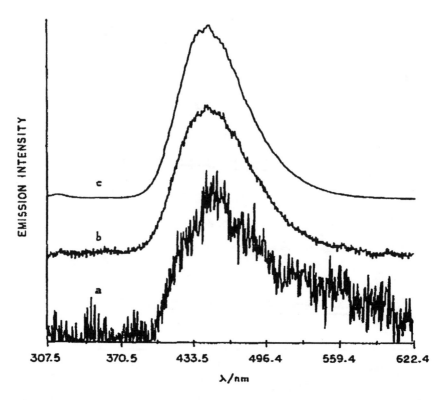

Figure 14 Ultraviolet-visible emission spectra in MeCN of (a) ECL from $S_2O_8^{2-}$-catalyzed reduction of PPD in the presence of TH [3.0 mM PPD, 3.0 mM TH, 20 mM $(TBA)_2S_2O_8$, 0.1 M TBABF$_4$, $E = 2.4$ V vs. SCE], (b) double potential step ECL from PPD/TH (3.0 mM PPD, 3.0 mM TH, 0.1 M TBABF$_4$, $E = +1.5$ to -2.4 V vs. SCE), and (c) photoexcited fluorescence of TH ($\lambda_{ex} = 300$ nm). (From Ref. 59.)

On the other hand, the thin film ECL of the conjugated polymer 4-methoxy-(2-ethylhexoxyl)-2,5-polyphenylenevinylene (MEH-PPV) (Fig. 15) was observed when the potential was scanned to -2.0 V vs. Fc/Fc$^+$ in a deoxygenated MeCN solution containing 0.10 M Bu$_4$NBF$_4$ and 10 mM $(TBA)_2S_2O_8$ [61].

C. Amine-Related Systems

1. Tri-*n*-Propylamine System

Noffsinger and Danielson [62] first reported the chemiluminescence of Ru(bpy)$_3^{3+}$ with aliphatic amines. Following this study, Leland and Powell [63]

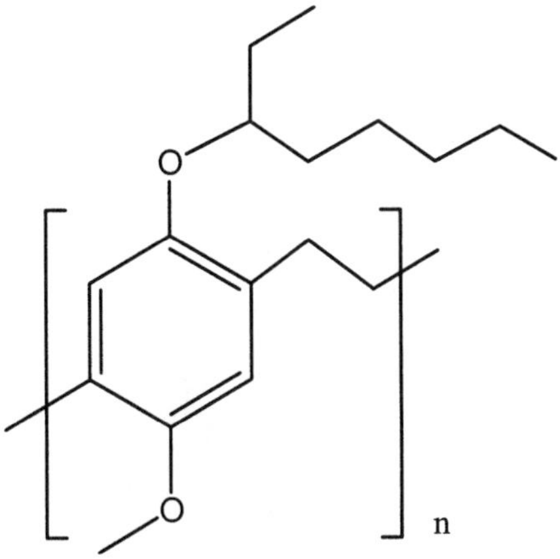

Figure 15 Structure of 4-methoxy-(2-ethylhexoxyl)-2,5-polyphenylenevinylene (MEH-PPV).

reported the ECL of $Ru(bpy)_3^{2+}$ with tri-n- propylamine (TPrA) as a coreactant. Since then, a wide range of ECL analytical applications involving $Ru(bpy)_3^{2+}$ or its derivatives as labels have been reported [7]. The $Ru(bpy)_3^{2+}$ (or its derivatives) with TPrA exhibit the highest ECL efficiency, and this system forms the basis of commercial systems for immunoassay and DNA analysis [7,63,64] (See Chapter 8). The ECL mechanism of this reaction is very complicated and has been investigated by many workers [62–68]. Generally, the ECL emission of this system as a function of applied potential consists of two waves (Fig. 5.16). The first occurs with the direct oxidation of TPrA at the electrode, and this wave is often merged into the foot of the second wave when relatively high concentrations of $Ru(bpy)_3^{2+}$ (in the millimolar range) are used. The second wave appears where $Ru(bpy)_3^{2+}$ is oxidized [65,68]. Both waves are associated with the emission from $Ru(bpy)_3^{2+}*$ [65]. The relative ECL intensity from the first wave is significant, particularly in dilute (less than micromolar) $Ru(bpy)_3^{2+}$ solutions containing millimolar TPrA. Thus, the bulk of the ECL signal obtained in this system with low concentrations of analytes, as in immunoassays and DNA probes with $Ru(bpy)_3^{2+}$ as an ECL label, probably originates from the first ECL wave. Scheme 4 summarizes the mechanism of the first ECL wave, where the cation radical species $TPrA^{\bullet+}$ formed during TPrA oxidation is a sufficiently stable intermediate (half-life of ~0.2 ms) that it can oxidize $Ru(bpy)_3^{+}$ (formed from the reduction of $Ru(bpy)_3^{2+}$ by $TPrA^{\bullet}$ free radical) to give $Ru(bpy)_3^{2+}*$ [68]. Direct evidence for $TPrA^{\bullet+}$ in neutral aqueous solution was obtained via flow cell electron spin resonance (ESR) experiments at ~20°C [68]. However, as indicated by digital simulations, fast scan cyclic voltammetry cannot verify the existence of $TPrA^{\bullet+}$. This is because of the cancelation of the cathodic current

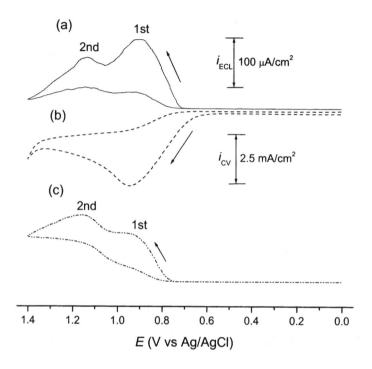

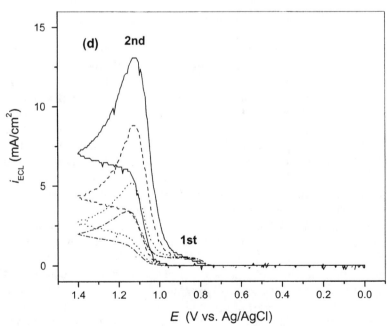

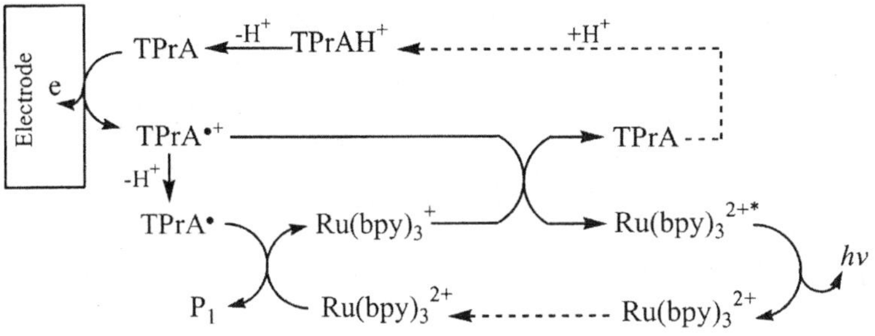

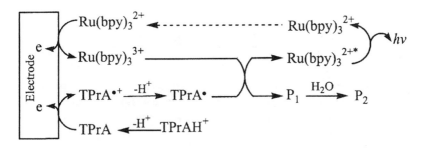

[where TPrA$^{\bullet+}$ = (CH$_3$CH$_2$CH$_2$)$_3$N$^{\bullet+}$, TPrAH$^+$ = Pr$_3$NH$^+$, TPrA$^{\bullet}$ =

Pr$_2$NC$^{\bullet}$HCH$_2$CH$_3$, P$_1$ = Pr$_2$N$^+$C = HCH$_2$CH$_3$]

Scheme 4   (From Ref. 68.)

[where P$_2$ = Pr$_2$NH + CH$_3$CH$_2$CHO]

Scheme 5   (From Ref. 68.)

Figure 16   (a) ECL and (b) cyclic voltammogram of 1.0 nM Ru(bpy)$_3^{2+}$ in the presence of 0.10 M TPrA with 0.10 M Tris/0.10 M LiClO$_4$ buffer (pH 8) at a 3 mm diameter glassy carbon electrode at a scan rate of 50 mV/s. (c) Same as (a) but with 1.0 μM Ru(bpy)$_3^{2+}$. The ECL intensity scale is given for (c) and should be multiplied by 100 for (a). (d) The first and second ECL responses in 100 mM TPrA (0.20 M PBS, pH 8.5) with different Ru(bpy)$_3^{2+}$ concentrations: 1 mM (solid line), 0.50 mM (dashed line), 0.10 mM (dotted line), and 0.05 mM (dash-dotted line), at a 3 mm diameter glassy carbon electrode at a scan rate of 100 mV/s. (From Ref. 68.)

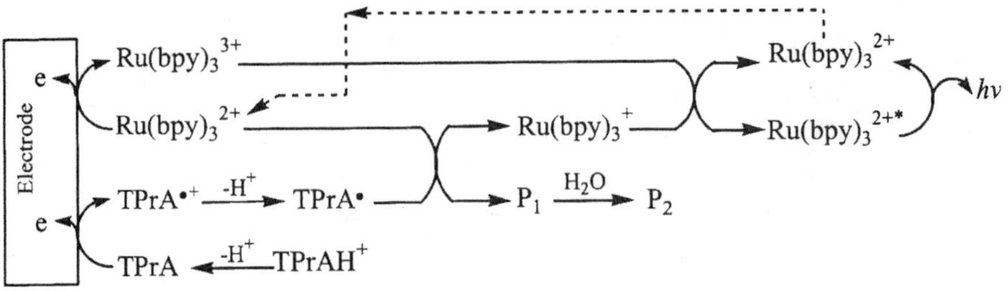

Scheme 6   (From Ref. 68.)

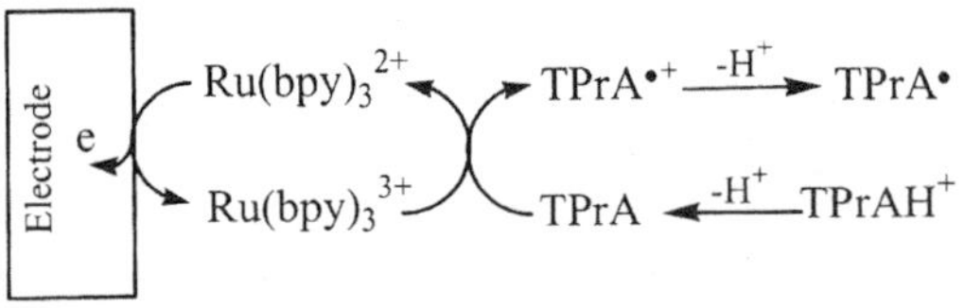

Scheme 7   (From Ref. 68.)

from the reduction of $TPrA^{\bullet+}$ by the larger anodic current contribution from the oxidation of $TPrA^{\bullet}$ [68].

The mechanism of the second ECL wave follows the classic oxidative reduction coreactant mechanism [2], where oxidation of TPrA generates a strongly reducing species $TPrA^{\bullet}$. Direct evidence for this free radical intermediate was obtained when the light was quenched upon addition of nitrobenzene derivatives [67]. This oxidation can occur via a "catalytic route" in which electrogenerated $Ru(bpy)_3^{3+}$ reacts with TPrA as well as by direct reaction of TPrA at the electrode described by both Scheme 5 and Scheme 6 [68].

The "catalytic route" involving homogeneous oxidation of TPrA with $Ru(bpy)_3^{3+}$ is shown in Scheme 7 [68]. The contribution of this process to the overall ECL intensity depends upon the $Ru(bpy)_3^{2+}$ concentration and is small when relatively low concentrations of $Ru(bpy)_3^{2+}$ are used [67].

As summarized in Scheme 8, the excited state of $Ru(bpy)_3^{2+}$ can be produced via three different routes: (1) $Ru(bpy)_3^{+}$ oxidation by $TPrA^{\bullet+}$ cation radicals, (2) $Ru(bpy)_3^{3+}$ reduction by $TPrA^{\bullet}$ free radicals, and (3) the $Ru(bpy)_3^{3+}$ and $Ru(bpy)_3^{+}$ annihilation reaction.

$$
\begin{array}{ll}
(1) & Ru(bpy)_3^{+} + TPrA^{\bullet+} \\
(2) & Ru(bpy)_3^{3+} + TPrA^{\bullet} \\
(3) & Ru(bpy)_3^{3+} + Ru(bpy)_3^{+}
\end{array}
\Biggr\} \longrightarrow Ru(bpy)_3^{2+*}
$$

Scheme 8

The direct oxidation of TPrA at an electrode plays an important role in the ECL process of the $Ru(bpy)_3^{2+}$/TPrA system and varies with the electrode material [65]. At a glassy carbon electrode, this kind of direct oxidation is evident. At Pt and Au electrodes, however, the formation of surface oxide could significantly block the direct oxidation of TPrA. As a result, these two types of electrodes exhibit much weaker ECL emission than that obtained at a glassy carbon electrode. Small amounts of halide species inhibit the growth of the surface oxide on Pt and Au electrodes and lead to an obvious increase in the TPrA oxidation current and hence ECL emission. Additionally, Au electrodes can be activated by anodic dissolution of Au in halide containing solutions. A 100-fold enhancement in ECL intensity by the addition of 0.10 mM bromide was reported [65].

The effect of electrode hydrophobicity on TPrA oxidation as well as on the ECL intensity of this system was also examined [69]. The oxidation rate of TPrA and hence the ECL intensity were found to be much larger at an electrode modified with a hydrophobic surface. Similar behavior was observed at Pt and Au electrodes in the presence of Triton X-100 [69,70] and some other nonionic surfactants [71]. These results were attributed to the adsorption of neutral TPrA species on this kind of hydrophobic electrode surface [69,70]. However, an earlier paper [72] also demonstrated a large increase in $Ru(bpy)_3^{2+}$ ECL with TPrA coreactant in the presence of three nonionic surfactants including Triton X-100 at a Pt electrode. The authors ascribed these increases to improved emission from $Ru(bpy)_3^{2+}$/surfactant or micelle species.

The ECL intensity for the first and second waves was found to be proportional to the concentration of both $Ru(bpy)_3^{2+}$ and TPrA species in a very large dynamic range [63–65,67,68], with limits of detection of 0.5 pM for $Ru(bpy)_3^{2+}$ [73] and 10 nM for TPrA [23] being reported.

The ECL intensity also depends on the solution pH [9,63] (Fig. 17), with dramatic increases at ~pH > 5.5 and a maximum value at pH 7.5. As revealed in Schemes 4–7, the solution pH must be sufficiently high to promote the deprotonation reactions of $TPrAH^+$ and $TPrA^{\bullet+}$; however, the maximum ECL signal is clearly obtained below the pK_a of $TPrAH^+$ (10.4) [64] but above that for $TPrA^{\bullet+}$ (~3.3) estimated on the basis of cyclic voltammetric digital simulations [68]; it is likely that the acidity of $TPrA^{\bullet+}$, and not the basicity of TPrA, is most important in determining the pH dependence of ECL [63]. Usually, pH values higher than 9 should not be used, because $Ru(bpy)_3^{3+}$ generated at the electrode could react with hydroxide ions to produce a significant ECL background signal [74].

On the basis of the ECL mechanism discussed above, it is evident why selection of highly efficient coreactants is difficult for the determination of low concentrations of the $Ru(bpy)_3^{2+}$ label. The coreactant must form both oxidant (e.g., $TPrA^{\bullet+}$) and reductant (e.g., $TPrA^{\bullet}$) with appropriate redox potentials and

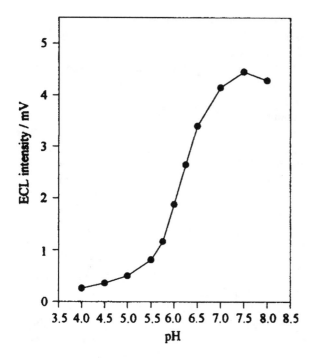

Figure 17 Effect of pH on the ECL intensity from the reaction of Ru(bpy)$_3^{2+}$ with TPrA. (From Ref. 9.)

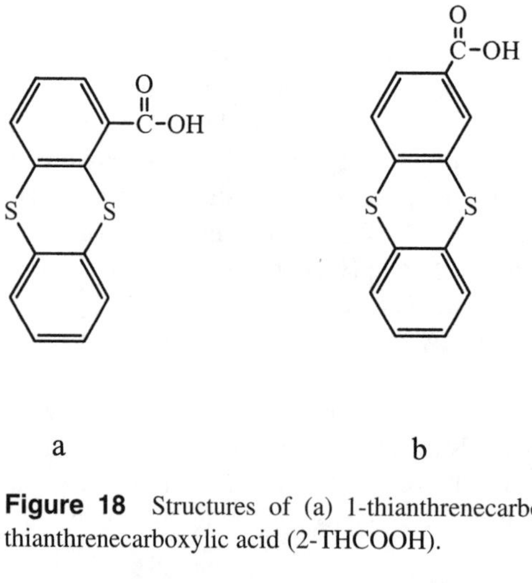

Figure 18 Structures of (a) 1-thianthrenecarboxylic acid (1-THCOOH) and (b) 2-thianthrenecarboxylic acid (2-THCOOH).

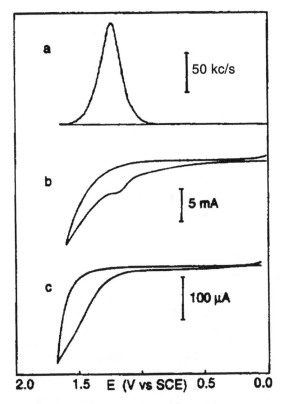

Figure 19 (a) Emission and (b) cyclic voltammogram of 1×10^{-5} M DPAS and 0.15 M TPrA in pH 7.5 sodium phosphate buffer at 6×9 mm Pt gauze (52 mesh) electrode. Scan rate, 100 mV/s. (c) Cyclic voltammogram of 0.05 M TPrA in pH 7.5 sodium phosphate buffer at a 1.5 mm diameter Pt electrode. Scan rate, 200 mV/s. (From Ref. 58.)

lifetimes. Thus the deprotonation rate of TPrA$^{\bullet+}$ must be just right to build up the needed concentrations of both species [68]. Moreover, in the commercial Origen analyzer (IGEN International, Inc., Gaithersburg, MD), the Ru(bpy)$_3^{2+}$-tagged species in an immunoassay are immobilized on 2.8 μm diameter nontransparent magnetic beads that are brought to an electrode by a magnetic field. The high sensitivity of the technique suggests that most of the labels on the beads participate in the ECL reaction, which can be well explained via Scheme 4, whereas the direct oxidation of Ru(bpy)$_3^{2+}$ on the beads would occur only for those within electron tunneling distance from the electrode, ~1–2 nm, and its contribution to the overall ECL intensity should be small or negligible.

In addition to Ru(bpy)$_3^{2+}$, many other metal chelate complexes (Chapter 7) and aromatic compounds or their derivatives can produce ECL in the presence of

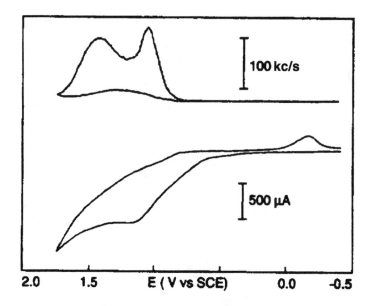

Figure 20 Cyclic voltammogram (bottom) and simultaneous emission (top) of 5.2 mM 2-THCOOH and 0.15 M TPrA in pH 7.5 sodium phosphate buffer at 6 mm diameter Pt electrode at a scan rate of 100 mV/s. (From Ref. 58.) The behavior of 1- THCOOH is similar except that the peak emission intensity is 6–7 times less intense that that of 2-THCOOH [58].

TPrA as a coreactant upon electrochemical oxidation. For example, the anodic oxidation of aqueous sodium 9,10-diphenylanthracene-2-sulfonate (DPAS) (Fig. 11a) and 1- and 2-thianthrenecarboxylic acid (1- and 2-THCOOH) (Fig. 18) in the presence of TPrA in sodium phosphate buffer solutions produce ECL emission (Figs. 19 and 20) [58]. The two distinct emission peaks as a function of potential in Figure 20 were attributed to reactions associated with the homogeneous oxidation of TPrA by 2-TH$^+$COO$^-$ and the direct oxidation of TPrA at the electrode, respectively, because at a much lower 2-THCOOH concentration only the first emission appears as a small shoulder in the dominant second emission [58]. For the DPAS/TPrA system, the ECL spectrum is characteristic of DPAS fluorescence, with peak intensity at 430 nm (Fig. 21). In contrast, the ECL spectra for 1- and 2-THCOOH are dramatically red-shifted from the fluorescence spectra (Fig. 22), suggesting that at least part of the ECL emission arises from a product of THCOOH rather than from the excited state of the intact acid.

The excited singlet state of DPAS (^{1}DPAS*) was thought to probably be formed via triplet–triplet annihilation (a T route [75]) [58]. However, the recently estimated standard potential of the Pr$_2$NC$^+$HEt/TPrA$^•$ couple enables

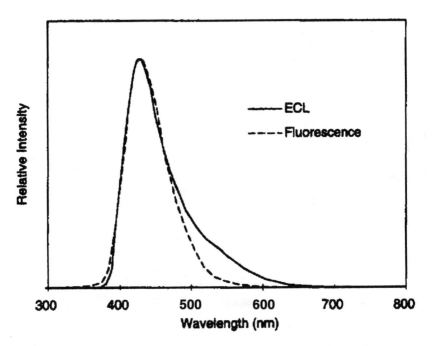

Figure 21 Scaled fluorescence and ECL spectra for a solution of 1×10^{-5} M DPAS and 0.15 M TPrA in pH 7.5 sodium phosphate buffer. The ECL emission was generated by repetitively pulsing a 6×9 mm Pt gauze (52 mesh) electrode from 0.0 V (vs SCE, 8 s) to +1.2 V (2 s) to -1.4 V (2 s) and back to 0.0 V. Light was generated on the positive pulse in this sequence (the negative pulse was for electrode-cleaning purposes) and was integrated for 2 min to produce the spectrum. Both peak intensities are at 430 nm. (From Ref. 58.)

one to reconsider this process. Because [58]

$$DPAS^{\bullet+} + e \longrightarrow DPAS \quad E^\circ \approx +1.3 \text{ V vs. SCE}$$
$$Pr_2NC^+HEt + e \longrightarrow TPrA^{\bullet} \quad E^\circ \approx -1.7 \text{ V vs. SCE [76]}$$
$$\quad\quad (-1.1 \text{ V vs. SCE was used [58]})$$
$$^1DPAS^* \longrightarrow DPAS + h\nu \quad E_s = 2.9 \text{ eV}$$

The enthalpy of the electron transfer reaction between $DPAS^{\bullet+}$ and $TPrA^{\bullet}$, correcting for an entropy of about 0.1 eV, is -2.9 eV, equal to that needed to produce $^1DPAS^*$. In considering the fact that the emission intensity for 1×10^{-5}M DPAS is about the same for the DPAS/TPrA oxidation as it is for the "energy-sufficient" DPAS/$S_2O_8^{2-}$ reduction [58], the oxidative DPAS/TPrA system could be "energy-sufficient" to produce $^1DPAS^*$ directly (an S route).

Scheme 9 summarizes the possible ECL mechanism of the DPAS/TPrA system.

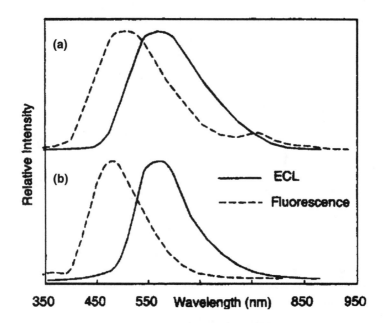

Figure 22 Fluorescence and ECL spectra (a) of 9.1 mM 1-THCOOH and (b) 20 mM 2-THCOOH taken in pH 8.5 sodium phosphate buffer containing 0.15 M TPrA. ECL emissions were generated by alternately pulsing a 6×9 mm Pt gauze (52 mesh) electrode between +1.4 (0.5 s) and -0.5 V (2 s) vs. SCE for a duration of 40 min for 1-THCOOH and 20 min for 2-THCOOH with a stirred solution. The peak fluorescence intensity is at 480 nm for 2-THCOOH and 505 nm for 1-THCOOH. The peak ECL intensities are at 570 nm for both isomers. The intensities have been scaled for comparison; those of the 2-THCOOH emissions are ~6–7 times those of 1-THCOOH. (From Ref. 58.)

$$TPrA - e \longrightarrow TPrA^{\bullet +}$$

$$DPAS - e \longrightarrow DPAS^{\bullet +}$$

$$DPAS^{\bullet +} + TPrA \longrightarrow DPAS + TPrA^{\bullet +}$$

$$TPrA^{\bullet +} - H^+ \longrightarrow TPrA^{\bullet}$$

$$DPAS^{\bullet +} + TPrA^{\bullet} \longrightarrow {}^1DPAS^* + Pr_2NC^+HEt$$

$$DPAS + TPrA^{\bullet} \longrightarrow DPAS^{\bullet -} + Pr_2NC^+HEt$$

$$DPAS^{\bullet +} + DPAS^{\bullet -} \longrightarrow {}^1DPAS^* + DPAS$$

$${}^1DPAS^* \longrightarrow DPAS + h\nu$$

Scheme 9

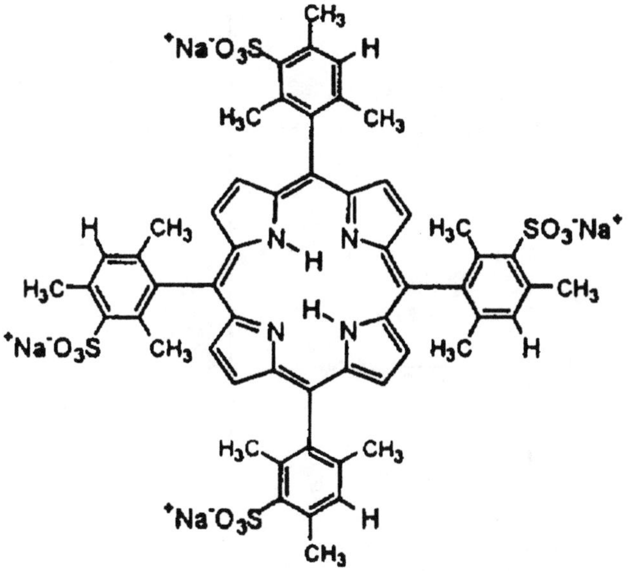

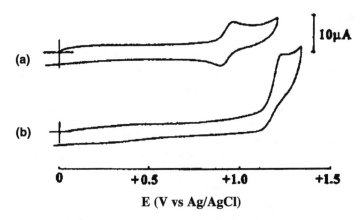

Figure 23 Structure of tetrakis(3-sulfonatomesityl)porphyrin (H$_2$TSMP). (From Ref. 77.)

Recently, a sterically hindered water-soluble porphyrin, tetrakis(3-sulfonatomesityl)porphyrin (H$_2$TSMP) (Fig. 23) was reported to be able to form stable cation species in aqueous media at pH of ~6.5 after one-electron electrochemical oxidation (Fig. 24) [77]. The anodic oxidation of H$_2$TSMP in

Figure 24 Cyclic voltammograms of 5×10^{-4} M H$_2$TSMP at pH values of (a) 6.5 and (b) 2.0. Scan rate = 0.05 V/s. (From Ref. 77.)

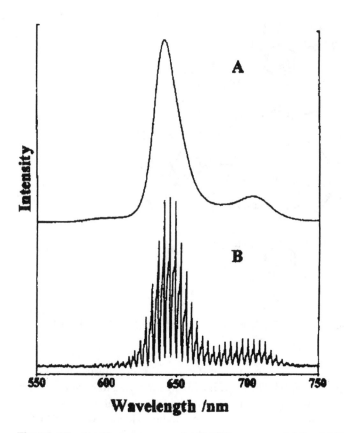

Figure 25 (A) Fluorescence and (B) ECL spectra of 1.0 mM H₂TSMP and 0.2 M Na₂SO₄ in pH 6.0 buffer solution. The ECL spectrum was obtained by pulsing the Pt electrode between $+1.00$ and $+0.45$ V vs Ag/AgCl at 0.2 Hz, and 50 mM TPrA was added as coreactant. (From Ref. 77.)

the presence of TPrA as a coreactant in aqueous solution produces ECL with maxima at 640 nm and 700 nm, which were assigned to the emission of the singlet state H₂TSMP* (Fig. 25) [77]. The formation of the excited state H₂TSMP should follow an S route rather than a T route [77] after adopting the new redox potential for the Pr₂NC⁺HEt/TPrA· couple ($E° \approx -1.7$ V vs. SCE [76]). Presumably, ECL results from a sequence of reactions analogous to those shown in Scheme 9, where DPAS·⁺/⁰/·⁻/⁎ is replaced by H₂TSMP·⁺/⁰/·⁻/⁎, although the homogeneous oxidation of TPrA with H₂TSMP·⁺ could be negligible due to the fact that the two redox couples have similar redox potential values of ~ 0.88 V vs. SEC [68,77]. Similar ECL behavior was observed when oxalate was used instead of TPrA as a coreactant [77].

In addition to solution-phase ECL as described above, in which both the luminophore and coreactant species are dissolved in a solution, ECL can be generated from systems in which either the luminophore is adsorbed/immobilized on the electrode in contact with a coreactant-containing electrolyte solution [23,61,78–82] or both the luminophore and the coreactant are encapsulated within a sol-gel matrix [83–86]. Xu and Bard [78] studied the ECL behavior that results from the anodic oxidation at HOPG (highly oriented pyrolytic graphite), Au, and Pt electrodes in the presence of TPrA after immersion of the electrodes in $Ru(bpy)_3^{2+}$ solutions followed by thorough washing, demonstrating that $Ru(bpy)_3^{2+}$ can be strongly adsorbed onto the surfaces of these electrodes. The ECL of $Ru(bpy)_3^{2+}$ immobilized in Nafion [23,82], Nafion-silica composite materials [82], silica gels [83–86], benzenesulfonic acid monolayer film [80], and poly(p-styrenesulfonate)–silica–Triton X-100 composite films [79,81] has been reported, and these systems have been used for ECL detection of a number of coreactants, including TPrA [23,79–81] and oxalate [23,79,81]. Solid-state ECL from gel-entrapped $Ru(bpy)_3^{2+}$/alkylamines and $Ru(bpy)_3^{2+}$/oxalate showed that when a microelectrode was used to generate the ECL the resultant emission was found to be remarkably stable over many hours, owing to the slow consumption of the coreactant at the small electrode as well as to the steady-state flux of reagents to the electrode surface [83–85].

The anodic oxidation of MEH-PPV polymer film (Fig. 15) immobilized on a Pt electrode in contact with $TPrA/Bu_4NBF_4$ in MeCN solution showed an orange luminescence [61]. As in the case of ECL generated via oxidative and reductive pulsing, the ECL of MEH-PPV polymer film in the presence of TPrA upon oxidation is also believed to be associated with electron and hole annihilation (Fig. 26). The Pt electrode serves as the hole injector (Fig. 26) while TPrA$^•$ (also formed at the Pt electrode) serves as a reductant strong enough to inject electrons into the conduction band, which leads to electron and hole annihilation and hence ECL (Fig. 26).

More recently, the ECL behavior of the aqueous $Ru(bpy)_3^{2+}$/TPrA system was studied at an as-deposited diamond electrode [87]. The anodic potential window for ECL at this kind of diamond electrode was found to be very wide and had a potential limit of 2.5 V vs. Ag/AgCl. Also, the ECL response was found to be extremely stable over potential cycling, which was attributed to the low degree of adsorption of the reaction products on the diamond surface.

2. Other Amine-Related Systems

Similar to the case of TPrA, a wide range of amine compounds can be used as coreactants and take part in ECL reactions with $Ru(bpy)_3^{2+}$ [4–12,88–93]. Several workers [5,8,9,62,63,88] have tried to correlate the coreactant efficiency with the amine structure. Although there are no strict rules governing ECL activity in

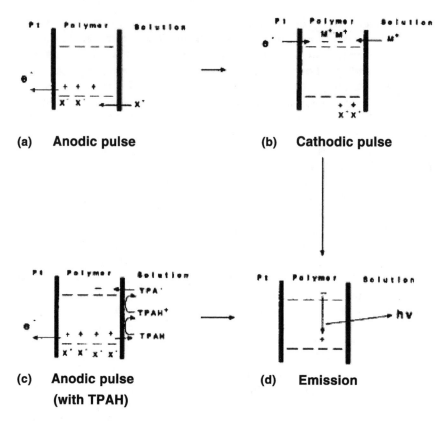

(a) **Anodic pulse**   (b) **Cathodic pulse**

(c) **Anodic pulse**   (d) **Emission**
     (with TPAH)

Figure 26 Schematic of ECL from MEH-PPV polymer (See Fig. 15) coated on Pt electrode via (a) oxidative and (b) reductive pulsing and (c) oxidation in the presence of TPrA (TPAH). (a) Oxidation of polymer (hole injection) at polymer/electrode surface and subsequent insertion of electrolyte anions. (b) Reduction of polymer (electron injection) at polymer/electrode surface and subsequent insertion of electrolyte cations. (c) Simultaneous oxidation of MEH-PPV on Pt surface to form holes and the oxidation of TPrA (TPAH) to TPrA˙ (TPA˙). (d) Luminescence resulting from the annihilation of holes and electrons. (From Ref. 61.)

amines, as a general rule the ECL intensity increases in the order primary < secondary < tertiary amines, with tertiary amines having the lowest detection limits [62,63]. The ECL intensity of primary amines can be improved by prior derivatization with divinylsulfone, $(CH_2CH)_2SO_2$, to form cyclic tertiary amines that then act as efficient coreactants [94]. As shown in Schemes 4–7, the amine should have a hydrogen atom attached to the α-carbon, so that upon oxidation newly formed radical cation species can undergo a deprotonation process to form a strongly reducing free radical species. Also, the nature of substituents attached to nitrogen or α-carbon on an amine molecule can affect the ECL activity [9]. In general, electron-withdrawing substituents, such as carbonyl, halogen, or hydroxyl groups, tend to cause a reduction of ECL activity, probably by making the lone pair of electrons on nitrogen less available for reaction [5], or by destabilizing the radical intermediates. Electron-donating groups such as alkyl chains have the opposite effect. Aromatic amines, aromatic substituted amines, and amines with a carbon–carbon double bond that can conjugate the radical intermediates consistently give a very low ECL response. This may be due to mesomeric or resonance stabilization of the radical intermediates, resulting in radicals that are less active toward the $Ru(bpy)_3^{3+}$ and hence hinder the ECL reaction. Note that aromatic amines may also quench the emission of the $Ru(bpy)_3^{2+*}$ complex, further reducing the ability of these compounds to give an ECL response [95]. Amines with a structure that inhibits the adoption of a more planar molecular geometry after electro-oxidation, such as quinuclidine and quinine, also hinder the formation of the nitrogen radical and hence tend to show lower ECL intensities than other amines that are able to tend toward a trigonal planar geometry, resulting in effective delocalization of charge. Furthermore, amines that have other functional groups that are more readily oxidized than the nitrogen atom may take part in alternative, ultimately non chemiluminescent, reactions. Finally, a linear correlation between the logarithm of the ECL signal and the first ionization potential of the amine [62] and a number of lupin alkaloids [91] was reported. In general, the ionization potentials of the alkylamines can be ordered primary > secondary > tertiary amines, which is inversely related to the order of the amine ECL signals. That is, the higher the ionization potential, the lower the ECL intensity. Clearly, these results are consistent with the proposed ECL mechanism, in which electrons are transferred from the amine to the electrode or $Ru(bpy)_3^{3+}$ species. The ionization potential of the amine is expected to be linearly proportional to the free energy change of the electron transfer reaction [96,97], and this free energy change is inversely proportional to the logarithm of the rate constant for the electron transfer [98,99], which is essentially directly related to the ECL intensity.

The interference of metal ions with $Ru(bpy)_3^{2+}$ ECL intensity was recently investigated when ethylenediaminetetraacetate (EDTA), a ditertiary amine compound, was used as a coreactant [90]. Fifteen metal ions were examined: Mg^{2+}, Ca^{2+}, Ce^{3+}, Nd^{3+}, Sm^{3+}, La^{3+}, Al^{3+}, Y^{3+}, Fe^{2+}, Co^{2+}, Ni^{2+}, Cu^{2+}, Zn^{2+}, Cd^{2+}, and Pb^{2+}. Metal ions that are preferentially bound to the oxygen atoms of EDTA (MI_O, from Mg^{2+} to Y^{3+}) were found to have a negligible effect on the intensity of ECL except for Al^{3+} and Y^{3+}, whereas metal ions that are preferentially bound to the nitrogen atoms of EDTA (MI_N, from Fe^{2+} to Pb^{2+}) can significantly reduce the ECL signal. This is because the tertiary amines of EDTA can react with MI_N and Al^{3+}, Y^{3+} ions to form stable complexes that are difficult to oxidize by electrogenerated $Ru(bpy)_3^{3+}$ or to be directly oxidized at the electrode [100–102].

There are many applications for ECL assays for amines because amine groups are prevalent in numerous biologically and pharmacologically important compounds and because amines absorb weakly in the UV/visible spectrum. A number of reviews have been recently published regarding these analytes [4–12].

Because the reduced form of nicotinamide adenine dinucleotide (NADH) or the phosphate NADPH contains a tertiary amine functional group, NAD(P)H can be used as a coreactant for $Ru(bpy)_3^{2+}$. However, in its oxidized form (NAD$^+$ or NADP$^+$) this group has been converted to an aromatic secondary amine (Fig. 27), which gives virtually no ECL response [23,103], probably because of the resonance stabilization effect of the radical intermediates [5,9]. The ECL mechanism involving NADH and $Ru(bpy)_3^{2+}$ is similar to that of the $Ru(bpy)_3^{2+}$/TPrA system and is illustrated in Fig. 28 [104]. Thus, species (substrates) that produce or consume NADH or NADPH in enzymatic reactions can be directly monitored by $Ru(bpy)_3^{2+}$ ECL (Fig. 29). For example, glucose was quantified by coupling two enzyme-catalyzed reactions, which results in stoichiometric formation of NADH [104] (Scheme 10). Other examples include the assays of ethanol, bicarbonate, and glucose phosphate dehydrogenase [104,105].

Glucose + ATP $\xrightarrow{\text{Hexokinase}}$ glucose-6-phosphate + ADP

Glucose-6-phosphate + NAD$^+$ $\xrightarrow{\text{G-6-P Dehydrogenase}}$ 6-phosphogluconate + H$^+$ + NADH

NADH + $Ru(bpy)_3^{2+}$ $\xrightarrow{\text{Oxidizing electrode}}$ ECL

Scheme 10   Schematic of reactions used to measure glucose by ECL. NADH, which is produced enzymatically, is used as a coreactant for $Ru(bpy)_3^{2+}$ ECL.

β-Lactam antibiotics, such as penicillin G shown in Scheme 11, can also act as the amine-containing coreactant in the ECL reaction with $Ru(bpy)_3^{2+}$ [106,107]. β-Lactamase-catalyzed hydrolysis of penicillin G forms a molecule with a secondary amine that causes dramatic changes in the ability of the antibiotic to promote $Ru(bpy)_3^{2+}$ light emission, permitting sensitive detection and quantification of β-lactams and β-lactamases. The efficiency of the ECL process can be further increased by direct covalent attachment of the β-lactamase substrate to a $Ru(bpy)_3^{2+}$ derivative [106].

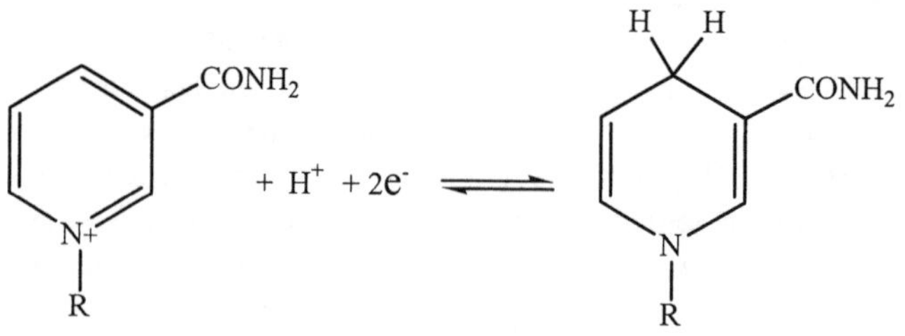

Figure 27   Structures of NAD$^+$ and NADH showing the part of the molecule where reversible reduction occurs, changing the ECL reactivity. (From Ref. 5.)

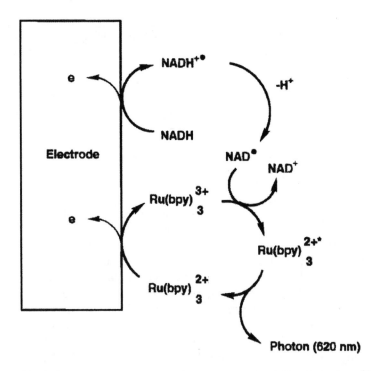

Figure 28 Proposed ECL reaction involving NADH and Ru(bpy)$_3^{2+}$. NADH and Ru(bpy)$_3^{2+}$ are both oxidized at the surface of an electrode. The one-electron oxidized cation radical NADH$^{\cdot+}$ loses a proton to become a strongly reducing radical, NAD$^{\cdot}$. NAD$^{\cdot}$ then transfers an electron to Ru(bpy)$_3^{3+}$ to form NAD$^+$ and the unstable excited state species, Ru(bpy)$_3^{2+}$*, which emits a photon when it decays to the ground state, Ru(bpy)$_3^{2+}$. Notably, the ECL reaction results in reformation of the functional reagents NAD$^+$ and Ru(bpy)$_3^{2+}$. (From Ref. 104.)

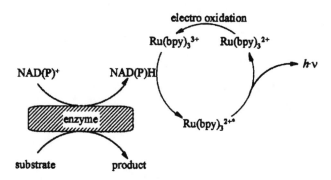

Figure 29 Enzymatic detection using Ru(bpy)$_3^{2+}$ ECL. NADH or NADPH is produced during a highly specific enzymatic reaction and detected with Ru(bpy)$_3^{2+}$ (From Ref. 10.)

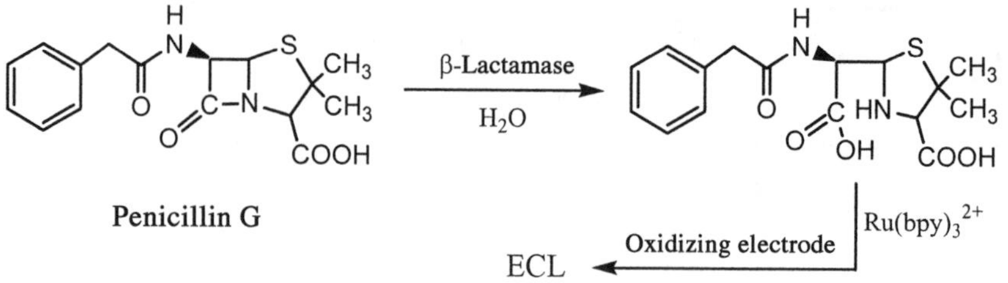

Scheme 11 Enzymatic hydrolysis of penicillin G by β-lactamase generates a coreactant that can be used to produce ECL from Ru(bpy)$_3^{2+}$.

Amino acids and peptides are all amine-containing compounds, therefore, they can be used as coreactants and detected by Ru(bpy)$_3^{2+}$ ECL [66,108–111]. The ECL intensity of the reaction is very dependent on the media pH, with the optimum signal-to-noise ratio being obtained when the amino acid is buffered at ~pH 10 using a 0.05 M boric acid buffer [108]. These pH values are higher than the N-terminal amino group pK_a of the amino acid, suggesting that the deprotonation at the amine site prior to oxidation is a very important process. Background reactions between Ru(bpy)$_3^{3+}$ and hydroxide ions at such a high pH range have been observed to be less pH-dependent and do not limit amino acid detection [108]. The ECL mechanism of the amino acid/Ru(bpy)$_3^{2+}$ system is similar to that for TPrA/Ru(bpy)$_3^{2+}$ and can be described as shown in Scheme 12 [108]. Not all amino acids react to yield ECL with the same efficiency [108,109,111]. Proline, the only secondary amine among all the amino acids studied (Fig. 30), gave the highest ECL signal. These data are consistent with the finding described previously that the ECL efficiency of amines follows the order primary < secondary < tertiary amines and that electron-withdrawing R groups (i.e., alcohols, serine, and threonine) tend to decrease ECL whereas electron-donating R groups (i.e., alkyl side chains, leucine, and valine) tend to enhance ECL.

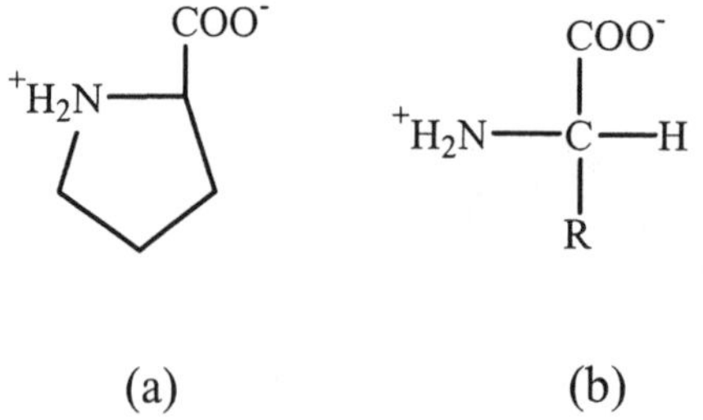

Figure 30 (a) Proline. (b) Common amino acids other than proline.

$$H_3N^+CHRCOO^- + OH^- \longrightarrow H_2NCHRCOO^- + H_2O$$

$$H_2NCHRCOO^- + Ru(bpy)_3{}^{3+} \longrightarrow H_2N^{\bullet+}CHRCOO^- + Ru(bpy)_3{}^{2+}$$

$$H_2N^{\bullet+}CHRCOO^- \longrightarrow H_2NC^{\bullet}RCOO^- + H^+$$

$$H_2NC^{\bullet}RCOO^- + Ru(bpy)_3{}^{3+} \longrightarrow HN{=}CRCOO^- +$$
$$Ru(bpy)_3{}^{2+}* + H^+$$

$$HN{=}CRCOO^- + H_2O \longrightarrow RCOCOO^- + NH_3$$

$$Ru(bpy)_3{}^{2+}* \longrightarrow Ru(bpy)_3{}^{2+} + h\nu$$

Scheme 12

D. Miscellaneous Systems

1. Pyruvate and Other Organic Acids or Salts

As in the case of oxalate described in Section II. A, the oxidation of pyruvate is suppressed by the presence of an oxide layer on the Pt electrode surface [2]. Because the oxidation potentials for pyruvate and $Ru(bpy)_3{}^{2+}$ are almost identical at a reduced Pt electrode, the $Ru(bpy)_3{}^{3+}$ produced electrochemically is insufficiently strong to oxidize the pyruvate in solution, and consequently there is no ECL generated. However, after addition of Ce(III) to an acidic solution containing $Ru(bpy)_3{}^{2+}$ and pyruvate, the ECL reaction proceeds upon anodic oxidation, because Ce(IV) produced from the oxidation of Ce(III) at the Pt electrode at higher potentials than $Ru(bpy)_3{}^{2+}$ oxidation is a strong oxidant and can oxidize pyruvate. This reacts to form the strongly reducing intermediate $CH_3CO^{\bullet}$ [2]. This species behaves like $CO_2{}^{\bullet-}$ and participates in electron transfer reactions with $Ru(bpy)_3{}^{3+}$ and $Ru(bpy)_3{}^{2+}$, as shown in Scheme 1 for oxalate, to produce ECL. The reaction mechanism of this system is summarized in Scheme 13.

$$Ru(bpy)_3^{2+} - e \rightarrow Ru(bpy)_3^{3+}$$

$$Ce^{3+} - e \rightarrow Ce^{4+}$$

$$Ce^{4+} + CH_3COCOO^- \rightarrow Ce^{3+} + CH_3COCOO^{\bullet}$$

$$CH_3COCOO^{\bullet} \rightarrow CH_3CO^{\bullet} + CO_2$$

$$Ru(bpy)_3^{3+} + CH_3CO^{\bullet} \xrightarrow{H_2O} Ru(bpy)_3^{2+*} + CH_3COOH + H^+$$

$$Ru(bpy)_3^{2+} + CH_3CO^{\bullet} \xrightarrow{H_2O} Ru(bpy)_3^{+} + CH_3COOH + H^+$$

$$Ru(bpy)_3^{3+} + Ru(bpy)_3^{+} \rightarrow Ru(bpy)_3^{2+*} + Ru(bpy)_3^{2+}$$

$$Ru(bpy)_3^{2+*} \rightarrow Ru(bpy)_3^{2+} + h\nu$$

Scheme 13

Many organic acids and salts, particularly those of the form R–CH(OH)COOH, such as lactic acid, citric acid, and tartaric acid, can also generate ECL much as pyruvate does in acidic conditions [2,112].

In the absence of Ce(III), ECL was also observed upon the electrochemical oxidation of Ru(bpy)$_3^{2+}$ and a number of organic acids in close to neutral [113] and strongly alkaline solutions [114,115]. However, in these cases, the role of the OH$^-$ ion in the ECL reaction mechanism appears to be crucial [114,115]. For citric acid, tartaric acid, malic acid, and D-gluconic acid, ECL emission (Fig. 31) resulted mainly from the reaction between alkoxide radical and Ru(bpy)$_3^{3+}$ in the potential range of 1.3–1.8 V vs. Ag/AgCl, as– illustrated in Scheme 14 [114]. At higher potentials, the ECL mechanism is very complicated and could involve a number of high-energy agents such as CO$_2^{\bullet-}$, OH$^\bullet$, O$^{\bullet-}$, $^\bullet$CHO, and HCOO$^{\bullet-}$ [114].

$$Ru(bpy)_3^{2+} - e \longrightarrow Ru(bpy)_3^{3+}$$
$$RCOH + OH^- \longrightarrow RCO^- + H_2O$$
$$RCO^- - e \longrightarrow RCO^\bullet$$
$$RCO^\bullet + Ru(bpy)_3^{3+} \longrightarrow Ru(bpy)_3^{2+*} + RCOO^-$$
$$2OH^- + RCO^\bullet + Ru(bpy)_3^{3+} \longrightarrow Ru(bpy)_3^{2+*} + RCOO^- + H_2O$$
$$Ru(bpy)_3^{2+*} \longrightarrow Ru(bpy)_3^{2+} + h\nu$$

$$R = -(CH_2COOH)_2COOH$$

Scheme 14

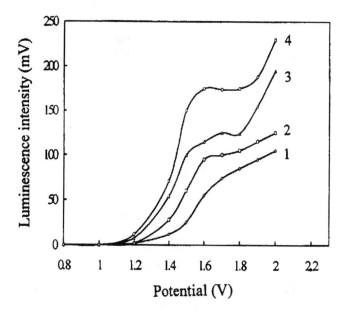

Figure 31 ECL intensity vs. applied potential profiles for (1) D-gluconic acid, (2) tartaric acid, (3) malic acid, and (4) citric acid. Carrier solution, 1.0 mM Ru(bpy)$_3^{2+}$, 0.1 M KOH, 0.1 M KF; flow rate, 0.35 mL/min; concentration of each acid, 50 μM; injection volume, 10 μL. (From Ref. 114.)

For the ascorbic acid/dehydroascorbic acid–$Ru(bpy)_3^{2+}$ system, Zorzi et al. [115] reported that the maximum light emission is at pH 10 and the ECL reaction between $Ru(bpy)_3^{3+}$ and the analyte does not occur significantly at pH values lower than 8, which is different from the reported of Chen et al. [113], who observed the maximum emission at pH 7. Zorzi et al. also pointed out that the real active coreactant species is dehydroascorbic acid or its related species and not ascorbic acid. The ECL reaction sequence could include the very reactive $OH^{\bullet}$ species produced from the reaction between $Ru(bpy)_3^{3+}$ and OH^- ions as well as an intermediate associated with dehydroascorbic acid and $OH^{\bullet}$ radicals [115].

2. Alcohols

In aqueous alkaline solutions (pH $\geq$ 10), anodic oxidation of alcohols or $Ru(bpy)_3^{2+}$ can generate weak ECL signals, presumably due to the formation of the excited state RCHO* [116–121] and the reaction between electrogenerated $Ru(bpy)_3^{3+}$ and hydroxyl ions [74], respectively. The ECL intensity was significantly enhanced after addition of the alcohols to $Ru(bpy)_3^{2+}$ alkaline solutions upon electrochemical oxidation [119–121] (Fig. 32). The alkoxy radical, $RCH_2O^{\bullet}$, generated by oxidation of the alcohols was proposed to be the reducing agent that reacts with electrogenerated $Ru(bpy)_3^{3+}$ to form excited state $Ru(bpy)_3^{2+}$* species. Scheme 15 summarizes the ECL mechanism of this system. The ECL intensity was found to be related to the alcohol structure [120]. Monohydric alcohols with shorter alkyl chain lengths produced stronger luminescence, and polyhydric alcohols generated increasing ECL intensities with increasing numbers of hydroxyl groups. Adjacent hydroxyl groups produced weaker ECL than separated hydroxyl groups.

$$Ru(bpy)_3^{2+} - e \longrightarrow Ru(bpy)_3^{3+}$$

$$RCH_2OH + OH^- \longrightarrow RCH_2O^- + H_2O$$

$$RCH_2O^- - e \longrightarrow RCH_2O^{\bullet}$$

$$RCH_2O^{\bullet} + OH^- \longrightarrow RCH_2O^- + H_2O$$

$$2RCH_2O^{\bullet} \longrightarrow RCHO^* + RCH_2OH$$

$$RCHO^* \longrightarrow RCHO + h\nu \quad (\lambda_{max} \sim 550 \text{ nm, weak})$$

$$RCHO^- + 3OH^- - 3e \longrightarrow RCOO^- + H_2O$$

$$4OH^- + RCH_2O^{\bullet} + 3\,Ru(bpy)_3^3 \longrightarrow {}^+RCOO^- + 3Ru(bpy)_3^{2+}* + 3H_2O$$

$$Ru(bpy)_3^{2+}* \longrightarrow Ru(bpy)_3^{2+} + h\nu \quad (\lambda_{max} \sim 608 \text{ nm})$$

Scheme 15

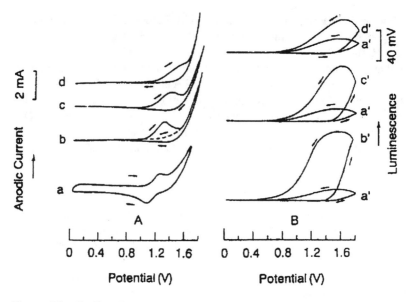

Figure 32 Cyclic voltammograms (A) and ECL intensity–potential profile curves (B) obtained at a glassy carbon electrode at a scan rate of 50 mV/s (reference electrode Ag/AgCl). (A) (a) 0.5 mM Ru(bpy)$_3^{2+}$ + electrolyte; (b) 50 mM methanol + electrolyte (---, electrolyte only); (c) 50 mM 1-propanol + electrolyte; (d) 50 mM 1-pentanol + electrolyte. (B)□(a′) 50 μM Ru(bpy)$_3^{2+}$ + electrolyte, (b′) 10 mM methanol + a′, (c′) 10 mM 1-propanol + a′, and (d′) 10 mM 1-pentanol + a′. Electrolyte: 0.20 M KOH-KCl, pH 12.0. (From Ref. 120.)

3. Hydrazine

Although numerous studies of hydrazine-related chemiluminescence have been reported, few papers have dealt with hydrazine as an ECL coreactant [122–124]. As proposed by Hercules and Lytle [125,126], the mechanism for CL reactions between Ru(bpy)$_3^{3+}$ and hydrazine is complicated; a number of high- energy intermediates such as N$_2$H$_3^\bullet$ and N$_2$H$_2$, and even HO$_2^\bullet$ and OH$^\bullet$, may be involved in it [115,125–127]. On the basis of the postulated CL reaction sequence (Scheme 16), Hercules and Lytle [125,126] were able to fit the experimentally observed CL intensity and Ru(bpy)$_3^{2+}$ vs. time profiles well with the data obtained from digital simulations at low concentrations [Ru(bpy)$_3^{3+}$ ~ 0.5 mM; N$_2$H$_4$ ~ 25 mM]. At higher hydrazine concentrations a qualitative fit was also observed. The kinetic data indicated that the reactions between Ru(bpy)$_3^{3+}$ and N$_2$H$_4$ and between Ru(bpy)$_3^{3+}$ and N$_2$H$_2$ are about 1% and about 99% efficient, respectively [126].

$$Ru(bpy)_3^{3+} + N_2H_4 \longrightarrow Ru(bpy)_3^{2+*} + N_2H_3^{\bullet}$$

$$2N_2H_3^{\bullet} \longrightarrow N_2 + NH_3$$

$$Ru(bpy)_3^{3+} + N_2H_3^{\bullet} \longrightarrow N_2H_2 + Ru(bpy)_3^{2+}$$

$$Ru(bpy)_3^{3+} + N_2H_2 \longrightarrow Ru(bpy)_3^{2+*} + N_2$$

$$Ru(bpy)_3^{2+*} \longrightarrow Ru(bpy)_3^{2+} + h\nu$$

Scheme 16

Both *ex situ* and *in situ*, electrochemically generated $Ru(bpy)_3^{3+}$ has been found to be a suitable reagent for the sensitive detection of gas-phase hydrazine and its derivatives dimethyl hydrazine and monomethyl hydrazine at pH ~6–7 [122]. Hydrazine can also be sensitively detected on the basis of the hydrazine–luminol ECL system upon the anodic oxidation at a preanodized Pt electrode [123]. The preanodized Pt electrode was found to have an electrocatalytic function for hydrazine oxidation. In this system, hydrazine is first oxidized at the preanodized Pt electrode (~0.25 V vs. SCE), and the electro-oxidation product of hydrazine, possibly a hydrazine radical, is produced in the diffusion layer of the electrode. Then this strongly reducing radical further reduces the dissolved oxygen in the solution to generate a superoxide anion radical. Finally, the reaction of the superoxide anion radical with the luminol present in the solution causes the emission of a strong light [123].

4. Oxygen

In the presence of oxygen, ECL was observed upon the reduction of bis(2,4,6-trischlorophenyl) oxalate (TCPO) solutions containing a fluorescer compound [DPA, rubrene, or $Ru(bpy)_3^{2+}$] at potentials corresponding to TCPO and O_2 reduction, where the ECL spectrum is typical of the fluorescer [128]. It was proposed that the electrochemical reduction products, $TCPO^{\bullet-}$ and $O_2^{\bullet-}$, at a Pt electrode in MeCN/benzene (2:1 v/v) solutions can rapidly react and form an intermediate, probably dioxetane, that reacts with the fluorescer to yield an excited state of the emitter and CO_2 (Scheme 17). Note that in the course of ECL generation it is the fluorescer itself and not its radical anion that is involved in the ECL reaction. Actually, at potentials where the radical anion (e.g., $DPA^{\bullet-}$) is produced, the ECL intensity decreases [128].

The effect of oxygen on the ECL of the ruthenium trisbipyridyl–viologen complexes has been recently reported [129]. The light emission observed previously [130] was found, in fact, to be from decomposition products formed during the ECL experiments in the presence of trace amounts of oxygen.

Another example of ECL resulting from the decomposition product due to an oxygen effect in the course of the ECL experiment was reported by Okajima

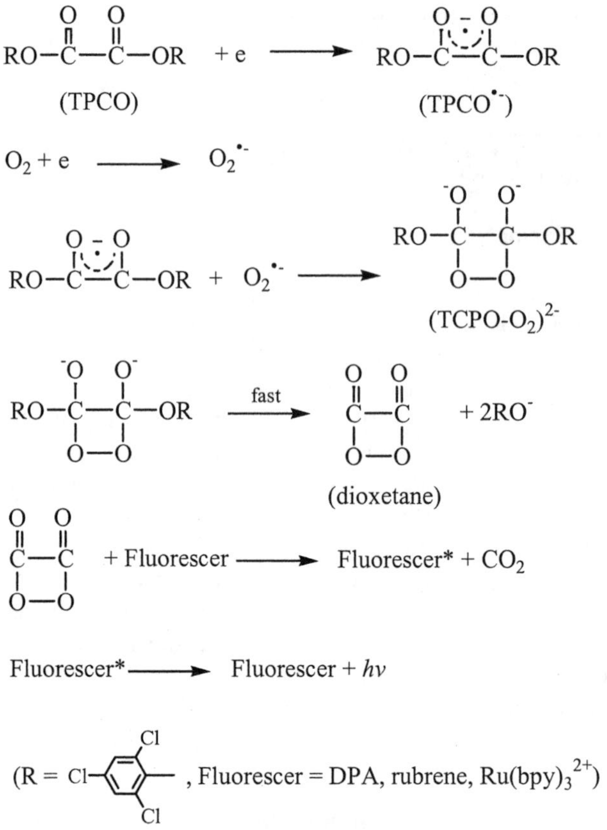

$$RO-\overset{\overset{O}{\|}}{C}-\overset{\overset{O}{\|}}{C}-OR \quad + e \quad \longrightarrow \quad RO-\overset{O \; - \; O}{\overset{\overset{\bullet}{\cdots}}{C}=C}-OR$$

(TPCO)   (TPCO·⁻)

$$O_2 + e \quad \longrightarrow \quad O_2^{\bullet-}$$

$$RO-\overset{O \; - \; O}{\overset{\overset{\bullet}{\cdots}}{C}=C}-OR \quad + \; O_2^{\bullet-} \quad \longrightarrow \quad RO-\overset{\overset{O^-}{|}}{\underset{\underset{O-O}{|}}{C}}-\overset{\overset{O^-}{|}}{C}-OR$$

(TCPO-O$_2$)$^{2-}$

$$RO-\overset{\overset{O^-}{|}}{\underset{\underset{O-O}{|}}{C}}-\overset{\overset{O^-}{|}}{C}-OR \quad \overset{fast}{\longrightarrow} \quad \overset{\overset{O}{\|}}{\underset{\underset{O-O}{|}}{C}}-\overset{\overset{O}{\|}}{C} \quad + \; 2RO^-$$

(dioxetane)

$$\overset{\overset{O}{\|}}{\underset{\underset{O-O}{|}}{C}}-\overset{\overset{O}{\|}}{C} \quad + \; Fluorescer \quad \longrightarrow \quad Fluorescer^* + CO_2$$

$$Fluorescer^* \longrightarrow \quad Fluorescer + h\nu$$

$$(R = Cl-\underset{Cl}{\overset{Cl}{\bigcirc}}- \; , Fluorescer = DPA, rubrene, Ru(bpy)_3^{2+})$$

Scheme 17

et al [131]. In this case, significant light emission was observed only on the reverse scan of the cyclic voltammetric experiment for the reduction of 3-methylindole (3-MIH) in air- saturated MeCN solutions (Fig. 33). The experimentally measured ECL spectrum in MeCN (λ_{max} ~475 nm) was similar to the fluorescence spectrum observed for o-formamidoacetophenone anion in DMSO solution (λ_{max} ~490 nm) [132], but not close to the fluorescence spectrum of 3-MIH in an air-saturated MeCN solution containing 0.1 M TEAP (λ_{max} ~343 nm). The proposed ECL mechanism of this system is shown in Scheme 18.

Previously, Okajima et al. [133,134] had observed the ECL of the MCLA (a *Cypridina luciferin* analog, Scheme 19) at a basal plane pyrolytic graphite electrode in air-saturated MeCN solutions (Fig. 34). In contrast to the ECL behavior of 3-MIH described above, this system can produce light upon the electroreduction of O_2 and MCLA and also on the reverse scan, when MCLA⁻ is

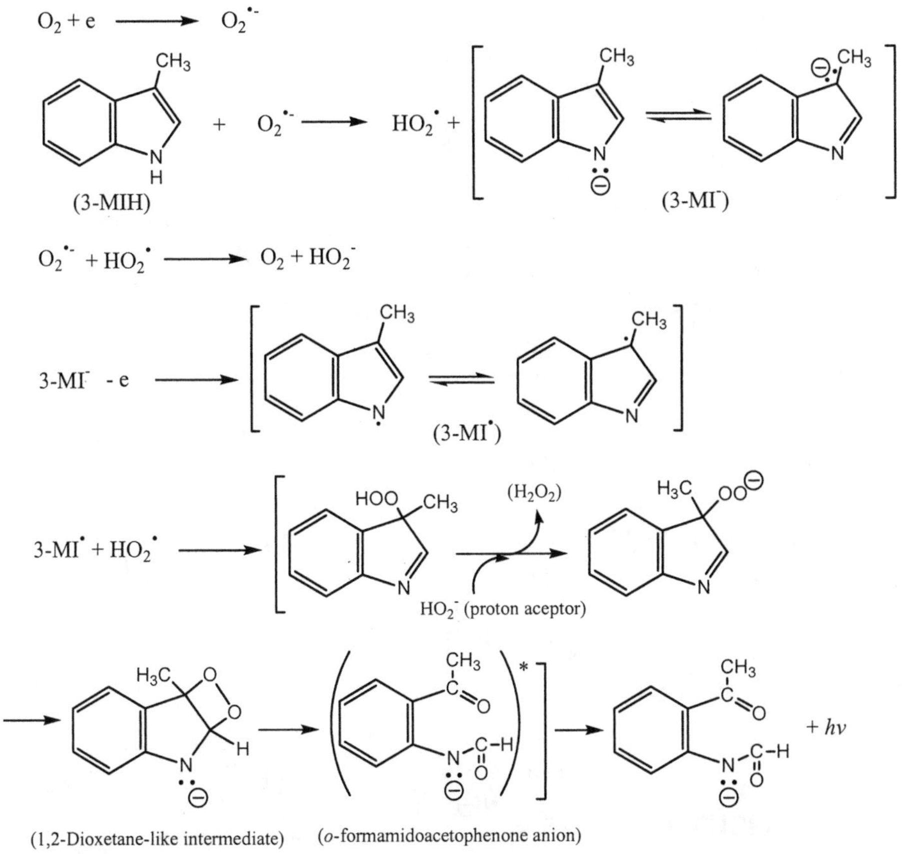

Scheme 18

believed to be mainly oxidized by O_2 regenerated at the electrode from O_2^- to MCLA$^\bullet$. Direct oxidation of MCLA$^-$ at the electrode occurs at ~ -0.25 V vs. Ag/AgCl [Fig. 34a(4)]. Scheme 19 shows the proposed ECL reaction sequence of this system.

Cathodic ECL of Ru(bpy)$_3^{2+}$ was recently reported at a glassy carbon electrode in aqueous solutions containing traditional alkylamine or organic acid coreactant in the presence of oxygen [135]. The ECL appears upon the electroreduction of oxygen at potentials of approximately -0.4 V vs. Ag/AgCl. Although the exact mechanism of this kind of ECL is still unclear, it is believed that a "reactive oxygen species" such as $O_2^{\bullet-}$, H_2O_2, or OH$^\bullet$ generated after oxygen reduction may play an important role during this process.

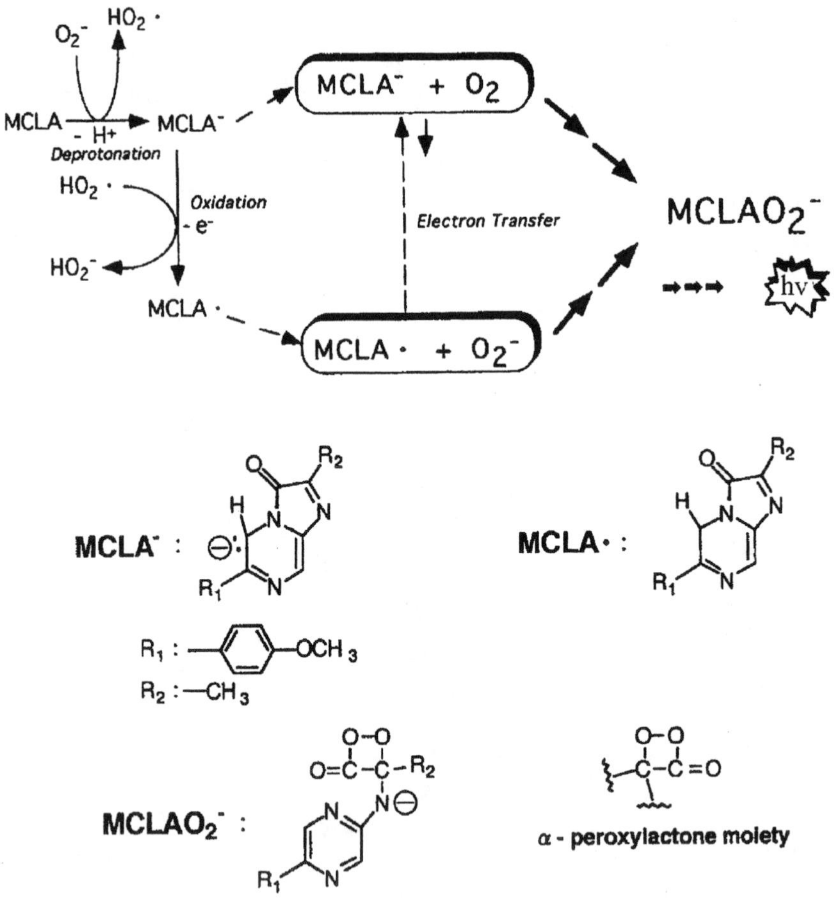

Scheme 19   (From Ref. 133.)

5.   Peroxide Systems

Peroxides can be used as coreactants because they produce reactive oxidizing agents when they are reduced. The reactivity of the produced oxidizing agents strongly depends on the electron density of the oxygen–oxygen (O–O) bond. Usually, peroxides with a low electron density on the O–O bond produce strong reactive oxidizing agents. Because the electron density is strongly affected by the functional groups attached to the O–O bond, oxidizing agents generated from peroxides that contain electron-withdrawing groups are more reactive than those formed from peroxides with electron-donating groups.

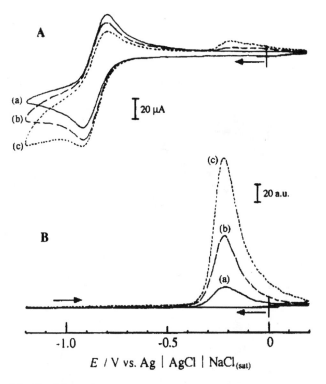

Figure 33 (A) Cyclic voltammograms and (B) light emission (λ_{max} ~475 nm) vs. potential profiles recorded simultaneously during the reduction of oxygen ($E°O_2/O_2^{\bullet-}$ ~ −0.86 V vs. Ag/AgCl) at a basal plane pyrolytic graphite electrode (area ~0.08 cm^2) in air-saturated MeCN solutions containing 0.1 M TEAP in the presence of (a) 0.5, (b) 1.7, and (c) 5.0 mM 3-MIH. Scan rate = 50 mV/s. (From Ref. 131.)

Benzoyl peroxide (BPO) (Fig. 35) is a good coreactant and has been used as an oxidizing agent to generate CL. For example, Chandross and Sonntag [136] demonstrated the generation of chemiluminescence by mixing DPA$^{\bullet-}$ and BPO. The proposed CL mechanism is as shown in Scheme 20.

$$DPA^{\bullet-} + BPO \longrightarrow DPA + BPO^{\bullet-}$$
$$BPO^{\bullet-} \longrightarrow C_6H_5CO_2^- + C_6H_5CO_2^{\bullet}$$
$$DPA^{\bullet-} + C_6H_5CO_2^{\bullet} \longrightarrow {}^1DPA^* + C_6H_5CO_2^-$$

Scheme 20

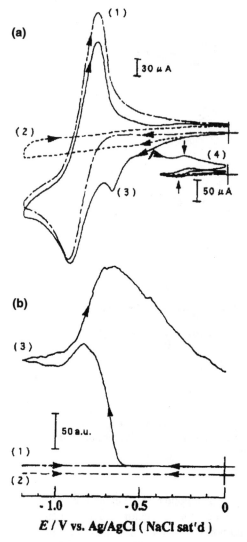

Figure 34 (a) Typical cyclic voltammograms and (b) light emission vs potential profiles (λ_{max} ~470 nm) obtained during the redox reaction of the O_2/O_2^- couple at basal plane pyrolytic graphite electrodes in (1,3,4) air-saturated and (2) N_2-saturated MeCN solutions containing 0.1 M TEAP (2,3,4) in the presence of and (1) in the absence of 0.41 mM MCLA. The scan rate used for 1–3 is 50 mV/s and for 4, 500 mV/s. (From Ref. 133.)

Likewise, BPO can be electrochemically reduced via an overall two-electron ECL reaction as shown in Scheme 21. BPO is first reduced to BPO$^{\bullet-}$ at the electrode by accepting one electron, and then this newly formed BPO$^{\bullet-}$ species immediately decomposes to benzoate anion and benzoate radical. The benzoate radical can be further reduced to benzoate anion by one electron.

$$BPO + e \longrightarrow BPO^{\bullet -}$$
$$BPO^{\bullet -} \longrightarrow C_6H_5CO_2^- + C_6H_5CO_2^{\bullet}$$
$$C_6H_5CO_2^{\bullet} + e \longrightarrow C_6H_5CO_2^-$$
$$\text{or } BPO + 2e \longrightarrow 2C_6H_5CO_2^-$$

Scheme 21

The benzoate radical formed is energetic enough to react with some cation radicals to form the excited states, because the redox potential value for the $C_6H_5CO_2^{\bullet}/C_6H_5CO_2^-$ couple ($> +1.5$ V vs. SCE) [136], which was estimated by using the CL reaction enthalpy of the emitter with benzoyl peroxide and the excited singlet state energy of an emitter, is quite positive. Akins and Coworkers [137,138] also demonstrated that ECL of various aromatic hydrocarbons, such as 9,10-DPA, rubrene, fluoranthene, and anthracene, can be generated in the presence of benzoyl peroxide after the electrode is brought to a potential of -1.90 V vs. SCE (Fig. 36). In contrast to the mechanism shown in Scheme 20, the ECL mechanism proposed by Akins and coworkers was considered to be energy-deficient (Scheme 22).

$$R + e \longrightarrow R^{\bullet -}$$
$$BPO + ne \longrightarrow \text{products}$$
$$R^{\bullet -} + BPO \longrightarrow R + BPO^{\bullet -}$$
$$BPO^{\bullet -} \longrightarrow C_6H_5CO_2^- + C_6H_5CO_2^{\bullet}$$
$$R^{\bullet -} + C_6H_5CO_2^{\bullet} \longrightarrow {}^3R^* + C_6H_5CO_2^-$$
$${}^3R^* + {}^3R^* \longrightarrow {}^1R^* + R$$
$${}^3R^* + BPO \longrightarrow R^{\bullet +} + BPO^{\bullet -}$$
$${}^3R^* + C_6H_5CO_2^{\bullet} \longrightarrow R^{\bullet +} + C_6H_5CO_2^-$$
$$R^{\bullet +} + R^{\bullet -} \longrightarrow {}^1R^* + R$$
$${}^1R^* \longrightarrow R + h\nu$$

R = aromatic hydrocarbons.

Scheme 22

The excited singlet state is produced by the triplet–triplet annihilation as well as by some anion–cation annihilation, where the cation is formed via a triplet quenching reaction. When the BPO concentration is relatively high

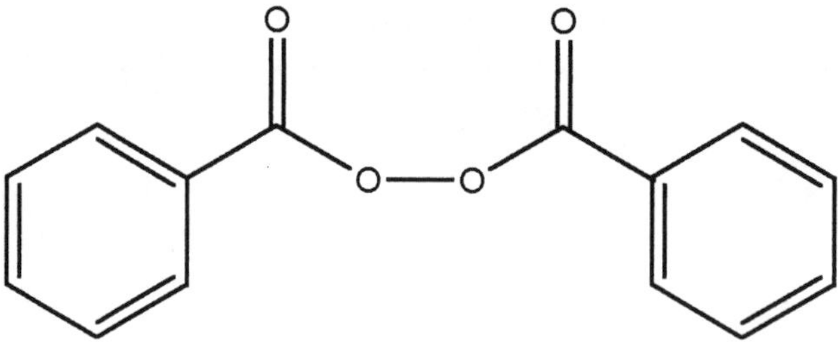

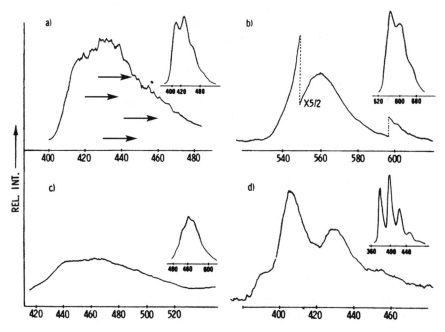

Figure 35 Molecular structure of benzoyl peroxide.

Figure 36 Chemiluminescence from systems of electrogenerated aromatic hydrocarbon anions and benzoyl peroxide (BPO). In all cases the supporting electrolyte used was tetraethylammonium perchlorate (TEAP) with N,N-dimethyl formamide (DMF) as the solvent. (a) 0.88 mM DPA and 1.3 mM BPO in DMF containing 0.1 M TEAP: potentiostated at −1.9 V vs. SCE, slit width 700 μm and RC time constant (TC) 7.5 s; (b) 0.43 mM rubrene and 0.41 mM BPO in 0.1 M TEAP in DMF: potentiostated at −1.4 V vs. SCE, slit width 500 μm, and TC = 14.7 s; (c) 1.06 mM fluoranthene and 4.2 mM BPO in 0.1 M TEAP in DMF: potentiostated at –1.9 V vs. SCE, slit width 1 mm, TC = 14.7 s; and (d) 1.0 mM anthracene and 1 mM BPO in 0.1 M TEAP in DMF. All wavelengths in nanometers. Inserts show fluorescence spectra, generally at higher resolution. (From Ref. 137.)

compared to that of aromatic hydrocarbons, the rate of the triplet quenching reaction ultimately becomes fast and the generation of cation species is predominant over the triplet–triplet annihilation. In this case, ECL is generated mainly through the anion–cation annihilation reaction. On the basis of this consideration, these authors [137,138] claimed that a redox potential value of +0.8 V vs. SCE for the $C_6H_5CO_2/C_6H_5CO_2^-$ couple would be energetic enough to produce ECL.

Hydrogen peroxide has also been used to generate light. Its electrochemical reduction is considered to be a two-step EE process [45]:

$$H_2O_2 + e \longrightarrow OH^- + OH^\bullet$$
$$OH^\bullet + e \longrightarrow OH^-$$

During the electrochemical reduction process, H_2O_2 produces a very strong oxidizing agent, hydroxyl radical, which has a redox potential value of +2.7 V vs SEC [45]. Because hydroxyl radicals undergo not only an energetic electron transfer reaction but also a chemical reaction to form adduct species, most ECL reactions involving H_2O_2 can, in fact, include decomposition reactions of hydroxyl radical adducts. The decomposition products can then produce luminescence [10,139,140]. Because the related mechanism is beyond our current coreactant concept, we do not discuss the details here. However, it is worthwhile to point out that H_2O_2 could be a promising cathodic ECL coreactant, because hydroxyl radical has such a high positive potential.

III. CONCLUSIONS

As we have seen in the previous sections, coreactant ECL does not require the generation of both oxidized and reduced forms of a luminophore. This can be a significant advantage, e.g., in analysis, compared to ECL produced via the annihilation reaction. First, the use of a coreactant can make ECL possible even for some fluorescent compounds that have only a reversible electrochemical reduction or oxidation. Second, even with solvents for ECL that have a narrow potential window so that only a reduced or oxidized form of a luminophore can be produced, it is still possible to generate ECL by use of a coreactant. Finally, when the annihilation reaction between oxidized and reduced species is not efficient, the use of a coreactant may produce more intense ECL.

To be a good coreactant candidate, the following criteria are generally required:

1. Solubility. The coreactant should be reasonably soluble in the reaction media, because the ECL intensity is generally proportional to the concentration of the coreactants.

2. Stability.   The intermediate species generated electrochemically and chemically should be sufficiently stable to allow appreciable reaction with the ECL precursor.

3. Electrochemical Properties.   The coreactant should be easily oxidized or reduced with the luminophore species at or near the electrode and undergo a rapid chemical reaction to form an intermediate that has sufficient reducing or oxidizing energy to react with the oxidized or reduced luminophore to form the excited state.

4. Kinetics.   The reaction rate between the intermediate and the oxidized or reduced luminophore species must be rapid [88].

5. Quenching Effect.   The coreactant and its redox products should not be good quenchers of the ECL compound's luminescence [88].

6. ECL Background.   The coreactant itself should not give any ECL signal over the potential range scanned.

REFERENCES

1.   Bard, A.J.; Faulkner, L.R. Electrochemical Methods: Fundamentals and Applications. 2nd ed.; Wiley: New York, 2001; pp 736–745.

2.   Rubinstein, I. Bard, A.J. Electrogenerated chemiluminescence. 37. Aqueous ECL systems based on tris(2,2'-bipyridine)ruthenium (2+) and oxalate or organic acids. J. Am. Chem. Soc. **1981**, *103*, 512–516.

3.   White, H.S.; Bard, A.J. Electrogenerated chemiluminescence. 41. Electrogenerated chemiluminescence and chemiluminescence of the tris(2,2'-bipyridine)ruthenium(2+)-peroxydisulfate(2−) system in acetonitrile-water solutions. J. Am. Chem. Soc. **1982**, *104*, 6891–6895.

4.   Andersson, A.-M.; Schmehl, R.H. Sensors based on electrogenerated chemiluminescence. Mol. Supramol. Photochem. **2001**, *7*, 153–187.

5.   Knight, A.W. Electrogenerated chemiluminescence. In *Chemiluminescence in Analytical Chemistry*; Grarcia-Campana, A.M., Baeyens, W.R.G., Eds.; Marcel Dekker: New York, 2001; pp 211–247.

6.   Richter, M.M. Electrochemiluminescence. In *Optical Biosensors: Present and Future*; Ligler, F.S., Taitt, C.A.R., Eds.; Elsevier: New York, 2002; pp 173–205.

7.   Bard, A.J.; Debad, J.D.; Leland, J.K.; Sigal, G.B.; Wilbur, J.L.; Wohlsatdter, J.N. Chemiluminescence: electrogenerated. In *Encyclopedia of Analytical Chemistry: Applications, Theory and Instrumentation*; Meyers, R.A., Ed.; Wiley: New York, 2000; Vol. 11, pp 9842–9849.

8.   Knight, A.W. A review of recent trends in analytical applications of electrogenerated chemiluminescence. TrAC, Trends Anal. Chem. **1999**, *18*, 47–62.

9.   Knight, A.W.; Greenway, G.M. Relationship between structural attributes and observed electrogenerated chemiluminescence (ECL) activity of tertiary amines as potential analytes for the tris(2,2-bipyridine)ruthenium(II) ECL reaction—a review. Analyst (Cambridge, UK) **1996**, *121*, 101R–106R.

10. Fahnrich, K.A.; Pravda, M.; Guilbault, G.G. Recent applications of electrogenerated chemiluminescence in chemical analysis. Talanta **2001**, *54*, 531–559.

11. Gerardi, R.D.; Barnett, N.W.; Lewis, A.W.. Analytical applications of tris(2,2'-bipyridyl)ruthenium (III) as a chemiluminescent reagent. Anal. Chim. Acta **1999**, *378*, 1–41.

12. Lee, W.Y. Tris-(2,2'-bipyridyl)ruthenium(II) electrogenerated chemiluminescence in analytical science. Mikrochim. Acta **1997**, *127*, 19–39.

13. Juris, A.; Balzani, V.; F Barigelletti, Campagna, S.; P Belser, Von Zelewsky, A. Ruthenium(II) polypyridine complexes: photophysics, photochemistry, electrochemistry, and chemiluminescence. Coord. Chem. Rev. **1988**, *84*, 85–277.

14. Hadd, A.G.; Birks, J.W. Peroxyoxalate chemiluminescence: mechanism and analtical detection. In *Selective Detectors: Environmental, Industrial, and Biomedical Applications*; Sievers, R.E., Ed.; Wiley: New York, 1995; Vol. 131, pp 209–240.

15. Chang, M.-M.; Saji, T.; Bard, A.J. Electrogenerated chemiluminescence. 30. Electrochemical oxidation of oxalate ion in the presence of luminescers in acetonitrile solutions. J. Am. Chem. Soc. **1977**, *99*, 5399–5403.

16. Butler, J.; Henglein, A. Elementary reactions of the reduction of thallium(1+) in aqueous solution. Rad. Phys. Chem. **1980**, *15*, 603–612.

17. Lingane, J.J. Chronopotentiometry with platinum and gold anodes. II. Oxidation of oxalic acid. J. Electroanal. Chem. **1960**, *1*, 379–395.

18. Giner, J. Anodic oxidation of oxalic acid on platinum. I. Passivation effects on electrodes of bright platinum. Electrochim. Acta **1961**, *4*, 42–54.

19. Anson, F.C.; Schultz, F.A. Effect of adsorption and electrode oxidation on the oxidation of oxalic acid at platinum electrodes. Anal. Chem. **1963**, *35*, 1114–1116.

20. Johnson, J.W.; Wroblowa, H.; Bockris, J.O.M. The mechanism of the electrochemical oxidation of oxalic acid. Electrochim. Acta **1964**, *9*, 639–651.

21. Keszthelyi, C.P.; Tachikawa, H.; Bard, A.J. Electrogenerated chemiluminescence. VIII. Thianthrene-2,5-diphenyl-1,3,4-oxadiazole system. Mixed energy-sufficient system. J. Am. Chem. Soc. **1972**, *94*, 1522–1527.

22. Ege, D.; Becker, W.G.; Bard, A.J. Electrogenerated chemiluminescent determination of tris(2,2'-bipyridine)ruthenium ion $(Ru(bpy)_3^{2+})$ at low levels. Anal. Chem. **1984**, *56*, 2413–2417.

23. Downey, T.M.; Nieman, T.A. Chemiluminescence detection using regenerable tris(2,2'-bipyridyl)ruthenium(II) immobilized in Nafion. Anal. Chem. **1992**, *64*, 261–268.

24. Kanoufi, F.; Bard, A.J. Electrogenerated chemiluminescence. 65. An investigation of the oxidation of oxalate by tris(polypyridine) ruthenium complexes and the effect of the electrochemical steps on the emission intensity. J. Phys. Chem. B **1999**, *103*, 10469–10480.

25. *CRC Handbook of Chemistry and Physics*; Weast, R.C., Astle,M.J., Beyer, W.H., Ed.; 67. CRC Press: Boca Raton, FL, 1987; p D162.

26. Malins, C.; Vandeloise, R.; D Walton, Donckt, E.V. Ultrasonic modification of light emission from electrochemiluminescence processes. J. Phys. Chem. A **1997**, *101*, 5063–5068.

27. Rubinstein, I.; Bard, A.J. Polymer films on electrodes. 4. Nafion-coated electrodes and electrogenerated chemiluminescence of surface-attached tris(2,2′-bipyridine) ruthenium(2+). J. Am. Chem. Soc. **1980**, *102*, 6641–6642.

28. Rubinstein, I.; Bard, A.J. Polymer films on electrodes. 5. Electrochemistry and chemiluminescence at Nafion-coated electrodes. J. Am. Chem. Soc. **1981**, *103*, 5007–5013.

29. Egashira, N.; Kumasako, H.; Ohga, K. Fabrication of a fiber-optic-based electrochemiluminescence sensor and its application to the determination of oxalate. Anal. Sci. **1990**, *6*, 903–904.

30. Zhang, X.; Bard, A.J. Electrogenerated chemiluminescent emission from an organized (L-B) monolayer of a tris(2,2′-bipyridine)ruthenium(2+)-based surfactant on semiconductor and metal electrodes. J. Phys. Chem. **1988**, *92*, 5566–5569.

31. Obeng, Y.S.; Bard, A.J. Electrogenerated chemiluminescence. 53. Electrochemistry and emission from adsorbed monolayers of a tris(bipyridyl)ruthenium(II)-based surfactant on gold and tin oxide electrodes. Langmuir **1991**, *7*, 195–200.

32. Sato, Y.; Uosaki, K. Electrochemical and electrogenerated chemiluminescence properties of tris(2,2′-bipyridine)ruthenium(II)-tridecanethiol derivative on ITO and gold electrodes. J. Electroanal. Chem. **1995**, *384*, 57–66.

33. Andersson, A.-M.; Isovitsch, R.; Miranda, D.; Wadhwa, S.; Schmehl, R.H. Electrogenerated chemiluminescence from Ru(II) bipyridylphosphonic acid complexes adsorbed to mesoporous TiO_2/ITO electrodes. Chem. Commun. (Cambridge) **2000**, 505–506.

34. Sykora, M.; Meyer, T.J. Electrogenerated chemiluminescence in SiO_2 sol-gel polymer composites. Chem. Mater. **1999**, *11*, 1186–1189.

35. Zu, Y.; Fan, F.-R.F.; Bard, A.J. Inverted region electron transfer demonstrated by electrogenerated chemiluminescence at the liquid/liquid interface. J. Phys. Chem. B **1999**, *103*, 6272–6276.

36. Kanoufi, F.; Cannes, C.; Zu, Y.; Bard, A.J. Scanning electrochemical microscopy. 43. Investigation of oxalate oxidation and electrogenerated chemiluminescence across the liquid-liquid interface. J. Phys. Chem. B **2001**, *105*, 8951–8962.

37. Lee, C.W.; Ouyang, J.; Bard, A.J. Electrogenerated chemiluminescence. Part 51. The tris(2,2′-bipyrazine)osmium(II) system. J. of Electroanal. Chem. Interfacial Electrochem. **1988**, *244*, 319–324.

38. Rodman, G.S.; Bard, A.J. Electrogenerated chemiluminescence. 52. Binuclear iridium(I) complexes. Inorg. Chem. **1990**, *29*, 4699–4702.

39. Egashira, N.; Kumasako, H.; Y Kurauchi, Ohga, K. Determination of oxalate in vegetables with a fiber-optic electrochemiluminescence sensor. Anal. Sci. **1992**, *8*, 713–714.

40. Skotty, D.R.; Nieman, T.A. Determination of oxalate in urine and plasma using reversed-phase ion-pair high-performance liquid chromatography with tris(2,2′-bipyridyl)ruthenium(II)-electrogenerated chemiluminescence detection. J. Chromatogr. B: Biomed. Appl. **1995**, *665*, 27–36.

41. Rubinstein, I.; Martin, C.R.; Bard, A.J. Electrogenerated chemiluminescent determination of oxalate. Anal. Chem. **1983**, *55*, 1580–1582.

42. Skotty, D.R.; Lee, W.-Y.; Nieman, T.A. Determination of dansyl amino acids and oxalate by HPLC with electrogenerated chemiluminescence detection using tris(2,2'-bipyridyl)ruthenium(II) in the mobile phase. Anal. Chem. **1996**, *68*, 1530–1535.

43. Barnett, N.W.; Bowser, T.A.; Russell, R.A. Determination of oxalate in alumina process liquors by ion chromatography with post column chemiluminescence detection. Anal. Proc. **1995**, *32*, 57–59.

44. Bolletta, F.; Ciano, M.; V Balzani, Serpone, N. Polypyridine transition metal complexes as light emission sensitizers in the electrochemical reduction of the persulfate ion. Inorg. Chim. Acta **1982**, *62*, 207–213.

45. Memming, R. Mechanism of the electrochemical reduction of persulfates and hydrogen peroxide. J. Electrochem. Soc. **1969**, *116*, 785–790.

46. Fabrizio, E.F.; Prieto, I.; Bard, A.J. Hydrocarbon cation radical formation by reduction of peroxydisulfate. J. Am. Chem. Soc. **2000**, *122*, 4996–4997.

47. Bolletta, F.; Juris, A.; Maestri, M.; Sandrini, D. Quantum yield of formation of the lowest excited state of tris(bipyridine)ruthenium(II) Ru(bpy)$_3^{2+}$ and tris(phenanthroline) ruthenium(II) Ru(phen)$_3^{2+}$. Inorg. Chim. Acta **1980**, *44*, L175–L176.

48. Kang, S.C.; Oh, S.; Kim, K.J. Enhanced electrogenerated chemiluminescence of tris(2,2'-bipyridyl)ruthenium(II)-S$_2$O$_8^{2-}$ system by sodium dodecyl sulfate. Bull. Korean Chem. Soc. **1990**, *11*, 505–508.

49. Xu, G.; Dong, S. Electrochemiluminescence of the Ru(bpy)$_3^{2+}$/S$_2$O$_8^{2-}$ system in purely aqueous solution at carbon paste electrode. Electroanalysis **2000**, *12*, 583–587.

50. Xu, G.; Dong, S. Electrochemiluminescent determination of peroxydisulfate with Cr(bpy)$_3^{3+}$ in purely aqueous solution. Anal. Chim. Acta **2000**, *412*, 235–240.

51. Yamazaki-Nishida, S.; Harima, Y.; Yamashita, K. Direct current electrogenerated chemiluminescence observed with tris(2,2'-bipyrazine)ruthenium(2+) in fully aqueous solution. J. Electroanal. Chem. Interfacial Electrochem. **1990**, *283*, 455–458.

52. Yamashita, K.; S Yamazaki-Nishida, Harima, Y.; Segawa, A. Direct current electrogenerated chemiluminescent microdetermination of peroxydisulfate in aqueous solution. Anal. Chem. **1991**, *63*, 872–876.

53. Ouyang, J.; Zietlow, T.C.; Hopkins, M.D.; Fant, F.R.F.; Gray, H.B.; Bard, A.J. Electrochemistry and electrogenerated chemiluminescence of tetrachlorotetrakis(trimethylphosphine) dimolybdenum. J. Phys. Chem. **1986**, *90*, 3841–3844.

54. Richter, M.M.; Bard, A.J. Electrogenerated chemiluminescence. 58. Ligand-sensitized electrogenerated chemiluminescence in europium labels. Anal. Chem. **1996**, *68*, 2641–2650.

55. Richter, M.M.; Bard, A.J.; Kim, W.; Schmehl, R.H. Electrogenerated chemiluminescence. 62. Enhanced ECL in bimetallic assemblies with ligands that bridge isolated chromophores. Anal. Chem. **1998**, *70*, 310–318.

56. Wang, H.; Xu, G.; Dong, S. Electrochemiluminescence of dichlorotris(1,10-phenanthroline)ruthenium(II) with peroxydisulfate in purely aqueous solution at carbon paste electrode. Microchem. J. **2002**, *72*, 43–48.

57. Tokel-Takvoryan, N.E.; Hemingway, R.E.; Bard, A.J. Electrogenerated chemiluminescence. XIII. Electrochemical and electrogenerated chemiluminescence studies of ruthenium chelates. J. Am. Chem. Soc. **1973**, *95*, 6582–6589.

58. Richards, T.C.; Bard, A.J. Electrogenerated chemiluminescence. 57. Emission from sodium 9,10-diphenylanthracene-2-sulfonate, thianthrenecarboxylic acids, and chlorpromazine in aqueous media. Anal. Chem. **1995**, *67*, 3140–3147.

59. Becker, W.G.; Seung, H.S.; Bard, A.J. Electrogenerated chemiluminescence. Part XLIII. Aromatic hydrocarbon/peroxydisulfate systems in acetonitrile-benzene solutions. J. Electroanal. Chem. Interfacial Electrochem. **1984**, *167*, 127–140.

60. Haapakka, K.; Kankare, J.; Lipiainen, K. Determination of polynuclear aromatic hydrocarbons in micellar solution by electrogenerated chemiluminescence. Anal. Chim. Acta **1988**, *215*, 341–345.

61. Richter, M.M.; Fan, F.-R.F.; Klavetter, F.; Heeger, A.J.; Bard, A.J. Electrochemistry and electrogenerated chemiluminescence of films of the conjugated polymer 4-methoxy-(2-ethylhexoxyl)-2,5-polyphenylenevinylene. Chem. Phys. Lett. **1994**, *226*, 115–120.

62. Noffsinger, J.B.; Danielson, N.D. Generation of chemiluminescence upon reaction of aliphatic amines with tris(2,2′-bipyridine)ruthenium(III). Anal. Chem. **1987**, *59*, 865–868.

63. Leland, K.; Powell, M.J. Electrogenerated chemiluminescence: an oxidative-reduction type ECL reaction sequence using tripropylamine. J. Electrochem. Soc. **1990**, *137*, 3127–3131.

64. Kanoufi, F.; Zu, Y.; Bard, A.J. Homogeneous oxidation of trialkylamines by metal complexes and its impact on electrogenerated chemiluminescence in the trialkylamine/Ru(bpy)$_3^{2+}$ system. J. Phys. Chem. B **2001**, *105*, 210–216.

65. Zu, Y.; Bard, A.J. Electrogenerated chemiluminescence. 66. The role of direct coreactant oxidation in the ruthenium tris(2,2′)bipyridyl/tripropylamine system and the effect of halide ions on the emission intensity. Anal. Chem. **2000**, *72*, 3223–3232.

66. He, L.; Cox, K.A.; Danielson, N.D. Chemiluminescence detection of amino acids, peptides, and proteins using tris(2,2′-bipyridine)ruthenium(III). Anal. Lett. **1990**, *23*, 195–210.

67. Gross, E.M.; Pastore, P.; Wightman, R.M. High-frequency electrochemiluminescent investigation of the reaction pathway between tris(2,2′-bipyridyl)ruthenium(II) and tripropylamine using carbon fiber microelectrodes. J. Phys. Chem. B **2001**, *105*, 8732–8738.

68. Miao, W.; Choi, J.-P.; Bard, A.J. Electrogenerated chemiluminescence 69: The tris(2,2′-bipyridine)ruthenium(II), (Ru(bpy)$_3^{2+}$)/tri-n-propylamine (TPrA) system revisited—a new route involving TPrA•$^+$ cation radicals. J. Am. Chem. Soc. **2002**, *124*, 14478–14485.

69. Zu, Y.; Bard, A.J. Electrogenerated chemiluminescence. 67. Dependence of light emission of the tris(2,2′-bipyridyl)ruthenium(II)/tripropylamine system on electrode surface hydrophobicity. Anal. Chem. **2001**, *73*, 3960–3964.

70. Bruce, D.; McCall, J.; Richter, M.M. Effects of electron withdrawing and donating groups on the efficiency of tris(2,2′-bipyridyl)ruthenium(II)/tri-n-propylamine electrochemiluminescence. Analyst (Cambridge, UK) **2002**, *127*, 125–128.

71. Factor, B.; Muegge, B.; Workman, S.; Bolton, E.; Bos, J.; Richter, M.M. Surfactant chain length effects on the light emission of tris(2,2′-bipyridyl) ruthenium(II)/tripropylamine electrogenerated chemiluminescence. Anal. Chem. **2001**, *73*, 4621–4624.

72. Workman, S.; Richter, M.M. The effects of nonionic surfactants on the tris(2,2′-bipyridyl)ruthenium(II)-tripropylamine electrochemiluminescence system. Anal. Chem. **2000**, *72*, 5556–5561.

73. Arora, A.; de Mello, A.J.; Manz, A. Sub-microliter electrochemiluminescence detector—a model for small volume analysis systems. Anal. Commun. **1997**, *34*, 393–395.

74. Hercules, D.M.; Lytle, F.E. Chemiluminescence from reduction reactions. J. Am. Chem. Soc. **1966**, *88*, 4745–4746.

75. Faulkner, L.R.; Bard, A.J. Techniques of electrogenerated chemiluminescence. In *Electronalytical Chemistry*; Bard, A.J., Ed.; Marcel Dekker: New York, 1977; Vol. 10, pp 1–95.

76. Lai, R; Bard, A.J. Electrogenerated chemiluminescence 70. The applications of ECL to determine electrode potentials of tri-n-propylamine, its radical cation, and intermediate free radical in MeCN/benzene solutions. J. Phys. Chem. A. 2003, 107, 3335–3340.

77. Chen, F.C.; Ho, J.H.; Chen, C.Y.; Su, Y.O.; Ho, T.I. Electrogenerated chemiluminescence of sterically hindered porphyrins in aqueous media. J. Electroanal. Chem. **2001**, *499*, 17–23.

78. Xu, X.-H.; Bard, A.J. Electrogenerated chemiluminescence. 55. Emission from adsorbed $Ru(bpy)_3^{2+}$ on graphite, platinum, and gold. Langmuir **1994**, *10*, 2409–2414.

79. Wang, H.; Xu, G.; Dong, S. Electrochemiluminescence of tris(2,2′-bipyri-dine)ruthenium(II) immobilized in poly(p-styrenesulfonate)-silica-Triton X-100 composite thin-films. Analyst (Cambridge, UK) **2001**, *126*, 1095–1099.

80. Wang, H.; Xu, G.; Dong, S. Electrochemistry and electrochemiluminescence of stable tris(2,2′-bipyridyl)ruthenium(II) monolayer assembled on benzene sulfonic acid modified glassy carbon electrode. Talanta **2001**, *55*, 61–67.

81. Wang, H.; Xu, G.; Dong, S. Electrochemiluminescence of tris(2,2′-bipyridyl)ruthenium(II) ion-exchanged in polyelectrolyte-silica composite thin-films. Electroanalysis **2002**, *14*, 853–857.

82. Khramov, A.N.; Collinson, M.M. Electrogenerated chemiluminescence of tris(2,2′-bipyridyl)ruthenium(II) ion-exchanged in Nafion-silica composite films. Anal. Chem. **2000**, *72*, 2943–2948.

83. Collinson, M.M.; Martin, S.A. Solid-state electrogenerated chemiluminescence in sol-gel derived monoliths. Chem. Commun. (Cambridge) **1999**, 899–900.

84. Collinson, M.M.; Taussig, J.; Martin, S.A. Solid-state electrogenerated chemiluminescence from gel-entrapped ruthenium(II) tris(bipyridine) and tripropylamine. Chem. Mater. **1999**, *11*, 2594–2599.

85. Collinson, M.M.; Novak, B.; Martin, S.A.; Taussig, J.S. Electro-chemiluminescence of ruthenium(II) tris(bipyridine) encapsulated in sol-gel glasses. Anal. Chem. **2000**, *72*, 2914–2918.

86. Collinson, M.M.; Novak, B. Diffusion and reactivity of ruthenium(II) tris(bipyridine) and cobalt(II) tris(bipyridine) in organically modified silicates. J. Sol-Gel Sci. Tech. **2002**, *23*, 215–220.

87. Honda, K.; Yoshimura, M.; Rao, T.N.; Fujishima, A. Electrogenerated chemiluminescence of the ruthenium tris(2,2′)bipyridyl/amines system on a boron-doped diamond electrode. J. Phys. Chem. B **2003**, *107*, 1653–1663.

88. Knight, A.W.; Greenway, G.M. Occurrence, mechanisms and analytical applications of electrogenerated chemiluminescence. A review. Analyst (Cambridge, UK) **1994**, *119*, 879–890.

89. Greenway, G.M.; Knight, A.W.; Knight, P.J. Electrogenerated chemiluminescent determination of codeine and related alkaloids and pharmaceuticals with tris(2,2′-bipyridine)ruthenium(II). Analyst (Cambridge, UK) **1995**, *120*, 2549–2552.

90. Xu, G.; Dong, S. Effect of metal ions on Ru(bpy)$_3^{2+}$ electrochemiluminescence. Analyst (Cambridge, UK) **1999**, *124*, 1085–1087.

91. Chen, X.; Yi, C.; M Li, Lu, X.; Z Li, Li, P.; Wang, X. Determination of sophoridine and related lupin alkaloids using tris(2,2′-bipyridine)ruthenium electrogenerated chemiluminescence. Anal. Chim. Acta **2002**, *466*, 79–86.

92. Song, Q.; Greenway, G.M.; McCreedy, T. Tris(2,2′-bipyridine)ruthenium(II) electrogenerated chemiluminescence of alkaloid type drugs with solid phase extraction sample preparation. Analyst (Cambridge, UK) **2001**, *126*, 37–40.

93. Xu, G.; Dong, S. Electrochemiluminescent detection of chlorpromazine by selective preconcentration at a lauric acid-modified carbon paste electrode using tris(2,2′-bipyridine)ruthenium(II). Anal. Chem. **2000**, *72*, 5308–5312.

94. Uchikura, K.; Kirisawa, M.; Sugii, A. Electrochemiluminescence detection of primary amines using tris(bipyridine)ruthenium(III) after derivatization with divinylsulfone. Anal. Sci. **1993**, *9*, 121–123.

95. Bock, C.R.; Connor, J.A.; Gutierrez, A.R.; Meyer, T.J.; Whitten, D.G.; Sullivan, B.P.; Nagle, J.K. Estimation of excited-state redox potentials by electron-transfer quenching. Application of electron-transfer theory to excited-state redox processes. J. Am. Chem. Soc. **1979**, *101*, 4815–4824.

96. Klingler, R.J.; Kochi, J.K. Heterogeneous rates of electron transfer. Application of cyclic voltammetric techniques to irreversible electrochemical processes. J. Am. Chem. Soc. **1980**, *102*, 4790–4798.

97. Gassman, P.G.; Mullins, M.J.; Richtsmeier, S.; Dixon, D.A. Effect of alkyl substitution on the ease of oxidation of bicyclo[1.1.0]butanes. Experimental verification of PRDDO calculations for the nature of the HOMO of bicyclo[1.1.0]butane. J. Am. Chem. Soc. **1979**, *101*, 5793–5797.

98. Marcus, R.A.; The theory of oxidation-reduction reactions involving electron transfer. V. Comparison and properties of electrochemical and chemical rate constants. J. Phys. Chem. **1963**, *67*, 853–857.

99. Amouyal, E.; Zidler, B.; P Keller, Moradpour, A. Excited-state electron transfer quenching by a series of water photoreduction mediators. Chem. Phys. Lett. **1980**, *74*, 314–317.

100. Reilley, C.N.; Scribner, W.G. Chronopotentiometric titrations. Anal. Chem. **1955**, *27*, 1210–1215.

101. Reilley, C.N.; Porterfield, W.W. Coulometric titrations with electrically released ethylenediaminetetraacetic acid (EDTA). Titrations of calcium, copper, zinc, and lead. Anal. Chem. **1956**, *28*, 443–447.

102. Reilley, C.N.; Scribner, W.G.; Temple, C. Amperometric titration of two- and three-component mixtures of cations with (ethylenedinitrilo)tetraacetic acid. Anal. Chem. **1956**, *28*, 450–454.

103. Martin, A.F.; Nieman, T.A. Glucose quantitation using an immobilized glucose dehydrogenase enzyme reactor and a tris(2,2'-bipyridyl)ruthenium(II) chemiluminescent sensor. Anal. Chim. Acta **1993**, *281*, 475–481.

104. Jameison, F.; Sanchez, R.I.; Dong, L.; Leland, J.K.; Yost, D.; Martin, M.T. Electrochemiluminescence-based quantitation of classical clinical chemistry analytes. Anal. Chem. **1996**, *68*, 1298–1302.

105. Yokoyama, K.; Sasaki, S.; K Ikebukuro, Takeuchi, T.; I Karube, Tokitsu, Y.; Masuda, Y. Biosensing based on NADH detection coupled to electrogenerated chemiluminescence from ruthenium tris(2,2'-bipyridine). Talanta **1994**, *41*, 1035–1040.

106. Liang, P.; Dong, L.; Martin, M.T. Light emission from ruthenium-labeled penicillins signaling their hydrolysis by β-lactamase. J. Am. Chem. Soc. **1996**, *118*, 9198–9199.

107. Liang, P.; Sanchez, R.I.; Martin, M.T. Electrochemiluminescence-based detection of beta-lactam antibiotics and beta-lactamases. Anal. Chem. **1996**, *68*, 2426–2431.

108. Brune, S.N.; Bobbitt, D.R. Effect of pH on the reaction of tris(2,2'-bipyridyl)ruthenium(III) with amino-acids: implications for their detection. Talanta **1991**, *38*, 419–424.

109. Brune, S.N.; Bobbitt, D.R. Role of electron-donating/withdrawing character, pH, and stoichiometry on the chemiluminescent reaction of tris(2,2'-bipyridyl)ruthenium(III) with amino acids. Anal. Chem. **1992**, *64*, 166–170.

110. Lee, W.-Y.; Nieman, T.A. Determination of dansyl amino acids using tris(2,2'-bipyridyl)ruthenium(II) chemiluminescence for post-column reaction detection in high-performance liquid chromatography. J. Chromatogr. **1994**, *659*, 111–118.

111. Lee, W.-Y.; Nieman, T.A. Evaluation of use of tris(2,2'-bipyridyl)ruthenium(III) as a chemiluminescent reagent for quantitation in flowing streams. Anal. Chem. **1995**, *67*, 1789–1796.

112. Knight, A.W.; Greenway, G.M. Electrogeneration chemiluminescent determination of pyruvate using tris(2,2'-bipyridine)ruthenium(II). Analyst (Cambridge, UK) **1995**, *120*, 2543–2547.

113. Chen, X.; Sato, M. High-performance liquid chromatographic determination of ascorbic acid in soft drinks and apple juice using tris(2,2'-bipyridine)ruthenium(II) electrochemiluminescence. Anal. Sci. **1995**, *11*, 749–754.

114. Chen, X.; Chen, W.; Y Jiang, Jia, L.; Wang, X. Electrogenerated chemiluminescence based on tris(2,2'-bipyridyl)ruthenium(II) with hydroxyl carboxylic acid. Microchem. J. **1998**, *59*, 427–436.

115. Zorzi, M.; Pastore, P.; Magno, F. A single calibration graph for the direct determination of ascorbic and dehydroascorbic acids by electrogenerated luminescence based on $Ru(bpy)_3^{2+}$ in aqueous solution. Anal. Chem. **2000**, *72*, 4934–4939.

116.   Karatani, H.; Kojima, M.; H Minakuchi, Soga, N.; Shizuki, T. Development and characterization of anodically initiated luminescent detection for alcohols and carbohydrates. Anal. Chim. Acta **1997**, *337*, 207–215.

117.   Karatani, H.; An electrochemically triggered chemiluminescence from polyhydric alcohols. J. Photochem. Photobiol. A: Chem. **1994**, *79*, 71–80.

118.   Karatani, H.; Shizuki, T. Luminescent electrooxidation of methanol in aqueous alkaline media. Electrochim. Acta **1996**, *41*, 1667–1676.

119.   Chen, X.; Jia, L.; X Wang, Hu, G.; Study of the electrochemiluminescence based on the reaction of hydroxyl compounds with ruthenium complex, Anal. Sci. **1997**, *13*(suppl., Asianalysis IV), 71–75.

120.   Chen, X.; Sato, M.; Lin, Y. Study of the electrochemiluminescence based on tris(2,2'-bipyridine)ruthenium(II) and alcohols in a flow injection system. Microchem. J. **1998**, *58*, 13–20.

121.   Chen, X.; Jia, L.; Sato, M. Study on the electrochemiluminescence based on tris(2,2'-bipyridine)ruthenium and methanol. Huaxue Xuebao **1998**, *56*, 238–243.

122.   Collins, G.E. Gas-phase chemical sensing using electrochemiluminescence. Sensors. Actuators. B: Chem. B **1996**, *35*, 202–206.

123.   Zheng, X.; Zhang, Z.; Z Guo, Wang, Q. Flow-injection electrogenerated chemiluminescence detection of hydrazine based on its in-situ electrochemical modification at a pre-anodized platinum electrode. Analyst (Cambridge, UK) **2002**, *127*, 1375–1379.

124.   Li, B.; Zhang, Z.; X Zheng, Xu, C. Flow-injection chemiluminescence determination of hydrazine using on-line electrogenerated BrO^- as the oxidant. Chem. Anal. (Warsaw) **2000**, *45*, 709–717.

125.   Hercules, D.M.; Lytle, F.E. Chemiluminescence from the reduction of aromatic amine cations and ruthenium(III) chelates. Photochem. Photobiol. **1971**, *13*, 123–133.

126.   Hercules, D.M. Chemiluminescence from electron-transfer reactions. Acc. Chem. Res. **1969**, *2*, 301–307.

127.   Martin, J.E.; Hart, E.J.; Adamson, A.W.; Gafney, H.; Halpern, J. Chemiluminescence from the reaction of the hydrated electron with tris(bipyridyl)ruthenium(III). J. Am. Chem. Soc. **1972**, *94*, 9238–9240.

128.   Brina, R.; Bard, A.J. Electrogenerated chemiluminescence. Part XLVII. Electrochemistry and electrogenerated chemiluminescence of bis(2,4,6-trichlorophenyl)oxalate-luminescer systems. J. Electroanal. Chem. Interfacial Electrochem. **1987**, *238*, 277–295.

129.   Clark, C.D.; Debad, J.D.; Yonemoto, E.H.; Mallouk, T.E.; Bard, A.J. Effect of oxygen on linked $Ru(bpy)_3^{2+}$-viologen species and methylviologen: A reinterpretation of the electrogenerated chemiluminescence. J. Am. Chem. Soc. **1997**, *119*, 10525–10531.

130.   Xu, X.; Shreder, K.; Iverson, B.L.; Bard, A.J. Generation by electron transfer of an emitting state not observed by photoexcitation in a linked $Ru(bpy)_3^{2+}$-methyl viologen. J. Am. Chem. Soc. **1996**, *118*, 3656–3660.

131.   Okajima, T.; Ohsaka, T. Chemiluminescence of 3-methylindole based on electrogeneration of superoxide ion in acetonitrile solutions. J. Electroanal. Chem. **2002**, *523*, 34–39.

132. Berger, A.W.; Driscoll, J.N.; Driscoll, J.S.; Pirog, J.A.; Linschitz, H. Chemiluminescence of indole derivatives in dimethyl sulfoxide. Photochem. Photobiol. **1968**, *7*, 415–420.

133. Okajima, T.; Tokuda, K.; Ohsaka, T. Chemiluminescence of a *Cypridina luciferin* analog by electrogenerated superoxide ion. Bioelectrochem. Bioenerg. **1996**, *41*, 205–208.

134. Okajima, T.; Tokuda, K.; Ohsaka, T. Chemiluminescence of a *Cypridina luciferin* analog by electrogenerated superoxide ion in the presence of a lipophilic ascorbic acid derivative. Denki Kagaku oyobi Kogyo Butsuri Kagaku **1996**, *64*, 1276–1279.

135. Cao, W.; Xu, G.; Z Zhang, Dong, S. Novel tris(2,2′-bipyridine)ruthenium(II) cathodic electrochemiluminescence in aqueous solution at a glassy carbon electrode. Chem. Commun. (Cambridge, UK) **2002**, 1540–1541.

136. Chandross, E.A.; Sonntag, F.I. Chemiluminescent electron-transfer reactions of radical anions. J. Am. Chem. Soc. **1966**, *88*, 1089–1096.

137. Akins, D.L.; Birke, R.L. Energy transfer reactions of electrogenerated aromatic anions and benzoyl peroxide. Chemiluminescence and its mechanism. Chem. Phys. Lett. **1974**, *29*, 428–435.

138. Santa Cruz T.D.; Akins, D.L.; Birke, R.L. Chemiluminescence and energy transfer in systems of electrogenerated aromatic anions and benzoyl peroxide. J. Am. Chem. Soc. **1976**, *98*, 1677–1682.

139. Sakura, S.; Electrochemiluminescence of hydrogen peroxide-luminol at a carbon electrode. Anal. Chim. Acta **1992**, *262*, 49–57.

140. Chen, G.N.; Zhang, L.; Lin, R.E.; Yang, Z.C.; Duan, J.P.; Chen, H.Q.; Hibbert, D.B. The electrogenerated chemiluminescent behavior of hemin and its catalytic activity for the electrogenerated chemiluminescence of lucigenin. Talanta **2000**, *50*, 1275–1281.

6

Organic ECL Systems

Samuel P. Forry and R. Mark Wightman
University of North Carolina at Chapel Hill, Chapel Hill, North Carolina, U.S.A.

I. INTRODUCTION

There have been several reviews of organic ECL [1–6]. Although current research in organic ECL continues to unravel the fundamentals of these light-emitting processes, much of the modern interest lies in using ECL systems for such applications as developing light-emitting devices and allowing sensitive detection [7–10]. However, the development of these applications of organic ECL systems requires a thorough understanding of the variables that contribute to ECL emission. Toward this end, we provide a summary of the parameters necessary for organic ECL. In addition, we have tabulated important characteristics of the most common organic ECL systems.

Two types of organic ECL systems can be described. In the first, termed single systems, the annihilating radical ions have the same neutral precursor. The second type, termed mixed systems, involves annihilating radical ions that have different precursors (see Chapter 1). Single systems are simpler to understand theoretically are considered first.

II. REQUIREMENTS FOR FORMING EXCITED STATES FROM SINGLE SYSTEMS

A. Electrochemistry

The first requirement for successful ECL with single systems is the generation of both a radical anion and a radical cation from the same parent (R) at an electrode surface:

$$R - e^- \rightarrow R^{+\bullet} \tag{1}$$

$$R + e^- \rightarrow R^{-\bullet} \tag{2}$$

Therefore, parent species in this type of reaction must undergo both oxidation and reduction within the potential range that is accessible in the solvent employed.

Once formed, the radical ions must be sufficiently long-lived to allow them to diffuse, encounter each other, and undergo homogeneous electron transfer. By their very nature, radical ions are reactive species and can undergo a variety of reactions in addition to the annihilation reaction that leads to light production. Radical cations are acidic and subject to nucleophilic attack. Radical anions tend to be more stable but can react with available proton donors, including the solvent or trace water they contain. Even the neutral ECL molecule (R) can be vulnerable to attack from the radical ions, quenching the ECL emission [11,12]. The chemically inert nature of aprotic solvents such as acetonitrile, butyronitrile, and dimethoxyethane minimizes these unwanted reactions. For optimal results, the solvents should be purified and dried [13]. Of course, the lifetime required for the radical species depends on the time scale of observation. When reagents are generated and react on a microsecond time scale, the requirements for a long-lived radical ion are considerably less than when operating on a time scale of seconds [14]. Before undertaking an ECL experiment, it is important to evaluate the radical ion stability with an electrochemical technique. Typically cyclic voltammetry is employed because it allows simultaneous evaluation of the stability and the $E_{1/2}$ values.

B. Excited States

Molecules of interest for single-system ECL must also have accessible electronic excited states. Following heterogeneous formation, the radical ions diffuse together to form an encounter complex:

$$R^{+\bullet} + R^{-\bullet} \rightarrow [R^{+\bullet} + R^{-\bullet}] \tag{3}$$

Homogeneous electron transfer within this encounter complex yields two neutral molecules:

$$[R^{+\bullet} + R^{-\bullet}] \rightarrow [^1R^* + R] \tag{4a}$$

$$[R^{+\bullet} + R^{-\bullet}] \rightarrow [^3R^* + R] \tag{4b}$$

$$[R^{+\bullet} + R^{-\bullet}] \rightarrow [2R] \tag{4c}$$

The driving force for ground-state formation [reaction (4c)] is large, and this pathway typically lies in the Marcus inverted region (Chapter 4). Thus when energetically accessible, excited state formation dominates [reactions (4a) and (4b)]. The type and number of excited states that are formed determine the amount of ECL observed. Although most organic molecules do not emit directly from triplet excited states, triplet–triplet annihilation can often provide an indirect pathway to emissive singlets (Chapter 3). A variety of methods have been used to probe the reaction mechanism for excited state formation, including magnetic field effects [15], addition of triplet quenchers [16], and the reaction order dependencies [17] (Chapter 4).

The free energy for the formation of excited states from radical ion annihilation within the encounter complex (ΔG_s and ΔG_t for formation of singlets and triplets) can be calculated [18–23]:

$$\Delta G = (E^\circ_{R/R-\bullet} - E^\circ_{R/R+\bullet}) + E_{0,0} - w_{a,u} \tag{5}$$

where $E^\circ_{R/R-\bullet}$ and $E^\circ_{R/R+\bullet}$ represent the formal potentials for radical ion formation, $E_{0,0}$ is the excited state energy (E_S for a singlet, E_T for a triplet), and $w_{a,u}$ is the work required to form the encounter complex from the electrochemically generated radical ions. The work term, which is primarily Coulombic, is affected by electrostatic screening from electrolyte in solution and can be calculated from Debye–Hückel theory [22–24],

$$w_{a,u} = \frac{z_{R+} z_{R-} e^2}{\in a} \left[\frac{\exp(\beta r \mu^{1/2})}{1 + \beta r \mu^{1/2}} \right] \exp(-\beta a \mu^{1/2}) \tag{6}$$

Here, z indicates the charge of the radical ions, e is the charge on an electron, $\in$ is the static dielectric constant, a is the separation between the ions in the encounter complex, r is the radius of the radical ion and associated counter ion [Eq. (6) is appropriate only for $r_{R+\bullet} = r_{R-\bullet}$], μ is the ionic strength, and

$$\beta = \left(\frac{8\pi N e^2}{1000 k_B T} \right)^{1/2} \tag{7}$$

where N is Avogadro's number, k_B is Boltzmann's constant, and T is temperature. The radius r can be calculated from the Stokes–Einstein relation

$$r = \frac{k_B T}{6\pi D_\eta} \tag{8}$$

where D is the diffusion coefficient and η is the solution viscosity. However, when electrolyte concentration is high, as in most electrochemical experiments, the effect of the work term $w_{a,u}$ is negligible [22], and Eq. (5) yields a simple comparison between the separation in formal potentials (ΔE°) and the excited state energy ($E_{0,0}$), which can be used to predict the reaction mechanism [17]. Systems where $\Delta G_s < 0$ are said to be "energy-sufficient," while systems with $\Delta G_s > 0$ are called "energy-insufficient." Because excited singlets are relatively stable and can emit directly whereas excited triplets can be quenched by radicals and typically must undergo triplet–triplet annihilation to yield emission, energy-sufficient systems proceed predominantly via singlet formation [reactions (3)–(4a)]. Energy-deficient systems produce ECL via formation of excited triplet [reactions (3)–(4b)] and triplet–triplet annihilation.

Calculation of ΔG is also important in predicting the kinetics of electron transfer [reaction (4)]. The apparent electron transfer rate constant $k'_{et,s}$ can be determined by using the Arrhenius–Eyring equation [18,20,23,25]

$$k'_{et} = Z \exp\left(\frac{-\Delta G^*}{k_B T}\right) \tag{9}$$

where Z is the liquid-phase collision number, ΔG^* is the free energy of activation, k_B is Boltzmann's constant, and T is temperature. ΔG^* can be related to ΔG by simple Marcus theory [18,20,23,25]:

$$\Delta G^* = \frac{(\Delta G + \lambda)^2}{4\lambda} \tag{10}$$

where λ is is the reorganizational energy required for the electron transfer. Thus the kinetics for electron transfer to form the excited states singlet, triplet, or ground state ($k'_{et,s}$, $k'_{et,t}$, and $k'_{et,g}$, respectively) can be calculated and compared. However, these rates must be corrected to account for the spin distribution of the radical ions. The annihilation of radical ion doublets, for example, gives rise to four possible electron-pairing arrangements, one indicating a singlet product and three indicating triplet degeneracy [7,20,23,26]. Thus, the efficiency of excited singlet state formation, ϕ_s, would be calculated [20] as

$$\phi_s = \frac{k'_{et,s}}{k'_{et,s} + 3k'_{et,t} + k'_{et,g}} \tag{11}$$

and a comparable equation could be written for ϕ_t. In energy-sufficient systems where emission is predominantly from the formation of excited singlets, the ECL efficiency is simply the product of the efficiencies for excited singlet formation and emission from the excited singlet excited ($\phi_{ECL} = \phi_s * \phi_f$, where ϕ_f is the fluorescence efficiency). When the electron transfer to form excited states is fast, it can become limited by the rate of formation of the encounter complex [reaction (3)], which depends on the frequency of radical ion encounters in solution, and is said to be "diffusion-limited."

The ECL emission intensity can be evaluated based on the ECL efficiency (ϕ_{ECL}). This term is defined as the product of the fraction of radical ions that annihilate to produce emissive excited states and the fluorescence quantum efficiency that describes the fate of those excited states [27]. Most common organic single systems involve strongly fluorescent molecules.

III. SINGLE SYSTEMS FOR ORGANIC ECL

Polyaromatic hydrocarbons (PAHs) exhibit many of the features required to achieve ECL from single systems. Typically they can be both reduced to form moderately stable radical anions and oxidized to form radical cations, and they have excited states that are accessible with the energy arising through radical ion annihilation reactions. Notably, their fluorescent quantum efficiencies are generally quite high.

9,10-Diphenylanthracene (DPA) is a model compound for ECL from organic single systems. DPA can be reversibly oxidized ($E^\circ_{R/R-\bullet}$ = 1.37 V vs. SCE) or reduced ($E^\circ_{R/R-\bullet}$ = -1.79 V vs. SCE) to give radical cations and anions, respectively [28]. The radical annihilation releases sufficient energy with ΔE° = 3.16 eV [Eq. (5), assuming high electrolyte concentration] to directly populate both the DPA excited singlet state (E_S = 3.06 eV) and the excited triplet state (E_T = 1.8 eV) [28]. The free energy for excited singlet formation is negative, indicating an energy-sufficient process, and the predicted mechanism leading to emission involves primarily reactions (1)–(3) and (4a) [29]. Calculation of the electron transfer rates indicates that the formation of both excited state singlets and triplets [reactions (4a) and (4b)] will be diffusion-limited and ground-state formation [reaction (4c)] will be exceedingly slow. DPA has unit fluorescence efficiency [18] and yields extremely bright ECL emission with an ECL efficiency that can approach 25% [23]. Because both pathways to form excited states are diffusion-limited, the ratio of singlets to triplets formed is governed by the spin statistics of the annihilating doublet radical ions [reaction (11)]. Thus, 25% efficiency represents the theoretical maximum for ECL from DPA singlets [23].

The efficiency of ECL varies with solvent and ionic strength [18,21,24]. In relatively polar solvents (acetonitrile, acetonitrile–toluene mixtures), encountering radical ions form radical ion pairs within which electron transfer occurs. The rate of dissociation of this encounter complex to re-form the radical ions is slowed at low electrolyte concentrations [18]. Thus, low ionic strength promotes passage into a radiative pathway, leading to an increase in ϕ_{ECL} [18]. In contrast, in nonpolar solvents, electrochemically formed radical ions do not exist alone but must be stabilized by electrolyte counter ions. Thus, formation of radical ions becomes difficult at low ionic strengths, leading to an increase in $\Delta E_{1/2}$. This also leads to a significant increase in ϕ_{ECL} [24].

Table 1 summarizes the electrochemical and excited state energetics of PAHs that have been used for ECL in single systems. In addition, the efficiency of their ECL emission relative to DPA is given. Note that the DPA system is among the most efficient ECL systems for any solution conditions [18,24]. As can be seen in the table, PAHs substituted with methyl or phenyl groups tend to give brighter ECL than their unsubstituted counterparts. These substituents block reactive sites in the radical cation and thus lower the competitive nucleophilic

Table 1 Single ECL Systems

Compound	$E^\circ_{R/R+\bullet}$ vs. SCE (eV)	$-E^\circ_{R/R-\bullet}$ vs. SCE (eV)	$E(^1R^*)$ (eV)	$E(^3R^*)$ (eV)	ϕ_{ECL}/ϕ_{DPA}
Anthracene	1.09[a] [60]	1.96[c] [60]	3.3 [26]	1.8 [61]	0.2 [62]
Tetracene	0.95[b] [63]	1.62[c] [64]	1.9 [34]	1.3 [65]	Weak [66]
Phenanthrene	1.5[a] [67]	2.4[c] [67]	3.57 [68]	2.66 [68]	0.005 [62]
Pyrene	1.16[a] [60]	2.08[c] [60]	3.4 [61]	2.1 [61]	0.002 [62]
Perylene	0.85[c] [69]	1.67[c] [69]	2.83 [34]	1.55 [34]	Weak [66]
Fluoranthene	1.48[c] [69]	1.74[c] [69]	3.0 [61]	2.3 [61]	0.005 [62]
Dibenz[a,h]anthracene	1.71[a,e] [70]	2.19[a] [36]	3.14 [36]	2.26 [36]	
Chrysene	1.35[a] [67]	2.25[c] [67]	3.43 [34]	2.44 [34]	0.005 [62]
Coronene	1.23[c] [69]	2.04[c] [69]	2.95 [69]	2.37 [69]	0.1 [62]
9,10-Dimethylanthracene	0.87[c] [69]	1.82[c] [69]	3.1 [71]	1.8 [71]	0.4 [14]
1,3-Dimethylbenzo[a]pyrene	1.04[a] [72]	1.97[a] [72]	2.97 [72]		0.0004 [72]
1,6-Dimethylbenzo[a]pyrene	1.05[a] [72]	1.97[a] [72]	2.97 [72]		0.2 [72]
2,3-Dimethylbenzo[a]pyrene	1.07[a] [72]	1.96[a] [72]	3 [72]		0.0003 [72]
3,6-Dimethylbenzo[a]pyrene	1.06[a] [72]	1.94[a] [72]	2.94 [72]		0.02 [72]
4,5-Dimethylbenzo[a]pyrene	1.11[a] [72]	1.97[a] [72]	3.02 [72]		0.006 [72]
9-Phenylanthracene	1.86[b] [63]	0.56[c] [63]	3.1 [30]		0.4 [14]
9,10-Diphenylanthracene	1.37[d] [28]	1.79[d] [28]	3.06 [28]	1.8 [28]	1
Rubrene	0.82[c] [69]	1.41[c] [69]	2.2 [61]	1.2 [69]	1 [62]
1,3,6,8-Tetraphenylpyrene	1.25[c] [15]	1.83[c] [15]	2.95 [15]	2 [15]	0.01 [62]
Phenothiazine	0.59[a] [73]	2.85[c] [74]			
1-Methyl-2,5-diphenylindene	1.25[a] [35]	2.36[a] [35]	3.18 [35]		5.8E-04 [35]
2-(p-tert-Butylphenyl)indene	1.27[a] [35]	2.57[a] [35]	3.34 [35]		4.5E-05 [35]
2-Phenylindene	1.36[a] [35]	2.45[a] [35]	3.36 [35]		1.8E-06 [35]
2-(4'-Biphenylyl)indene	1.29[a] [35]	2.24[a] [35]	3.18 [35]		1.3E-03 [35]

2-(2'-Naphthylyl)indene	1.29[a] [35]	2.25[a] [35]	3.22 [35]	1.6E-05 [35]
2-(1'-Naphthyl)indene	1.29[a] [35]	2.22[a] [35]	3.07 [35]	2.0E-05 [35]
2-(1-Naphthyl)-5-phenylindene	1.28[a] [35]	2.18[a] [35]	2.98 [35]	2.0E-05 [35]
3,4-Dihydro-2,6-diphenylnaphthalene	1.31[a] [35]	2.26[a] [35]	3.12 [35]	2.3E-05 [35]
3-(4-Biphenyl)-1,2-dihydronaphthalene	1.3[a] [35]	2.28[a] [35]	3.13 [35]	2.8E-04 [35]
3,4-Dihydro-2-(2-naphthyl)-6-naphthalene	1.23[a] [35]	2.18[a] [35]	3.02 [35]	2.1E-04 [35]
3,4-Dihydro-2-(4-biphenyl)phenanthrene	1.19[a] [35]	2.15[a] [35]	3.07 [35]	6.9E-05 [35]
3,4-Dihydro-2-phenylphenanthrene	1.25[a] [35]	2.25[a] [35]	3.07 [35]	3.9E-05 [35]
1,8-Diphenyl-1,3,5,7-octatetraene				0.005 [62]
Brucine				weak [66]
Thebaine				weak [66]
2-Phenylnaphthalene				weak [66]
2-Fluoronaphthalene				weak [66]
α,β,γ,δ-Tetraphenylporphin	0.75[b] [75]	0.96[b] [75]	1.57[b] [75]	<0.01[b] [75]
1,4-Dimethoxynaphthalene	1.1[a] [76]	2.69[c] [76]		
9-Methoxyanthracene	1.05[a] [76]	1.92[c] [76]		
9,10-Dimethoxyanthracene	0.98[a] [76]	1.9[c] [76]		
9,10-Bis(2,6-dimethoxyphenyl)anthracene	1.18[a] [76]	2.08[c] [76]		
10,10'-Dimethoxy-9,9'-bianthracenyl	1.1[a] [76]	1.8[c] [76]		
9,10-Diphenoxyanthracene	1.2[a] [76]	1.71[c] [76]		
1,4-Bis(methylthio)naphthalene	1.07[a] [76]	2.1[c] [76]		
2,6-Bis(methylthio)naphthalene	1.1[a] [76]	2.24[c] [76]		
9,10-Bis(methylthio)anthracene	1.11[a] [76]	1.55[c] [76]		
1-Dimethylaminonaphthalene	0.75[a] [76]	2.58[c] [76]		
2-Dimethylaminonaphthalene	0.67[a] [76]	2.63[c] [77]		
1,5-Bisdimethylaminonaphthalene	0.58[a] [77]	2.64[c] [76]		
2,6-Bisdimethylaminonaphthalene	0.26[a] [76]	2.71[c] [76]		

(continued)

Table 1 Continued

Compound	$E°_{R/R•+}$ vs. SCE (eV)	$-E°_{R/R•-}$ vs. SCE (eV)	$E(^1R*)$ (eV)	$E(^3R*)$ (eV)	ϕ_{ECL}/ϕ_{DPA}
2,7-Bisdimethylaminonaphthalene	0.57[a] [76]	2.77[c] [76]			
N,N-Tetramethylbenzidine	0.43[a] [76]				
1,4,5,8-Tetraphenylnaphthalene	1.39[a] [76]	1.98[c] [76]			
9,10-Bis(phenylethynyl)anthracene	1.16[f] [77]	1.29[c] [76]			
1,6-Bis(dimethylamino)pyrene	0.49[a] [76]	2.16[c] [76]			
1,6-Bis(methylthio)pyrene	0.96[a] [76]	1.83[c] [76]			
1,3,4,7-Tetraphenylisobenzofuran	0.89[c] [78]	1.86[c] [78]			
1,3,5,6-Tetraphenylisobenzofuran	0.85[c] [78]	1.79[c] [78]			
1,3-Di-p-anisyl-4,7-diphenylisobenzofuran	0.84[c] [78]	1.92[c] [78]			
N-Methyl-1,3,4,7-tetraphenylisoindole	0.67[c] [78]	2.35[c] [78]			
N-Methyl-1,3-di-p-anisyl-4,7-diphenylisoindole	0.59[c] [78]	2.42[c] [78]			
Tetracene-5,12-quinone	1.75[c] [78]	1.02[c] [78]			
1-p-Anisyl-3,5-diphenyl-2-pyrazoline	0.61[f] [79]	2.38[f] [79]		1.88 [79]	
Tetraanisylpyrrole	0.66[a] [80]	<2.9[a] [80]	3.49 [80]		
Tetraphenylpyrrole	0.90[f] [80]	2.78[f] [80]	3.47 [80]		
Tetratolylpyrrole	0.81[a] [80]	2.85[a] [80]	3.38 [80]		
Tetra-p-chlorophenyl-pyrrole	1.06[f] [80]	2.52[f] [80]	3.45 [80]		
Tetraphenylfuran	1.19[a] [80]	2.47[a] [80]	3.46 [80]		
Tetraphenylthiophene	1.38[a] [80]	2.30[a] [80]			
2-Methylnaphthalene	1.73[f] [81]	2.33[f] [81]			
1,4-Diphenylnaphthalene	1.17[f] [81]	1.94[f] [81]			
2-(2-Fluorenyl)-5-naphthyloxazole	1.34[f] [81]	1.78[f] [81]			
1-(6-Phenylindenyl)naphthalene	0.96[f] [81]	2.32[f] [81]			
5,6-Dihydro-5H-dinaphtho[2,1-g:1',2'-h][1,5]-dioxocin	1.09[f] [81]	2.04[f] [81]			

4,5,6,7-Tetrahydrodinaphtho[2,1-g:1',2'-i]-[1,6]dioxecin	1.09^f [81]	2.31^f [81]		
Dibenzo[[ff]-4,4',7,7'-tetraphenyl]diindeno[1,2,3-cd:1',2',3'-lm]perylene	0.946^g [82]	1.20^g [82]	2.1 [82]	0.2 [82]
N,N'-Bis(2,6-diisopropylphenyl)-3,4,9,10-perylenetetracarboxylic diimide	1.65^h [83]	0.52^h [83]		0.4 [83]
N-(2,6-Diisopropylphenyl)-N'-octylterrylenetetracarboxylic diimide	1.11^h [83]	0.63^h [83]	1.84 [83]	0.3 [83]
N,N'-Bis(2,6-diisopropylphenyl)-3,4,9,10-quaterrylenetetracarboxylic diimide	0.83^h [83]	0.57^h [83]	1.58 [83]	Weak [83]
3,7-[Bis[4-phenyl-2-quinolyl]]-10-methylphenothiazine	0.69^f [84]	1.96^f [84]	2.3 [84]	0.8 [84]
9,10-Dimethylsulfone-7,12-diphenylbenzo[k]fluoranthene	1.5^i [85]	1.8^i [85]	2.97 [85]	1 [85]
9,10-Dimethyl-7,12-diphenylbenzo[k]fluoranthene	1.36^i [85]	1.95^i [85]	2.97 [85]	

^{a}In MeCN.
^{b}In MeCl.
^{c}In DMF.
^{d}In PhCN.
^{e}Versus NHE.
^{f}In 1:1 benzene:MeCN
^{g}In 9:1 benzene:MeCN
^{h}In 3:2 CHCl$_3$:MeCN
^{i}In 4:1 benzene:MeCN

substitution or addition reactions (e.g., DPA yields considerably brighter ECL than anthracene). Functionalization can also serve to reduce intermolecular interactions that lead to alternative emissive states (such as excimer emission as discussed in Section VI). In general, increased aromaticity correlates with lower energy excited states and also with the formation of radical ion at lower potentials.

IV. REQUIREMENTS FOR FORMING EXCITED STATES FROM MIXED SYSTEMS

"Mixed systems" refers to the ECL mechanism where the radical cation and radical anion are formed from different molecules:

$$D + e^- \rightarrow D^{+\bullet} \tag{12}$$

$$A + e^- \rightarrow A^{-\bullet} \tag{13}$$

Excited states can be formed by annihilation reactions that are similar to reactions (3) and (4):

$$D^{+\bullet} + A^{-\bullet} \rightarrow [D^{+\bullet} + A^{-\bullet}] \tag{14}$$

$$[D^{+\bullet} + A^{-\bullet}] \rightarrow [^1D^* + A] \quad \text{or} \quad [D + {}^1A^*] \tag{15a}$$

$$[D^{+\bullet} + A^{-\bullet}] \rightarrow [^3D^* + A] \quad \text{or} \quad [D + {}^3A^*] \tag{15b}$$

$$[D^{+\bullet} + A^{-\bullet}] \rightarrow [D + A] \tag{15c}$$

where the energy levels of the A and D excited states determine which complex is formed. Equations (5)–(11), shown above for single systems can be applied to mixed systems as well, allowing calculation of the available energy for forming excited states and the speed of competing electron transfers (see, e.g., Refs. 17, 22, 30). To simplify interpretation, most investigations of mixed ECL systems couple one species for which the reaction energetics indicate no excited state formation with a second species that can be excited. These are called, respectively, the nonemissive and emissive reaction partners.

A. Nonemissive Reaction Partners

1. Excited States

As can be seen in reactions (15), many different excited states can conceivably arise from electron transfer in mixed systems. These various electron transfer processes are competitive, however, and in most actual systems the spectra of the ECL emission indicate the formation of only a single excited state. This can be understood theoretically by calculating the rate for each of the electron transfers based on Marcus theory [Eqs. (5), (9), and (10)]. Typically this leads to the conclusion that one of the excited states will be formed rapidly and the other mechanistic pathways will be negligible. Nonemissive reaction partners typically exhibit highly energetic excited states, leading to positive ΔG and slow electron transfer to the excited state. Alternatively, nonemissive reaction partners may form excited states but exhibit a very low emission quantum efficiency and remain undetected due to stronger emissions from other compounds in the system. It is important to note that a nonemissive reaction partner in one chemical system may be emissive when coupled with a different radical ion that yields a larger $\Delta E°$ or no dominating light-producing pathway.

2. Electrochemistry

As with single systems, electrochemical formation of stable radical ions is central to mixed system ECL. The radical ions of nonemissive reaction partners must be sufficiently long-lived to find an emissive reaction partner with which to undergo homogeneous electron transfer [reactions (15)].

The formal potential of the reaction partner offers a degree of control not present in single systems. When the excited state energetics for the emissive partner, $E_{0,0}$, are known, the nonemissive partner can be chosen on the basis of its electron transfer potential to change the free energy for excited state formation [Eq. (5)] and thermodynamically control the reaction pathways leading to ECL between energy-sufficient and energy-insufficient mechanisms [17]. Alternatively, series of nonemissive reaction partners exhibiting electrochemistry across a range of potentials can be considered to systematically vary ΔG [28].

3. Solubility

One use of reaction partners is to increase the ECL light emission by providing high concentrations of one radical ion to facilitate formation of the encounter complex [reaction (14)]. This allows more sensitive detection and easier quantification [9]. Toward this end the solubility of the reaction partner in the particular solvent being used is important [31].

4. Example

One of the most commonly studied nonemissive reaction partners is Würster's Blue (N,N,N',N'-tetramethyl-p-phenylenediamine, TMPD). TMPD exhibits stable radical cation formation ($E°=0.24$ V vs. SCE) but no reduction [32]. The triplet and singlet excited states of TMPD are highly energetic (2.83 eV and 3.52 eV, respectively) and are not populated during ECL [95]. TMPD can be paired with DPA (discussed in Section III) for mixed system ECL. In this system, the potential is controlled to generate the radical cation of TMPD and the radical anion of only DPA. Calculation of the free energy available for the formation of an excited state from the encounter complex indicates that this system will proceed via the energy-insufficient pathway. Thus, annihilation of the radical ions generates only DPA triplets that produce light via triplet–triplet annihilation. Because the rate limiting step for light generation is triplet–triplet annihilation, the ECL intensity is expected to be second order with respect to the concentration of DPA, and this is seen experimentally [17].

A mixed system that allows ECL via the energy-sufficient pathway is the combination of benzophenone (BP) and DPA. BP is readily reduced ($E° = -1.91$) but not readily oxidized [28]. When combined with DPA, the DPA radical cation must be generated and react with the BP radical anion. The energetics of the triplet and singlet excited states of BP are 3 eV and 3.2 eV, respectively [28]. Considering Eqs. (5)–(11) for a mixed system, it is predicted that excited triplets and singlets of both BP and DPA will be formed by electron transfer within the encounter complex. Nevertheless, BP can be considered a nonemissive reaction partner in this system because of its inefficient excited state quantum efficiency. BP excited singlets undergo intersystem crossing with nearly unity conversion to generate triplets [33]. The BP triplets typically exhibit a low luminescence quantum efficiency because of quenching by solvent, polyaromatic hydrocarbons, and radicals. Thus, very little emission from BP is seen experimentally. By comparison, emission from excited DPA singlets is particularly efficient [23]. Further, the energetics of excited DPA singlets are well matched to the energetics of the electron transfer between DPA radical cations and BP radical anions. This yields light generation through the energy-sufficient mechanism and light intensity that is first-order with respect to the concentration of DPA [17].

These two examples, DPA radical anions mixed with TMPD radical cations and BP radical cations mixed with DPA radical anions, illustrate energy-insufficient and energy-sufficient pathways in mixed system ECL. They also show the complexity of mixed systems, because the nonemitting species can be either the radical cations or the radical anions. In both cases, the spectrum of the generated light must be compared to a fluorescence spectrum of the emissive species, DPA, to demonstrate that the ECL signal does indeed arise entirely from DPA excited singlet emission [17].

B. Emissive Reaction Partners

Often, species capable of single-system ECL are coupled to a nonemissive reaction partner that is more easily oxidized or reduced (see, e.g., Refs. 17 and 34). The potentials applied are then controlled to generate only the desired radical cation and radical anion.

However, mixed systems are particularly interesting because they allow a greater range of emissive species to participate for ECL. Emissive molecules that are capable of forming only a single stable radical ion and thus incapable of single-system ECL can still participate in mixed ECL systems. For example, when considering ECL from *trans*-stilbene derivatives, Wilson et al. [35] reported weak or undetectable single-system ECL due to unstable radical cations. However, a mixed system with tri-p-tolylamine radical cations allowed stable ECL.

V. MIXED SYSTEMS FOR ORGANIC ECL

A. Mixed System Nonemitters

Table 2 summarizes common nonemitting organic molecules used in investigations of mixed ECL systems. Substituted aniline and aromatic ketones make up the bulk of mixed system nonemitters [28,36]. Other molecules include nitrogen-, oxygen-, or sulfur containing aromatic heterocycles, alkenes, and even solvents that can be electrolyzed to yield radical ions [35,37,38].

1. Aniline Derivatives

By far the most common reaction partners for mixed system ECL are substituted anilines. These compounds form radical cations ($D^{+\bullet}$) whose stability depends on their particular structure [39]. The formal potential for oxidation of these compounds is modest (typically <1 V), facilitating radical cation formation with various solvents and electrode materials. Aniline derivatives typically exhibit high electronic excited state energies that do not get populated during annihilation reactions. This simplifies mechanistic interpretation of the ECL emission.

2. Aromatic Ketones

Benzophenone and other related ketones can be reduced to form stable radical anions. The reduction potentials of aromatic ketones vary widely. This allows a large range of ΔE° values to be considered in mechanistic studies [17,28].

B. Mixed System Emitters

As noted above, any of the molecules involved in single-system ECL can also be studied in mixed systems. These have already been summarized in Table 1. Other emissive molecules have been considered only in mixed systems, though their participation in ECL from single systems may be possible. These are listed in Table 3.

VI. EXCIPLEX EMISSION, CHARACTERISTICS AND REQUIREMENTS

In addition to forming singlet and triplet excited states as in reactions (4) and (12), radical ion annihilation can lead directly from the contact radical ion pair to the emissive exciplex excited state (Chapter 4) [40,41].

Table 2 Nonemitters in Mixed systems

Compound	$E^{\circ}_{R/R+\bullet}$ vs. SCE (eV)	$E^{\circ}_{R/R-\bullet}$ vs. SCE (eV)	$E\,(^{1}R^{*})$ (eV)	$E\,(^{3}R^{*}$ (eV)
Würster's Blue (*N,N,N′,N′*-tetramethyl-*p*-phenylenediamine)	0.24[a] [32]	n/a	3.52 [71]	2.83 [71]
Würster's Red (*N,N*-dimethyl-*p*-phenylenediamine)	0.45[a,d] [61]	n/a		2.7 [61]
Tri-*p*-tolylamine	0.74[b] [86]	n/a	3.51 [86]	2.96 [86]
Tri-p-anisylamine	0.51[b] [36]	n/a	3.26 [36]	
Triphenylamine	0.95[b] [36]	n/a	3.5 [36]	
N,N′-Diphenylbenzidine	0.65[b] [36]	n/a	3.4 [36]	
4-Methyltriphenylamine	0.84[b] [36]	n/a		
4,4′-Dimethoxytriphenylamine	0.57[b] [36]	n/a		
Benzophenone	n/a	1.91[b] [28]	3.2 [28]	3.00 [28]
1,4-Benzoquinone	n/a	0.53[c] [28]	5.2 [28]	
1,4-Naphthaquinone	n/a	0.75[c] [28]	5.1 [28]	
9,10-Anthraquinone	n/a	0.89[c] [28]	3.07 [28]	
4-Methylbenzophenone	n/a	1.90[c] [28]		
4-Methoxybenzophenone	n/a	1.94[c] [28]		
9-Fluorenone	n/a	0.81[c] [28]		
2-Naphthyl phenyl ketone	n/a	1.18[c] [28]		
Benzil	n/a	1.21[c] [28]		2.67 [28]
trans-Stilbene	1.46[b] [35]	2.37[b] [35]	3.8 [61]	2 [61]
cis-Stilbene	n/a	2.13[a] [87]		2.5 [87]
10-Methylphenothiazine	0.69[b] [36]	n/a	3.4 [61]	2.4 [61]
10-Phenylphenothiazine	0.88[a] [87]	n/a		2.4 [87]
2,5-Diphenyl-1,3,4-oxadiazole	n/a	2.17[b] [89]	3.9 [89]	2.5 [89]
Benzonitrile	n/a	2.74[a,e] [88]		
Thianthrene	1.25[b] [89]	2.54[a] [69]	3.9 [89]	2.6 [89]
Bis-1,8-(*N,N*-dimethylamino)naphthalene	0.7[b] [41]	n/a	2.4	
N-p-Tolylphthalimide	n/a	1.40[a] [79]		
7,14-Diphenyl-acenaphtho[1,2-*k*]fluoranthene	n/a	1.47[a] [79]		
1,4-Bis[5-phenyl-oxazolyl-(2)]-benzene	n/a	1.76[a] [79]		
2-*p*-Biphenyl-5-phenyl-1,3,4-oxadiazole	n/a	1.93[a] [79]		
2,5-Di-*p*-biphenylyl-oxazole	n/a	2.01[a] [79]		
2,5-Diphenyl-oxazole	n/a	2.21[a] [79]		
2-*p*-Anisyl-5-phenyl-oxazole	n/a	2.33[a] [79]		

[a]In DMF.
[b]In MeCN.
[c]In PhCN.
[d]Vs. SSCE.
[e]Vs. Ag/ClO$_4$.

Table 3 Emitters in Mixed Systems

Compound	$E^{\circ}_{R/R+\bullet}$ vs. SCE (eV)	$E^{\circ}_{R/R-\bullet}$ vs. SCE (eV)	$E\,(^{1}R^{*})$ (eV)	$E\,(^{3}R)$ (eV)
Naphthalene	1.54[a] [60]	2.54[b] [60]	3.9 [61]	[61]
1,2-Benzanthracene	1.18[a] [28]	2.00[a] [28]	3.22 [28]	2.08 [28]
Benzophenone		1.91[a] [28]	3.2 [28]	3 [28]
Benzo[a]pyrene		1.98[a] [36]	3.07 [36]	1.81 [36]
Benzo[e]pyrene		2.26[a] [36]	3.39 [36]	2.29 [36]
Bibenz[a,c]anthracene		2.17[a] [36]	3.4 [36]	2.2 [36]
7,12-Dimethylbenz[a]anthracene		2.21[a] [26]	2.96 [36]	1.92 [36]
3-Methylcholanthrene		2.29[a] [36]	3.42 [36]	
Picene		2.28[c] [34]	3.31 [34]	2.58 [34]
1,12-Benzoperylene		1.92[c] [71]	3.07 [69]	2 [69]
5,10-Dihydroindeno[2, 1-a]indene	1.14[a] [35]	2.62[a] [35]	3.34 [35]	
2-(p-Tolyl)indene	1.25[a] [35]	2.48[a] [35]	3.34 [35]	
3,4-Dihydro-2-(2-naphthyl)naphthalene	1.25[a] [35]	2.25[a] [35]	3.18 [35]	
Thianthrene	1.25[a] [89]	2.54[b] [69]	3 [89]	2.6 [89]
Acetophenone		2.18[a] [41]		3.2 [41]
Dibenzoylmethane		1.55[a] [41]		2.77 [41]
Benz[a]anthracene		2.17[a] [41]	3.35 [41]	2.08 [41]
1,1,4,4-Tetraphenyl-1,3- butadiene		2.2[a] [41]	3.14 [41]	
1,4-Diphenyl–1,3-butadiene		2.12[a] [41]	3.6 [41]	
trans-Stilbene	1.46[b] [35]	2.37[b] [35]	3.8 [61]	2 [61]
9-Methylanthracene		2.19[b] [35]	3.2 [41]	1.8 [41]
2,5-Diphenyloxazole		2.38[b] [35]	3.61 [41]	
3-[1,1′-Biphenyl]-4-yl-4,5- dihydro-1-(4-methoxyphenyl)-5-phenyl-1H-pyrazole	0.59[a] [79]	2.17[b] [79]		1.83 [79]
4,5-Dihydro-1-(4-methoxyphenyl)-5-phenyl-3-(2-phenylethenyl)-1H-pyrazole	0.58[a] [79]	2.16[b] [79]		1.75 [79]
3,5-diphenyl-1-p-tolyl-2-pyrazoline	0.75[a] [79]	2.37[b] [79]		2.07 [79]
1,1′-[1,1′-biphenyl]-4,4′-diyl-bis[4,5-dihydro-3,5-diphenyl-1H-pyrazole	0.56[a] [79]	2.36[b] [79]		1.88 [79]
Bi(1-phenyl-2-pyrazolin-3-yl)	0.49[a] [79]	2.36[b] [79]		1.75 [79]

[a]In MeCN.
[b]In DMF.
[c]In DME.

$$[D^{+\bullet} + A^{-\bullet}] \rightarrow {}^1(AD)^* \qquad\qquad (16)$$

$${}^1(AD)^* \rightarrow A + D + h\nu \qquad\qquad (17)$$

[When a single system exhibits this behavior, the dimer exciplex, ${}^1(R_2)^*$, is termed an "excimer."] Although exciplexes can be formed through spectroscopic excitation, its formation is particularly likely in ECL because of the close proximity of the radical ions in the contact radical ion pair. Exciplexes that are not seen following spectroscopic excitation can dominate in the ECL spectra [42]. Additionally, exciplex generation in polar solvents is possible through ECL, though not with spectroscopic excitation [43].

Exciplex emission is characterized by a broad featureless emission redshifted from the singlet emission of the individual molecules (${}^1A^*$ or ${}^1D^*$). Further, the wavelength of exciplex emission varies with solvent polarity [41]. In general, singlet and exciplex emission are observed simultaneously [44]. The result is that emission from these systems is quite broad (several hundred nanometers) and can appear white [30].

The main requirement for exciplex formation is structural [45]. The participating molecules must be able to align such that there is significant π-orbital overlap. For steric reasons, this occurs most often with planar PAHs. Table 4 lists excimer and exciplex systems along with the characteristic emission assignments for the ECL spectrum.

VII. TWISTED INTRAMOLECULAR CHARGE TRANSFER EMISSION

In exciplex emission, the proximity of participating species to one another is of primary importance. Some researchers have used synthetic procedures to include pairs of exciplex-forming species in a single molecule (A–D, though A and D are not necessarily different [46]). The excited state emission from the resulting molecule shows emission only from the twisted intramolecular charge transfer (TICT) excited state [46–50], sometimes called an intramolecular exciplex [51,52]. For example, 4-(9-anthracenyl)-N,N-dimethylaniline (ADMA) combines

ADMA

Table 4 Systems Exhibiting Excimer and Exciplex Emission

D/D$^{+\bullet}$	A/A$^{-\bullet}$	Emissive state
Excimer		
Perylene	Perylene	^{1}A*, ^{1}A$_2$* [42,90,91]
Pyrene	Pyrene	^{1}A*, ^{1}A$_2$* [42,92,93]
4,5,6,7-Tetrahydrodinaphtho[2,1-g:1′,2′-i]–[1,6]dioxecin	4,5,6,7-Tetrahydrodinaphtho[2,1-g:1′,2′-i]–[1,6]dioxecin	^{1}A*, ^{1}A$_2$* [45]
Benzo[a]pyrene	Benzo[a]pyrene	^{1}A*, ^{1}A$_2$* [42]
Naphthalene	Naphthalene	^{1}A$_2$*, [93]
1,6-Bis(dimethylamino)pyrene	1,6-Bis(dimethylamino)pyrene	^{1}A$_2$* [76]
1,6-bis(methylthio)pyrene	1,6-Bis(methylthio)pyrene	^{1}A$_2$* [76]
Exciplex		
Würster's Blue (N,N,N′,N′-tetramethyl-p-phenylenediamine)	Pyrene	^{1}A*, ^{1}AD* [71]
	9,10-Dimethylanthracene	^{1}A*, ^{1}A$_2$* [71]
	9,10-Diphenylanthracene	^{1}A*, ^{1}A$_2$* [71]
	1,12-Benzoperylene	^{1}A*, ^{1}A$_2$* [71]
	Benzo[a]pyrene	^{1}A*, ^{1}A$_2$* [71]
	Benzo[e]pyrene	^{1}A*, ^{1}A$_2$* [71]
	Dibenz[a,h]anthracene	^{1}A*, ^{1}A$_2$* [71]
	Benz[a]anthracene	^{1}A*, ^{1}A$_2$* [71]
	9-Methylanthracene	^{1}A*, ^{1}A$_2$* [71]
	Anthracene	^{1}A*, ^{1}A$_2$* [71]
	N-Ethylcarbazole	^{1}D*, ^{1}A*, ^{1}AD* [44]
	4,4′-Dimethyl-biphenyl	^{1}D*, ^{1}A*, ^{1}AD* [44, 94]
	4-Methylbiphenyl	^{1}D*, ^{1}A*, ^{1}AD* [44]
	Biphenyl	^{1}D*, ^{1}A*, ^{1}AD* [44]
	Naphthalene	^{1}D*, ^{1}A*, ^{1}AD* [44]
	m-Terphenyl	^{1}D*, ^{1}A*, ^{1}AD* [44]
	Phenanthrene	^{1}D*, ^{1}A*, ^{1}AD* [44]
	Bitolyl	^{1}D*, ^{1}A*, ^{1}AD* [94]
	Chrysene	^{1}A*, ^{1}AD* [94]
Tri-p-tolylamine	9,10-Dimethylanthracene	^{1}A*, ^{1}AD* [40, 95]
	9-Methylanthracene	^{1}A*, ^{1}AD* [40, 43, 94, 95]
	Anthracene	^{1}A*, ^{1}AD* [40, 95]
	Benzo[a]pyrene	^{1}A*, ^{1}AD* [36]
	Benzo[e]pyrene	^{1}A*, ^{1}AD* [36]
	Dibenz[a,h]anthracene	^{1}A*, ^{1}AD* [36]
	Dibenz[a,c]anthracene	^{1}A*, ^{1}AD* [36]
	Benz[a]anthracene	^{1}A*, ^{1}AD* [36]

(Continued)

Table 4 Continued

D/D^{+}*	A/A$^{-\bullet}$	Emissive state
	7,12-Dimethylbenz[*a*]anthracene	^{1}A*, ^{1}AD* [36]
	3-Methylcholanthrene	^{1}A*, ^{1}AD* [36]
	9,10-Diphenylanthracene	^{1}A*, ^{1}AD* [94]
	9,10-Di(1-naphthyl)anthracene	^{1}A*, ^{1}AD* [94]
	9,9′-Bianthracene	^{1}A*, ^{1}AD* [94]
	Benzophenone	^{1}A*, ^{1}AD* [43, 96, 97]
	Naphthalene	^{1}D*, ^{1}AD* [45, 96]
	Acetophenone	^{1}AD* [43]
	Dibenzoylmethane	^{1}AD* [43]
	1,1,4,4-Tetraphenyl-1,3-butadiene	^{1}AD* [43]
	1,4-Diphenyl-1,3-butadiene	^{1}AD* [43]
	trans-Stilbene	^{1}AD* [43, 35]
	2,5-Diphenyloxazole	^{1}AD* [43]
	Naphthalene	^{1}AD* [43]
	1,4-Dicyanobenzene	^{1}AD* [97]
	1-Methylnaphthalene	^{1}A*, ^{1}AD* [45]
	2-Methylnaphthalene	^{1}A*, ^{1}AD* [45]
	1,4-Diphenylnaphthalene	^{1}A*, ^{1}AD* [45]
	1,5-Diphenylnaphthalene	^{1}A*, ^{1}AD* [45]
	1,4-Dinaphthylbenzene	^{1}A*, ^{1}AD* [45]
	2-(2-Fluorenyl)-5-naphthyloxazole	^{1}A*, ^{1}AD* [45]
	1-(6-Phenylindenyl)-naphthalene	^{1}A*, ^{1}AD* [45]
	2,2′-Binaphthyl	^{1}A*, ^{1}AD* [45]
	6-Phenyl-2,2′-binaphthyl	^{1}A*, ^{1}AD* [45]
	Dinaphtho[2,1-*d*:1′,2′-f][1,3]dioxepin	^{1}A*, ^{1}AD* [45]
	5,6-Dihydro-4*H*-dinaphtho[2,1-*g*:1′, 2′-*h*][1,5]-dioxocin	^{1}AD* [45]
	4,5,6,7-Tetrahydrodinaphtho[2,1-*g*:1′,2′-*i*]-[1,6]dioxecin	^{1}A*, ^{1}AD* [45]
	5,10-Dihydroindeno[2,1-*a*]indene	^{1}A*, ^{1}AD* [35]
	1-Methyl-2,5-diphenylindene	A*, ^{1}AD* [35]

(Continued)

Table 4 Continued

D/D+*	A/A−•	Emissive state
	2-(p-tert-butylphenyl)indene	¹¹AD* [35]
	2-Phenylindene	¹AD* [35]
	2-(p-tolyl)indene	¹AD* [35]
	2-(4′-Biphenyl)indene	¹A*, ¹AD* [35]
	2-(2′-Naphthyl)indene	¹A*, ¹AD* [35]
	2-(1′-Naphthyl)indene	¹A*, ¹AD* [35]
	2-(1-Naphthyl)-5-phenylindene	¹A*, ¹AD* [35]
	3,4-Dihydro-2,6-diphenylnaphthalene	¹A*, ¹AD* [35]
	3-(4-Biphenyl)-1,2-dihydronaphthalene	¹A*, ¹AD* [35]
	3,4-Dihydro-2-(2-naph-thyl)naphthalene	¹A*, ¹AD* [35]
	3,4-Dihydro-2-(2-naph-thyl)-6-naphthalene	¹A*, ¹AD* [35]
	3,4-Dihydro-2-(4-biphenyl)-phenanthrene	¹AD* [35]
	3,4-Dihydro-2-phenylphenanthrene	¹A*, ¹AD [35]*
Tris(p-dimethylaminophenyl)amine	4,4′-Dimethylbiphenyl	¹D*, ¹AD* [44]
	4-Methylbiphenyl	¹D*, ¹AD* [44]
	Biphenyl	¹D*, ¹AD* [44]
	Naphthalene	¹D*, ¹A*, ¹AD* [44]
	m-Terphenyl	¹D*, ¹A*, ¹AD* [44]
	Phenanthrene	¹D*, ¹A*, ¹AD* [44]
	2,4,6-Trimethyl benzoni-trile	¹D*, ¹AD* [44]
	Picene	¹A*, ¹AD* [44]
	trans-Stilbene	¹AD* [44]
	Benzo[c]phenanthrene	¹A*, ¹AD* [94]
	p-Terphenyl	¹A*, ¹AD* [94]
	p-Quaterphenyl	¹A*, ¹AD* [94]
	p-Quinquiphenyl	¹A*, ¹AD* [94]
	p-Sexiphenyl	¹A*, ¹AD* [94]
	Chrysene	¹A*, ¹AD* [94]
Triphenylamine	Chrysene	¹A*, ¹AD* [36]
N,N′-Diphenylbenzidin		¹A*, ¹AD* [36]
Tri-p-tolylamine		¹A*, ¹AD* [36]
Tri-p-anisylamine		¹A*, ¹AD* [36]
4,4′-Dimethoxytriphenylamine		¹A*, ¹AD* [36]

(*Continued*)

Table 4 Continued

D/D⁺*	A/A⁻•	Emissive state
4-Methyltriphenylamine		^{1}A*, ^{1}AD* [36]
10-Methylphenothiazine		^{1}A*, ^{1}AD* [36]
Triphenylamine	Naphthalene	^{1}AD*, [43]
N,N'-Diphenylbenzidine		^{1}AD*, [43]
Bis-1,8-(N,N-dimethylamino)naphthalene		^{1}AD* [43]
Tri-p-anisylamine		^{1}AD* [43]
Perylene	Pyrene	^{1}D*, ^{1}AD* [91]

anthracene and aniline moieties. Heterogeneous electron transfer of TICT emitters yields radical ions localized within the molecule,

$$A—D + e^- \rightarrow A^{-•}—D \tag{18}$$

$$A—D - e^- \rightarrow A—D^{+•} \tag{19}$$

When excited spectroscopically, these species yield a faint emission band at short wavelengths assigned to $^1A^*$ and a more intense emission at lower energy corresponding to $^1(A—D)^*$, the TICT excited state. In ECL emission, only the longer wavelength TICT emission is seen:

$$A^{-•}—D + A—D^{+•} \rightarrow {}^1(A—D)^* + A—D \tag{20}$$

$$^1(A—D)^* \rightarrow A—D + h\nu \tag{21}$$

Synthetic procedures allow the physical separation between the moieties of interest to be controlled as well as the steric hindrance to the molecular folding necessary to form the TICT excited state [47,52]. This allows exploration of Marcus theory for electron transfer by yielding well-defined geometries for the electron transfer [47,49,50,52]. Table 5 lists emissive TICT molecules studied to date, the $\Delta E°$ for radical cation and anion formation, and their reported ECL efficiency.

VIII. PREANNIHILATION EMISSION

When ECL was first being investigated, there were several reports of preannihilation ECL [53–58]. This referred to a weak emission seen for single systems in some solvents when one radical ion was generated but the potential was insufficient to generate the second radical ion. Although many theories were proposed to explain this emission [53,55,56], the eventual conclusion was that electrolysis was responsible for generating species in solution that were oxidized or

Table 5 Emissive TICT Systems

Compound	ΔE° (eV)[a]	ϕ_{ECL} (%)[a]
4-(9-Anthracenyl)-N,N-dimethylaniline	2.77 [47, 51, 52]	1.1 [47, 51]
4-(9-Anthracenylmethyl)-N,N-dimethylaniline	2.76 [51]	
4-[2-(9-Anthracenyl)ethyl]-N,N-dimethylaniline	2.77 [51]	
4-(9-Anthracenyl)-N,N-di-p-tolylaniline	2.77 [52]	2.8[b] [52]
4-(9-Anthracenyl)-N,N-di-p-anisylaniline	2.63 [52]	0.84[b] [52]
4-[9-(10-Phenylanthracenyl)]-N,N-dimethylaniline	2.74 [52]	2.4[b] [52]
4-[9-(10-Phenylanthracenyl)]-N,N-tolylaniline	2.77 [52]	3.6[b] [52]
4-[9-(10-Phenylanthracenyl)]-N,N-anisylaniline	2.61 [52]	0.81[b] [52]
4-(1-Pyrenyl)-N,N-dimethylaniline	2.84 [52]	4.4[b] [52]
4-(1-Pyrenyl)-N,N-tolylaniline	2.89 [52]	3.0[b] [52]
4-(1-Pyrenyl)-N,N-anisylaniline	2.72 [52]	0.32[b] [52]
4-(9-Anthacenyl)-N,N-diethylaniline	2.78 [47]	0.14 [47]
4-(9-Anthacenyl)-3,5-dimethyl-N,N-dimethylaniline	2.8 [47]	0.57 [47]
4-(9-Anthacenyl)-3-methyl-N,N-dimethylaniline	2.82 [47]	0.44 [47]
4-(9-Anthacenyl)-3-methoxy-N,N-dimethylaniline	2.79 [47]	
1-Methyl-5-(9-anthracenyl)-indoline	2.66 [47]	0.22 [47]
4-(1-Naphthyl)-N,N-dimethylaniline	3.31 [49]	
4-(3-Fluoranthenyl)-N,N-dimethylaniline	2.57 [49]	2.7 [49]
4-(1-Pyrenyl)-N,N-dimethylaniline	2.86 [49]	4.7 [49]
9,9′-Bianthryl	3.17 [46]	3 [46]
10,10′-Dimethoxy-9,9′-bianthryl	3.00 [46]	3 [46]

[a]In MeCN unless otherwise noted.
[b]In 3:1 MeCN:PhCN.

reduced at less extreme potentials than the second radical ion [54,57–59]. Thus preannihilation ECL represents a mixed ECL system where the nonemissive reaction partner is a breakdown product of the radical ions of the emissive reaction partner.

IX. ORGANIC COREACTANT SYSTEMS

Although many coreactant ECL systems involve organic molecules, these were covered in Chapter 5 and will not receive further consideration here.

ACKNOWLEDGMENT

Support for this work was provided by NSF grant CHE-0096837.

REFERENCES

1. Hercules, D.M. Chemiluminescence from electron-transfer reactions. Acc. Chem. Res. **1969**, *2*, 301–307.
2. Weller, A., Zachariasse, K. Chemiluminescence from radical ion recombination. VI. Reactions, yields, and energies. Chemiluminescence Biolumin., Pap. Int. Conf. **1973**, Plenum, NY, NY 169–180.
3. Faulkner, L.R. Chemiluminescence from electron-transfer processes. Methods Enzymol. **1978**, *57*(Biolumin. Chemilumin.), 494–526.
4. Park, S.-M., Tryk, D.A. Excited state intermediates probed by electrogenerated chemiluminescence. Rev. Chem. Intermed. **1980**, *4*(1–4), 43–79.
5. Faulkner, L.R., Glass, R.S. Electrochemiluminescence. Chem. Biol. Gener. Excited States **1982**, Academic Press, NY, NY 191–227.
6. Knight, A.W., Greenway, G.M. Occurrence, mechanisms and analytical applications of electrogenerated chemiluminescence. A review. Analyst **1994**, *119*(5), 879–890.
7. Rothberg, L.J., Lovinger, A.J. Status of and prospects for organic electroluminescence. J. Mater. Res. **1996**, *11*(12), 3174–3187.
8. Armstrong, N.R., Anderson, J., Lee, P., McDonald, E., Wightman, R.M., Hall, H.K., Hopkins, T., Padias, A., Thayumanavan, S., Barlow, S., Marder, S. Electrochemical models for the radical annihilation reactions in organic light-emitting diodes. Proc SPIE **1998**, *3476*, 178–187.
9. Forry, S., Wightman, R.M. Electrogenerated chemiluminescence detection in reversed-phase liquid chromatography. Anal. Chem. **2002**, *74*(3), 528–532.
10. Maus, R.G., McDonald, E.M., Wightman, R.M. Imaging of nonuniform current density at microelectrodes by electrogenerated chemiluminescence. Anal. Chem. **1999**, *71*(21), 4944–4950.
11. Werner, T.C., Chang, J., Hercules, D.M. The electrochemiluminescence of anthracene and 9,10-dimethylanthracene. The role of direct excimer formation. J. Am. Chem. Soc. **1970**, *92*(4), 763–768.
12. Faulkner, L.R., Bard, A.J. Electrogenerated chemiluminescence. I. Mechanism of anthracene chemiluminescence in N,N-dimethylformamide solution. J. Am. Chem. Soc. **1968**, *90*(23), 6284–6290.
13. Brilmyer, G.H., Bard, A.J. Electrogenerated chemiluminescence. XXXVI. The production of steady direct current ECL in thin layer and flow cells. J. Electrochem. Soc. **1980**, *127*(1), 104–110.
14. Collinson, M.M., Wightman, R.M. High-frequency generation of electrochemiluminescence at microelectrodes. Anal. Chem. **1993**, *65*(19), 2576–2582.
15. Faulkner, L.R., Tachikawa, H., Bard, A.J. Electrogenerated chemiluminescence. VII. Influence of an external magnetic field on luminescence intensity. J. Am. Chem. Soc. **1972**, *94*(3), 691–699.
16. Freed, D.J., Faulkner, L.R. Mechanisms of chemiluminescent electron-transfer reactions. II. Triplet yield of electron transfer in the fluoranthene-10-methylphenothiazine system. J. Am. Chem. Soc. **1971**, *93*(15), 3565–3568.
17. Ritchie, E.L., Pastore, P., Wightman, R.M. Free energy control of reaction pathways in electrogenerated chemiluminescence. J. Am. Chem. Soc. **1997**, *119*(49), 11920–11925.

18. Meyer, T.J. Excited-state electron transfer. In: S.J. Lippard, ed. Progress in Inorganic Chemistry. New York: Wiley, 1983, pp 389–441.

19. Weller, A. Electron-transfer and complex formation in the excited state. Pure Appl. Chem. **1968**, *16*, 115–123.

20. Maness, K.M., Bartelt, J.E., Wightman, R.M. Effects of solvent and ionic strength on the electrochemiluminescence of 9,10-diphenylanthracene. J. Phys. Chem. **1994**, *98*(15), 3993–3998.

21. Kapturkiewicz, A. Solvent and temperature control of the reaction mechanism and efficiency in the electrogenerated chemiluminescence of rubrene. J. Electroanal. Chem. **1994**, *372*, 101–116.

22. Gross, E.M., Anderson, J.D., Slaterbeck, A.F., Thayumanavan, S., Barlow, S., Zhang, Y., Marder, S.R., Hall, H.K., Nabor, M. Flore, Wang, J.-F., Mash, E.A., Armstrong, N.R., Wightman, R.M. Electrogenerated chemiluminescence from derivatives of aluminum quinolate and quinacridones: cross-reactions with triarylamines lead to singlet emission through triplet-triplet annihilation pathways. J. Am. Chem. Soc. **2000**, *122*(20), 4972–4979.

23. Maness, K.M., Wightman, R.M. Electrochemiluminescence in low ionic strength solutions of 1,2-dimethoxyethane. J. Electroanal. Chem. **1994**, *396*, 85–95.

24. Sutin, N. Nuclear, electronic, and frequency factors in electron-transfer reactions. Acc. Chem. Res. **1982**, *15*, 275–282.

25. Marcus, R.A. On the theory of chemiluminescent electron-transfer reactions. J. Chem. Phys. **1965**, *43*(8), 2654–2657.

26. Hoytink, G.J. Electrochemiluminescence of aromatic hydrocarbons. Discuss. Faraday Soc. **1968**, *No. 45*, 14–22.

27. Keszthelyi, C.P., Tokel-Takvoryan, N.E., Bard, A.J. Electrogenerated chemiluminescence: determination of the absolute luminescence efficiency in electrogenerated chemiluminescence; 9,10-diphenylanthracene-thianthrene and other systems. Anal. Chem. **1975**, *47*(2), 249–256.

28. Beldeman, F.E., Hercules, D.M. Electrogenerated chemiluminescence from 9,10-diphenylanthracene cations reacting with radical anions. J. Phys. Chem. **1979**, *83*(17), 2203–2209.

29. Collinson, M.M., Wightman, R.M., Pastore, P. Evaluation of ion-annihilation reaction kinetics using high-frequency generation of electrochemiluminescence. J. Phys. Chem. **1994**, *98*(46), 11942–11947.

30. Slaterbeck, A.F., Meehan, T.D., Gross, E.M., Wightman, R.M. Selective population of excited states during electrogenerated chemiluminescence with 10-methylphenothiazine. J. Phys. Chem. B **2002**, *106*(23), 6088–6095.

31. Wightman, R.M., Curtis, C.L., Flowers, P.A., Maus, R.G., McDonald, E.M. Imaging microelectrodes with high-frequency electrogenerated chemiluminescence. J. Phys. Chem. B **1998**, *102*(49), 9991–9996.

32. Faulkner, L.R., Bard, A.J. Electrogenerated chemiluminescence. I. Mechanism of anthracene chemiluminescence in N,N-dimethylformamide solution. J. Am. Chem. Soc. **1968**, *90*(23), 6284–90.

33. Park, S.M., Bard, A.J. Electrogenerated chemiluminescence. The production of benzophenone phosphorescence in fluid solution by radical ion reaction. Chem. Phys. Let. **1976**, *38*(2), 257–262.

34. Weller, A., Zachariasse, K. Chemiluminescence from chemical oxidation of aromatic anions. J. Chem. Phys. **1967**, *46*(12), 4984–4985.

35. Wilson, J.R., Park, S.-M., Daub, G.H. Electrogenerated chemiluminescence of trans-stilbene derivatives. J. Electrochem. Soc. **1981**, *128*(10), 2085–2089.

36. Sharifian, H.A., Park, S.-M. Electrogenerated chemiluminescence of several polycyclic aromatic hydrocarbons. Photochem. Photobiol. **1982**, *36*(1), 83–90.

37. Keszthelyi, C.P., Tachikawa, H., Bard, A.J. Electrogenerated chemiluminescence. VIII. Thianthrene-2,5-diphenyl-1,3,4-oxadiazole system. Mixed energy-sufficient system. J. Am. Chem. Soc. **1972**, *94*(5), 1522–1527.

38. Wightman, R.M., Curtis, C.L., Flowers, P.A., Maus, R.G., McDonald, E.M. Imaging microelectrodes with high-frequency electrogenerated chemiluminescence. J. Phys. Chem. B **1998**, *102*(49), 9991–9996.

39. Seo, E.T., Nelson, R.F., Fritsch, F.M., Marcoux, L.S., Leedy, D.W., Adams, R.N. Anodic oxidation pathways of aromatic amines. Electrochemical and electron paramagnetic resonance studies. J. Am. Chem. Soc. **1966**, *88*(15), 3498–3503.

40. Weller, A., Zachariasse, K. Direct hetero-excimer formation from radical ions. Chem. Phys. Lett. **1971**, *10*(5), 590–594.

41. Park, S.-M., Bard, A.J. Electrogenerated chemiluminescence. XXII. Generation of exciplexes in the radical ion reaction. J. Am. Chem. Soc. **1975**, *97*(11), 2978–2985.

42. Chandross, E.A., Longworth, J.W., Visco, R.E. Excimer formation and emission via the annihilation of electro-generated aromatic hydrocarbon radical cations and anions. J. Am. Chem. Soc. **1965**, *87*(14), 3259–3260.

43. Park, S.-M., Bard, A.J. Electrogenerated chemiluminescence. XXII. Generation of exciplexes in the radical ion reaction. J. Am. Chem. Soc. **1975**, *97*(11), 2978–2985.

44. Weller, A., Zachariasse, K. Chemiluminescence from radical ion recombination. VII. Heteroexcimer chemiluminescence yields. Chemiluminescence Biolumin., Pap. Int. Conf. **1973**, Plenum, NY, NY 181–191.

45. Park, S.-M., Paffett, M.T., Daub, G.H. Electrogenerated chemiluminescence of naphthalene derivatives. Steric effects on exciplex emissions. J. Am. Chem. Soc. **1977**, *99*(16), 5393–5399.

46. Kapturkiewiez, A. Electrochemical generation of excited twisted intramolecular charge transfer states. Part VI. J. Electroanal Chem. **1993**, *348*, 283–302.

47. Kapturkiewiez, A., Grabowski, Z.R., Jasny, J. Electrochemical generation of excited TICT states. Part I. J. Electroanal Chem. **1990**, *279*, 55–65.

48. Kapturkiewiez, A. Electrochemical generation of excited TICT states. Part II. Effect of the supporting electrolyte. J. Electroanal Chem. **1990**, *290*, 135–143.

49. Kapturkiewiez, A. Electrochemical generation of excited TICT* states. Part III. Aryl derivatives of N,N-dimethylaniline. J. Electroanal Chem. **1991**, *302*, 131–144.

50. Kapturkiewiez, A. Electrochemical generation of excited TICT states. Part IV. Effect of the solvent. J. Electroanal Chem. **1991**, *170*, 87–105.

51. Itaya, K., Toshima, S. Formation of intramolecular exciplexes in electrogenerated chemiluminescence. Chem. Phys. Lett. **1977**, *51*(3), 447–452.

52. Kawai, M., Itaya, K., Toshima, S. Formation of intramolecular exciplexes in electrogenerated chemiluminescence. 2. J. Phys. Chem. **1980**, *84*(19), 2368–2374.

53. Maricle, D.L., Maurer, A. Pre-annihilation electrochemiluminescence of rubrene. J. Am. Chem. Soc. **1967**, *89*(1), 188–189.

54. Malbin, M.D., Mark, H.B. The reduction of aromatic hydrocarons. IV. The effects of a proton donor on preannihilation electrochemiluminescence. J. Phys. Chem. **1969**, *73*(9), 2992–2995.

55. Maricle, D.L., Zweig, A., Maurer, A.H., Brinen, J.S. Pre-annihilation electrochemiluminescence. Electrochim. Acta **1986**, *13*, 1209–1220.

56. Zweig, A., Maricle, D.L. On the mechanism of preannihilative electrochemiluminescence. J. Phys. Chem. **1968**, *72*(1), 377–378.

57. Chandross, E.A., Visco, R.E. On preannihilative electrochemiluminescence and the heterogeneous electrochemical formation of excited states. J. Phys. Chem. **1968**, *72*(1), 378–379.

58. Zweig, A., Hoffmann, A.K., Maricle, D.L., Maurer, A.H. An investigation of the mechanism of some electrochemiluminescent processes. J. Am. Chem. Soc. **1968**, *90*(2), 261–268.

59. Hercules, D.M., Lansbury, R.C., Roe, D.K. Chemiluminescence from the reduction of rubrene radical cations. J. Am. Chem. Soc. **1966**, *88*(2), 4578–4583.

60. Weig, A., Maurer, A.H., Roberts, B.G. Oxidation, reduction, and electrochemiluminescence of donor-substituted polycyclic aromatic hydrocarbons. J. Org. Chem. **1967**, *32*(5), 1322–1329.

61. Faulkner, L.R., Freed, D.J. Mechanisms of chemiluminescent electron-transfer reactions. I. Role of the triplet state in energy-deficient systems. J. Am. Chem. Soc. **1971**, *93*(9), 2097–2102.

62. Fleet, B., Keliher, P.N., Kirkbright, G.F., Pickford, C.J. Analytical usefulness of electrochemiluminescence for the determination of microgram amounts of aromatic hydrocarbons. Analyst **1969**, *94*(1123), 847–854.

63. Bard, A.J., Santhanam, K.S.V., Maloy, J.T., Phelps, J., Wheeler, L.O. Steric effects and the electrochemistry of phenyl-substituted anthracenes and related compounds. Discuss. Faraday Soc. **1968**, *No. 45*, 167–174.

64. Tachikawa, H., Bard, A.J. Electrogenerated chemiluminescence. XII. Magnetic field effects on ECL [electrogenerated chemiluminescence] in the tetracene-TMPD [N,N′-tetramethyl-p-phenylenediamine] system. Evidence for triplet-triplet annihilation of tetracene. Chem. Phys. Lett. **1973**, *19*(2), 287–289.

65. Siegel, T.M., Mark, H.B. Mechanism of the chemiluminescent reaction of certain alkyl halides with electrogenerated aromatic hydrocarbon radical anions. J. Am. Chem. Soc. **1972**, *94*(26), 9020–9027.

66. Hercules, D.M. Chemiluminescence resulting from electrochemically generated species. Science **1964**, *145*(3634), 808–809.

67. Loutfy, R.O., Loutfy, R.O. The interrelation between polarographic half-wave potentials and the energies of electronic excited states. Can. J. Chem. **1976**, *54*(9), 1454–1463.

68. Abdel-Shafi, A.A., Wilkinson, F. Charge transfer effects on the efficiency of singlet oxygen production following oxygen quenching of excited singlet and triplet states of aromatic hydrocarbons in acetonitrile. J. Phys. Chem. A. **2000**, *104*(24), 5747–5757.

69. Santa Cruz, T.D., Akins, D.L., Birke, R.L. Chemiluminescence and energy transfer in systems of electrogenerated aromatic anions and benzoyl peroxide. J. Am. Chem. Soc. **1976**, *98*(7), 1677–1682.

70. Harada, M., Watanabe, I., Watarai, H. Calculation of electrostatic solvation energies for polycyclic aromatic hydrocarbon mono-cations in acetonitrile. Chem. Phys. Lett. **1999**, *301*(3,4), 270–274.

71. Weller, A., Zachariasse, K. Chemiluminescence from radical ion recombination. Experimental evidence for triplet-triplet annihilation mechanism. Chem. Phys. Lett. **1971**, *10*(2), 197–200.

72. Park, S.-M. Fluorescence and electrogenerated chemiluminescence studies of several dimethyl-substituted benzo[a]pyrenes. Photochem. Photobiol. **1978**, *28*(1), 83–90.

73. Kowert, B.A., Marcoux, L., Bard, A.J. Homogeneous electron-transfer reactions of several aromatic anion and cation radicals. J. Am. Chem. Soc. **1972**, *94*(16), 5538–5550.

74. Fakhr, A., Mugnier, Y., Laviron, E. Electrochemical reduction of phenothiazine and fluorobenzene at low temperature. Electrochim. Acta **1983**, *28*(12), 1897–1898.

75. Tokel, N., Keszthelyi, C.P., Bard, A.J. Electrogenerated chemiluminescence. X. α, β, γ, δ-Tetraphenylporphin chemiluminescence. J. Am. Chem. Soc. **1972**, 94(14), 4872–4877.

76. Zweig, A., Maurer, A.M., Roberts, B.G. Oxidation, reduction and electrochemilumi-nescence of donor-substituted polycyclic aromatic hydrocarbons. J. Org. Chem. **1967**, *32*(5), 1322–1329.

77. Zweig, A., Lehnsen, J.H. Cumulative influence of conjugated substituents on the p-system properties of aromatic hydrocarbons. X. Cumulative influence of methylthio groups on the p-system properties of aromatic hydrocarbons. J. Am. Chem. Soc. **1965**, *87*(12), 2647–2657.

78. Zweig, A., Hoffmann, A.K., Maricle, D.L., Maurer, A.H. An investigation of the mechanism of some electrochemiluminescent processes. J. Am. Chem. Soc. **1968**, *90*(2), 261–268.

79. Pragst, F., Fabian, G., Ziebig, R., Schmidt, D., Jugelt, W. Direct formation of excited singlet states in the electrogenerated chemiluminescence of mixed systems involving N-aryl-2-pyrazolines. Chem. Phys. Lett. **1975**, *36*(5), 630–634.

80. Libert, M., Bard, A.J. Electrogenerated chemiluminescence. XXVI. Systems involv-ing tetraarylpyrroles, tetraphenylfuran, and tetraphenylthiophene. J. Electrochem. Soc. **1976**, *123*(6), 814–818.

81. Park, S.-M., Paffett, M.T., Daub, G.H. Electrogenerated chemiluminescence of naphthalene derivatives. Steric effects on exciplex emissions. J. Am. Chem. Soc. **1977**, 99(16), 5393–5399.

82. Debad, J.D., Morris, J.C., Lynch, V., Magnus, P., Bard, A.J. Dibenzotetra-phenylperiflanthene: synthesis, photophysical properties, and electrogenerated chemiluminescence. J. Am. Chem. Soc. **1996**, *118*(10), 2374–2379.

83. Lee, S.K., Zu, Y., Herrmann, A., Geerts, Y., Muellen, K., Bard, A.J. Electrochemistry, spectroscopy and electrogenerated chemiluminescence of pery-lene, terrylene, and quaterrylene diimides in aprotic solution. J. Am. Chem. Soc. **1999**, *121*(14), 3513–3520.

84. Lai, R.Y., Fabrizio, E.F., Lu, L., Jenekhe, S.A., Bard, A.J. Synthesis, cyclic voltam-metric studies, and electrogenerated chemiluminescence of a new donor-acceptor molecule: 3,7-[bis[4-phenyl-2-quinolyl]]-10-methylphenothiazine. J. Am. Chem. Soc. **2001**, *123*(37), 9112–9118.

85. Fabrizio, E.F., Payne, A., Westlund, N.E., Bard, A.J., Magnus, P.P. Photophysical, electrochemical, and electrogenerated chemiluminescent properties of 9,10-dimethyl-7,12-diphenylbenzo[k]fluoranthene and 9,10-dimethylsulfone-7,12-diphenylbenzo[k]fluoranthene. J. Phys. Chem. A **2002**, *106*(10), 1961–1968.

86. Keszthelyi, C.P., Tokel-Takvoryan, N.E., Tachikawa, H. Bard, A.J. Electrogenerated chemiluminescence. XIV. Effect of supporting electrolyte concentration and magnetic field effects in the 9,10-dimethylanthracene-tri-p-tolylamine system in tetrahydrofuran. Chem. Phys. Lett. **1973**, *23*(2), 219–222.

87. Faulkner, L.R. Techniques of electrogenerated chemiluminescence. In: A.J. Bard, Ed. Electroanalytical Chemistry. New York: Marcel Dekker, 1977, 1–95.

88. Rieger, P.H., Bernal, I., Reinmuth, W.H., Fraenkel, G.K. Electron spin resonance of electrolytically generated nitrile radicals. J. Am. Chem. Soc. **1963**, *85*, 683–693.

89. Michael, P.R., Faulkner, L.R. Electrochemiluminescence from the thianthrene-2,5-diphenyl-1,3,4-oxadiazole system. Evidence for light production by the T route. J. Am. Chem. Soc. **1977**, *99*(24), 7754–7761.

90. Werner, T.C., Chang, J., Hercules, D.M. Electrochemiluminescence of perylene. The role of direct excimer formation. J. Am. Chem. Soc. **1970**, *92*(19), 5560–5565.

91. Oyama, M., Mitani, M., Okazaki, S. Formation of π-excimer or π-exciplex in electrogenerated chemiluminescence involving perylene molecule revealed using a dual-electrolysis stopped-flow method. Electrochem Commun **2000**, *2*(5), 363–366.

92. Fleet, B., Keliher, P.N., Kirkbright, G.F,; Pickford, C.J. Analytical usefulness of electrochemiluminescence for the determination of microgram amounts of aromatic hydrocarbons. Analyst **1969**, *94*(1123), 847–854.

93. Weller, A. Singlet- and triplet-state exciplexes. Exciplex, Proc. Meet. **1975**, 23–38.

94. Zachariasse, K. Exciplexes in chemiluminescent radical ion recombination. Exciplex, Proc. Meet. **1975**, 275–303.

95. Weller, A. Determination of chemiluminescence quantum yields of radical ion reactions. Chem. Phys. Lett. **1971**, *10*(4), 424–427.

96. Bard, A.J., Park, S.-M. Exciplexes in electrogenerated chemiluminescence. Exciplex, Proc. Meet. **1975**, 305–326.

97. Tachikawa, H., Bard, A.J. Electrogenerated chemiluminescence. Effect of solvent and magnetic field on ECL [electrogenerated chemiluminescence] of rubrene systems. Chem. Phys. Lett. **1974**, *26*(2), 246–251.

7
Metal Chelate Systems

Mark M. Richter
Southwest Missouri State University, Springfield, Missouri, U.S.A.

I. INTRODUCTION

In little more than three decades, electrogenerated chemiluminescence (ECL) has moved from being a "laboratory curiosity" to being a useful analytical technique. Metal chelates, in particular the dication of tris(bipyridine)ruthenium(II), $Ru(bpy)_3^{2+}$ (bpy = 2,2′-bipyridine), have played a pivotal role in this transformation. This is not surprising, considering that many metal chelates display the electrochemical and spectroscopic qualities required of ECL luminophores (Chapter 1). Representative systems include main group metals (e.g., Si and Al) [1,2], transition metal complexes incorporating Ru, Os, and Pt [3–8], and rare earth chelates [9,10], to name a few. There has been particular emphasis on characterizing the nature of the excited state, discerning the mechanisms by which this state is formed, and determining the efficiency of excited state formation. Various techniques were used and are still being used, including detailed electrochemical studies, spectroscopic and spin-resonance measurements, and magnetic field effects (Chapters 1–4) [3,11,12].

With the development in the early 1980s of a method for the binding of $Ru(bpy)_3^{2+}$ to biological molecules of interest (e.g., antibodies, proteins, nucleic acids) [13] came renewed interest in discovering additional ECL luminophores and understanding the fundamental properties of ECL phenomena. For example, although $Ru(bpy)_3^{2+}$ has many properties that make it an ideal ECL luminophore for sensitive and selective analytical methods, it would be useful to have other ECL labels that span a wide range of wavelengths so that simultaneous determination of several analytes in a single sample is possible. Metal chelate systems have been the focus of many of these studies. Therefore, this chapter offers the reader a survey of developments within metal chelate ECL.

There are two obvious choices when discussing metal chelates, organizing systems by the metal or by the chelate(s) used to bind the metal. In this chapter the former is used. The advantages of organizing by the metal are that systems containing the same metal ion can be grouped together for comparison, and we pay homage (in a humble way) to the periodic table. The main disadvantage is that several papers discuss the ECL of different metal ions with the same chelate, so some repetition within different sections is inevitable. Hopefully the advantages outweigh the disadvantages. Therefore, Sections II–VI deal with 'ruthenium and osmium polypyridine-based ligand complexes,' the 'main group elements,' the 'first row transition metals', 'the second and third row transition metals', and the 'rare earth elements', respectively. Individual metals and their chelates are organized alphabetically by metal within these groups. The ECL of ruthenium and osmium polypyridine complexes are discussed separately from the second and third row transition metals because of the pivotal role $Ru(bpy)_3^{2+}$ has played, and continues to play, in the development of ECL. Osmium is included in this section owing to the many similarities between ruthenium and osmium (and the important differences that, to date, have made ruthenium the choice for practical applications).

The photophysics and photochemistry of the various systems are discussed on an as needed basis to complement the electrochemiluminescence studies. No separate sections or introduction are presented. Several excellent monographs exist that discuss the rich excited state chemistry of metal systems, and the reader is encouraged to consult them [14,15].

II. RUTHENIUM AND OSMIUM POLYPYRIDINE-BASED LIGAND COMPLEXES

Electrochemical, photophysical, and ECL data under both annihilation and core-actant conditions are presented in Tables 1 and 2.

A. Ruthenium

1. Monometallics

Since the discovery that $Ru(bpy)_3^{2+}$ is photoluminescent [16], a large body of literature has appeared aimed at understanding both the ground and excited state properties of $Ru(bpy)_3^{2+}$, $Os(bpy)_3^{2+}$ and their polyazine derivatives [14,15]. Therefore, it is not surprising that these compounds have also played an important role in the development of ECL.

The first report of ECL in a metal chelate was in 1972; the excited state of $Ru(bpy)_3^{2+}$ had been generated in aprotic media by annihilation of the reduced

Table 1 Electrochemical, Spectroscopic, and ECL Properties of Ruthenium and Osmium Polypyridine Complexes. ECL Generated via Annihilation

Complex	Solvent	$E_{1/2}$(ox) (V)	$E_{1/2}$(red) (V)	λ_{em} (nm)	λ_{ecl} (nm)	ϕ_{em}	ϕ_{ecl}
Ru(bpy)$_3$$^{2+}$	MeCN	+1.35[a]	-1.32[a]	608[a]	608[a]	0.062[b]	0.05[c]
Ru(phen)$_3$$^{2+}$	MeCN	+1.40[d]	-1.40[d]	590[d]	590[d]	—	—
Ru(terpy)$_3$$^{2+}$	MeCN	+1.28[d]	-1.43[d]	—	660 nm	—	—
Ru(TPTZ)$_3$$^{2+}$	MeCN	+1.52[d]	-0.84[d]	—[e]	—[e]	—[e]	—[e]
Ru(bpz)$_3$$^{2+}$	MeCN	+1.90[f,g]	-0.76[f,g]	585[f]	585[f]	~-0.03[h,i]	~-0.04[h]
Ru(dp-bpy)$_3$$^{2+}$	MeCN	+1.2[j,k]	-1.4[j,k]	635[j]	635[j]	0.26[j]	0.14[j]
Ru(dp-phen)$_3$$^{2+}$	MeCN	+1.3[j,k]	-1.3[j,k]	615[j]	615[j]	0.31[j]	0.24[j]
(bpy)$_2$Ru(bphb)$^{2+}$	MeCN	+1.569[l]	-1.291[l]	624[l]	624[l]	0.08[l]	0.0066[l]
[(bpy)$_2$Ru]$_2$(bphb)$^{4+}$	MeCN	+1.337[l]	-1.287[l]	624[l]	624[l]	0.11[l]	0.16[l]
Os(bpy)$_3$$^{2+}$	MeCN	+0.82[m]	-1.26[m]	724[m]	720[m]	0.00462[b]	—
Os(bpy)$_2$(diphos)$^{2+}$	MeCN	+1.34[m]	-1.23[m]	612[m]	610[m]	0.0550[b]	~0.7[m]
Os(bpy)$_2$(dppene)$^{2+}$	MeCN	+1.38[m]	-1.23[m]	605[m]	606[m]	0.0699[b]	—
Os(bpy)$_2$(DMSO)$_2$$^{2+}$	MeCN	+1.79[m]	-1.00[m]	575[m]	—[e]	—	—
Os(bpy)$_2$dpae^{2+}	MeCN	+1.21[m]	-1.25[m]	632[m]	633[m]	—	—
Os(bpz)$_3$$^{2+}$	MeCN	+1.52[n]	-0.67[n]	700[n]	700[n]	—	—
Os(phen)$_3$$^{2+}$	DMF	+0.86[m]	-1.25[m]	690[m]	691[m]	0.0159[b]	—
Os(phen)$_3$$^{2+}$	MeCN	+0.82[o]	-1.21[o]	—	740[o]	0.32[o]	0.40[o]
Os(phen)$_2$(diphos)$^{2+}$	MeCN	+1.29[m]	-1.29[m]	601[m]	600[m]	0.138[b]	—
Os(phen)$_2$(dppene)$^{2+}$	MeCN	+1.42[m]	-1.21[m]	597[m]	599[m]	0.239[b]	—
Os(phen)$_2$dpae^{2+}	MeCN	+1.21[m]	-1.26[m]	622[m]	—[e]	0.121[b]	—[e]

Please see text for ligand definitions.

All potentials are in volts vs. SSCE (0.2360 vs. NHE) unless otherwise noted.

Photoluminescence efficiencies (ϕ_{em}) and ECL efficiencies (ϕ_{ecl}) are relative to Ru(bpy)$_3$$^{2+}$ at 0.062 and 0.05, respectively.

[a]Ref. 5.
[b]Ref. 21.
[c]Refs. 18 and 20.
[d]Ref. 17.
[e]No observed ECL.
[f]Ref. 32.
[g] vs. Ag QRE.
[h]Ref. 33.
[i]Relative to Rhodamine B (ϕ_{em} = 0.61 in ethanol).
[j]Ref. 41.
[k]Cathodic (Epc) and anodic (Epa) peak potentials.
[l]Ref. 8.
[m]Ref. 57.
[n]Ref. 59.
[o]Ref. 56.

Table 2 Electrochemical, Spectroscopic, and ECL Properties of Ruthenium and Osmium Polypyridine Complexes. ECL Generated via Coreactants

Complex/coreactant	Solvent (electrolyte)	$E_{1/2}$ (V)	λ_{em} (nm)	λ_{ecl} (nm)	ϕ_{em}	ϕ_{ecl}
Ru(bpy)$_3$$^{2+}$/C$_2O_4$$^{2-}$	H$_2$O (0.1 M H$_2$SO$_4$)	—	610[a]	610[a]	—	0.02[a,b]
Ru(bpy)$_3$$^{2+}$/S$_2O_8$$^{2-}$	MeCN–H$_2$O (50:50 v/v)	−1.35[c]	625[c]	625[c]	—	0.025[b,c]
Ru(bpy)$_3$$^{2+}$/TPrA	H$_2$O (0.2 M KH$_2$PO$_4$)	—	610[d]	610[d]	—	1.0[e]
Ru(bpz)$_3$$^{2+}$/ S$_2O_8$$^{2-}$	H$_2$O (0.1 M Na$_2$SO$_4$)	~−0.8V[f]	585[f]	590[f]	—	—
Ru(dp-bpy)$_3$$^{2+}$/TPrA	H$_2$O (0.2 M phosphate buffer)	+1.2[g]	—	—	0.26[g]	2.4[g]
Ru(dp-phen)$_3$$^{2+}$/TPrA	H$_2$O (0.2 M phosphate buffer)	+1.3[g]	—	—	0.31[g]	0.8[g]
(bpy)$_2$Ru(AZA-bpy)$^{2+}$/TPrA	MeCN–H$_2$O (0.1 M KH$_2$PO$_4$)	+1.32[h]	603[h]	603[h]	0.062[h]	0.84[h]
(bpy)$_2$Ru(AZA-bpy)$^{2+}$/TPrA	H$_2$O (0.2 M KH$_2$PO$_4$)	+1.32[i]	613[i]	613[i]	0.062[i]	0.51[i]
(bpy)$_2$Ru(CE-bpy)$^{2+}$/TPrA	H$_2$O (0.1 M Tris)	1.28[j,k]		650[j]	—	1.0[l]
(bpy)$_2$Ru(CE-bpy)$^{2+}$/TPrA	MeCN (0.1 M TBAClO$_4$)	1.28[j,k]		655[j]	—	~0.5[m]
(bpy)$_2$Ru(bpy-C$_{19}$)$^{2+}$/C$_2$O$_4$$^{2-}$	H$_2$O(0.5 M Na$_2$SO$_4$)	+1.06[n]	600[n,o]	680[n]	—	—
Ru(bpy)$_3$$^{2+}$/C$_2O_4$$^{2-}$	H$_2$O (0.1 M NaCl+0.1 M KH$_2$PO$_4$)	+1.26[p]		591[p]	0.062[p]	0.0011[p]
Ru(phen)$_3$$^{2+}$/ C$_2O_4$$^{2-}$	H$_2$O (0.1 M NaCl+0.1 M KH$_2$PO$_4$)	+1.25[p]		585[p]	0.065[p]	0.0015[p]
Ru(dmbp)$_3$$^{2+}$/ C$_2O_4$$^{2-}$	H$_2$O (0.1 M NaCl+0.1 M KH$_2$PO$_4$)	+1.21[p]		594[p]	0.045[p]	5×10^{-3}[p]
Ru(dmphen)$_3$$^{2+}$/ C$_2O_4$$^{2-}$	H$_2$O (0.1 M NaCl+0.1 M KH$_2$PO$_4$)	+1.12[p]		591[p]	—	—
(bpy)$_2$Ru(bphb)$^{2+}$/TPrA	MeCN (0.1 M TBAPF$_6$)	+1.333[q]	624[q]	624[q]	0.08[q]	1.5[q]
	MeCN–H$_2$O (50:50 v/v; 0.1 M TBAPF$_6$)					1.6[q]
	H$_2$O (0.2 M KH$_2$PO$_4$)					0.058[q]
(bpy)$_2$Ru(bphb)$^{2+}$/S$_2$O$_8$$^{2-}$	MeCN	−1.291[q]	624[q]	624[q]	0.08[q]	0.4[q,r]
	MeCN–H$_2$O (50:50 v/v)					0.7[q]
[(bpy)$_2$Ru]$_2$(bphb)$^{4+}$/TPrA	MeCN	+1.333[q]	624[q]	624[q]	0.11[q]	2.6[q]
	MeCN–H$_2$O (50:50 v/v)					2.8[q]
	H$_2$O (0.2 M KH$_2$PO$_4$)					2.0[q]

(continued)

Table 2 Continued

Complex/coreactant	Solvent (electrolyte)	$E_{1/2}$ (V)	λ_{em} (nm)	λ_{ecl} (nm)	ϕ_{em}	ϕ_{ecl}
$[(bpy)_2Ru]_2(bphb)^{4+}/S_2O_8{}^{2-}$	MeCN	-1.291^q	624^q	624^q	0.11^q	$0.6^{q,r}$
	MeCN–H$_2$O (50:50 v/v)	—	—	—	—	$0.89^{q,r}$
$(bpy)_2Ru(bpy\text{-}O\text{-}C_8)^{2+}$	MeCN (0.1 M TBAPF$_6$)	$+1.214^s$	618^s	618^s	0.75^s	$1.0^{s,t}$
Den-8-Ru	MeCN (0.1 M TBAPF$_6$)	-1.207^s	618^s	618^s	0.75^s	$\sim5^{s,u}$
$Os(bpy)_2(dppene)^{2+}/TPrA$	CH$_3$CN–H$_2$O (50:50 v/v)	$+1.02^v$	368^v	585^v	0.088^v	0.95^v
$Os(bpy)_2(dppene)^{2+}/TPrA$	H$_2$O (0.2 M KH$_2$PO$_4$)	$+1.16^v$	368^v	589^v	0.157^v	2.0^v

All electrochemical and ECL experiments were referenced with respect to SSCE (0.2360 vs. NHE).

Photoluminescence efficiency with respect to Ru(bpy)$_3{}^{2+}$ ($\phi_{em} = 0.082$) unless otherwise noted.

Relative ECL efficiency with respect to Ru(bpy)$_3{}^{2+}$/TPrA ($\phi_{ecl} = 1$) unless otherwise noted.

[a]Ref. 22.

[b]Relative to Ru(bpy)$_3{}^{2+}$ annihilation ECL (0.05) in MeCN.

[c]Ref. 24.

[d]Ref. 25.

[e]Assigned a value of 1.0 for comparison purposes.

[f]Ref. 34.

[g]Ref. 41.

[h]Ref. 45.

[i]Ref. 46.

[j]Ref. 44.

[k]MeCN (0.1 M TBAP).

[l]Set to 1.0 for comparison to CE-bpy in MeCN. Relative ϕ_{ecl} vs. Ru(bpy)$_3{}^{2+}$/TPrA not reported.

[m]Relative to CE-bpy in H$_2$O ($\phi_{ecl} = 1$).

[n]Ref. 36.

[o]In MeCN.

[p]Ref. 50.

[q]Ref. 8.

[r]Relative to Ru(bpy)$_3{}^{2+}$/S$_2$O$_8{}^{2-}$ at 1.0.

[s]Ref. 52.

[t]Set to 1.0 for comparison with Den-8-Ru.

[u]Relative to (bpy)$_2$Ru(bpy-O-C$_8$)$^{2+}$ at 1.0.

[v]Ref. 60.

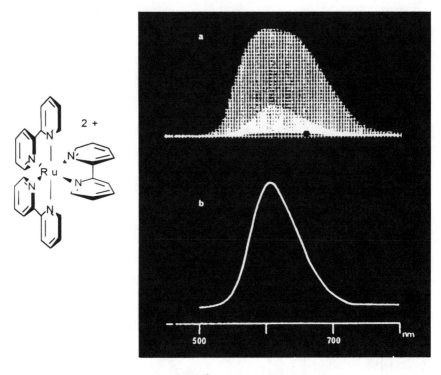

Figure 1   (Left) Structure of Ru(bpy)$_3^{2+}$. (a) ECL emission spectrum of Ru(bpy)$_3$Cl$_2$ from 450 to 800 nm using a cyclic square wave at 0.2 Hz between +1.75 and –1.60 V vs. Ag reference electrode. (b) Fluorescence emission spectrum of Ru(bpy)$_3$Cl$_2$ in MeCN with excitation at 500 nm. (From Ref. 5. Copyright 1972 American Chemical Society.)

(Ru(bpy)$_3^{1+}$) and oxidized (Ru(bpy)$_3^{3+}$) species (Fig. 1) [5]. Subsequently, Ru(bpy)$_3^{2+}$ has become the most thoroughly studied ECL active molecule [17–20]. This is for a number of reasons, including its strong luminescence and solubility in both aqueous and nonaqueous media at room temperature and its ability to undergo reversible one-electron transfer reactions at easily attainable potentials, leading to stable reduced and oxidized species (Fig. 2). The synthetic versatility of polypyridine ligands is also an asset. Furthermore, the overall ECL efficiency (photons produced per redox event) is a product of the photoluminescence quantum yield and the efficiency of production of the excited state. Ru(bpy)$_3^{2+}$ has a photoluminescent quantum efficiency (ϕ_{em}) of 0.0682 [21] and an ECL efficiency (ϕ_{ecl}) of 0.0500 [22,24]. Under certain conditions, the annihilation reaction between Ru(bpy)$_3^{1+}$ and oxidized Ru(bpy)$_3^{3+}$ produces the emitting charge transfer triplet with an efficiency approaching 100% [17] and is

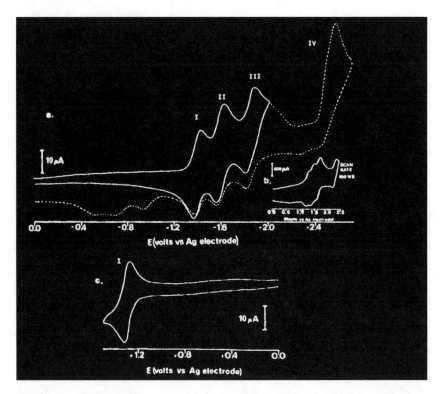

Figure 2 Cyclic voltammograms of 1 mM $Ru(bpy)_3(ClO_4)_2$/MeCN/0.1 M $TBABF_4$ at a Pt disk microelectrode. (a) Scan from 0.0 to –2.6 V, scan rate = 200 mV/s; (b) scan as in (a), scan rate = 100 V/s; (c) scan from 0.0 to +1.4 V, scan rate = 200 mV/s. (From Ref. 31. Copyright 1973 American Chemical Society.)

comparable with photoluminescence data [23] showing that about 5% of the excited states produce luminescence.

Many other studies on $Ru(bpy)_3^{2+}$ have followed, and much of this work has been reviewed [3,11,12]. For example, the first report of ECL in aqueous solution involved $Ru(bpy)_3^{2+}$ and the oxalate ion ($C_2O_4^{2-}$) [22]. Subsequently, other species were shown to act as coreactants, among them peroxydisulfate [24] and tri-*n*-propylamine (TPrA) [25], and $Ru(bpy)_3^{2+}$ was the ECL luminophore in each study. The discovery of $Ru(bpy)_3^{2+}$/TPrA [25] allowed efficient ECL not only in aqueous media but also at physiological pH. By attachment of a suitable group to the bipyridine moieties (e.g., *n*-hydroxysuccinimide), $Ru(bpy)_3^{2+}$ can be linked to biologically interesting molecules, such as antibodies or DNA, where it serves as a label for analysis in an analogous manner to radioactive or fluorescent labels (Chapter 9) [13,26]. These developments have resulted in a wide range of analytical applications for ECL in clinical diagnostic assays [13,26] in which $Ru(bpy)_3^{2+}$ plays a key role. Added to this is the study of $Ru(bpy)_3^{2+}$ for use in liquid chromatography detection [27], the measurement of biologically and pharmacologically important compounds [e.g., alkylamines, antibiotics, antihistamines, opiates, and the reduced form of NADH (i.e., adenine dinucleotide)] [4,28], and in environmental applications [29,30], to name a few. In fact, with the interest in using ruthenium chelates for multiple applications (e.g., solar energy conversion, catalysis, oxygen sensing, and ECL), one has to hope the world's supply of ruthenium doesn't run out!

Obviously, the wealth of studies on $Ru(bpy)_3^{2+}$ ECL have eclipsed those of other ruthenium chelates. However, several papers exist on both mono- and multimetallic systems that examine ECL from fundamental and applied angles. For example, the ECL of four ruthenium(II) chelates, including $Ru(bpy)_3^{2+}$ as a standard, were reported in acetonitrile solution [31]. The compounds were $Ru(bpy)_3^{2+}$, $Ru(phen)_3^{2+}$ (phen = 1,10-phenanthroline), $Ru(TPTZ)_2^{3+}$ (TPTZ = 2,4,6-tripyridyl-2-triazine), and $Ru(terpy)_2^{2+}$ (terpy = 2,2',2''-terpyridine) (Fig. 3).

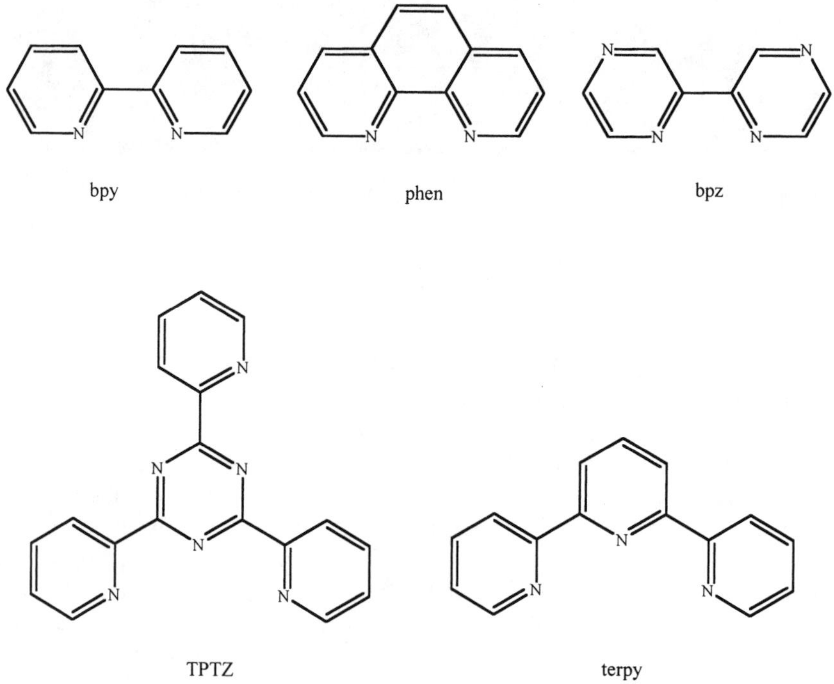

Figure 3 2,2'-Bipyridine (bpy); 1,10-phenanthroline (phen); 2,2'-bipyrazine (bpz); 2,4,6-tripyridyl-2-triazine (TPTZ); 2,2',2''-terpyridine (terpy).

All compounds displayed one-electron oxidation and reduction waves using cyclic voltammetry (e.g., Fig. 2). The bpy, phen, and terpy compounds showed ECL via redox reaction of the oxidized and reduced forms (i.e., annihilation), identified as the lowest triplet MLCT state. $Ru(TPTZ)_2^{3+}$ displayed no measurable ECL under these conditions. Experiments at a rotating ring-disk electrode (RRDE) were used to generate steady-state ECL for $Ru(bpy)_3^{2+}$ and yielded an ECL efficiency of 5–6%.

Another ruthenium chelate that has been studied is $Ru(bpz)_3^{2+}$ (bpz =2,2'-bipyrazine) (Fig. 3) using both annihilation and coreactant methodologies. Of particular interest in these studies is that the oxidation and reduction potentials occur at about 0.5 V more positive than $Ru(bpy)_3^{2+}$ (Fig. 4). Such a positive potential shift may facilitate electrochemical, photochemical, and ECL studies in aqueous solutions. The first report on $Ru(bpz)_3^{2+}$ noted ECL characteristic of $Ru(bpz)_3^{2+*}$ in acetonitrile upon formation of the +3 and +1 states [32]. It was also noted that the shift in the +3 state compared to $Ru(bpy)_3^{2+}$ resulted in greater stabilization of the +2 versus +3 species, with correspondingly weaker ECL in the bpz versus the bpy complex. Another study [33] measured the temperature dependence of ϕ_{ecl} in acetonitrile, with the aim of further understanding the mechanism of $Ru(bpz)_3^{2+}$ annihilation ECL. A bright orange luminescence characteristic of $Ru(bpz)_3^{2+*}$ ECL was also observed in aqueous solution using $S_2O_8^{2-}$ as a reductive–oxidative coreactant [34], suggesting that $Ru(bpz)_3^{2+}/S_2O_8^{2-}$ ECL might be useful for the determination of persulfate in trace amounts. This suggestion was realized in subsequent work with nanomolar (nM) detection limits of $S_2O_8^{2-}$ achieved [35].

Electrogenerated chemiluminescence has also been measured from a $Ru(bpy)_3^{2+}$-based surfactant adsorbed on semiconductor and metal electrodes [36]. ECL was observed from an organized monomolecular layer of $(bpy)_2Ru(bpy-C_{19})^{2+}$ (Fig. 5) adsorbed on indium-doped tin oxide (ITO), Pt, and Au electrodes. The surfactant monolayer was coated on the substrate using Langmuir–Blodgett methods, and ECL was generated in aqueous oxalate solution upon application of a positive potential. The ECL of the adsorbed monolayers was easily detectable, with an emission maximum of ~680 nm (Fig. 5). Interestingly, the emission from the monolayer on ITO electrodes was 100–1000 times more intense than that on Pt and Au.

A number of studies of metal chelate ECL in aqueous surfactant solution have appeared [37–42]. This is due, in part, to the widespread use of surfactants in both fundamental and applied studies and the observation that changes in PL and, subsequently, ECL intensities and excited state lifetimes are observed for many systems in surfactant media. For example, increased ECL efficiencies (eightfold and greater) and duration of the ECL signal were observed in surfactant media upon oxidation of the coreactant TPrA and $Ru(bpy)_3^{2+}$ (Fig. 6) [37,38,42]. However, the mechanism of the surfactant effect is still under study.

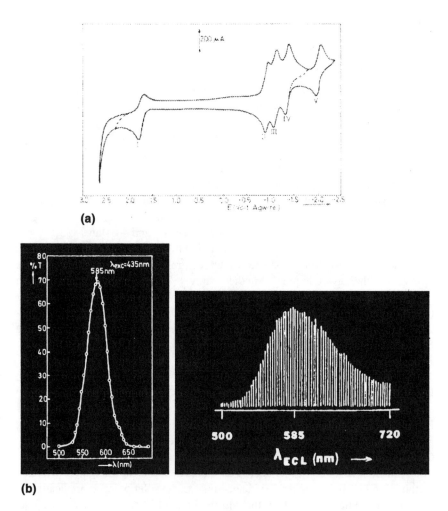

Figure 4 (a) Cyclic voltammogram of 1 mM Ru(bpz)$_3^{2+}$(PF$_6^-$)$_2$/MeCN/0.1 M TBAPF$_6$ at a Pt electrode (scan rate 100 mV/s, $T = 25°C$, Ag wire reference electrode). (b) Luminescence spectrum of a 10^{-5} M solution of Ru(bpz)$_3^{2+}$ in MeCN at room temperature and ECL spectrum of a 1 mM Ru(bpz)(PF$_6$)$_2$/MeCN/0.1 M TBAPF$_6$ solution (pulsing limits –0.8 to +1.85 vs. Ag wire reference electrode at +0.5 Hz). (From Ref. 32. Copyright 1983 American Chemical Society.)

The effects of surfactants on Ru(bpy)$_3^{2+}$/TPrA, [42] and the ruthenium derivatives Ru(dp-bpy)$_3^{2+}$ and Ru(dp-phen)$_3^{2+}$ (dp-bpy = 4,4′-biphenyl-2,2′-bipyridyl; dp-phen = 4,7-diphenyl-1,10-phenanthroline) [41] were attributed to strong hydrophobic interactions between the ECL luminophore and micellized surfac-

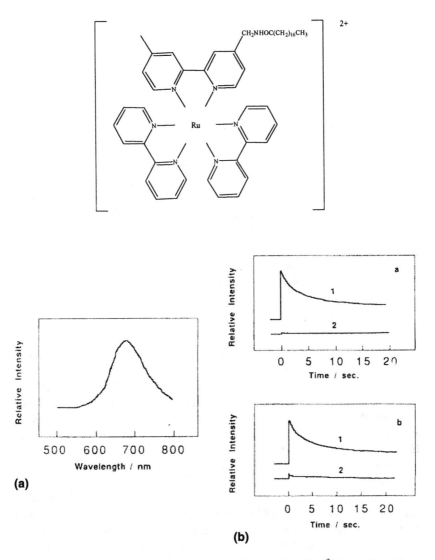

Figure 5 (A) ECL spectrum of monolayer of $(bpy)_2Ru(bpy-C_{19})^{2+}$ (see structure) on an In-doped SnO_2 electrode in 0.02 M $Na_2C_2O_4$, 0.4 M Na_2SO_4 aqueous solution (pH 5.5) with the electrode at +1.25 V vs. SCE. (B) (1) (ECL emission spectra from a monolayer of $(bpy)_2Ru(bpy-C_{19})^{2+}$ (see structure) on metal electrodes and (2) background emissions of the same electrodes after washing off the monolayer with chloroform, in 0.02 M $Na_2C_2O_4$, 0.4 M Na_2SO_4 aqueous solution (pH 5.5) with a potential step of +1.25 V vs. SC applied to the electrode at time 0. (a) Pt/mica electrode; (b) Au foil electrode. (From Ref. 36. Copyright 1988 American Chemical Society.)

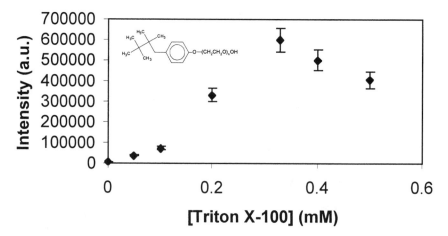

Figure 6 Intensity versus concentration of Triton X-100 plot for Ru(bpy)$_3^{2+}$ (0.1 μM), C$_9$H$_{21}$N (0.05 M), and potassium phosphate buffer (0.2 M). Error bars are ±10%. Triton X-100 = polyethylene glycol *tert*-octylphenyl ether, n = 9,10. (From Ref. 42. Copyright 2000 American Chemical Society.)

tant. However, recent work [38] on the effects of electrode hydrophobicity on ECL indicate that adsorption of Triton X-100 on Pt and Au electrodes renders the surface more hydrophobic, facilitating coreactant oxidation and leading to increased ECL intensities in the Ru(bpy)$_3^{2+}$/TPrA system. Studies of the effects of nonionic chain lengths on Ru(bpy)$_3^{2+}$ / TPrA ECL [37] confirm these results. Recently, the effect of Triton X-100 at concentrations both below and above the critical micelle concentration on (bpy)$_2$Ru(DC-bpy) and (bpy)$_2$Ru(DM-bpy) (DC = 4,4′-dicarboxy-2,2′-bipyridine; DM = 4,4′-dimethyl-2,2′-bipyridine) ECL was reported relative to Ru(bpy)$_3^{2+}$ in aqueous solution and aqueous surfactant solution [39]. ECL was generated using TPrA as a coreactant upon potential sweep from +0.0 to +1.5 V vs. Ag/AgCl reference electrode. This study clearly showed that the surfactant effect on the ECL of ruthenium complexes is not limited to Ru(bpy)$_3^{2+}$ and that adsorption of surfactant plays a role in the ECL process, allowing for increased ruthenium and/or TPrA oxidation. Surprisingly, the hydrophobicity of the ECL luminophore did not lead to dramatic increases or decreases in ECL efficiency, suggesting that adsorption of TPrA or other factors contribute to the surfactant effect [5]. Although the effects of micelles and discrete complexation of the surfactants with Ru(bpy)$_3^{2+}$ and TPrA cannot be ruled out, these studies indicate that increases in ECL intensity are probably due to changes in electrode hydrophobicity upon formation of a surfactant adsorption layer and less likely due to micelle interactions. Such dramatic increases in ECL intensity, coupled with work on more efficient ECL labels and coreactants, may have profound impacts on the sensitivity of ECL for a variety of applications.

Ruthenium (II) diimine complexes that have phosphonic acid substituents have been shown to adsorb (via the phosphonic acid substituents) to TiO_2-modified ITO electrodes and undergo ECL [43]. Oxalate ion was used as the coreactant in these systems. It was shown that optically transparent electrodes containing TiO_2 can serve as solid supports for the generation of ECL from adsorbed ruthenium chelates and that the stability of the modified surfaces for sustained ECL depends on the number of substituents per chromophore.

Recently, the ECL of monometallic ruthenium complexes was extended to systems containing a crown ether moiety covalently bonded to a bipyridyl ligand. The ECL of $Ru(bpy)_2(CE-bpy)^{2+}$ [CE-bpy is a bipyridine ligand in which a crown either (15-crown-5) is bound to the bpy ligand in the 3 and 3' positions] [44] and $(bpy)_2Ru(AZA-bpy)^{2+}$ [bpy = 2,2'-bipyridine; AZA-bpy = 4-(N-aza-18-crown-6-methyl-2,2'-bipyridine)] [45,46] have been reported. $Ru(bpy)_2(CE-bpy)^{2+}$ is sensitive to sodium ions in aqueous buffered solution (Fig. 7), and $(bpy)_2Ru(AZA-bpy)^{2+}$ has been shown to be sensitive to Pb^{2+}, Hg^{2+}, Cu^{2+}, Ag^+, and K^+ in 50:50 (v/v) $CH_3CN:H_2O$ (0.1 M KH_2PO_4 as electrolyte) and aqueous (0.1 M KH_2PO_4) solution (Fig. 8). These studies clearly show the versatility of ECL for sensing metal ions in solution. Also, systems that are capable of sensing metal ions not directly involved in redox reactions would be useful in a variety of tests, e.g., in the determination of electrolytes and metal ions in clinical and environmental analyses.

Another application of ECL in metal chelates is the detection of β-lactamase activity [47]. Pencillin and its derivatives do not act as coreactants with $Ru(bpy)_3^{2+}$ to produce ECL. However, β-lactamase-catalyzed hydrolysis of pencillin forms a molecule with a secondary amine that can act as a coreactant. The efficiency of the ECL process has been increased by covalent attachment of a β-lactamase substrate to a $Ru(bpy)_3^{2+}$ derivative [48]. The ECL of aminopeptidase and esterase cleavage products has also been reported to have been achieved by covalently attaching such species as ligands to bis(bipyridine)ruthenium(II). $(bpy)_2Ru^{2+}$ has little to no intrinsic ECL, but attachment of a third ligand leads to enhanced ECL [49].

In an effort to understand the structure–activity relationships for ECL in coreactant systems, a series of ruthenium bipyridine complexes were used to probe the homogeneous oxidation of oxalate and the effect of electrochemical steps on the ECL emission intensity [50]. ECL was generated by oxidation of $Ru(bpy)_3^{2+}$, $Ru(phen)_3^{2+}$, $Ru(bpy)_2(dmbp)^{2+}$, $Ru(dmphen)_3^{2+}$, or $Ru(dmbp)_3^{2+}$ (dmbp = 4,4'-dimethyl-2,2'-bipyridine; dmphen = 4,7-dimethyl-1,10-phenanthroline) and oxalate in aqueous solution. The luminescent emission was related to the driving force for the electron transfer reactions, and the different pathways for

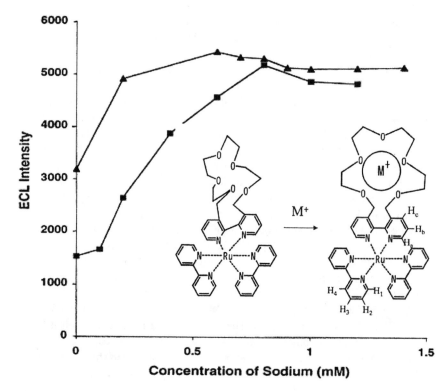

Figure 7 Electrogenerated chemiluminescence signal intensity as a function of sodium concentration for solutions containing 0.3 mM Ru(bpy)$_2$(CE-bpy)$^{2+}$ and 30 mM TPrA in 0.1 M TBAClO$_4$, MeCN; and 0.1 M, pH 7.0 Tris buffer. (From Ref. 44. Copyright 2002 American Chemical Society.)

coreactant (i.e., CO$_2^{\bullet-}$) reaction. Understanding structure–activity relationships may prove useful in both designing new ECL systems and improving the performance of existing systems.

2. Multimetallics

Solution-phase coreactant ECL is quite sensitive, with subpicomolar detection limits achieved [25]. When the ECL luminophore is bound to a magnetic particle (the particle can then be captured on the surface of the electrode prior to electrochemical stimulation), detection limits as low as 10^{-18} M are attainable [26]. However, there are many systems where even greater sensitivity is needed, such as in environmental (where preconcentration of samples is often necessary) and molecular diagnostics applications, where the detection of as few as 10 molecules

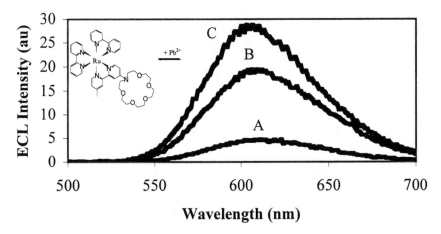

Figure 8 Perturbation of ECL emission spectrum of $(bpy)_2Ru(AZA-bpy)(PF_6)_2$ (0.1 mM) in 50:50 (v/v) $CH_3CN:H_2O$ upon addition of Pb^{2+}. (A) 0 mM Pb^{2+}, (B) 0.5 mM Pb^{2+} (five fold excess), and (C) 1 mM Pb^{2+} (ten fold excess). (From Ref. 45. Copyright 2002 American Chemical Society.)

would eliminate the need for sample amplification (e.g., via the polymerase chain reaction). One approach has been to vary the properties of the ECL luminophore. For example, $Ru(bpy)_3^{2+}$ has an ECL efficiency of 0.050 [22], or ~5% of the $Ru(bpy)_3^{2+}$ molecules that undergo electron transfer generate emission. With the goal of increasing the magnitude of ECL emission, and therefore increasing ECL sensitivity and lowering detection limits, the ECL of the bimetallic ruthenium system $[(bpy)_2Ru]_2(bphb)^{4+}$ [bphb = 1,4-bis(4′-methyl-2,2′-bipyridin-4-yl)benzene] was studied (Fig. 9) [8]. The ligand bphb is capable of binding two independent metal centers through a "bridging ligand" framework. This bimetallic species produced more intense emission (two- to threefold) than $Ru(bpy)_3^{2+}$ in aqueous and nonaqueous solution using annihilation and coreactant methods (Tables 1 and 2). A key point to this study was that for enhanced ECL to be possible in multimetallic assemblies, there must be small electronic coupling between metal centers via the bridging ligand so that the metal centers are electronically isolated or "valence trapped" (Robin and Day Class I systems) [51].

This work has recently extended to dendrimeric systems containing eight $Ru(bpy)_3^{2+}$ units at the periphery (Fig. 10) of a carbosilane dendrimer platform [52]. Preliminary experiments indicated that the ECL of the $Ru(bpy)_3^{2+}$ dendrimer is five times that of the reference monometallic species. As with the bimetallic study, spectroscopic and electrochemical studies show that the $Ru(bpy)_3^{2+}$ units do not interact in either the ground or excited state and that ECL (and photoluminescence) emission can be amplified by using multimetallic

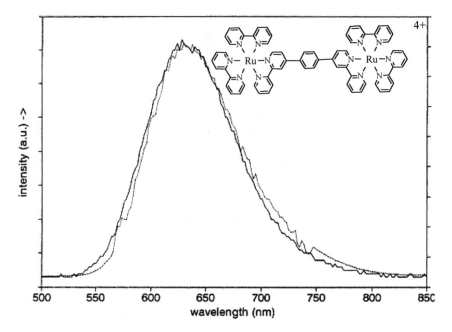

Figure 9 Electrogenerated chemiluminescence emission spectra of 1 mM $[(bpy)_2Ru]_2(bphb)(PF_6)_4$ generated via annihilation (—) and in the presence of TPrA (—). MeCN solutions were 1 μM in complex, 0.1 M in TPrA, and 0.1 M in Bu_4NPF_6, where appropriate. The annihilation spectrum was generated by alternate pulsing the Pt electrode potential between +1.65 and –1.10 V. The TPrA spectrum was generated by pulsing electrode potentials between 0 and +1.65 V via an oxidative–reductive coreactant sequence. Pulse length 0.1 s. Annihilation spectrum (—) offset 5 nm to the red for clarity. (From Ref. 8. Copyright 1998 American Chemical Society.)

species. Multimetallic compounds such as these show much promise for use in analytical applications. However, it has yet to be shown whether these types of labels will change nucleic acid hybridization or affinity binding of antigens and antibodies in diagnostic applications.

B. Osmium

Extension of $Ru(bpy)_3^{2+}$ ECL to osmium systems has been somewhat limited owing to the larger spin–orbit coupling in osmium systems that results in shorter excited state lifetimes and weaker emission efficiencies [53,54]. The development of osmium based sensors would be advantageous, because osmium systems are more photostable than their ruthenium analogs. Also, polypyridyl osmium

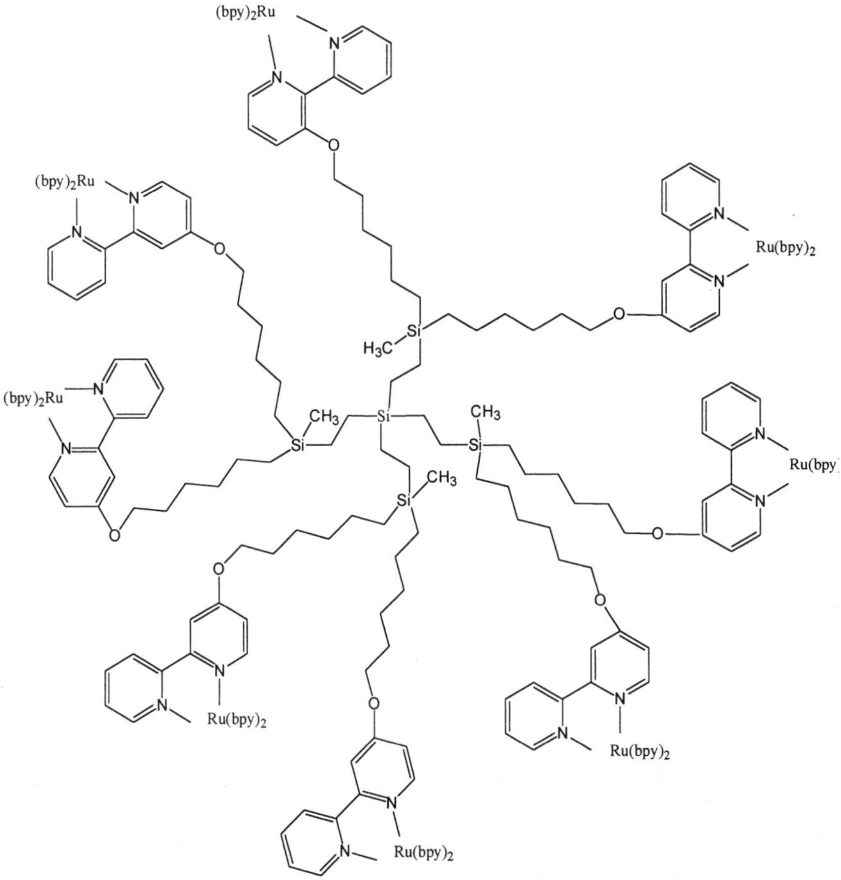

Figure 10 Dendritic supramolecular assembly with tris(bipyridyl)ruthenium(II) units. (From Ref. 52. Copyright 2001 American Chemical Society.)

complexes usually oxidize at less positive potentials than analogous ruthenium systems, and this could be important in designing DNA-labeling agents [55].

The first report of ECL in an osmium complex was of $Os(phen)_3^{2+}$ in DMF and MeCN using $S_2O_8^{2-}$ reduction to generate the excited state [56]. ECL intensities of 40% were obtained in DMF using $Ru(bpy)_3^{2+}$ at 100% as a relative standard. The mechanism of the reaction was believed to be electron transfer between electrogenerated $Os(phen)_3^{+}$ and $SO_4^{-\bullet}$. However, it was noted that the relative ECL intensities do not correlate well with photoluminescence intensities [~32% for $Os(phen)_3^{2+}$], indicating that other factors related to the heterogeneous or homogeneous electron transfer processes may play a role.

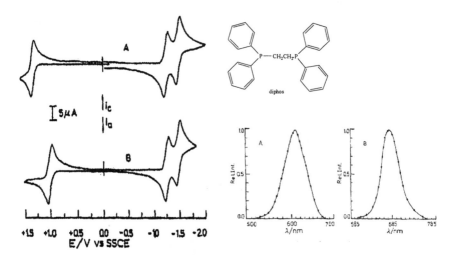

Figure 11 Structure of 1,2-bis(diphenylphosphino)ethane (diphos) (Left) Cyclic voltammograms at 0.2 V/s (0.1 M TBAClO₄) for (A) Os(bpy)₂(diphos)²⁺ and (B) Os(bpy)₂(MeCN)₂²⁺. (Right) Normalized photoluminescence and ECL emission (—) for (A) Os(bpy)₂(diphos)²⁺ and (B) Os(bpy)₂(MeCN)₂²⁺. (From Ref. 57. Copyright 1984 Elsevier Science).

Much higher ECL intensities were observed via annihilation in a series of osmium complexes containing bpy and phen ligands [57]. The complexes were of the general form $Os(bpy)_2L^{2+}$ or $Os(phen)_2L^{2+}$, with L = 1,2-bis(diphenylphos-phino)ethane (diphos, Fig. 11), 1,2-bis(diphenylphosphino)methane (dppm), CH_3CN, dimethylsulfoxide (DMSO), 1,2-*cis*-bis-2-diphenylphosphinoethy-lene (dppene), and 1,2-bis(diphenylarsinoethane) (dpae). Unlike $Os(bpy)_3^{2+}$ and $Os(phen)_3^{2+}$, many of these complexes show very intense photoluminescence with efficiencies two to three orders of magnitude higher than $Os(bpy)_3^{2+}$. The complexes undergo reversible oxidation and reduction with behavior analogous to that of $Ru(bpy)_3^{2+}$ [i.e., a metal-centered ($Os^{2+/3+}$) oxidation and sequential ligand-based reductions] (Fig. 11). Therefore, ECL was generated when a platinum elec-trode in MeCN containing metal complex and electrolyte was pulsed between the first anodic and cathodic peaks. In many instances, bright orange luminescence characteristic of the osmium chelate could be observed in a semi-darkened room. ECL spectra confirmed the identity of the ECL emission as due to an MLCT state (Fig. 11) [57]. A subsequent study on several of these complexes probed the mech-anism of ECL and found that there was a good correlation between the observed ECL intensity and the photoluminescence quantum yield [58]. This indicates that the ECL excited state may form directly upon annihilation without intervening deactivation pathways. ECL was also observed when [Os(4,4′-distyryl-2,2′-bipyridine)$_2$(bis-1,2-phenylphoninoethane)]$^{2+}$ was adsorbed on the surface of an electrode via electroreductive polymerization [58]. Although the ECL emission efficiency was weaker than that of electropolymerized Ru(v-bpy)$_3$$^{2+}$ (v-bpy = 4-vinyl-4'-methyl-2,2'-bipyridine) the emission intensity was much longer lived (~2 h), suggesting possible uses in display device technology.

Experiments were designed to study the microenvironmental effects of micelles on the electrochemical and ECL behavior of Os(bpy)$_3$$^{2+}$ [40]. The one-electron oxidation of Os(bpy)$_3$$^{2+/3+}$ near +0.6 V vs. SCE allowed the studies to be carried out in aqueous solution using oxalate to generate Os(bpy)$_3$$^{2+*}$. Anionic [sodium dodecyl sulfate (SDS)], cationic [cetyltrimethylammonium bromide (CTAB)], and neutral (Triton X-100, Fig. 6) species were chosen as representative surfactants. Measurements of changes in oxidation peak current and ECL intensities at different electrolyte concentrations indicated a strong interaction of Os(bpy)$_3$$^{2+}$ with SDS. Above the cmc, strong hydrophobic interactions between the osmium chelate and SDS micelles suppress both peak current and ECL intensity. Owing to electrostatic effects, the interaction of Os(bpy)$_3$$^{2+}$ and CTAB was much weaker than that of Os(bpy)$_3$$^{2+}$ and SDS, and Triton X-100 does not appear to have any effect.

The ECL of Os(bpz)$_3$$^{2+}$ (Fig. 5), produced by alternate generation of Os(bpz)$_3$$^+$ and Os(bpz)$_3$$^{3+}$ [59], was more intense than previously studied osmium(II) tris chelates [56]. An ECL emission maximum was observed at a wavelength of 700 nm in MeCN, very similar to the photoluminescence spectrum, leading to the conclusion that the emitting state species is Os(bpz)$_3$$^{+*}$. ECL was also observed when the electrode potential was scanned from +0.0 to +1.4 V in an aqueous phosphate buffer solution containing Os(bpz)$_3$$^{2+}$ and oxalate [59].

Os(phen)$_2$(dppene)$^{2+}$ (Fig. 12) exhibits electrochemiluminescence in aqueous and mixed aqueous/nonaqueous solutions using TPrA as coreactant [60]. In fact, the ECL emission quantum efficiency is twofold that of Ru(bpy)$_3$$^{2+}$ in aqueous buffered solution. The ECL spectra were identical to photoluminescence spectra (Fig. 12), indicating formation of the same metal-to-ligand (MLCT) excited states in both ECL and PL. The ECL is also linear over several orders of magnitude in aqueous and mixed solution with theoretical detection limits (blank plus three times the standard deviation of the noise) of 16.9 nM in H$_2$O and 0.29 nM in MeCN–H$_2$O (50:50 v/v). These observations may prove useful in diagnostic or environmental applications where greater sensitivity and detection limits are required than Ru(bpy)$_3$$^{2+}$ can provide. The lower potentials required to excite osmium systems compared to Ru(bpy)$_3$$^{2+}$ may also prove useful in DNA diagnostic applications. However, it will still be necessary to develop other osmium phosphine systems that can oxidize at even lower potentials, because it has been well documented that oligonucleotide sequences undergo irreversible oxidative damage at potentials of 1 V or greater [55].

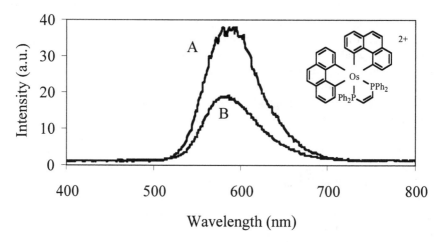

Figure 12 (A) Photoluminescence spectra of 0.01 mM Os(phen)$_2$(dppene)$^{2+}$ in aqueous solution (0.1 M KH$_2$PO$_4$). Excitation wavelengths were at 368 nm with slit widths of 5 nm. (B) ECL spectra of 0.01 mM Os(phen)$_2$(dppene)$^{2+}$ in aqueous solution (0.05M TPrA, 0.1M KH$_2$PO$_4$). (From Ref. 60. Copyright 2002 American Chemical Society.)

III. MAIN GROUP ELEMENTS

Electrochemical, photophysical, and ECL data for the main group elements under both annihilation and coreactant conditions are presented in Table 3.

A. Aluminum

Annihilation ECL has been used to study solutions of aluminum quinolate/triary-lamine and related organic complexes that are used in organic light-emitting diodes (OLEDs) [2,61]. The motivation behind these studies was not only to probe the mechanism of light emission but also to develop a method to rapidly screen candidates for OLEDs. For example, some of the most successful OLEDs created to date are based on tris(8-hydroxyquinoline)aluminum (Alq$_3$) and a triarylamine such as 4,4′-bis(m-tolylphenylamino)biphenyl (TPD) (Fig. 13) [62]. Solution ECL was generated by using "cross-reactions" between radical cations of triarylamines and radical anions of small-molecule OLED materials (i.e., derivatives of Alq$_3$ and quinacridones).

$$Alq_3 + e^- \rightarrow Alq_3^{\bullet-} \tag{1}$$

$$TPD \rightarrow TPD^{\bullet+} + e^- \tag{2}$$

$$TPD^{\bullet+} + Alq_3 \rightarrow Alq_3^{\bullet+} + TPD \tag{3}$$

Table 3 Electrochemical, Spectroscopic, and ECL Properties of Main Group Metal Complexes

Complex/reactant	Solvent	$E_{1/2}$(ox) (V)	$E_{1/2}$(red) (V)	λ_{em} (nm)	λ_{ecl} (nm)	ϕ_{em}	ϕ_{ecl} (%)
Alq$_3$/TPD[a]	50:50 v/v MeCN–toluene (0.1 M TBAPF$_6$)	+0.73[b,c]	−2.30[b,c]	527[b]	510[b]	—	0.09[b,d]
Alq$_3$/TPDF$_2$[a]	50:50 v/v MeCN–toluene (0.1 M TBAPF$_6$)	+0.73[b,c]	−2.30[b,c]	527[b]	510[b]	—	0.18[b,d]
Al(qs)$_3$/TPD[a]	50:50 v/v MeCN–toluene (0.1 M TBAPF$_6$)	+1.04[b,c]	−1.98[b,c]	505[b]	510[b]	—	0.03[b,d]
Al(qs)$_3$/TPDF$_2$[a]	50:50 v/v MeCN–toluene (0.1 M TBAPF$_6$)	+1.04[b,c]	−1.98[b,c]	505[b] 4	510[b]	—	0.10[b,d]
Al(HQS)$_3$/TPrA[e]	H$_2$O (0.2 M KH$_2$PO$_4$)	0.56[f,g] 1.21[g]	499[f]	99[f]	499[f]	0.06[f,h]	0.002[f,i]
SiPc(OR)$_2$	CH$_2$Cl$_2$ (0.1 M TBAP)	+1.00[j]	−0.90,−1.48[j]	~684[j]	725[j]	—	—
SiNc(OR)$_2$	CH$_2$Cl$_2$ (0.1 M TBAP)	+0.58,+1.24[j]	−1.02,−1.56[j]	~792[j]	828[j]	—	—
RO(SiPcO)$_2$R	CH$_2$Cl$_2$ (0.1 M TBAP)	+0.72,+1.20[j]	−0.82,−1.22[j]	—[k]	—[k]	—	—

Please see text for ligand definitions.

Potentials are in volts vs. SSCE (0.2360 vs. NHE) unless otherwise noted.

[a] ECL generated via cross-reaction annihilation methodology (e.g., reaction between Alq$_3$$^{•-}$ and TPD$^{•+}$).

[b] Ref. 61.

[c] Volts vs. Fc/Fc$^+$ (0.631 V vs. NHE).

[d] Relative to diphenylanthracene (ϕ_{ecl} = 6.3%).

[e] ECL generated via coreactant methodology.

[f] Ref. 63.

[g] Anodic peak potentials (Epa).

[h] Relative to Ru(bpy)$_3$$^{2+}$ at 0.082.

[i] Relative to Ru(bpy)$_3$$^{2+}$/TPrA at 1.

[j] Ref. 66.

[k] No observed photoluminescence or ECL.

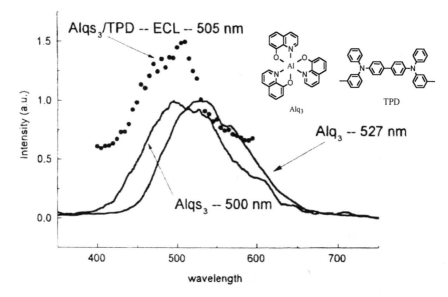

Figure 13 Data points: ECL response for a typical double potential step experiment involving $Alqs_3^{-\bullet}/TPD^{+\bullet}$ (qs = sulfonamide derivative of q). Curves of Electroemissive response from OLEDs involving either Alq_3 or $Al(qs)_3$. (From Ref. 2. Copyright 1998 American Chemical Society.)

$$Alq_3^{\bullet+} + Alq_3^{\bullet-} \rightarrow Alq_3^{*s} + Alq_3 \tag{4}$$

$$Alq_3^{\bullet-} + TPD^{\bullet+} \rightarrow Alq_3^{*s} + TPD \tag{5}$$

$$Alq_3^{*s} \rightarrow Alq_3 + h\nu \tag{6}$$

where *s refers to the excited singlet state. Voltammetric and ECL studies clearly show that the excited state formed in ECL is the same as that formed using electroluminescence (i.e., in solid-state OLED devices) (Fig. 13).

More recently, tris(8-hydroxyquinoline-5-sulfonic acid)aluminum(III) [$Al(HQS)_3$] was generated in situ in aqueous buffered solutions, and its ECL was measured [63]. The chelating ligand 8-hydroxyquinoline (Fig. 13) and its analogs occupy an important place in analytical chemistry, perhaps second only to EDTA and its derivatives. Of special interest is the intense luminescence exhibited by many metal complexes containing 8-hydroxyquinoline-type ligands, including $Al(HQS)_3$ [64,65], such that new applications for these chelates are continually being developed (e.g., chromatography and OLEDs [61,62]). Of particular interest are derivatives such as HQS that are water-soluble and retain the luminescence properties of the 8-hydroxyquinoline parent complex. Therefore, the ECL of $Al(HQS)_3$ was measured using TPrA in aqueous buffered solution (Fig. 14).

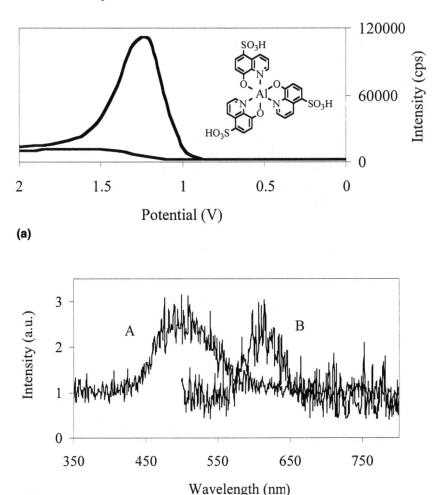

(a)

(b)

Figure 14 (a) ECL intensity versus potential for 4.1×10^{-4} M (250 ppm) Al(HQS)$_3$$^{2+}$ and 0.05 M TPrA in aqueous solution (0.2M KH$_2$PO$_4$). (b) (A) Photolumiminescence of 0.4 mM solution of Al(HQS)$_3$ (slit widths 3 nm) in 0.2 M KH$_2$PO$_4$ and (B) electrochemiluminescence of 0.4 mM solution of Al(HQS)$_3$ (slit widths 10 nm) in 0.2 M KH$_2$PO$_4$/0.05M TPrA. ECL spectrum offset by 100 nm to the red for clarity.

Conditions for ECL emission were optimized with the ECL correlating to aluminum concentration over several orders of magnitude. The ECL of several metal ions other than aluminum with HQS and effects on Al(HQS)$_3$ ECL were also examined. Only two, cadmium(II) and zinc(II), showed any significant effect, and both led to ~50% enhancement of the ECL emission.

B. Silicon

Although they are technically semimetals or metalloids, the ECL of silicon compounds [66] and silicon nanocrystal quantum dots [67] have been reported. The ECL of the latter is discussed in Chapter 11, so we focus on the former here. Phthalocyanine compounds often show high thermal and chemical stability as well as interesting optical and electrical properties. Therefore, the synthesis, spectral characterization, and electrochemical behavior of bis(tri-*n*-hexyl-siloxy)(2,3-phthalocyanato)silicon SiPc(OR)$_2$ and its naphthocyanine analog SiNc(OR)$_2$ were reported [OR = OSi(*n*-C$_6$H$_{13}$)$_3$] (Fig. 15) in dichloromethane (CH$_2$Cl$_2$)[66]. There were two reversible oxidations for SiPc(OR)$_2$ and one reversible reduction, whereas SiNc(OR)$_2$ generated two reversible reductions and one reversible oxidation. Both compounds emit upon electrochemical generation of the oxidized and reduced forms (annihilation ECL) with emission maxima in the visible (684 nm) for [SiPc(OR)$_2$] and in the near-infrared (792 nm) for SiNc(OR)$_2$. Interestingly, no fluorescence or ECL was observed from the dimer of the phthalocyanine derivative (*n*-C$_6$H$_{13}$)$_3$SiO(SiPcO)$_2$Si(*n*-C$_6$H$_{13}$)$_3$.

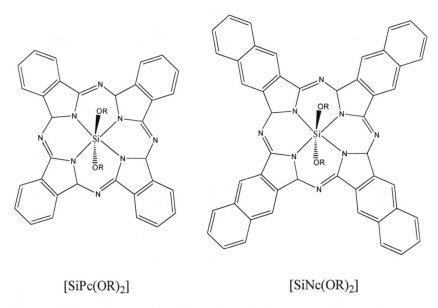

[SiPc(OR)$_2$]   [SiNc(OR)$_2$]

Figure 15 Structures of SiPc(OR)$_2$ and SiNc(OR)$_2$, where OR = OSi(*n*-C$_6$H$_{13}$)$_3$. (From Ref. 66. Copyright 1984 American Chemical Society.)

C. Thallium

As discussed in Chapter 1, light can be emitted at certain oxide-covered aluminum cathodes during the reduction of persulfate, oxygen, or hydrogen peroxide. In this manner emission has been obtained from thallium(I) ions adsorbed in the electrode's oxide layer [68]. This "cathodic" luminescence is often termed "electrogenerated chemiluminescence" in the literature and may prove useful in practical applications. However, the mechanism is distinctly different than that of most other types of ECL reactions (e.g., annihilation, coreactant), so it will not be discussed in more detail.

To the best of our knowledge, only one other report of light emission from thallium via electrochemical methods has been reported. This emission comes from a somewhat surprising source. Low-level ECL has been obtained from blood extracted from tunicates ("sea-squirts"), a class of marine invertebrates [69]. Although the exact source of the low-level ECL is not known, it is believed to emanate from metal ion tunichrome complexes, because a synthetic analog of the tunichrome chromophore showed a tenfold ECL enhancement in the presence of Tl^+ (also see mercury in Section IV.D).

IV. FIRST ROW TRANSITION METALS

Electrochemical, photophysical, and ECL data for the first row transition metals under both annihilation and coreactant conditions are presented in Table 4.

A. Chromium Complexes

The electrochemical reduction of $Cr(bpy)_3^{3+}$ and $Cr(4,4'-Me_2bpy)_3^{3+}$ (4,4'-Me_2bpy = 4,4'-dimethyl-2,2'-bipyridine) in aqueous solution containing persulfate ($S_2O_8^{2-}$; reductive–oxidative coreactant) resulted in emission characteristic of the chromium metal-to-ligand charge transfer (MLCT) state [56]. The ECL emission intensity of $Cr(bpy)_3^{3+}$ in water was approximately 25% that of $Ru(bpy)_3^{2+}$ in dimethylformamide (DMF), and that of $Cr(4,4'-Me_2bpy)_3^{3+}$ was approximately 75%. These values correlate well with photoluminescence efficiencies. The following mechanism was proposed for ECL emission in the chromium chelates with $S_2O_8^{2-}$:

$$Cr(bpy)_3^{3+} + e^- \rightarrow Cr(bpy)_3^{2+} \tag{7}$$

$$S_2O_8^{2-} + e^- \rightarrow SO_4^{2-} + SO_4^{-\bullet} \tag{8}$$

$$Cr(bpy)_3^{2+} + SO_4^{-\bullet} \rightarrow {}^*Cr(bpy)_3^{3+} + SO_4^{2-} \tag{9}$$

$${}^*Cr(bpy)_3^{3+} \rightarrow Cr(bpy)_3^{3+} + h\nu \tag{10}$$

Table 4   Electrochemical, Spectroscopic, and ECL Properties of First Row Transition Metal Complexes

Complex	Solvent (electrolyte)	$E_{1/2}$(ox) (V)	$E_{1/2}$(red) (V)	λ_{em} (nm)	λ_{ecl} (nm)	ϕ_{em}	ϕ_{ecl}
$Cr(bpy)_3^{3+}$[a]	MeCN (0.1 M TBAP)	—	—	730[b]	730[b]	—	—
$Cr(CN)_6^{3-}$[a]	MeCN (0.1 M TBAP)	≥2.0 V[b,c]	-1.75[b,c]	800[b]	800[b]	—	$\sim$3×10^{-4}[b,d]
$Cr(bpy)_3^{3+}/S_2O_8^{2-}$[e]	H_2O (0.1 M NaCl)	—	-0.26[f]	727[f]	727[f]	0.02[f]	0.25[f,g]
$Cr(4,4'\text{-}Me_2bpy)_3^{3+}/S_2O_8^{2-}$[e]	H_2O (0.1 M NaCl)	—	-0.46[f]	727[f]	727[f]	0.06[f]	0.45[f,g]
$Cr(phen)_3^{3+}/S_2O_8^{2-}$[e]	H_2O (0.1 M NaCl)	—	-0.28[f]	727[f]	727[f]	0.10[f]	0.08[f,g]
$Cr(5\text{-}Cl\text{-}phen)_3^{3+}/S_2O_8^{2-}$[e]	H_2O (0.1 M NaCl)	—	-0.17[f]	727[f]	727[f]	0.08[f]	0.07[f,g]
$[Cu(pyridine)I]_4$[a]	CH_2Cl_2 (0.1 M TBABF$_4$)	+0.8[h]	$-0.8, -1.6$[h]	698[h]	698[h]	0.05[h]	—
$Cu(dmp)_2(PF_6)_2/TPrA$[e]	MeCN (0.1 M TBAPF$_6$)	+0.52[i]	—	519,642[i]	—	0.04[i,j]	0.004[i,k]
$Cu(dmp)_2(PF_6)_2/TPrA$[e]	MeCN–H_2O (50:50 v/v; 0.1 M KH_2PO_4)	+0.52[i]	—	519,642[i]	—	0.04[i,j]	0.002[i,k]
$Cu(dmp)_2(PF_6)_2/TPrA$[e]	H_2O (0.1 M KH_2PO_4)	+0.52[i]	—	519,642[i]	—	0.04[i,j]	0.001[i,k]

Please see text for ligand definitions.

Potentials are in volts vs. SSCE (0.2360 vs. NHE) unless otherwise noted.

Photoluminescence efficiencies relative to Ru(bpy)$_3^{2+}$ (ϕ_{em} = 0.082) unless otherwise noted.

[a]ECL generated via annihilation.

[b]Ref. 71.

[c]Vs. Ag wire quasi-reference electrode.

[d]Relative to Ru(bpy)$_3^{2+}$ at 0.05.

[e]ECL generated via the listed coreactant.

[f]Ref. 56.

[g]Relative to Ru(bpy)$_3^{2+}$/S$_2$O$_8^{2-}$ at 1.

[h]Ref. 72.

[i]Ref. 75.

[j]In CH$_2$Cl$_2$.

[k]Relative to Ru(bpy)$_3^{2+}$/TPrA at 1.

Very weak ECL was also observed for $Cr(phen)_3^{3+}$ and $Cr(5\text{-}Cl\text{-}phen)_3^{3+}$ (5-Cl-phen = 5-chloro-1,10-phenanthroline) [56].

Recently, reductive–oxidative ECL of $Cr(bpy)_3^{3+}$ was used to detect $S_2O_8^{2-}$ at glassy carbon electrodes [70]. A linear detection range of 7×10^{-6} to 1×10^{-4} M with a theoretical detection limit of 1×10^{-6} M were obtained at a potential of –0.5 V (vs. Ag/AgCl). Although these detection limits were lower than those of $Ru(bpy)_3^{2+}$ and $Ru(bpz)_3^{2+}$ (2,2'-bipyrazine) (Fig. 5) the applied potential was considerably more positive in $Cr(bpy)_3^{3+}$, indicating that the reduction of water may have been avoided with the chromium system.

The ECL of a non-polypyridyl chromium(III) species has also been reported. ECL in $Cr(CN)_6^{3-}$ was generated via an annihilation pathway in nonaqueous solvents (Fig. 16) [71]. This is somewhat surprising, because the Cr(IV) oxidation state is usually inaccessible in aprotic solution (i.e., DMF, MeCN, DMSO). ECL spectra characteristic of the ^2E_g excited state of $Cr(CN)_6^{3-}$ were obtained in all three-solvents when the working electrode was pulsed between +2.2 and –2.0 V (vs. an Ag quasi-reference electrode). ECL was not observed when anodic limits of less than +2.0 V were applied, indicating that the oxidant produced at these potentials is from background processes. The identity of the oxidant is unclear, but from electrochemical evidence it is not believed to be $Cr(CN)_6^{4-}$, indicating that "cross-reaction" ECL is taking place between $Cr(CN)_6^{2-}$ and another electrogenerated species in solution. This methodology was also used to generate ECL in $Cr(bpy)_3^{3+}$ [71].

B. Copper Complexes

Several studies have appeared on the ECL of copper systems. The first, to our knowledge, involved the polynuclear complex $[Cu(pyridine)I]_4$ [72]. $[Cu(pyridine)I]_4$ is one of a rare group of complexes containing metal–metal bonds that display photoluminescence in fluid solution at room temperature [other examples include $Pt_2(diphosphonate)_2^{4-}$ (ECL discussed below) and Mo_6Cl_{12}]. ECL was generated using a terminal voltage of 5 V and a frequency of 1 Hz. Although the ECL was very weak compared to photoluminescence intensities, it was possible to assign the emission to the lowest excited state of the complex. The poor ECL efficiencies were attributed to the instability of the reduced and oxidized electro-generated products (e.g., quasi reversible to irreversible cyclic voltammetric waves). The solvent (CH_2Cl_2) was also believed to play a role in ECL generation because it reduces at –2.33 V vs. SCE, much lower than the 5 V needed to generate the $[Cu(pyridine)I]_4$ excited state [72].

Cathodically generated luminescence was also observed for Cu(II) ion adsorbed onto an aluminum oxide electrode, and Cu(II) complexes containing bpy and ethylenediamine-type ligands [73].

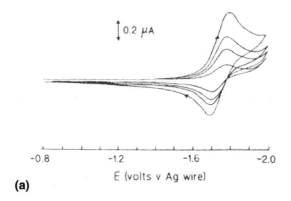

(a)

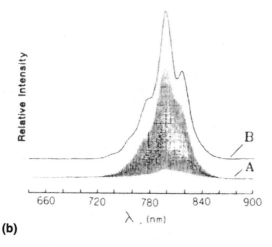

(b)

Figure 16   (a) Cyclic voltammogram of 2 mM $Cr(CN)_6^{3-}$/Me_2SO/0.05M $TBAClO_4$, for scan rates of 20, 100, 200, and 400 mV/s (hanging mercury drop working electrode). (b) (A) ECL spectrum of a 2.5 mM $Cr(CN)_6^{3-}$/MeCN/0.1 M $TBAClO_4$ solution (pulsing limits +2.2 and –2.0 V vs. Ag quasi-reference electrode at 0.5 Hz; Pt working and auxiliary electrodes). (B) Photoluminescence spectrum of a 2.5 mM $Cr(CN)_6^{3-}$/MeCN solution on 365 nm excitation. (From Ref. 71. Copyright 1987 American Chemical Society.)

A diaminotoluene isomer produces weak ECL in the presence of Cu^{2+} ions with the use of TPrA [74]. Because of the weak nature of the ECL it was not possible to characterize the nature of the emitting state (e.g., ECL spectra), so extreme care must be taken when making mechanistic arguments. However, the

ECL was believed to generate from copper diaminotoluene chelates. Despite the weak ECL, it was possible to detect the ligand using Cu(II) in the parts per million range.

In the previous examples, ECL was generated from copper in the +2 oxidation state. More recently, ECL from Cu(I) chelates has been reported [75]. Cu(I) bis-phenanthroline (i.e., [Cu(NN)$_2$]$^+$) complexes have long been known to show interesting photophysical properties. For example, excitation into the visible MLCT band of Cu(dmp)$_2$(PF$_6$) (dmp = 2,9-dimethyl-1,10-phenanthroline, Fig. 17)

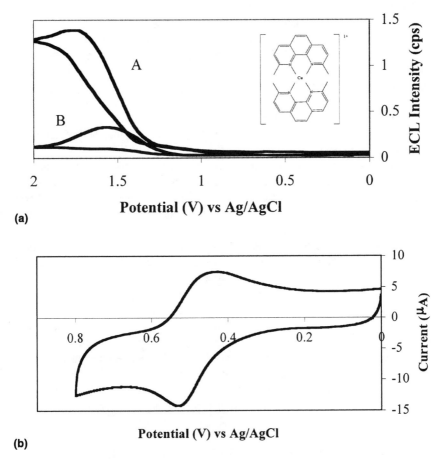

(a)

(b)

Figure 17 (a) ECL intensity versus potential for 3 μM Cu(dmp)$_2$$^+$ in (A) CH$_3$CN and (B) 50:50 (v/v) CH$_3$CN:H$_2$O. (b) Cyclic voltammogram of Cu(dmp)$_2$$^+$ in 50:50 (v/v) CH$_3$CN:H$_2$O (0.2 M KH$_2$PO$_4$). (From Ref. 75. Copyright 2001 American Chemical Society.)

results in $\phi_{em}(CH_2Cl_2) = 0.02\%$ and $\tau \cong 85$ ns [76,77]. They also display strong visible absorptions and ground-state and excited state redox potentials that, like their Ru(II) counterparts, make them potentially useful in applications such as solar energy conversion, molecular sensing, and photocatalysis. They are also easier to synthesize and purify than their ruthenium analogs. ECL was observed for $Cu(dmp)_2^+$ in aqueous, nonaqueous, and mixed solvent solutions using tri-*n*-propylamine as an "oxidative–reductive" coreactant (Fig. 17) [75]. The ECL intensity peaks a potential corresponding to oxidation of both the coreactant and $Cu(dmp)_2^+$. The peak potential corresponding to maximum ECL emission was approximately 500 mV more positive than corresponding oxidative peak potentials, indicating that the ECL emission may be due to the formation of either the $*Cu(dmp)_2^+$ metal-to-ligand charge transfer excited state or an excited state product of $Cu(dmp)_2^+$ oxidation. ECL efficiencies (ϕ_{ecl} = photons generated per redox event) were solvent-dependent [(ϕ_{ecl} (CH_3CN) > ϕ_{ecl} (50:50 (v/v) $CH_3CN:H_2O$) > ϕ_{ecl} (H_2O)] and corresponded fairly well with photoluminescence efficiencies. Increased ECL efficiencies ($\geq$50-fold) were observed in the presence of the nonionic surfactant Triton X-100.

The work described above on $Cu(dmp)_2^+$/TPrA ECL has been extended to the detection of copper ions in aqueous sample solutions. ECL has been observed from aqueous solutions of $Cu(dmp)_2^+$ generated in situ from copper ions and dmp [78]. In fact, the ligand dmp has been known for more than 50 years to be a specific chelator for Cu^+ [79]. ECL is generated by reducing Cu^{2+} ions to Cu^+ with hydroxylamine hydrochloride and then complexing with the chelating agent 2,9-dimethyl-1,10-phenanthroline to form $Cu(dmp)_2^+$ (Fig. 17), followed by oxidation in the presence of tri-*n*-propylamine. The ECL intensity peaks at a potential corresponding to oxidation of both TPrA and $Cu(dmp)_2^+$, indicating that the emission is from a $Cu(dmp)_2^+$ MLCT excited state (e.g., much like that in Fig. 17a). Conditions for ECL emission were optimized and used to generate a calibration curve that was linear over the 0.1–5 mg/L (ppm) range. The theoretical limit of detection was 6 µg/L (ppb), with a practical limit of detection of 0.1 ppm, well below the U.S. EPA federal standard of 1.3 ppm for copper in drinking water. The ECL of several metal ions other than copper with dmp and effects on $Cu(dmp)_2^+$ ECL were also examined. This work suggests that Cu/dmp ECL may find use as a detection method in analytical applications such as liquid chromatography, flow injection analysis, or the determination of copper in aqueous samples.

V. SECOND AND THIRD ROW TRANSITION METALS

Electrochemical, photophysical, and ECL data for the second and third row transition metals under both annihilation and coreactant conditions are presented in Tables 5 and 6, respectively.

Table 5 Electrochemical, Spectroscopic, and ECL Properties of Second and Third Row Transition Metal Complexes. ECL Generated Via Annihilation.

Complex	Solvent (electrolyte)	$E_{1/2}$(ox) (V)	$E_{1/2}$(red) (V)	λ_{em} (nm)	λ_{ecl} (nm)	ϕ_{em}	ϕ_{ecl}
$Ir(ppy)_3$	MeCN (0.1 M TBABF$_4$)	+0.7[a]	~−1.9[a]	495[a]	—	—	—
$Ir(ppy)_3$	Benzonitrile	—	—	530[b]	530[b]	—	—
$Ir(ppy)_3$	MeCN (0.1 M TBABPF$_6$)	+1.06[c,d]	−2.34[c,d]	510[c]	510[c]	—	0.14[c,e]
$[Ir(COD)(\mu\text{-pz})]_2$	THF (0.3 M TBABF$_4$)	+0.342[f]	−2.492[f]	682[f]	694[f]	—	—
$[Ir(COD)\,(\mu\text{-mpz})]_2$	THF (0.3 M TBABF$_4$)	+0.318[f]	−2.487[f]	—	—	—	—
$[Ir(COD)\,(\mu\text{-dmpz})]_2$	THF (0.3 M TBABF$_4$)	+0.279[f]	−2.510[f]	714[f]	719[f]	—	—
$Mo_6Cl_{14}{}^{2-}$	MeCN (0.1 M TBAP)	+1.56[g]	−1.53[g]	700[g]	700[g]	0.19[g]	—
$Mo_6Cl_{14}{}^-/D^{-h}$	CH$_2$Cl$_2$ (0.1 M TBAP)	+1.56[i]	j	~700[i]	~700[i]	—	—
$Mo_6Cl_{14}{}^{3-}/A^{+k}$	CH$_2$Cl$_2$ (0.1 M TBAP)	1	−1.53[i]	~700[i]	~700[i]	—	—
$Mo_2Cl_4(PMe_3)_4$	THF (0.1 M TBABF$_4$)	+0.69[m]	−1.78[m]	680[m]	680[m]	0.13[m]	0.002[m,n]
Pt(TPP)	CH$_2$Cl$_2$ (0.1 M TBAP)	+0.82[d,o]	−1.46[d,o]	654[o]	656[o]	—	—
Pd(TPP)	CH$_2$Cl$_2$ (0.1 M TBAP)	+1.01[d,o]	−1.45[d,o]	700[o]	700[o]	—	—
$Pt_2(P_2H_2O_5)_4{}^{4-}$	MeCN (0.1 M TBABF$_4$)			517[p]	517[p]	—	—
$Pt_2(P_2H_2O_5)_4{}^{4-}/Bu_4N^{+q}$	MeCN (0.1 M TBABF$_4$)	+0.5[d,r,s]		512[r]	512[r]	—	—
$Pt(thpy)_2$	DMF (0.1 M TBAP)	+0.82[s,t]	−1.77[t]	580[t]	580[t]	1.2[t]	0.005[n,t]
$Pt(q)_2$	MeCN (0.005 M TBABF$_4$)	+0.9[a]	−1.7[a]	650[a]	>530[a]	0.01[a]	—
$Pt_2(dba)_3$	CH$_2$Cl$_2$ (0.1 M TBAP)			782[u]	~780[u]	0.95[u,v]	—
$Pt_3(tbaa)_3$	CH$_2$Cl$_2$ (0.1 M TBAP)	+1.05[d,u]	−1.1[d,u]	782[u]	~780[u]	1.05[u,v]	—
$Pt_2(dba)_3$	CH$_2$Cl$_2$ (0.1 M TBAP)			726[u]	—	0.29[u,v]	—
$Pt_3(tbaa)_3$	CH$_2$Cl$_2$ (0.1 M TBAP)			724[u]	—	0.28[u,v]	—
Pt(OEP)	CH$_2$Cl$_2$ (0.1 M TBAPF$_6$)	~+0.8[c,d]	~−1.6[c,d]	619[c]	619[c]	—	0.02[c,d]
$Re(CO)_3Cl(phen)$	MeCN (0.1 M TBAClO$_4$)	+1.33[w]	−1.34[w]	598[w]	598[w]	—	—
$Re(CO)_3Cl(4,7\text{-}$ diphenylphen)	MeCN (0.1 M TBAClO$_4$)	+1.32[w]	−1.30[w]	610[w]	610[w]	—	—
$Re(CO)_3Cl$ (phen)/THPO* (phen)	MeCN (0.1 M TBABF$_4$)	—		610[y]	610[y]	—	—
$Re(CO)_3Cl(phen)$	MeCN (0.1 M TBAPF$_6$)	+0.98[s,z,aa]	−1.74[z,aa]	601[z]	601[z]	0.0026[z]	$6.4 10^{-4}$ [z]

(*continued*)

Table 5 Electrochemical, Spectroscopic, and ECL Properties of Second and Third Row Transition Metal Complexes. ECL Generated Via Annihilation.Coreactants

Complex	Solvent (electrolyte)	$E_{1/2}$(ox) (V)	$E_{1/2}$(red) (V)	λ_{em} (nm)	λ_{ecl} (nm)	ϕ_{em}	ϕ_{ecl}
Re(CO)$_3$Cl (2,9-Me$_2$phen)	MeCN (0.1 M TBAPF$_6$)	+0.98^{z,aa}	−1.84^{z,aa}	595^z	595^z	0.012^z	5.1×10$^{-4\ z}$
Re(CO)$_3$Cl (5,6-Me$_2$phen)	MeCN (0.1 M TBAPF$_6$)	+0.96^{s,z,aa}	−1.76^{z,aa}	599^z	599^z	0.022^z	4.2×10$^{-4\ z}$
Re(CO)$_3$Cl (4,7-Me$_2$phen)	MeCN (0.1 M TBAPF$_6$)	+0.94^{s,z,aa}	−1.85^{z,aa}	588^z	588^z	0.044^z	0.0011^z
Re(CO)$_3$Cl(3,4, 7,8-Me$_2$phen)	MeCN (0.1 M TBAPF$_6$)	+0.95^{s,z,aa}	−2.01^{z,aa,bb}	599^z	599^z	0.0044^z	3.1×10$^{-5\ z}$
Re(CO)$_3$Cl(bpy)	MeCN (0.1 M TBAPF$_6$)	+0.98^{s,z,aa}	−1.75^{z,aa}	609^z	609^z	0.023^z	2.9×10$^{-4,z}$
Re(CO)$_3$Cl(4,4'-Me$_2$bpy)	MeCN (0.1 M TBAPF$_6$)	+0.93^{s,z,aa}	−1.84^{z,aa}	599^z	599^z	0.017^z	6.9×10$^{-4,z}$

Please see text for ligand definitions.

Potentials are in volts vs. SSCE (0.2360 vs. NHE) unless otherwise noted.

Photoluminescence efficiencies relative to Ru(bpy)$_3$$^{2+}$ (ϕ_{em} = 0.082) unless otherwise noted.

[a]Ref. 72.

[b]Ref. 86.

[c]Ref. 87.

[d]Ag wire quasi-reference electrode.

[e]Relative to diphenylanthracene at 0.063.

[f]Ref. 91.

[g]Ref. 93.

[h]ECL generated via cross-reaction between Mo$_6$Cl$_{14}$$^-$ and electroactive donor species (D$^-$; nitroaromatic radical anions).

[i]Refs. 95–97.

[j]Potential varies based on the nature of D$^-$.

[k]ECL generated via cross-reaction between Mo$_6$Cl$_{14}$$^{3-}$ and electroactive acceptor species (A$^+$; aromatic amine radical cations).

[l]Potential varies based on the nature of A$^+$.

[m]Ref. 99.

[n]Relative to Ru(bpy)$_3$$^{2+}$ at 0.05.

[o]Ref. 100.

[p]Ref. 101.

[q]ECL generated via cross-reaction between Pt$_2$(P$_2$H$_2$O$_5$)$_4$$^{5-}$ and tetrabutylammonium cation (Bu$_4$N$^+$).

[r]Ref. 102.

[s]Anodic peak potential (Epa).

[t]Ref. 105.

[u]Ref. 106.

[v]Relative to Cr(bpy)$_3$(ClO$_4$)$_2$.

[w]Ref. 107.

[x]ECL generated via cross-reaction between Re(CO)$_3$Cl(phen)$^+$ and α-tetralone (produced on decomposition of THPO$^-$).

[y]Refs. 72 and 108.

[z]Ref. 109.

[aa]Volts vs. Fc/Fc$^+$ (0.631 V vs. NHE).

[bb]Cathodic peak potential (Epc).

Table 6 Electrochemical, Spectroscopic, and ECL Properties of Second and Third Row Transition Metal Complexes. ECL Generated via Coreactants.

Complex/coreactant	Solvent (electrolyte)	$E_{1/2}$ (V)	λ_{em} (nm)	λ_{ecl} (nm)	ϕ_{em}	ϕ_{ecl}
[Ir(COD)(μ-pz)]₂/C₂O₄²⁻	THF (0.3 M TBABF₄)	+0.342[a]	682[a]	694[a]	—	—
[Ir(COD)(μ-dmpz)]₂/ C₂O₄²⁻	THF (0.3 M TBABF₄)	+0.279[a]	714[a]	719[a]	—	—
Ir(ppy)₃/TPrA	MeCN (0.1 M TBAPF₆)	+0.71[b]	517[b]	517[b]	0.14[b]	0.33[b,c]
Ir(ppy)₃/TPrA	MeCN:H₂O (50:50 v/v; 0.1 M KH₂PO₄)	+0.53[b]	507,532[b]	517[b]	0.10[b]	0.0044[b,c]
Ir(ppy)₃/TPrA	H₂O (0.1 M KH₂PO₄)	+0.51[b]	507,532[b]	517[b]	0.08[b]	0.00092[b,c]
Mo₂Cl₄(PMe₃)₄/S₂O₈²⁻	THF (0.1 M TBABF₄)	−1.78[d]	680[d]	680[d]	0.13[d]	—
Mo₂Cl₄(PMe₃)₄/C₂O₄⁻	THF (0.1 M TBABF₄)	−1.78[d]	680[d]	—[e]	0.13[d]	—
Pt(thpy)₂/S₂O₈²⁻	DMF (0.1 M TBAP)	−1.77[f]	580[f]	—[e,f]	—	—
Re(CO)₃Cl(phen)/ S₂O₈²⁻	MeCN (0.1 M TBAPF₆)	−1.74[g,h,i]	601[g]	601[g]	0.0026[g]	0.0011[g,j]
Re(CO)₃Cl(2,9-Me₂phen)/ S₂O₈²⁻	MeCN (0.1 M TBAPF₆)	−1.84[g,h,i]	595[g]	595[g]	0.012[g]	0.0012[g,j]
Re(CO)₃Cl(5,6-Me₂phen)/ S₂O₈²⁻	MeCN (0.1 M TBAPF₆)	−1.76[g,h,i]	599[g]	599[g]	0.022[g]	0.0054[g,j]
Re(CO)₃Cl(4,7-Me₂phen)/ S₂O₈²⁻	MeCN (0.1 M TBAPF₆)	−1.85[g,h,i]	588[g]	588[g]	0.044[g]	0.030[g,j]
Re(CO)₃Cl(3,4,7,8-Me₄phen)/ S₂O₈²⁻	MeCN (0.1 M TBAPF₆)	−2.01[g,h,i]	599[g]	599[g]	0.0044[g]	0.0049[g,j]
Re(CO)₃Cl(bpy)/ S₂O₈²⁻	MeCN (0.1 M TBAPF₆)	−1.75[g,h,i]	609[g]	609[g]	0.023[g]	0.0031[g,j]
Re(CO)₃Cl(4,4'-Me₂bpy)/ S₂O₈²⁻	MeCN (0.1 M TBAPF₆)	−1.84[g,h,i]	599[g]	599[g]	0.017[g]	0.0043[g,j]
Re(CO)₃Cl(phen)/ TPrA	MeCN (0.1 M TBAPF₆)	+0.98[g,h,k]	601[g]	601[g]	0.0026[g]	0.087[c,g]
Re(CO)₃Cl(2,9-Me₂phen)/TPrA	MeCN (0.1 M TBAPF₆)	+0.98[g,h,k]	595[g]	595[g]	0.012[g]	0.12[c,g]
Re(CO)₃Cl(5,6-Me₂phen)/TPrA	MeCN (0.1 M TBAPF₆)	+0.96[g,h,k]	599[g]	599[g]	0.022[g]	0.12[c,g]
Re(CO)₃Cl(4,7-Me₂phen)/TPrA	MeCN (0.1 M TBAPF₆)	+0.94[g,h,k]	588[g]	588[g]	0.044[g]	0.22[c,g]
Re(CO)₃Cl(3,4,7,8-Me₄phen)/TPrA	MeCN (0.1 M TBAPF₆)	+0.95[g,h,k]	599[g]	599[g]	0.0044[g]	0.0027[c,g]
Re(CO)₃Cl(bpy)/TPrA	MeCN (0.1 M TBAPF₆)	+0.98[g,h,k]	609[g]	609[g]	0.023[g]	0.52[c,g]
Re(CO)₃Cl(4,4'-Me₂bpy)/TPrA	MeCN (0.1 M TBAPF₆)	+0.93[g,h,k]	599[g]	599[g]	0.017[g]	0.40[c,g]

Please see text for ligand definitions.
Potentials are in volts vs. SSCE (0.2360 vs. NHE) unless otherwise noted.
Photoluminescence efficiencies relative to Ru(bpy)₃²⁺ (ϕ_{em} = 0.082) unless otherwise noted.
[a]Ref. 91.
[b]Ref. 88.
[c]Relative to Ru(bpy)₃²⁺/TPrA at 1.0.
[d]Ref. 99.

[e]Weak ECL such that no spectrum was obtained.
[f]Ref. 105.
[g]Ref. 109.
[h]Volts vs. Fc/Fc⁺ (0.631 V vs. NHE).
[i]Cathodic peak potential (Epc).
[j]Relative to Ru(bpy)₃²⁺/S₂O₈²⁻ at 1.0.
[k]Cathodic peak potential (Epa).

A. Cadmium

An ECL system has been developed to detect cadmium ions in aqueous solution using TPrA as coreactant [80]. Various organic compounds were screened and shown to give ECL with Cd(II), with 1,10-phenanthroline giving the highest sensitivities and lowest detection limits. A linear range of 0.05–1 ppm with a detection limit [signal-to-noise ratio (S/N) = 3] of 4.9 ppb was achieved. The ECL of numerous metal ions with cadmium(II) and 1,10-phenanthroline in solution showed that many divalent ions interfere quite strongly with the cadmium ECL process [80]. This is not surprising, because 1,10-phenanthroline is known to bind many other metal ions, including those that have been shown to display ECL (e.g., Ru, Os, and Re). In an effort to improve both the efficiency of ECL emission (i.e., sensitivity) and the selectivity of this process, the effects of nonionic surfactant and ligand modification on the cadmium/TPrA emission system was explored [81]. Of the ligands studied (phenanthroline, bathophenanthroline salt, terpyridine, dimethylphenanthroline, and bipyridine), phenanthroline has the highest specificity for cadmium(II) ECL. The surfactant effect on ECL is currently an area of active research and has been shown to increase the sensitivity of $Ru(bpy)_3^{2+}$ ECL by up to tenfold [37–39]. Much like $Ru(bpy)_3^{2+}/$ TPrA ECL, Triton X-100 had the greatest effect on Cd(II)-phen/TPrA with an approximately twofold increase in ECL intensity at 2% Triton X-100 by weight [81]. This may reflect greater adsorption of Cd-phen complexes in the hydrophobic Cd-Phen micelle or, as is seen in ruthenium [37–39], iridum [82], and copper systems [75], adsorption of a surfactant layer on the electrode.

B. Gold

2,4-Diaminotoluene forms a weakly electrochemiluminescent compound with Au^+ [74]. As with copper(II) in the presence of 3,4-diaminotoluene (Section IV.B), it was possible to detect the ligand in the ppm range. Because many aminoaromatic ligands and compounds are associated with the degradation of explosives, this work suggests that solution ECL can be used to detect both ligands and metals in aqueous solution.

C. Iridium

Ortho-metalated complexes of Ir(III) show interesting photophysical properties. For example, excitation into the visible metal-to-ligand charge transfer (MLCT) band of Ir(ppy)$_3$ (ppy = 2-phenylpyridine) (Fig. 18) results in an excited state lifetime of approximately 100 ns in CH_2Cl_2 and 5 μs in toluene [83]. They also display strong visible absorptions and ground-state and excited state redox potentials that, like their Ru(II) counterparts, make them potentially useful in applications such as solar energy conversion, molecular sensing, and ECL.

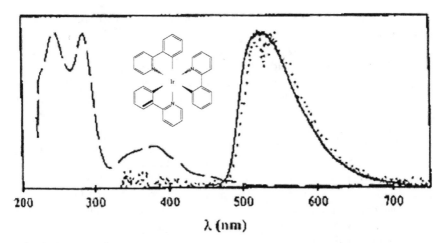

Figure 18 Absorbance of 10 μM Ir(ppy)₃ in CH₂Cl₂ (dashed line), photoluminescence (solid line) of 10 μM Ir(ppy)₃ in MeCN with an excitation wavelength of 380 nm, and ECL spectral response (dotted line) of 0.5 mM Ir(ppy)₃ in 50:50 MeCN:benzene containing 0.1 M TBAPF₆. (From Ref. 87.)

Weak ECL was reported for Ir(ppy)₃ in acetonitrile using the following reaction sequence to generate the excited state [72]:

$$\text{Ir(ppy)}_3 + e^- \rightarrow \text{Ir(ppy)}_3^- \quad (E_{1/2} \approx -1.9 \text{ V vs. SCE}) \tag{11}$$

$$\text{Ir(ppy)}_3 \rightarrow \text{Ir(ppy)}_3^+ + e^- \quad (E_{1/2} = +0.7 \text{ V vs. SCE}) \tag{12}$$

$$\text{Ir(ppy)}_3^+ + \text{Ir(ppy)}_3^- \rightarrow \text{Ir(ppy)}_3^* + \text{Ir(ppy)}_3 \tag{13}$$

$$\text{Ir(ppy)}_3^* \rightarrow \text{Ir(ppy)}_3 + h\nu \tag{14}$$

However, no ECL efficiencies and spectra were reported, making assignments of the ECL emission difficult.

Despite the weak ECL efficiency, the corresponding electroluminescent emission (EL) is quite intense (≤7%), so Ir(ppy)₃ has been incorporated into numerous EL devices. For example, the recent demonstration [84] of highly efficient green electrophosphorescence (i.e, electroluminescence) from Ir(ppy)₃ doped into a dicarbazole-biphenyl host, and subsequent work on a series of ortho-metalated iridium complexes with fluorinated aromatic ligands that emit over a wide color range [85] opens up the possibility of using these types of systems in organic light-emitting devices (OLEDs) such as flat panel displays. This has prompted a reinvestigation of the solution ECL of ortho-metalated iridium systems. In one study solution ECL was observed for Ir(ppy)₃ in benzonitrile via annihilation [86]. In this work, a thin-layer cell composed of glass/indium-tin oxide (ITO)/emitting solution/ITO/glass was used, and an ECL emission spectrum characteristic of Ir(ppy)$_3$* was observed.

The role of the excited triplet in the ECL of Ir(ppy)$_3$ was explored (Fig. 18) [87]. Radical cations and anions of Ir(ppy)$_3$ were generated in acetonitrile and acetonitrile–benzene solutions and recombined to produce ECL. The light generated in this manner was stable, and the emission spectrum was the same as that reported for OLED devices based on this compound. Also, the ECL from Ir(ppy)$_3$ was brighter than that from a 9,10-diphenylanthracene standard.

The ECL of Ir(ppy)$_3$ has also been extended to aqueous and partially aqueous solutions using coreactant ECL [88]. More specifically, the ECL of Ir(ppy)$_3$ was reported in acetonitrile (MeCN), mixed MeCN–H$_2$O (50:50 v/v), and aqueous (0.1 M KH$_2$PO$_4$) solutions with tri-n-propylamine (TPrA) as an oxidative–reductive coreactant. Photoluminescence (PL) efficiencies of 0.039, 0.050, and 0.069 were obtained in aqueous, mixed, and acetonitrile solutions, respectively, compared to Ru(bpy)$_3^{2+}$ ($\phi_{em} = 0.042$). The high photoluminescence efficiencies of Ir(ppy)$_3$, coupled with stable oxidative redox chemistry and the ability to distinguish between Ir(ppy)$_3$ and Ru(bpy)$_3^{2+}$ based on photoluminescence emission spectra makes this system of interest in fundamental and applied ECL studies. Tri-n-propylamine was used as an oxidative–reductive coreactant to generate ECL owing to the reversible to quasi-reversible nature of the Ir(ppy)$_3^{0/+}$ anodic redox couple. The ECL intensity peaks at potentials of ~+0.8 V. At these potentials, oxidation of both TPrA (E_a ~ +0.5 V vs. Ag/AgCl) and Ir(ppy)$_3$ (E^0 ~ +0.7 V) has occurred. ECL efficiencies (ϕ_{ecl} = photons emitted per redox event) of 0.00092 in aqueous, 0.0044 in mixed, and 0.33 in MeCN solutions for Ir(ppy)$_3$ were obtained using Ru(bpy)$_3^{2+}$ as a relative standard ($\phi_{ecl} = 1$). ECL emission spectra were obtained (Fig. 19) that were identical to photoluminescence spectra, indicating that the same MLCT excited state was formed in both experiments. The ECL was also linear over several orders of magnitude in mixed and acetonitrile solutions with theoretical detection limits (blank plus three times the standard deviation of the noise) of 1.23 nM in CH$_3$CN and 0.23 μM in CH$_3$CN–H$_2$O (50:50 v/v).

These studies illustrate that Ir(ppy)$_3$ [and, by analogy, other ortho-metalated iridium(III) complexes] exhibits ECL in aqueous and nonaqueous solutions. Although the ECL emission quantum efficiency was weaker than that of Ru(bpy)$_3^{2+}$ under similar conditions, the green ECL emission maximum of Ir(ppy)$_3$ and the red/orange emission of Ru(bpy)$_3^{2+}$ are far enough removed that it is possible to distinguish both signals in a single ECL experiment (Fig. 19) [88]. This ability may prove useful in applications where an ECL internal standard or multianalyte determination is desired. The lower potentials required to excite Ir(ppy)$_3$ (i.e., ≤ 1 V) compared to Ru(bpy)$_3^{2+}$ (i.e., ~1.2–1.6 V) may also prove useful in DNA diagnostic applications because it has been well documented that oligonucleotide sequences undergo irreversible oxidative damage at potentials of 1 V and higher [89].

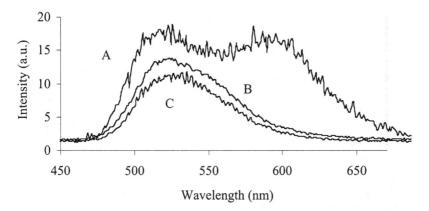

Figure 19 ECL spectra of (A) a solution of 10 μM Ir(ppy)₃ and 10 μM Ru(bpy)₃²⁺ in CH₃CN containing 0.05 M TPrA (0.1 M Bu₄NPF₆ as electrolyte), (B) 10 μM Ir(ppy)₃ (0.05 M TPrA) in CH₃CN (0.1 M Bu₄NPF₆), and (C) 10 μM Ir(ppy)₃ (0.05 M TPrA) in CH₃CN:H₂O 50:50 v/v, 0.1 M KH₂PO₄). (From Ref. 88. Copyright 2002 American Chemical Society.)

The effects of the nonionic surfactant Triton X-100 (Triton X-100 = polyethylene glycol *tert*-octylphenyl ether) (Fig. 6) on the properties of tris(2-phenylpyridine)iridium(III) electrochemiluminescence have also been investigated [90]. ECL in aqueous surfactant solution is currently an area of active study [37–42]. Solubilization of Ru(bpy)₃²⁺ (bpy = 2,2′-bipyridine) and analogous compounds in aqueous nonionic surfactant solutions leads to significant, and potentially useful, changes in the ECL properties [37,39,41]. For example, increases in both ECL efficiency (≥ eightfold) and duration of the ECL signal were observed in surfactant media upon oxidation of Ru(bpy)₃²⁺ and TPrA (Fig. 6) [38,42]. Because of the analytical importance of ECL in immunoassays and DNA probe analysis [26], the improved detection sensitivity of ruthenium polypyridyl compounds' ECL in the presence of nonionic surfactants (Section II.A), [38,39,42] and the widespread use of surfactants in clinical assays, an understanding of the nature of Ir(ppy)₃–surfactant interactions is critical to their use in practical applications. Oxidation of Ir(ppy)₃ produced ECL in the presence of tri-*n*-propylamine in aqueous surfactant solution. Increases in ECL efficiency (≥ tenfold) and TPrA oxidation current (≥ twofold) were observed in surfactant media. The mechanism of the surfactant effect was discussed in detail in Section II.A. Briefly, data support adsorption of surfactant on the electrode surface, facilitating TPrA and Ir(ppy)₃ oxidation and leading to higher ECL efficiencies [90]. Regardless of the mechanism of the surfactant effect, the fact that the emission maxima of Ir(ppy)₃ and Ru(bpy)₃²⁺ are far enough removed that it is possible to distinguish both signals in a single ECL experiment (Fig. 19) [88] and the similar reactivities of both in the presence of surfactants (e.g., pH and increased ECL efficiencies) improve the likelihood that both ruthenium- and iridium-based compounds can be incorporated into a single ECL assay.

The electrochemistry and ECL of a series of bimetallic iridium complexes, $[Ir(COD)(\mu\text{-}L)]_2$ (abbreviated Ir_2), where COD is 1,5-cyclooctadiene and L is the anion of pyrazole (pz), and substituted derivatives 3-methylpyrazole (mpz) and 3,5-dimethylpyrazole (dmpz) (Fig. 20) were studied in tetrahydrofuran (THF)/0.3 M tetra-n-butylammonium hexafluorophosphate [91]. Reversible one-electron oxidation waves were observed for all complexes, with the potential shifting to more negative values with increasing substitution of the bridging ligand (i.e., L). Irreversible to quasi-reversible reduction waves were also observed, with more reversible behavior observed with increasing substitution of the bridging ligands. ECL characteristic of the $^3B_2$ state of Ir_2 was produced upon sequential generation of the Ir_2^+ and Ir_2^- species, showing that the excited state formed electrochemically is the same as that formed in photoluminescence. The excited state species, Ir_2^* can also be produced by oxidizing Ir_2 in the presence of tetrabutylammonium oxalate (an oxidative–reductive coreactant) (Fig. 20).

D. Mercury

Weak ECL has been reported for mercury(II) in the presence of a synthetic tunichrome chromophore using TPrA as an oxidative–reductive coreactant [69]. Although the tunichrome chromophore itself exhibited intrinsic ECL, a nine- to tenfold increase in ECL was observed in the presence of Hg^{2+}, suggesting the formation of a metaltunichrome complex (also thallium–tunichrome in Section II.C).

Electrogenerated chemiluminescence was also observed for the chemiluminescent (CL) polymer diazoluminomelanin (DALM) with a luminol pendant group [92]. In the absence of mercury, both anodic and cathodic ECL were observed when no coreactant was present. Similar "background" ECL has been observed for other metal chelates, most notably $Ru(bpy)_3^{2+}$, and has been traced to a light-emitting reaction between the ECL luminophore [e.g., $Ru(bpy)_3^{3+}$] and OH^- ions [22]. In the presence of Hg(II), however, an approximately twofold increase in ECL for DALM was observed upon application of an anodic electrochemical sequence. If TPrA is present in a DALM-Hg^{2+} solution, a tenfold enhancement has been reported, again suggesting the formation of a mercury-DALM chelate complex.

As with many systems that produce relatively weak ECL, emission spectra were not obtained for Hg-tunichrome or Hg-DALM. This makes mechanistic arguments (as well as excited state assignments of the emission) difficult, so care must be taken when interpreting data. Nevertheless, this work coupled with

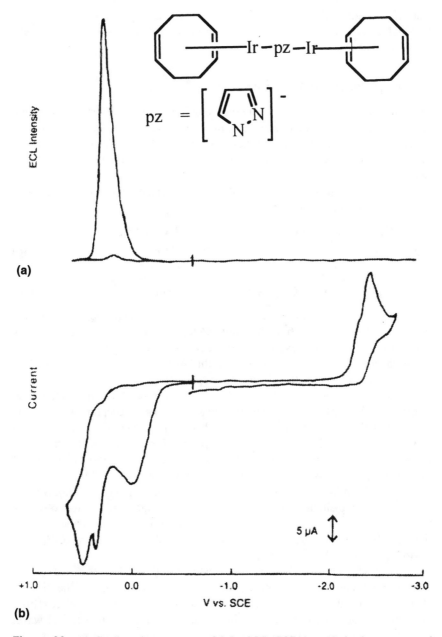

Figure 20 (a) Cyclic voltammogram of 2.5 mM Ir(COD)(μ-pz)]₂ in the presence of fivefold excess of (TBA)₂oxalate in THF/0.3 M TBAPF₆. (b) ECL vs. potential profile for solution in (a). (From Ref. 91. Copyright 1990 American Chemical Society.)

$Cu(dmp)_2^+$ and $Al(HQS)_3$ (Sections IV.B and III.A, respectively) demonstrates that ECL can be produced in situ without prior isolation and purification of metal ligand complexes.

E. Molybdenum and Tungsten

Though technically not chelates, a considerable amount of effort has gone into studying the ECL of molybdenum and tungsten clusters. This is due, in part, to the rich excited state chemistry of these complexes [93] and the ability of ECL to probe the kinetic and mechanistic aspects of highly exergonic electron transfer reactions.

The ECL of $Mo_6Cl_{14}^{2-}$ was produced by alternate oxidation and reduction at a Pt electrode in acetonitrile (CH_3CN) solution [94]:

$$Mo_6Cl_{14}^{2-} + e^- \rightarrow Mo_6Cl_{14}^{3-} \quad (E_- = -1.53 \text{ V vs. SCE}) \tag{15}$$

$$Mo_6Cl_{14}^{2-} \rightarrow Mo_6Cl_{14}^- + e^- \quad (E_- = +1.56 \text{ V vs. SCE}) \tag{16}$$

$$Mo_6Cl_{14}^- + Mo_6Cl_{14}^{3-} \rightarrow {}^*Mo_6Cl_{14}^{2-} + Mo_6Cl_{14}^{2-} \tag{17}$$

$${}^*Mo_6Cl_{14}^{2-} \rightarrow Mo_6Cl_{14}^{2-} + h\nu \quad (\lambda_{ecl} \approx 700 \text{ nm}) \tag{18}$$

Although the photoluminescence efficiency was quite high ($\phi_{em} = 0.19$), the measured ECL efficiency (photons emitted/electrons transferred), ϕ_{ecl}, was very low ($\leq 10^{-5}$). Although these ECL efficiencies are weak compared to those of metal chelate systems [e.g., $Ru(bpy)_3^{2+}$], this work clearly showed that ECL is possible in highly symmetric all-inorganic species.

Electrogenerated $Mo_6Cl_{14}^-$ and $Mo_6Cl_{14}^{3-}$ were also shown to react with electroactive donor (D^-; nitroaromatic radical anions) and acceptor (A^+; aromatic amine radical cations) species. The excited states generated in this way and the dependence of the ECL on the free energy driving force of the annihilation reactions were used to probe the energy dynamics (e.g., reorganizational barriers) of the $Mo_6Cl_{14}^{2-}$ ECL system [95]. Subsequent investigations on $Mo_6Cl_{14}^{2-}$ were focused on the effects of long-distance electron transfer in the context of Marcus theory to help explain the observed weak ECL efficiencies [96], the role of solvation on $Mo_6Cl_{14}^{2-}$ ECL [97], and the partitioning of the electrochemical excitation energy in molybdenum and tungsten clusters [$Mo_6Cl_{14}^{2-}$ and $M_6X_8Y_6^{2-}$ (M = Mo, W; X, Y = Cl, Br, I)] [98].

The high photoemission quantum yield of $Mo_2Cl_4(PMe_3)_4$ ($\phi_{em} = 0.26$) and long excited state lifetime (140 ns) prompted an investigation of the electrochemistry and ECL of this compound in THF and MeCN solutions [99]. Formation of the +1 cation resulted in a quasireversible electrochemical wave, with the product of oxidation undergoing subsequent chemical reactions. The $Mo_2Cl_4(PMe_3)_4^{0/-}$ couple is reversible, indicating that the reduction product is stable (Fig. 21). ECL was generated via annihilation and coreactant mechanisms.

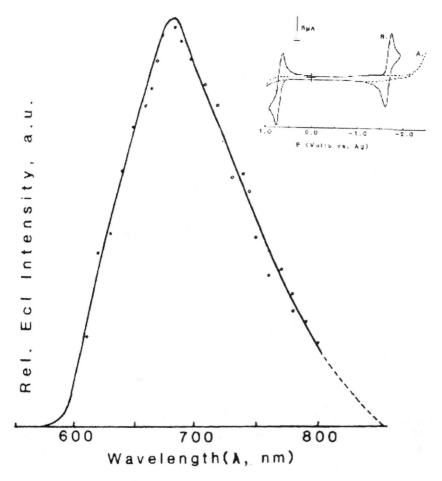

Figure 21 ECL spectrum of $Mo_2Cl_4(PMe_3)_4$ in THF and cyclic voltammograms of (A) 0.5 M TBABF$_4$ in THF and (B) 2 mM Mo $Mo_2Cl_4(PMe_3)_4$, 0.5 M TBABF$_4$ in THF. (From Ref. 99. Copyright 1986 American Chemical Society.)

ECL was observed at a Pt electrode whose potential was pulsed between -1.95 and 0.70 V. The ECL spectrum showed a peak at 680 nm, with an efficiency of 0.002 (Fig. 21). Owing to the stability of the -1 anion, $S_2O_8^{2-}$ was used as a reductive–oxidative coreactant to generate ECL. ECL was generated by pulsing between -0.5 and -2.0 V in an CH_3CN solution of $Mo_2Cl_4(PMe_3)_4$ and $S_2O_8^{2-}$. The ECL spectra generated with $S_2O_8^{2-}$ and via annihilation were the same as PL spectra. However, the ECL generated by sweeping reductively in the presence of persulfate was more intense than that generated via annihilation, probably due to the greater stability of the reduced form of the complex compared to the oxidized $[Mo_2Cl_4(PMe_3)_4^+]$ form. ECL was also detected using oxalate $(C_2O_4^{2-})$ as an oxidative–reductive coreactant and pulsing the potential between 0.0 and 0.7 V. However, the emission intensity was much weaker than the ECL generated by both annihilation and $S_2O_8^{2-}$, confirming the instability of the oxidized form, because $Mo_2Cl_4(PMe_3)_4^+$ is involved in $C_2O_4^{2-}$ systems [99].

F. Platinum and Palladium Complexes

The electrochemistry of Pt(II) and Pd(II) $\alpha,\beta,\gamma,\delta$-tetraphenylporphyrin complexes [Pt(TPP) and Pd(TPP)] in dichloromethane solution (0.1 M tetra-n-butylammonium perchlorate as electrolyte) are characterized by one-electron reduction and oxidation waves [100]. Electron transfer reactions between these electrogenerated species leads to low level light emission characteristic of the lowest triplet states of Pt(TPP) and Pd(TPP):

$$Pd(TPP)^+ + Pd(TPP)^- \rightarrow Pd(TPP)^* + Pd(TPP) \tag{19}$$

$$Pt(TPP)^+ + Pt(TPP)^- \rightarrow Pt(TPP)^* + Pt(TPP) \tag{20}$$

The ECL emission intensities are much lower than those observed for $Ru(bpy)_3^{2+}$ and may be low because of initial formation of the singlet excited state on electron transfer coupled with inefficient crossing to the emitting triplet, or quenching of the triplet state by ground-state species via the formation of a triplet excimer.

Two independent reports on the ECL of tetrakis(pyrophosphito)diplatinate(II), $Pt_2(P_2O_5H_2)_4^{4-}$, have appeared [101,102]. Interest in this compound can be traced to the intense photoluminescence found in both solution and the solid state, as well as its structure, which includes face-to-face square planes. Also, the excited states involve metal–metal bond excitation [103] and are significantly different from those of many luminescent metal complexes, such as $Ru(bpy)_3^{2+}$, that involve MLCT states. In the first study [101], ECL was generated in MeCN solutions by applying 4 V alternating current (potential sweep from +2.0 to –2.0 V vs. SCE) at 280 Hz. Green emission ($\lambda \approx 517$ nm) characteristic of the $Pt_2(P_2O_5H_2)_4^{4-*}$ excited state was observed, which was attributed to electron transfer between $Pt_2(P_2O_5H_2)_4^{3-}$ and $Pt_2(P_2O_5H_2)_4^{5-}$. Subsequent work [102] using controlled potential experiments (potential sweep from +0.7 to –3.0 V and +0.0 V to –3.0 V vs. Ag quasi-reference electrode) found that ECL characteristic of $Pt_2(P_2O_5H_2)_4^{4-*}$ was generated at potentials corresponding to reduction of solvent and electrolyte media. Generation of $Pt_2(P_2O_5H_2)_4^{3-}$ was not necessary to produce ECL. This led the authors to suggest a mechanism involving electron transfer between $Pt_2(P_2O_5H_2)_4^{5-}$ and Bu_4N^+ (i.e., tetra-n-butylammonium ion). Interestingly, when a mercury pool electrode was used in place of platinum, the characteristic emission at 512 nm and a new, as yet unidentified, peak at 607 nm appeared in the ECL spectrum [102].

Emission of ECL from intraligand (IL) states was observed for Pt(q)$_2$ where q = 8-quinolato [72]. 8-Quinolato is often called quinolin-8-olate or 8-hydroxyquinoline and is shown in Fig. 13 for the aluminum complex. Pt(q)$_2$ exhibits intense PL ($\phi_{em} \approx 0.01$) at $\lambda_{em} = 650$ nm under ambient conditions with the lowest energy emitting state assigned to the IL triplet of the chelating ligand [104]. Weak ECL characteristic of the IL emission of Pt(q)$_2$ was generated by using a terminal voltage of 4 V and a frequency of 30 Hz [72]. It was assumed that the source of the ECL was annihilation of the Pt(q)$_2^+$ and Pt(q)$_2^-$ species. The low ECL intensity was traced to quasireversible to irreverisible electrochemical oxidation and reduction processes corresponding to formation of Pt(q)$_2^+$ and Pt(q)$_2^-$.

The ortho-metalated platinum(II) complex, Pt(thpy)$_2$ (Fig. 22), is photoluminescent ($\lambda_{max} = 580$ nm) in fluid solution at room temperature, undergoes two reversible reductions ($E_{pc} = -1.80$ and -2.11 V vs. SCE) and one irreversible oxidation ($E_{pa} \approx 0.82$ V) in DMF solution, and displays ECL for both annihilation and coreactant (i.e., S$_2$O$_8^{2-}$) pathways [105]. The ECL spectra in all cases

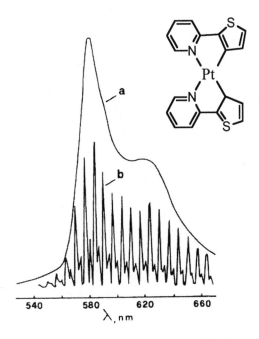

Figure 22 (a) Photoluminescence and (b) electrochemiluminescence spectra of 0.1 mM Pt(thpy)$_2$ in DMF. (From Ref. 105. Copyright 1986 American Chemical Society.)

were identical with the photoluminescence spectrum (Fig. 22), indicating that the chemical reactions that follow the electrochemical processes lead to the same MLCT state that is generated by photoexcitation.

Electrogenerated chemiluminescence has also been reported for the d^{10} transition metal complexes $M_2(dba)_3$ and $M_3(tbaa)_3$, where M = Pd(0) or Pt(0), dba = dibenzylideneacetone, and tbaa = tribenzylideneacetylacetone [106]. All four complexes exhibited strong photoluminescence at room temperature in a variety of solvents (e.g., CH_2Cl_2, CH_3CN, toluene). The relative emission quantum yields compared to $Cr(bpy)_3(ClO_4)_3$ were substantially greater for the Pt(0) systems, with those for $Pt_2(dba)_3$ and $Pt_3(tbaa)_3$ being 95% and 105%, respectively. In cyclic voltammetric studies, two reversible one-electron reduction steps and an irreversible one-electron oxidation step were reported for the Pd and Pt systems. Therefore, ECL was attempted by pulsing between the anodic and cathodic limits sufficiently to produce the oxidized [e.g., $M_2(dba)_3^+$] and reduced [e.g., $M_2(dba)_3^-$] forms of the parent complex. Surprisingly, no ECL was obtained for room temperature solutions of any of the complexes when the working electrode (Pt or Au) was stepped between +1.2 and −1.3 V (or −1.9 V vs. Ag quasi-reference electrode). However, ECL was obtained when the cell temperature was lowered to −50°C for both $Pt_2(dba)_3$ and $Pt_3(dba)_3$, indicating that cell temperature may be employed as a variable in studying weak ECL in transition metal systems. The ECL spectra for the platinum complexes matched the photoluminescence spectra, indicating population of the same excited states in both experiments.

The role of triplet states in the ECL of Pt(II) octaethyl porphyrin, PtOEP, was reported [87]. Radical cations and anions of PtOEP were generated in solution and recombined to produce ECL. The ECL from PtOEP was stable, and its emission spectrum was the same as those reported for both photoluminescence and organic light-emitting diode (OLED) devices based in PtOEP. However, the ECL emission efficiency was lower than that of a 9,10-diphenylanthracene standard.

G. Rhenium

The first excited state electron transfer and ECL studies of metal carbonyl complexes were carried out on *fac*-Re(CO)$_3$Cl(phen) [107]. Cyclic voltammetry established the ground-state potentials in acetonitrile at +1.3 V and −1.3 V vs. SCE, with a lowest excited energy state of ~2.3 eV. ECL can be observed for *fac*-Re(CO)$_3$Cl(phen) if a Pt electrode is cycled from −1.3 to +1.3 V vs. SCE in a potential step fashion. The ECL spectrum was identical to that formed by photoluminescence in the same medium.

fac-Re(CO)$_3$Cl(phen) also shows an intense chemiluminescence during the catalytic decomposition of tetralin hydroperoxide (THPO) in boiling tetraline [108]. The following mechanism was proposed:

$$Re(CO)_3Cl(phen) + THPO \rightarrow Re(CO)_3Cl(phen)^+ + THPO^- \quad (21)$$

$$THPO^- \rightarrow \alpha\text{-tetralone} + H_2O \quad (22)$$

$$Re(CO)_3Cl(phen)^+ + \alpha\text{-tetralone} \rightarrow Re(CO)_3Cl(phen)^*$$
$$+ \alpha\text{-tetralone} \quad (23)$$

$$Re(CO)_3Cl(phen)^* \rightarrow Re(CO)_3Cl(phen) + h\nu \quad (24)$$

Subsequent work showed that the same sequence of reactions could be initiated in acetonitrile via ECL, using a voltage greater than 2.6 V (e.g., potential pulse from +1.3 V to −1.3 V) and a frequency of 30 Hz [72]. THPO is apparently reduced during the cathodic cycle ($E_- = 0.973$ V vs. SCE), whereas the complex is oxidized during the anodic sweep, leading to the formation of $Re(CO)_3Cl(phen)^*$ via reactions (22)–(24).

The work described above was extended to other $Re(NN)(CO)_3Cl$ complexes (where NN is 1,10-phenanthroline, 2,2′-bipyridine, or a phenanthroline and bipyridine derivative containing methyl groups) (Fig. 23) [109]. These complexes are photoluminescent at room temperature and in acetonitrile display one chemically reversible one-electron reduction process and one chemically irreversible oxidation process. The wavelength maximum of emission (λ_{em}) was dependent on the nature of L, and a linear relationship between λ_{em} and the difference in electrode potentials for oxidation and reduction was evident (Fig. 23).

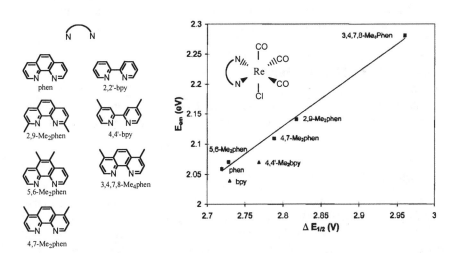

Figure 23 (Left) Schematic of phenanthroline, bipyridine, and substituted derivatives. (Right) and Correlation of emission energy (E_{em}) and difference in redox potentials [$\Delta E_{1/2} = E_{1/2}(Re^{II/I}) - E_{1/2}(L/L^-)$] in MeCN at room temperature. The line represents the best fit for the phenanthroline data only and has a correlation coefficient of 0.994. (From Ref. 109. Copyright 1996 American Chemical Society.)

ECL was observed in acetonitrile by stepping the potential of a platinum disk working electrode between values sufficient to form the radical anionic and cationic species. The relative amount of light produced during the pulses was dependent on the potential limits and pulse duration. ECL was also generated in the presence of the coreactants TPrA (potential sweep from +0.0 to +2.0 V) and $S_2O_8^{2-}$ (potential sweep from +0.0 to −1.8 V vs. Ag quasi-reference). In most cases, the ECL spectrum was identical to the PL spectrum, indicating that the chemical reactions following electrochemical oxidation or reduction form the same MLCT states that are generated via PL.

H. Silver

As with the mercury-DALM system discussed in Section IV.D, ECL was observed for the chemiluminescent (CL) polymer diazoluminomelanin (DALM) with a luminol pendant group in the presence and absence of silver, indicating the formation of an Ag^+-DALM chelate complex [92]. However, it was not possible to confirm the formation of such a complex because of the low level of the ECL emission.

A method for the trace determination of Ag(I) in basic aqueous solution has also been developed. This method relies on the enhanced ECL of 6-(2-hydroxy-4-diethylaminophenylazo)-2,3-dihydro-1,4-phthalazine-1,4-dione (HDEA) in the presence of Ag(I) ions [110]. Whether this system involves chelation between the metal ion and HDEA was not reported.

VI. RARE EARTH ELEMENTS

Electrochemical, photophysical, and ECL data for the rare earth elements under annihilation and coreactant conditions are presented in Tables 7 and 8, respectively.

A. Europium

Many trivalent lanthanide complexes display high photoluminescence efficiencies, large separations (~300 nm) between absorption and emission bands, and narrow emission spectra [111], all of which make them attractive as potential ECL labels. Therefore, it is not surprising that several reports have appeared on the ECL of lanthanide complexes.

The first report of ECL in a europium chelate described intermolecular energy transfer from an exciplex (formed by the radical cation of tri-*p*-tolylamine and either the benzophenone or dibenzoylmethane radical anions) directly to Eu(DBM)$_3$•piperidine or Eu(DNM)$_3$•piperidine (DBM = dibenzylmethide and

Table 7 Electrochemical, Spectroscopic, and ECL Properties of Rare Earth Metal Complexes. ECL Generated via Annihilation.

Complex	Solvent (Electrolyte)	$E_{1/2}$(ox) (V)	$E_{1/2}$(red) (V)	λ_{em} (nm)	λ_{ecl} (nm)	ϕ_{em}	ϕ_{ecl}
Eu(dibenzoylmethide)$_3$ piperidine[a]	MeCN (0.1 M TBAClO$_4$)	—	−1.94[b]	612[b]	—[c]	—	—
Eu(dinaphthoylmethide)$_3$ piperidine[a]	MeCN (0.1 M TBAClO$_4$)	—	−1.68[b]	610[b]	—[c]	—	—
Eu(TTFA)$_3$(phen)	MeCN (0.05 M TBABF$_4$)	—	−1.3,−1.63[d]	610[d]	—[c]	—	—
Tb(TTFA)$_3$(phen)	MeCN (0.05 M TBABF$_4$)	—	−1.5[d]	565[d]	—[c]	—	—
Tb(TTFA)$_4^-$	MeCN (0.05 M TBABF$_4$)	—	−1.75[d]	565[d]	—[c]	—	—

Please see text for ligand definitions.

Potentials are in volts vs. SSCE (0.2360 vs. NHE) unless otherwise noted.

Photoluminescence efficiencies relative to Ru(bpy)$_3^{2+}$ (ϕ_{em} = 0.082) unless otherwise noted.

[a] ECL generated via exciplex energy transfer between europium complex and ^{1}(A$^-$D$^+$)*, where D = tri-p-tolylamine radical cation and A = benzophenone or dibenzoylmethane radical anion.

[b] Ref. 9.

[c] Weak ECL such that no spectrum was obtained.

[d] Ref. 72.

Table 8 Electrochemical, Spectroscopic, and ECL Properties of Rare Earth Metal Complexes. ECL Generated via Coreactants

Complex/coreactant	Solvent (electrolyte)	$E_{1/2}$ (V)	λ_{em} (nm)	λ_{ecl} (nm)	ϕ_{em}	ϕ_{ecl}
$(pipH^+)Eu(DBM)_4/S_2O_8{}^{2-}$	MeCN (0.1 M TBAPF$_6$)	$-1.68,{}^a -2.41^b$	611[b]	611[b]	0.099[b]	0.052[b]
$(TBA)Eu(DBM)_4/S_2O_8{}^{2-}$	MeCN (0.1 M TBAPF$_6$)	-2.42^b	611[b]	611[b]	0.20[b]	0.06[b]
$(TBA)Eu(TTA)_4/S_2O_8{}^{2-}$	MeCN (0.1 M TBAPF$_6$)	$-2.12^{a,b}$	612[b]	612[b]	1.01[b]	0.006[b]
$(TBA)Eu(HFAC)_4/S_2O_8{}^{2-}$	MeCN (0.1 M TBAPF$_6$)	$-2.72^{a,b}$	611[b]	611[b]	0.095[b]	0.01[b]
$(TBA)Eu(BA)_4/S_2O_8{}^{2-}$	MeCN (0.1 M TBAPF$_6$)	$-2.82^{a,b}$	612[b]	612[b]	0.13[b]	0.02[b]
$(TBA)Eu(BTA)_4/S_2O_8{}^{2-}$	MeCN (0.1 M TBAPF$_6$)	$-2.36^{a,b}$	611[b]	611[b]	1.2[b]	0.01[b]
$Eu[2.2.2](NO_3)_3/S_2O_8{}^{2-}$	MeCN (0.1 M TBAPF$_6$)	$-0.21^{b,c}$	612[b]	—	0.0004[b]	—[d]
$Eu[2.2.1](NO_3)_3/S_2O_8{}^{2-}$	MeCN (0.1 M TBAPF$_6$)	$-0.44^{b,c}$	612[b]	—	—	—[d]
$Eu[2.2.1](DBM)(NO_3)_2/S_2O_8{}^{2-}$	MeCN (0.1 M TBAPF$_6$)	-0.56^b	612[b]	612[b]	0.0002[b]	$2\times10^{-4,b}$
$Eu[2.2.1](BA)(NO_3)_2/S_2O_8{}^{2-}$	MeCN (0.1 M TBAPF$_6$)	-0.58^b	612[b]	612[b]	0.0013[b]	$5\times10^{-5,b}$
$Eu[2.2.1](TTA)(NO_3)_2/S_2O_8{}^{2-}$	MeCN (0.1 M TBAPF$_6$)	-0.64^b	614[b]	—	0.0011[b]	—[d]
$Eu[2.2.1](BTA)(NO_3)_2/S_2O_8{}^{2-}$	MeCN (0.1 M TBAPF$_6$)	-0.83^b	612[b]	—	0.046[b]	—[d]
$Eu[2.2.1](HFAC)(NO_3)_2/S_2O_8{}^{2-}$	MeCN (0.1 M TBAPF$_6$)	—	613[b]	—	0.0014[b]	—[d]

Ligand definitions: DBM = dibenzoylmethide; TTA = thenoyltrifluoroacetate; BA = benzoylacetate; HFAC = hexafluoroacetylacetate; pipH$^+$ = piperidine cation; BTA = benzoyltrifluoroacetate; [2.2.2] = 4,7,13,16,21,24-hexaoxa-1,10-diazacyclo[8.8.8]hexacosane; [2.2.1] = 4,7,13,16,21-hexaoxa-1,10-diazacyclo[8.8.5]tricosane.

Potentials are in volts vs. Fc/Fc$^+$ (0.40 vs. NHE) unless otherwise noted.

Photoluminescence efficiencies relative to Ru(bpy)$_3{}^{2+}$ (ϕ_{em} = 0.082) unless otherwise noted.

ECL efficiencies are relative to Ru(bpy)$_3{}^{2+}$/S$_2$O$_8{}^{2-}$ (ϕ_{em} = 1).

[a] Cathodic peak potential, Epa.

[b] Ref. 10.

[c] Potentials in volts vs. SSCE (0.236 vs. NHE).

[d] No observed ECL.

DNM = dinaphthoylmethide) [9]. ECL resulted from "sensitization" of the europium complexes, with the donor formed electrochemically, not optically. However, no direct ECL of the Eu(III) chelates was found in that study.

Weak ECL from Eu(TTFA)$_3$(phen) [TTFA = 4,4,4-trifluoro-1-(2-thienyl)-1,5-butanediono/thenoyltrifluoroacetonato] was detected by using a photon-counting apparatus [72]. This ECL was attributed to intramolecular energy transfer from a nonemitting excited state of the intraligand orbitals to the emitting state of the metal-centered f orbitals, analogous to events proposed in photochemical experiments.

The ECL of a series of europium chelates, cryptates, and mixed-ligand chelate/cryptate complexes was studied [10]. The complexes were of the following general forms: EuL$_4^-$, where L = β-diketonate, a bis-chelating ligand (such as dibenzoylmethide), added as salts (A)EuL$_4$ (where A = tetrabutylammonium or piperidinium ion (pipH$^+$), Eu(crypt)$^{3+}$, where crypt = a cryptand ligand, e.g., 4,7,13,16,21-pentaoxa-1,10-diazabicyclo[8,8,5]tricosane; and Eu(crypt)L^{2+} for the mixed-ligand systems. ECL was generated in acetonitrile solution using S$_2$O$_8^{2-}$ (+0.0 V to –2.0 V vs. Ag quasi-reference) and appears to occur by a different mechanism than in transition metal systems via a "ligand-sensitization" route, where ECL occurs in the organic ligands with subsequent transfer to the f orbitals of the metal centers. ECL spectra (Fig. 24) matched the photoluminescence spectra with a narrow emission band around 612 nm, corresponding to a metal centered 4f–4f transition. Although it was clear from this work that the ligands play an integral role in rare earth ECL, very low ECL efficiencies were observed in nonaqueous solvents and little to no ECL was observed in aqueous media.

B. Terbium

As with Eu(TTFA)$_3$(phen), weak ECL was observed for Tb(TTFA)$_4^-$ and Tb(TTFA)$_3$(phen), which was attributed to intramolecular energy transfer from a nonemitting ligand-centered state to the emitting state centered on the metal f orbitals [72].

Electrochemically induced light emission involving Tb(III) was also generated via "cathodic luminescence", where Tb(III) ions and chelates were adsorbed onto an aluminum oxide electrode with subsequent generation of excited states [112].

VII. CONCLUSIONS

This chapter has focused on developments within metal chelate ECL. Considering the number of metal complexes that display the electrochemical and spectroscopic qualities required of ECL, it is not surprising that they have played

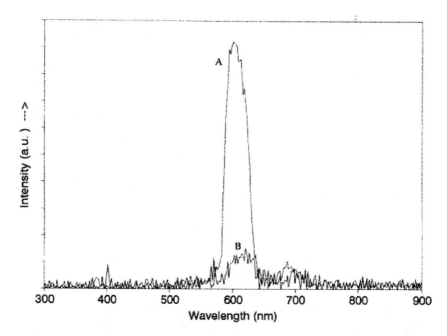

Figure 24 ECL spectra of (A) (TBA)Eu(dibenzoylmethide)$_4$ and (B) (TBA)Eu(benzoy-lacetonate)$_4$ in $CH_3CN/S_2O_8^{2-}$. Solutions were 4 mM in chelate, 10 mM in $S_2O_8^{2-}$, and 0.1 M in TBABF$_4$. The potential was continuously pulsed between 0.0 and 2.5 V vs. AgQRE with a pulse duration of 1 s and an acquisition time of 30 min. (From Ref. 10. Copyright 1996 American Chemical Society.)

an important role in transforming ECL from a laboratory curiosity to a useful analytical technique. There are also many metal chelates yet to be explored! One wonders what the future holds for ECL, but whatever the improvements and new aspects of ECL that emerge, metal chelates will undoubtedly continue to play an important role.

ACKNOWLEDGMENTS

I thank Research Corporation, the American Chemical Society Petroleum Research Fund, the National Science Foundation, and Southwest Missouri State University for support. A number of undergraduate and graduate student cowork-ers have also been responsible for work from the Richter group reported here. Although most are listed in the references, I am especially grateful to Craig Alexander, Jeff McCall, Brigitte Factor, Scott Workman, Brian Muegge, and

David Bruce—without them, expanding the frontiers of ECL (at least in the Richter group) would be nearly impossible.

REFERENCES

1. Wheeler, B.L., Nagasubramanian, G., Bard, A.J., Schechtman, L.A., Dininny, D.R., Kenney, M.E. A silicon phthalocyanine and a silicon naphthalocyanine: synthesis, electrochemistry, and electrogenerated chemiluminescence. J. Am. Chem. Soc. **1984**, *106*, 7404–7410.
2. Anderson, J.D., McDonald, E.M., Lee, P.A., Anderson, M.L., Ritchie, E.L., Hall, H.K., Hopkins, T., Mash, E.A., Wang, J., Padias, A., Thayumanavan, S., Barlow, S., Marder, S.R., Jabbour, G.E., Shaheen, S., Kippelen, B., Peyghambarian, N., Wightman, R.M., Armstrong, N.R. Electrochemistry and electrogenerated chemiluminescence processes of the components of aluminum quinolate/triarylamine, and related organic light-emitting diodes. J. Am. Chem. Soc. **1998**, *120*, 9646–9655.
3. Faulkner, L.R., Glass, R.S. In: Electrochemiluminescence 1982, Chemical and Biological Generation of Excited States; Eds. A. Waldemar, and C. Giuseppe, Academic Press, New York, Ch. 6.
4. Knight, A.W., Greenway, G.M. Occurrence, mechanisms and analytical applications of electrogenerated chemiluminescence. Analyst **1994**, *119*, 879–890.
5. Tokel, N., Bard, A.J. Electrogenerated chemiluminescence. IX. Electrochemistry and emission from systems containing tris(2,2'-bipyridine)ruthenium(II) dichloride. J. Am. Chem. Soc. **1972**, *94*, 2862–2863.
6. Vogler, A., Kunkeley, H. Electrochemiluminescence of tetrakis(diphosphonato) diplatinate(II). Angew. Chem. Int. Ed. Engl. **1984**, *23*(4), 316–317.
7. Kim, J., Fan, F-R.F., Bard, A.J., Che, C-M., Gray, H.B. Electrogenerated chemiluminescence. On the electrogenerated chemiluminescence (ECL) of tetrakis(pyrophosphito)diplatinate(II), $Pt_2(P_2O_5H_2)_4^{4-}$. Chem. Phys. Lett. **1985**, *121*, 543–546.
8. Richter, M.M., Bard, A.J., Kim, W., Schmehl, R.S. Electrogenerated chemiluminescence. 62. Enhanced ECL in bimetallic assemblies with ligands that bridge isolated chromophores. Anal. Chem. **1998**, *70*, 310–318.
9. Hemingway, R.E., Park, S-M., Bard, A.J. Electrogenerated chemiluminescence. XXI. Energy transfer from an exciplex to a rare earth chelate. J. Am. Chem. Soc. **1975**, *95*, 200–201.
10. Richter, M.M., Bard, A.J. Electrogenerated chemiluminescence. 58. Ligand Sensitized electrogenerated chemiluminescence in europium labels. Anal. Chem. **1996**, *68*, 2641–2650.
11. Faulkner, L.R., Bard, A.J. 1977. In: Electroanalytical Chemistry, Vol. 10, Ed. A.J. Bard, Marcel Dekker, New York, pp 1–95.
12. Bard, A.J., Faulkner, L.R. Electrochemical Methods Fundamentals and Applications 2nd ed. Wiley, New York, pp 736–745, 2001.

13. (a) Bard, A.J., Whitesides, G.M. US Patent 5, 221, 605, 1993. (b) Bard, A.J., Whitesides, G.M. US Patent 5, 238, 808, 1993. (c) Bard, A.J., Whitesides, G.M. US Patent 5, 310, 687, 1994.

14. Roundhill, D.M. 1994, Photochemistry and Photophysics of Coordination Complexes, Plenum, New York, Ch. 5.

15. Kalyanasundaram, K. Photochemistry and Photophysics of Polypyridine and Porphyrin Complexes. New York: Academic Press, 1992.

16. Paris, J.P., Brandt, W.W. Charge transfer luminescence of a ruthenium(II) chelate. J. Am. Chem. Soc. **1959**, *81*(18), 5001–5002.

17. (a) Tokel-Takvoryan, N.E., Hemingway, R.E., Bard, A.J. Electrogenerated chemiluminescence. XIII. Electrochemical and electrogenerated chemiluminescence studies of ruthenium chelates. J. Am. Chem. Soc. **1973**, *95*, 6582–.

18. Wallace, W.L., Bard, A.J. Electrogenerated chemiluminescence. 35. Temperature dependence of the ECL efficiency of tris(2,2'-bipyridine)rubidium(2+) in acetonitrile and evidence for very high excited state yields from electron transfer reactions. J. Phys. Chem. **1979**, *83*, 1350–1357.

19. Luttmer, J.D., Bard, A.J. Electrogenerated chemiluminescence. 38. Emission intensity-time transients in the tris(2,2'-bipyridine)ruthenium(II) system. J. Phys. Chem., **1981**, *85*, 1155.

20. Glass, R.S., Faulkner, L.R. Electrogenerated chemiluminescence from the tris(2,2'-bipyridine)ruthenium(II) system. An example of S-route behavior. J. Phys. Chem., **1981**, *85*, 1160.

21. (a) Caspar, J.V., Meyer, T.J. Photochemistry of tris(2,2'-bipyridine)ruthenium(2+) ion (Ru(bpy)32+). Solvent effects. J. Am. Chem. Soc. **1983**, *105*, 5583–5590. (b) Van Houten, J., Watts, R.J. Temperature dependence of the photophysical and photochemical properties of the tris(2,2'-bipyridyl)ruthenium(II) ion in aqueous solution, J. Am. Chem. Soc. **1975**, *98*, 4853–4858.

22. Rubinstein, I. and Bard, A.J. Electrogenerated chemiluminescence. 37. Aqueous ECL systems based on Ru(2,2'-bipyridine)$_3{}^{2+}$ and oxalate or organic acids. J. Am. Chem. Soc. **1981**, *103*, 512–516.

23. (a) Demas, J.N., Crosby, G.A. Quantum efficiencies on transition metal complexes. II. Charge-transfer luminescence. J. Am. Chem. Soc. **1971**, *93*, 2841–2847. (b) Meyer, T.J. Optical and thermal electron transfer in metal complexes. Acc. Chem. Res. **1978**, *11*(3), 94–100.

24. White, H.S., Bard, A.J. Electrogenerated chemiluminescence. 41. Electrogenerated chemiluminescence and chemiluminescence of the Ru(2,2'-bpy)$_3{}^{2+}$-S$_2$O$_8{}^{2-}$ system in acetonitrile-water solutions. J. Am. Chem. Soc. **1982**, *104*(25), 6891–6895.

25. Leland, J.K., Powell, M.J. Electrogenerated chemiluminescence: An oxidative-reduction type ECL reaction using tripropyl amine. J. Electroanal. Chem. **1990**, *137*(10), 3127–3131.

26. Bard, A.J., Debad, J.D., Leland, J.K., Sigal, G.B., Wilbur, J.L., Wohlstadter, J.N. Electrochemiuminescence. In *Encycopedia of Analytical Chemistry*; R.A. Meyers, Ed., Wiley, Chichester, 2000, pp 9842–9849.

27. Downey, T.M., Nieman, T.A. Chemiluminescence detection using regenerable

tris(2,2′-bipyridyl)ruthenium(II) immobilized in Nafion. Anal. Chem. **1992**, *64*, 261–268.

28. (a) Danielson, N.D., He, L., Noffsinger, J.B., Trelli, L. Determination of erythromycin in tablets and capsules using flow injection analysis with chemiluminescence detection. J. Pharm. Biomed. Anal. **1989**, *7*, 1281–1285. (b) Knight, A.W. A review of recent trends in analytical applications of electrogenerated chemiluminescence. Trends Anal. Chem. **1999**, *18*, 47–62.

29. Dixon, S.B., Sanford, J., Swift, B.W., 1993, Principles and Practices for Petroleum Contaminated Soils, Eds. E.J. Calabrese and P.T. Kosteki, Chelsea, MI: Lewis, pp 85–99.

30. (a) Alexander, C., Richter, M.M. Measurement of fatty amine ethoxylate surfactants using electrochemiluminescence. Anal. Chim. Acta. **1999**, *402*, 105–112. (b) McCall, J., Alexander, C., Richter, M.M. Quenching of electrogenerated chemiluminescence by phenols, hydroquinones, catechols and benzoquinones. Anal. Chem. **1999**, *71*, 2523–2527. (c) McCall, J., Richter, M.M. Phenol substituent effects on electrogenerated chemiluminescence quenching. Analyst **2000**, *125*, 545–548.

31. Tokel-Takvoryan, N.E., Hemingway, R.E., Bard, A.J. Electrogenerated chemiluminescence. XIII. Electrochemical and electrogenerated chemiluminescence studies of ruthenium chelates. J. Am. Chem. Soc. **1973**, *95*(20), 6582–6589.

32. Gonzales-Velasco, J., Rubinstein, I., Crutchley, R.J., Lever, A.B.P., Bard, A.J. Electrogenerated chemiluminescence. 42. Electrochemistry and electrogenerated chemiluminescence of the tris(2,2′-bipyrazine)ruthenium(II) system. Inorg. Chem. **1983**, *22*, 822–825.

33. Gonzalez-Velasco, J. Temperature dependence of the electrogenerated chemiluminescence efficiency of Ru(bpz)$_3^{2+}$ in acetonitrile. A mechanistic interpretation. J. Phys. Chem. **1988**, *92*, 2202–2207.

34. Yamazaki-Nishida, S., Harima, Y., Yamashita, K. Direct current electrogenerated chemiluminescence observed with [Ru(bpz)$_3^{2+}$] in fully aqueous solution. J. Electroanal. Chem. **1990**, *283*, 455–458.

35. Yamashita, K., Yamazaki-Nishida, S., Harmia, Y., Segawa, A. Direct current electrogenerated chemiluminescent microdetermination of peroxydisulfate in aqueous solution. Anal. Chem., **1991**, *63*, 872–876.

36. Zhang, X., Bard, A.J. Electrogenerated chemiluminescent emission from an organized (L-B) monolayer of a Ru(bpy)$_3^{2+}$-based surfactant on semiconductor and metal electrodes. J. Phys. Chem. **1988**, *92*, 5566–5569.

37. Factor, B., Muegge, B., Workman, S., Bolton, E., Bos, J., Richter, M.M. Surfactant chain length effects on the light emission of tris(2,2′-bipyridine) ruthenium(II)/tripropylamine electrogenerated chemiluminescence. Anal. Chem. **2001**, *73*, 4621–4624.

38. Zu, Y., Bard, A.J. Electrogenerated chemiluminescence. 67. Dependence of light emission of the tris(2,2′)bipyridylruthenium(II)/tripropylamine system on electrode surface hydrophobicity. Anal. Chem. **2001**, *73*, 3960–3964.

39. Bruce, D., McCall, J., Richter, M.M. Effects of electron withdrawing and donating groups on the efficiency of tris(2,2′-bipyridyl)ruthenium(II)/tri-*n*-propylamine electrochemiluminescence. Analyst. **2002**, *127*, 125–128.

40. Ouyang, J., Bard, A.J. Electrogenerated chemiluminescence. 50. Electrochemistry and electrogenerated chemiluminescence of micelle solubilized Os(bpy)$_3^{2+}$. Bull. Chem. Soc. Jpn. **1988**, *61*, 17–24.

41. McCord, P., Bard, A.J. Electrogenerated chemiluminescence. 54. Electrogenerated chemiluminescence of ruthenium(II) 4,4'-diphenyl-2,2'-bipyridine and ruthenium(II) 4,7-diphenyl-1,10-phenanthroline systems in aqueous and acetonitrile solutions. J. Electroanal. Chem. **1991**, *318*, 91–99.

42. Workman, S., Richter, M.M. The effects of nonionic surfactants on the tris(2,2'-bipyridyl)ruthenium(II)–tripropylamine electrochemiluminescence system. Anal. Chem. **2000**, *72*, 5556–5561.

43. Andersson, A-M., Isovitsch, R., Miranda, D., Wadhwa, S., Schmehl, R.H. Electrogenerated chemiluminescence from Ru(II) bipyridylphosphonic acid complexes adsorbed to mesoporous TiO$_2$/TiO$_2$ electrodes. J. Chem. Soc. Chem. Commun. **2000**, 505–506.

44. Lai, R.Y., Chiba, M., Kitamura, N., Bard, A.J. Electrogenerated chemiluminescence. 68. Detection of sodium ion with a ruthenium(II) complex with crown ether moiety at the 3,3'-positions on the 2,2'-bipyridine ligand. Anal. Chem. **2002**, *74*, 551–553.

45. Muegge, B.D., Richter, M.M. Electrochemiluminescent detection of metal cations using a ruthenium(II) bipyridyl complex containing a crown ether moiety. Anal. Chem. **2002**, *74*, 547–550.

46. Bruce, D., Richter, M.M. Electrochemiluminescence in aqueous solution of a ruthenium(II) bipyridyl complex containing a crown ether moiety in the presence of metal ions. Analyst. 2002, *127*, 1492–1494.

47. Liang, P., Sanchez, L., Martin, M.T. Electrochemiluminescence-based quantitation of classical clinical chemistry analytes. Anal. Chem. **1996**, *68*, 1298–1302.

48. Liang, P., Dong, L., Martin, M.T. Light emission from ruthenium-labeled penicillins signaling their hydrolysis by β-lactamase J. Am. Chem. Soc. **1996**, *118*, 9198–9199.

49. Dong, L., Martin, M.T. Enzyme triggered formation of electrochemiluminescent ruthenium complexes. Anal. Biochem., **1996**, *236*, 344–347.

50. Kanoufi, F., Bard, A.J. Electrogenerated chemiluminescence. 65. An investigation of the oxidation of oxalate by tris(polypyridine) ruthenium complexes and the effect of the electrochemical steps on emission intensity. J. Phys. Chem. B. **1999**, *103*, 10469–10480.

51. Robin, M.B., Day, P. Mixed valence chemistry—a survey and classification. Adv. Inorg. Chem. Radiochem. **1967**, *10*, 247–405.

52. Zhou, M. Roovers, J. Dendritic supramolecular assembly with multiple Ru(II) tris(bipyridine) units at the periphery: synthesis, spectroscopic, and electrochemical study. Macromolecules. **2001**, *34*, 244–252.

53. Creutz, C., Chou, M., Netzel, T.L., Okumura, M., Sutin, N. Lifetimes, spectra, and quenching of the excited states of polypyridine complexes of iron(II), ruthenium(II), and osmium(II). J. Am. Chem. Soc. **1980**, *102*, 1309–1319.

54. Kober, E.M., Meyer, T.J. Concerning the absorption spectra of the ions M(bpy)$_3^{2+}$ (M = Fe, Ru, Os; bpy = 2,2'-bipyridine). Inorg. Chem. **1982**, *21*, 3967–3977.

55. Heller, A. Spiers Memorial Lecture. On the hypothesis of cathodic protection of genes. Faraday Disc. **2000**, *116*, 1–13.

56. Bolletta, B., Ciano, T., Balzani, V., Serpone, N. Polypyridine transition metal complexes as light emission sensitizers in the electrochemical reduction of the persulfate ion. Inorg. Chim. Acta **1982**, *62*, 207–213.

57. Abruna, H.D. Electrogenerated chemiluminescence of bipyridine and phenanthroline complexes of osmium. J. Electroanal. Chem. **1984**, *175*, 321–326.

58. Abruna, H.D. Electrochemiluminescence of osmium complexes—spectral, electrochemical and mechanistic studies. J. Electrochem. Soc.: Electrochem. Sci. Technol. **1985**, *132*(4), 842–849.

59. Lee, C-W., Ouyang, J., Bard, A.J. Electrogenerated chemiluminescence. 51. The tris(2,2′-bipyrazine)osmium(II) system. J. Electroanal. Chem. **1988**, *244*, 319–324.

60. Bruce, D., Richter, M.M., Brewer, K.J. Electrochemiluminescence from $Os(phen)_2(dppene)^{2+}$ (phen = 1,10-phenanthroline and dppene = bis(diphenylphosphinoethene)). Anal. Chem. **2002**, *74*, 3157–3159.

61. Gross, E.M., Anderson, J.D., Slaterbeck, A.F., Thayumanavan, S., Barlow, S., Zhang, Y., Marder, S.R., Hall, H.K., Flore, Nabor, M., Wang, J.-F., Mash, E.A., Armstrong, N.R., Wightman, R.M. Electrogenerated chemiluminescence from derivatives of aluminum quinolate and quinacridones: cross-reactions with triarylamines lead to single emission through triplet-triplet annihilation pathways. J. Am. Chem. Soc. **2000**, *122*, 4972–4979.

62. (a) Tang, C.W., Van Slyke, S.A. Organic electroluminescent diodes. Appl. Phys. Lett. **1987**, *51*, 913–915. (b) Jabbour, G.E., Kawabe, Y., Shaheen, S.E., Wang, J.F., Morrell, M.M., Kippelen, B. Peyghambarian N. Highly efficient and bright organic electroluminescent devices with an aluminum cathode. Appl. Phys. Lett. **1997**, *71*, 1762–1764.

63. Muegge, B.D., Richter, M.M. Electrochemiluminescence of tris(8-hydroxyquinoline-5-sulfonic acid)aluminum(III) in aqueous solution. Anal. Chem. 2003, *75,* 1102–1105.

64. Soroka, K., Vithanage, R.S., Phillips, D.A., Walker, B., Dasgupta, P.K. Fluorescence properties of metal complexes of 8-hydroxyquinoline-5-sulfonic acid and chromatographic applications. Anal. Chem. **1987**, *59*, 629–636.

65. Fernandez-Butierrez, A., Muoz de la Pena, A. In: Molecular Luminescence Spectroscopy. Methods and Applications: Part 1. S.G. Shulman, ed. Wiley: New York, 1985; pp 371–456.

66. Wheeler, B.L., Nagasubramanian, G., Bard, A.J., Schechtman, L.A., Dininny, D.R., Kenney, M.E. A silicon phthalocyanine and a silicon naphthalocyanine: synthesis, electrochemistry and electrogenerated chemiluminescence. J. Am. Chem. Soc. **1984**, *106*, 7404–7410.

67. Ding, Z., Quinn, B.M., Harma, S.K., Pell, L.E., Korgel, B.A., Bard, A.J. Electrochemistry and electrogenerated chemiluminescence from silicon nanocrystal quantum dots. Science. **2002**, *296*, 1293–1297.

68. Haapakka, K., Kankare, J., Kulmala, S. Feasibility of low-voltage cathodic electroluminescence at oxide-covered aluminum electrodes for trace metal determinations in aqueous solution. Anal. Chim. Acta **1985**, *171*, 259–267.

69. Bruno, J.G., Collard, S.B., Andrews, A.R.J. Further characterization of tunicate and tunichrome electrochemiluminescence. J. Biolum. Chemilumin. **1997**, *12*, 155–164.

70. Xu; G., Dong, S. Electrochemiluminescent determination of peroxydisulfate with Cr(bpy)$_3^{3+}$ in purely aqueous solution. Anal. Chim. Acta **2000**, *412*, 235–240.

71. Kane-Maguire, N.A.P., Guckert, J.A., O'Neill, P.J. Electrogenerated chemiluminescence of hexacyanochromate(III) and tris(2,2'-bipyridine)chromium(III) in aprotic solvents. Inorg. Chem. **1987**, *26*, 2340–2342.

72. Vogler, A., Kunkely, H. Electrochemiluminescence of organometallics and other transition metal complexes. In: K.S. Suslick, ed. ACS Symp Ser No. 333. High Energy Processes in Organometallic Chemistry. Washington D.C.: Am. Chem. Soc. **1987**, pp 155–168.

73. Haapakka, K., Kankare, J., Kulmala, S. Feasibility of low-voltage cathodic electroluminescence at oxide-covered aluminum electrodes for trace metal determinations in aqueous solutions. Anal. Chim. Acta **1985**, *171*, 259–267.

74. Bruno, J.G., Cornette, J.C. An electrochemiluminescence assay based on the interaction of diaminotoluene isomers with gold(I) and copper(II) ions. Microchem. J. **1997**, *56*, 305–314.

75. McCall, J., Bruce, D., Workman, S., Cole, C., Richter, M.M. Electrochemiluminescence of copper(I) bis(2,9-dimethyl-1,10-phenanthroline). Anal. Chem. **2001**, *73*, 4617–4620.

76. Blaskie, M.W., McMillin, D.R. Photostudies of copper(I) systems. 6. Room-temperature emission and quenching studies of [Cu(dmp)$_2$]$^+$. Inorg. Chem **1980**, *19*, 3519–3522.

77. Ruthkosky, M., Castellano, F.N., Meyer, G.J. Photodriven electron and energy transfer from copper phenanthroline excited states. Inorg. Chem. **1996**, *35*, 6406–6412.

78. High, B., Bruce, D., Richter, M.M. Determining copper ions in water using electrochemiluminescence. Anal. Chim. Acta **2001**, *449*, 17–22.

79. Case, F.H. Substituted 1,10-phenanthrolines. I. The synthesis of certain mono- and polymethyl-1,10-phenanthrolines. J. Am. Chem. Soc., **1948**, *70*, 3994–3996.

80. Taverna, P.J., Mayfield, H., Andrews, A.R.J. Determination of cadmium ions by a novel electrochemiluminescence method. Anal. Chim. Acta **1998**, *373*, 111–117.

81. Whitchurch, C., Andrews, A.R.J. Ligand and surfactant effects on a novel electrochemiluminescent reaction involving cadmium. Anal. Chim. Acta **2002**, *454*, 45–51.

82. Cole, C., Muegge, B.D., Richter, M.M. The effects of Triton X-100 (polyethylene glycol tert-octylphenyl ether) on the tris(2-phenylpyridine)iridium(III)-tripropylamine electrochemiluminescence system. Anal. Chem. 2003, *75,* 601–604.

83. King, K.A., Spellane, P.J., Watts, R.J. Excited-state properties of a triply ortho-metalated iridium(III) complex. J. Am. Chem. Soc. **1985**, *107*, 1431–1432.

84. Baldo, M.A., Lamansky, S., Burrows, P.E., Thompson, M.E., Forrest, S.R. Very high-efficiency green organic light-emitting devices based on electrophosphorescence. Appl. Phys. Lett. **1999**, *75*(1), 4–6.

85. Grushin, V.V., Herron, N., LeCloux, D.D., Marshall, W.J., Petrov, V.A., Wang, Y.

New, efficient electroluminescent materials based on organometallic Ir complexes. Chem. Commun **2001**, 1494–1495.

86. Nishimura, K., Hamada, Y., Tsujioka, T., Shibata, K., Fuyuki, T. Solution electrochemiluminescent cell using tris(phenylpyridine)iridium. Jpn. J. Appl. Phys. **2001**, *40*, L945–L947.

87. Gross, E.M., Armstrong, N.R., Wightman, R.M. Electrogenerated chemiluminescence from phosphorescent molecules used in organic light-emitting diodes. J. Electrochem. Soc. **2002**, *149*(5), E137–E142.

88. Bruce, D., Richter, M.M. Green electrochemiluminescence from ortho-metalated tris(2-phenylpyridine)iridium(III). Anal. Chem. **2002**, *74*, 1340–1342.

89. Palecek, E. In: Topics in Bioelectrochemistry and Bioenergetics, Vol. 5. G. Milazzo, ed. Wiley, London, 1983, p. 65.

90. Cole, C., Muegge, B.D., Richter, M.M. The effects of Triton X-100 (polyethylene glycol tert-octylphenyl ether) on the tris(2-phenylpyridine)iridium(III)-tripropylamine electrochemiluminescence system. Anal. Chem. 2003, *75*, 601–604.

91. Rodman, G.S., Bard, A.J. Electrogenerated chemiluminescence. 52. Binuclear iridium(I) complexes. Inorg. Chem. **1990**, *29*, 4699–4702.

92. Bruno, J.G., Parker, J.E., Holwitt, E., Alls, J.L., Kiel, J.L. Preliminary electrochemiluminescence studies of metal ion-bacterial diazoluminomelanin (DALM) interactions. J. Biolum. Chemilum. **1998**, *13*, 117–123.

93. (a) Maverick, A.W., Gray, H.B. Luminescence and redox photochemistry of the molybdenum(II) cluster $Mo_6Cl_{14}^{2-}$. J. Am. Chem. Soc. **1981**, *103*, 1298–1300. (b) Maverick, A.W., Najdzionek, J.S., MacKenzie, D., Nocera, D.G., Gray, H.B. Spectroscopic, electrochemical, and photochemical properties of molybdenum(II) and tungsten(II) halide clusters. J. Am. Chem. Soc. **1983**, *105*, 1878–1882.

94. Nocera, D.G., Gray, H.B. Electrochemical reduction of molybdenum(II) and tungsten(II) halide cluster ions. Electrogenerated chemiluminescence of $Mo_6Cl_{14}^{2-}$. J. Am. Chem. Soc. **1984**, *106*, 824–825.

95. Mussell, R.D., Nocera, D.G. Electrogenerated chemiluminescence of $Mo_6Cl_{14}^{2-}$: free-energy effects on chemiluminescence reactivity. Polyhedron. **1986**, *5*(12), 47–50.

96. Mussell, R.D., Nocera, D.G. Effect of long-distance electron transfer on chemiluminescence efficiencies. J. Am. Chem. Soc. **1988**, *110*, 2764–2772.

97. Mussell, R.D., Nocera, D.G. Role of solvation in the electrogenerated chemiluminescence of hexanuclear molybdenum cluster ion. J. Phys. Chem. **1991**, *95*, 6919–6924.

98. Mussell, R.D., Nocera, D.G. Partitioning of the electrochemical excitation energy in the electrogenerated chemiluminescence of hexanuclear molybdenum and tungsten clusters. Inorg. Chem. **1990**, *29*, 3711–3717.

99. Ouyang, J., Zietlow, T.C., Hopkins, M.D., Fan, F-R.F., Gray, H.B., Bard, A.J. Electrochemistry and electrogenerated chemiluminescence of $Mo_2Cl_4(PMe_3)_4$. J. Phys. Chem. **1986**, *90*, 3841–3844.

100. Tokel-Takvoryan, N.E., Bard, A.J. Electrogenerated chemiluminescence. X. ECL of palladium and platinum $\alpha,\beta,\gamma,\delta$-tetraphenylporphyrin complexes. Chem. Phys. Lett. **1974**, *25*(2), 235–238.

101. Vogler, A., Kunkely, H. Electrochemiluminescence of tetrakis(diphosphonato) diplatinate(II). Angew. Chem. Int. Ed. Engl. **1984**, *23*(4), 316–317.

102. Kim, J., Fan, F.F., Bard, A.J., Che, C-M., Gray, H.B. Electrogenerated chemiluminescence. On the electrogenerated chemiluminescence (ECL) of tetrakis(pyrophosphito)diplatinate(II), $Pt_2(P_2O_5H_2)_4{}^{4-}$. Chem. Phys. Letts. **1985**, *121*(6), 543–546.

103. (a) Fordyce, W.A., Brummer, J.G., Crosby, G.A. Electronic spectroscopy of a diplatinum(II) octaphosphite complex. J. Am. Chem. Soc. **1981**, *103*(24), 7061–7064. (b) Che, C.M., Butler, L.G., Gray, H.B. Spectroscopic properties and redox chemistry of the phosphorescent excited state of octahydrotetrakis(phosphorus pentoxide)diplatinate(4–) ion $(Pt_2(P_2O_5)_4H_8{}^{4-})$. J. Am. Chem. Soc. **1981**, *103*(24), 7796–7797.

104. Ballardini, R., Indelli, M.T., Varani, G., Bignozzi, C.A., Scandola, F. Bis(8-quinolinolato)platinum(II): a novel complex exhibiting efficient, long-lived luminescence in fluid solution. Inorg. Chim. Acta. **1978**, *31*, L423–L424.

105. Bonafede, S., Ciano, M., Bolletta, F., Balzani, V., Chassot, L., von Zelewsky, A. Electrogeneated chemiluminescence of an ortho-metalated platinum(II) complex. J. Phys. Chem. **1986**, *90*, 3836–3841.

106. Kane-Maguire, N.A.P., Wright, L.L., Guckert, J.A., Tweet, W.S. Photoluminescence and electrogenerated chemiluminescence of palladium(0) and platinum(0) complexes of dibenzyl and tribenzylideneacetylacetone. Inorg. Chem. **1988**, *27*(17), 2905–2907.

107. Luong, J.C., Nadjo, L., Wrighton, M.S. Ground and excited state electron transfer processes involving *fac*-tricarbonylchloro(1,10-phenanthroline)rhenium(I). Electrogenerated chemiluminescence and electron transfer quenching of the lowest excited state. J. Am. Chem. Soc. **1978**, *100*(18), 5790–5795.

108. Vogler, A., Kunkely, H. Chemiluminescence of tricarbonyl(chloro)(1,10-phenanthroline)rhenium(I) during the catalytic decomposition of tetralinyl hydroperoxide. Angew. Chem. Int. Ed. Engl. **1981**, *20*, 469–470.

109. Richter, M.M., Debad, J.D., Striplin, D.R., Crosby, G.A., Bard, A.J. Electrogenerated chemiluminescence. 59. Rhenium complexes. Anal. Chem. **1996**, *68*(24), 4370–4376.

110. Jungru, A., Jinming, L. Determination of trace silver(I) in basic aqueous solution by electrochemiluminescence of a new reagent 6-(2-hydroxy-4-diethylaminophenylazo)-2,3-dihydro-1,4-phthalazine-1,4-dione. Chem. Res. Chin. Univ. **1991**, *7*(1), 32–36.

111. (a) Bhaumik, M.L., El-Sayed, M.A. Studies on the triplet-triplet energy transfer to rare earth chelates. J. Phys. Chem. *69*, 275–280. (b) Wildes, P.D., White, E.H. Dioxetane-sensitized chemiluminescence of lanthanide chelates. Chemical source of "monochromatic" light. J. Am. Chem Soc. **1971**, *93*, 6286–6288.

112. Kankare, J., Falden, K., Kulmala, S., Haapakka, K. Cathodically induced time-resolved lanthanide(III) electroluminescence at stationary aluminum disk electrodes. Anal. Chim. Acta. **1992**, *256*, 17–28.

8

Clinical and Biological Applications of ECL

Jeff D. Debad, Eli N. Glezer, Jacob Wohlstadter, and George B. Sigal
Meso Scale Discovery, Gaithersburg, Maryland, U.S.A.

Jonathan K. Leland
IGEN International, Gaithersburg, Maryland, U.S.A.

I. INTRODUCTION

The application of electrochemiluminescence (ECL) to the detection of biologically important analytes has seen substantial success, as measured by the widespread use of the technology in basic research, in environmental and industrial testing, and in clinical laboratory analyzers. Initial interest in the technique was spurred by the ability of ECL measurements to detect analytes in extremely low concentrations, though other favorable attributes such as insensitivity to matrix effects and a high dynamic range have contributed to ECL's acceptance as a robust detection methodology. The flexibility of the technique permits the probing of nearly any type of molecular interaction. ECL-based assays have been developed to detect proteins, nucleic acids, and small molecules and to probe ligand–receptor interactions and enzyme activity. The commercial availability of ECL reagents and instruments has also played a large part in fueling interest in the technique. Relatively inexpensive commercial instrumentation and ECL labels with linker groups that are easily attached to biomolecules have allowed researchers to quickly and easily develop numerous assay formats for a variety of applications. Fields that currently use ECL in the detection of biological analytes include clinical diagnostics, food and water testing, biowarfare agent detection, and basic research within both academia and industry.

Assays based on ECL share many attributes with other techniques used to conduct biological assays. A common approach to the detection of an analyte present at low concentrations is to link the analytes to a "label", a material that emits a strong, easily measured, signal. This signal is used to measure the label and thereby the analyte with which it is associated. Specificity for a particular analyte can be achieved by attaching the label to a reagent that binds specifically to the analyte, for example, a labeled antibody directed against the analyte. Examples of labels that are commonly used in biological assays include radioactive isotopes, fluorescent or chemiluminescent materials, and enzymes. ECL-based assays are unique in that they employ electrochemiluminescent species as labels and measure the labels by inducing and measuring their electrochemiluminescence. The availability of efficient systems for generating ECL has made possible the development of highly sensitive ECL assays.

The overwhelming majority of ECL-based assays and all commercial systems use a $Ru(bpy)_3^{2+}$ derivative as the electrochemiluminescent label. This label is induced to emit ECL at an oxidizing electrode in the presence of a tertiary amine such as tripropylamine (TPrA, the coreactant). The mechanism of $Ru(bpy)_3^{2+}$ ECL generation (Fig. 1) involves a high-energy electron transfer reaction between oxidation products of the ruthenium label and TPrA [1]. The energy of the reaction is sufficient to promote the label to a luminescent excited state. Emission of a photon regenerates the ground state, and as a consequence a single label may participate in multiple reaction cycles to produce multiple photons (see Chapter 5 for more details on ECL generation). The $Ru(bpy)_3^{2+}$/TPrA system is especially well suited for use in biological assay because it is highly efficient for generating ECL and performs optimally in aqueous solutions at neutral pH [1].

This chapter is divided into several sections. The instrumentation required to produce and measure ECL is presented in Section II, and a discussion on the formatting of ECL-based assays follows in Section III. This should provide the reader with a general understanding of how the technique can be applied to biological molecule detection and offers references for more detailed reading. Sections IV–VIII present a review of applications in various fields, including the use of ECL-based assays in clinical testing, life science research, environmental applications including food and water testing, and the detection of biowarfare agents.

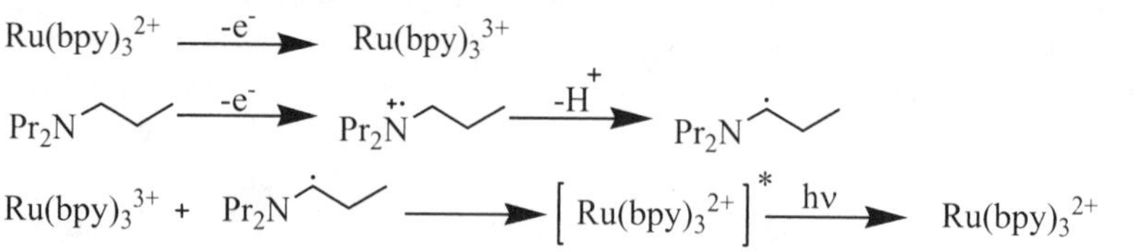

$$Ru(bpy)_3^{2+} \xrightarrow{-e^-} Ru(bpy)_3^{3+} \qquad (1)$$

$$Pr_2N\diagup\diagdown \xrightarrow{-e^-} Pr_2\overset{+\cdot}{N}\diagup\diagdown \xrightarrow{-H^+} Pr_2\overset{\cdot}{N}\diagup\diagdown \qquad (2)$$

$$Ru(bpy)_3^{3+} + Pr_2\overset{\cdot}{N}\diagup\diagdown \longrightarrow \left[Ru(bpy)_3^{2+}\right]^* \xrightarrow{h\nu} Ru(bpy)_3^{2+} \qquad (3)$$

Figure 1 Mechanism of $Ru(bpy)_3^{2+}$/TPrA ECL generation in aqueous solution.

II. INSTRUMENTATION

In its most basic form, an ECL measurement consists of supplying electrical energy for the ECL reaction and measuring the generated luminescence with an optical detector (Chapter 2). The reaction takes place in a solution that contains the reagents required for ECL and an electrolyte for conducting current between the working and counter electrodes. The generated luminescence is proportional to the amount of ECL label present on, or very near, the working electrode surface. Three types of measurements have been applied to biological assays: the label being measured is either free in solution, bound to a solid support such as microparticles, or bound directly to the electrode surface.

The basic equipment needed for ECL generation consists of an electrochemical cell and a light detector. ECL is typically detected with a photodetector such as a photomultiplier tube (PMT), a photodiode, or a charge-coupled device (CCD) that is positioned to collect light from the working electrode. In flow-based systems, part of the electrochemical cell is transparent so that the light generated at the electrode(s) can reach the detector. The electrodes and detector are contained within a lighttight housing to reduce background light.

Electrogenerated chemiluminescence instrumentation has been commercially available since 1994 and has been rapidly adopted by users in clinical, industrial, and research laboratories. More than 8000 ECL instruments have been placed worldwide, and the instrumentation has proved to be versatile and easily adapted to a wide range of applications. The first commercial instrument to employ ECL detection was the ORIGEN Analyzer [2] (IGEN International Inc.) (Fig. 2A) [3]. The ORIGEN Analyzer is adapted to measure ECL labels present on the surface of magnetically responsive beads. The beads are coated with binding reagents and are used as assay supports in solid-phase binding assays. As will be described in more detail in the section on assay formats, the assays are designed so that the amount of label on the beads is indicative of the amount of analyte in a sample.

The basic components of the instrument and other flow-based systems that use ORIGEN technology are shown in Figure 3. The essential element is the measurement module, which consists of a flow cell containing a reusable platinum electrode and a PMT for light detection. Typical assays use $Ru(bpy)_3^{2+}$ as the ECL label and are run offline in sample tubes that are manually loaded into the instrument's carousel. Each individual sample is drawn into the flow cell, where the paramagnetic beads are magnetically captured on the platinum electrode (Fig. 4). Highly sensitive assays can be realized by this procedure, because the labels bound to the beads are concentrated onto the electrode surface, where they are most efficiently induced to emit ECL. Any unbound labeled reagents remain dispersed in solution and can be washed away to reduce background

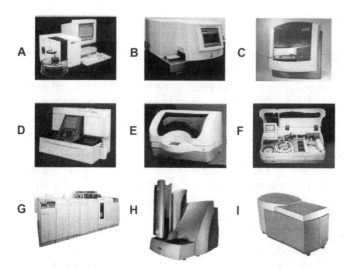

Figure 2 Commercial flow cell–based ECL instrumentation. (A) The ORIGEN 1.5, (B) M series M-384, and (C) M-1 analyzers by Igen International, (D) PicoLumi by Eisai, (E) Elecsys 1010, (F) Elecsys 2010, and (G) the MODULAR system containing the E-170 immunoassay module by Roche Diagnostics, (H) Sector HTS Imager and (I) Sector PR Reader by Meso Scale Discovery.

emission. A solution containing TPrA is then drawn through the flow cell to wash the beads and supply coreactant for ECL, which is initiated by applying a voltage between the working and counter electrodes. A PMT collects the light, and results are reported as light intensity for each sample. The beads are then washed from the cell, and a cleaning cycle is initiated before the next sample in the carousel is drawn. The instrument can also be used to detect solution-based ECL.

IGEN International recently developed a new generation of ECL-based instruments named the M-Series. These systems are also based on ORIGEN technology but are designed for higher throughput than the ORIGEN analyzer. The M-Series M-8 was the first to launch in 1999, followed by the M-384 in 2002 (Fig. 2B) and the M-1 in 2003 (Fig. 2C). These instruments are designed for a variety of applications in life sciences, high-throughput drug screening, and food, water, and animal health testing [4].

There are currently five other commercial flow-based ECL systems, and all use ORIGEN technology. Closely related to the ORIGEN analyzer is the NucliSens [5] reader and system (bioMérieux Inc.). This system was developed and marketed for the detection of nucleic acids; it is marketed along with the company's NASBA technology for the amplification of RNA. The assays detect amplified viral RNA using the $Ru(bpy)_3^{2+}$/TPrA ECL system and are used in clinical blood testing.

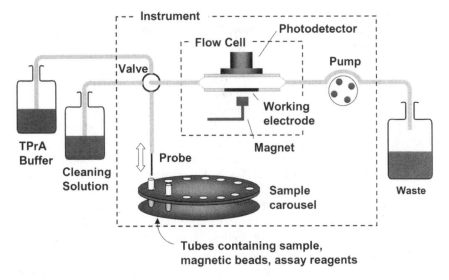

Figure 3 Components of a flow system based on ORIGEN technology.

Another instrument, the PicoLumi (Eisai Company), (Fig. 2D), is a fully automated immunoassay reader marketed in Japan. The system is also based on ORIGEN technology and uses PicoLumi series test kits with a test menu featuring primarily tumor marker assays.

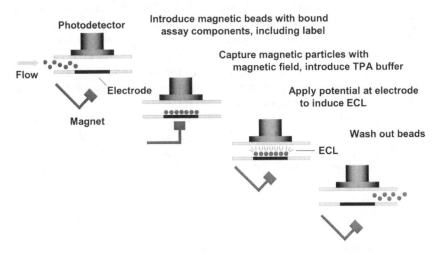

Figure 4 Generic process for measuring magnetic bead–based ECL assays in flow cells.

Roche Diagnostics markets three analyzers based on ORIGEN technology for clinical immunoassay testing. The Elecsys [6] 1010 is a routine and short-turnaround-time (STAT) analyzer for small- to medium-volume laboratories (Fig. 2E), and the Elecsys 2010 is a continuous random access analyzer for medium- to large-volume laboratories (Fig. 2F). Patient samples are loaded in vacutainer tubes, and the entire assay process is automated. These systems feature an expansive menu of over 50 immunoassay tests in the areas of cardiology, fertility, thyroid function, oncology, anemia, infectious disease, and osteoporosis. The introduction of the Elecsys systems has been one of the most successful launches in the clinical marketplace, with more than 7000 Elecsys systems in clinical laboratories worldwide. Roche Diagnostics recently released a new clinical product based on ECL detection: the E170 module for their high-volume MODULAR *ANALYTICS* clinical analyzer (Fig. 2G). This modular system incorporates a variety of clinical diagnostic systems in addition to immunodiagnostics and is capable of running up to 170 ECL-based immunoassay tests per hour with the same test menu as the Elecsys products.

The ORIGEN flow cell technology is amenable to miniaturization using microfabrication techniques. There has been a report of a microcell for DNA quantification using ECL detection, although this system has not been commercialized [7]. The microfabricated cell employs a built-in silicon diode detector and thin-film platinum electrodes and uses paramagnetic bead capture with the $Ru(bpy)_3^{2+}$/TPrA ECL system.

The other major class of instrumentation for conducting ECL assays is designed to measure labels bound directly to the surface of an electrode. In contrast to bead-based assays, the electrode surface itself is used as an assay support in solid-phase binding assays. In this case a binding reagent is immobilized directly on the electrode surface, and the ECL label is bound at the electrode surface during the assay. This approach enables an array or microarray format, where multiple binding sites are patterned on the electrode surface, and multiple measurements are performed in parallel using the same sample. This class of instrumentation typically uses disposable electrodes that are used for only one electrochemiluminescence measurement.

The Sector HTS Imager and Sector PR [8] Reader instruments (Meso Scale Discovery, a division of Meso Scale Diagnostics, LLC) are the first commercial ECL instruments for measuring ECL from solid-phase assays carried out on disposable electrodes (Figs. 2H and 2I). Assays are carried out on the surfaces of screen-printed carbon ink electrodes within the wells of multiwell plates called Multi-Array plates [8,9]. A variety of plate formats are available to suit the needs of drug discovery and biological research, including 96-, 384-, and 1536-well formats. Unique Multi-Spot [8] plates with patterned microarrays within each plate well are also available, enabling multiple assays to be performed in each well of the plates (Fig. 5). The instruments are targeted for use in high-throughput screening, where large numbers of chemical compounds are tested for their effects on biological sys-

Figure 5 Pictures of MSD Multi-Spot plates with seven binding domains in each well of the 96-well plates.

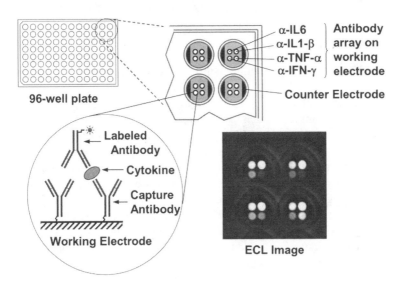

Figure 6 Schematic diagram of a Multi-Spot plate assay for four human cytokines. Each spot within each well of the 96-well plate contains capture antibody specific for one cytokine. Inset shows an image of the ECL emitted from assays in four wells using samples with varying concentrations of cytokines.

tems in the search for new drugs. Other uses include large-scale proteomics/genomics studies, multianalyte assays, and general laboratory use including assay development and low- to medium-throughput biological measurements. ECL is generated using $Ru(bpy)_3^{2+}$ analogs and TPrA or similar coreactants, and the light is collected with either a CCD camera (Sector HTS Imager) or a series of photodiodes (Sector PR Reader). An example of multiple simultaneous assays is shown in Figure 6, where immunoassays were used to simultaneously detect four human cytokines on the Sector HTS instrument using a Multi-Spot plate with four binding regions [9]. The instrument converts light intensity from each capture zone to a numerical value.

III. ASSAY FORMATS

The ECL instruments described in the preceding section measure ECL labels bound to a solid-phase support in proximity to an electrode. In the case of ORIGEN instrumentation, the label is bound to magnetic beads that are collected on the surface of the electrode. In the case of Multi-Array instrumentation, label is bound to the electrode itself. Due to the short lifetime of the electrogenerated species involved in the ECL process [10], only labels that are in close proximity to the electrode surface produce ECL. As a result of this surface selectivity, ECL labels bound to the solid phase can be sensitively and selectively measured in the presence of labels in solution, enabling highly sensitive assays with very few or no wash steps.

The use of solid-phase supports allows assay formats employed in other solid-phase assay techniques, such as enzyme-linked immunosorbent assay (ELISA) and radioimmunoassay (RIA), to be adapted easily to ECL-based instrumentation. The common feature of solid-phase assay measurements is the correlation of the analyte to be measured to the amount of label bound to the solid phase. In perhaps the simplest example, the direct binding assay, the binding of a labeled reagent to an immobilized reagent is determined by measuring the accumulation of the label on the solid phase. This format is particularly useful for studying receptor–ligand interactions in biology (Fig. 7A).

Binding interactions of unlabeled species can be detected by using slightly more complicated formats such as the sandwich binding assay or the competitive binding assay. These formats are most commonly used in immunoassays for clinical analytes. The sandwich immunoassay format employs two antibodies directed against the target molecule of interest; one antibody is immobilized on a solid phase, the other carries an ECL label (Fig. 7B). Binding of both antibodies to the target results in the attachment of the ECL label to the solid phase, where it can be subsequently measured, thus correlating the amount of label on the solid phase to the amount of target analyte in the sample. The competitive immunoassay employs an antibody and an analog of the analyte of interest, one immobilized on a solid phase and the other labeled with an ECL reporter. Analyte present in a sample competes with the analog for binding to the antibody and decreases the accumulation of labels on the solid phase (Fig. 7C).

Similar techniques have been employed in nucleic acid assays. Nucleic acid amplification products have been measured using a sandwich hybridization

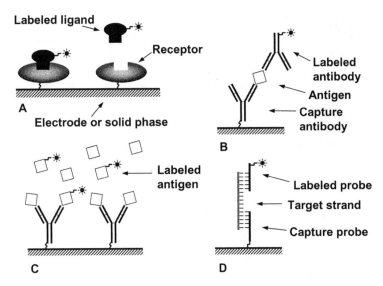

Figure 7 Examples of some binding assay formats using ECL detection. (A) A simple receptor–ligand assay using a labeled ligand; (B) a sandwich immunoassay; (C) a competition assay where analyte competes with labeled analyte for antibody-binding sites on the solid support; (D) capture of an oligonucleotide using a capture and label sequence specific to the target sequence.

format employing a labeled probe and an immobilized probe, each specific for a different sequence in the amplification product (Fig. 7D). Amplification products have also been measured by using direct binding assays in which labels are incorporated into the amplification products during amplification using labeled primers or nucleotides. Another alternative for the detection of double-stranded DNA bound near the electrode is to intercalate an ECL tag into the DNA, although this method has not yet found commercial applications [11,12].

Solid-phase ECL assays have also been used to measure the activity of enzymes. Often these employ binding assays, as described above, for detecting the products of the enzymatic reaction. For example, labeled antibodies specific against phosphotyrosine have been used to detect the tyrosine kinase–induced phosphorylation of amino acid–containing substrates immobilized on magnetic beads [4] or carbon electrodes [13] (Fig. 8A). Alternatively, the activity of an enzyme can be directly coupled to the accumulation or loss of labels from a solid phase. The activity of proteases can be determined by measuring the release of labels tethered to a solid phase via a peptide or protein linker (Fig. 8B). Conversely, the ligation activity of HIV integrase can be determined by measuring the ligation of labeled oligonucleotides to an immobilized oligonucleotide [14] (Fig. 8C).

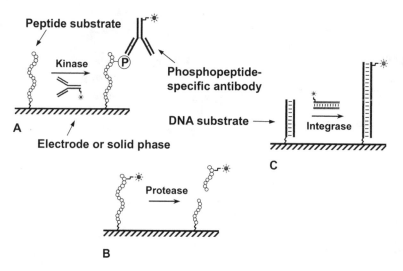

Figure 8 Examples of ECL assays using enzymes. (A) An assay for kinase activity using a labeled antibody to recognize the phosphorylated immobilized product. (B) A protease activity assay in which cleavage of the immobilized substrate removes the tag from the solid support. (C) An assay using immobilized and free labeled DNA strands to test for integrase activity.

Many of the above assay formats can be used in the search for disease therapies. For example, inhibition of HIV integrase by an added compound, signaled by a decrease in ECL in the above assay, identifies the compound as a potential antiviral drug candidate. Such studies are usually performed by screening large libraries of chemical and biological compounds for their effects on such model assays, aided by high-throughput ECL instrumentation that has recently become available. Follow-up studies on the compounds that affect the model system would determine the feasibility of the compound for use as a drug. This area represents a rapidly growing area in ECL applications, and further examples are discussed in Section V.

The development of many ECL assay formats has been aided by the wide variety of Ru(bpy)$_3^{2+}$ derivatives that are available for linking to biological molecules. Figure 9 shows some of the Ru(bpy)$_3^{2+}$ derivatives that have been developed. Proteins are most commonly labeled via reaction of lysine amino groups with Ru(bpy)$_3^{2+}$ NHS ester derivatives. Derivatives presenting maleimide groups, a functional group that reacts with thiols, have proved especially useful in labeling peptides and nucleic acids prepared by solid-phase synthesis because of the ability of these techniques to introduce thiols at selected locations in the oligomer chains [15]. Nucleic acids have also been labeled at the 5′ terminus using a phosphoramidite-containing derivative, a reagent that allows probes to be modi-

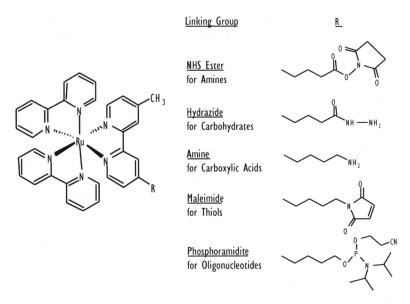

Figure 9 Functionalized derivatives of Ru(bpy)$_3{}^{2+}$.

fied with Ru(bpy)$_3{}^{2+}$ during automated synthesis [16]. Nucleic acids produced by enzymatic replication of nucleic acid templates have been labeled through the enzymatic incorporation of labeled nucleotides such as Ru(bpy)$_3{}^{2+}$-labeled dUTP [13].

A variety of approaches have been developed for immobilizing biological molecules on solid phases for use in ECL assays. Magnetic particles suitable for use in ORIGEN assays are available that have hydrophobic surfaces or charged surfaces suitable for adsorption of proteins and other macromolecules. Alternatively, reagents can be covalently bound to magnetic particles through the use of particles whose surfaces are coated with reactive species. To avoid optimizing the immobilization of different reagents for each assay, most ORIGEN assays use streptavidin-coated magnetic particles. Biotin-labeled reagents such as capture antibodies or nucleic acid probes are easily bound to the surface of these generic particles through streptavidin–biotin interactions. Similarly, biological molecules can be immobilized on carbon ink electrodes for use in Multi-Array assays via direct adsorption of the reagents on the electrodes or through the use of an adsorbed capture reagent such as streptavidin or avidin [9].

Although not commercially important at this time, approaches have been developed for carrying out ECL binding assays in homogeneous assay formats, i.e., assay formats that do not employ reagents immobilized on a solid phase. The binding in solution of small labeled antigens to antibodies has been shown to decrease the ECL signal at an electrode in contact with the antigen solution [17]. The decrease in ECL as a result of the binding event is most likely a result of the decrease in the rate of diffusion of the label to the electrode and a decrease in the accessibility of the label to the electrode. The effect can be amplified by coating the antibodies on particles so as to increase the effective size of the antibodies.

In the assays discussed above, the concentration of the analyte is determined by measuring the ECL emission of a label, usually $Ru(bpy)_3^{2+}$, in the presence of a high concentration of coreactant, typically TPrA. There are some assays for biological molecules or function that instead rely on the detection of coreactants in the presence of a constant $Ru(bpy)_3^{2+}$ concentration. A method for the detection of double-stranded DNA has been demonstrated in this manner, in which an intercalator acts as a coreactant for ECL generation with $Ru(bpy)_3^{2+}$ in solution [18]. Most other ECL assays for coreactants are based on the production or consumption of a coreactant by an enzyme when analyte is present, an example being the detection of the coreactant NADH formed by glucose dehydrogenase in the presence of glucose. More examples are provided in the following section.

IV. DETECTION OF CLINICALLY IMPORTANT ANALYTES

The largest area of biological molecule detection in which ECL has been applied is the clinical setting. Clinical assays are defined as methods for measuring constituents present within the human body. Examples of potential analytes include hormones, proteins secreted by cancer cells, enzymes, viral or bacterial proteins or toxins, and antibodies generated through immune responses. The presence or absence or the concentration of these constituents provides clinicians with information on the functioning of the body or the presence of specific disease states. With many clinical assays, it is common that the sample specimen is blood or one of the blood derivatives serum or plasma. Yet there are many clinical analytes not found in blood, making it necessary to perform clinical tests in a wide variety of sample specimens including urine, feces, mucus, tissue, and spinal chord fluid. Thus the clinical laboratory is challenged by a great number of analytes and specimen types spanning many physiological conditions and diseases. These labs are also burdened by ever-increasing demands for shorter turnaround times and lower costs and by strict requirements for very accurate results with low error rates.

Numerous detection methods are available for clinical assays; however, the fundamental attributes of the ECL detection method complement the rigorous demands of a clinical assay. ECL methods are very accurate, insensitive to matrix affects, and available in a variety of assay formats; allow the detection of analytes in very low concentrations; and exhibit a broad linear detection range to cover those analytes present in higher concentrations. Thus, ECL-based detection methods are capable of measuring a diverse set of analytes in many matrices.

Clinical applications of ECL have used two general methods in which either the label or coreactant concentration is related to analyte concentration [19]; however, the vast majority use a format in which the label concentration is measured. In fact, all commercially available ECL instruments use a Ru(bpy)$_3^{2+}$ derivative as the label and detect ECL in a solution with excess TPrA. This method couples very well to affinity binding reactions such as antibody/antigen or nucleic acid/complementary nucleic acid formats [20–22]. The ECL label is bound to one of the species participating in the affinity binding reaction, and the coreactant is present in vast excess over the other assay components or potentially interfering species.

Clinical assays generally require a calibration curve, typically generated by running a blank and a known concentration of the analyte of interest in the sample matrix. Establishment of the blank or zero analyte level is a critical aspect of the calibration and one for which ECL is well- suited. In the absence of an ECL label (and therefore zero analyte) there is a low-level emission fundamental to the ECL mechanism. In commercial ECL instrumentation used for clinical applications, this background emission is very reproducible and is independent of the sample matrix. Biological substances do not contribute to or alter the background emission, and thus the steady blank emission level provides a quiet background to detect analytes at very low levels.

The list of ECL-based assays developed for clinical analytes is impressive and constantly growing (Tables 1 and 2). The large number of analytes precludes a discussion of each, and the reader is directed to the referenced work or specific company websites and literature for further details.

Tumor markers make up a large portion of ECL assays for clinically important analytes. These include α-fetoprotein (AFP) [23,24], carcinoembryonic antigen (CEA) [24,25], cytokeratin 19 [26], des-γ-carboxy prothrombin [27,28], prostate specific antigen (PSA) [25,29–31], and also tests for mRNA coding for various cancer markers [32–35]. These tests can be used to diagnose cancer, to monitor the effectiveness of drug treatments or surgeries, and also to monitor for recurrence. In some cases, such as PSA testing of older men, the assays have been used in a screening fashion to detect cancer early, when treatment is typically most effective.

A number of fertility and related tests are also available. Estradiol, follitropin (FSH), prolactin, progesterone, luteinizing hormone (LH) [36], testosterone [37], and human chorionic gonadotropin (hCG) [25,38] levels can be used to judge and monitor fertility and to diagnose the source of problems associated with low fertility.

Thyroid function is also well represented among clinical ECL assays, reflecting the significant prevalence of thyroid diseases among the population.

Table 1 Summary of Clinical Assays that Use ECL Detection

Analyte	Application	Detection method	Ref.
I. Assays for ECL Labels			
A. Immunoassays			
AFP, α-fetoprotein	Tumor, fertility	TPrA, Ru(bpy)$_3^{2+}$	23, 24
Anti-Borna disease antibodies	Infectious disease	TPrA, Ru(bpy)$_3^{2+}$	71–73
β-Amyloid peptide	Alzheimer's disease	TPrA, Ru(bpy)$_3^{2+}$	74
CA (cancer antigen) 15–3	Tumor marker	TPrA, Ru(bpy)$_3^{2+}$	24, 75
CA (cancer antigen) 19–9	Tumor marker	TPrA, Ru(bpy)$_3^{2+}$	24
CA (cancer antigen) 72–4	Tumor marker	TPrA, Ru(bpy)$_3^{2+}$	76
CA (cancer antigen) 125 II	Tumor marker	TPrA, Ru(bpy)$_3^{2+}$	77, 78
CEA, carcinoembryonic antigen	Tumor marker	TPrA, Ru(bpy)$_3^{2+}$	24, 25
β-Crosslaps	Bone	TPrA, Ru(bpy)$_3^{2+}$	79, 80
C-Telopeptides	Bone	TPrA, Ru(bpy)$_3^{2+}$	81
Anti-CMV antibodies	Infectious disease	TPrA, Ru(bpy)$_3^{2+}$	56
Cytokeratin 19	Tumor marker	TPrA, Ru(bpy)$_3^{2+}$	26
CKMB, creatine kinase	Cardiac	TPrA, Ru(bpy)$_3^{2+}$	25, 82
Cytomegalovirus	Infectious disease	TPrA, Ru(bpy)$_3^{2+}$	83
Des-γ-carboxy prothrombin	Tumor marker	TPrA, Ru(bpy)$_3^{2+}$	27, 28
Estradiol	Reproductive endocrinology	TPrA, Ru(bpy)$_3^{2+}$	36
Ferritin	Anemia	TPrA, Ru(bpy)$_3^{2+}$	24
FSH, follitropin	Reproductive endocrinology	TPrA, Ru(bpy)$_3^{2+}$	36
HCG, human chorionic gonadotropin	Reproductive endocrinology	TPrA, Ru(bpy)$_3^{2+}$	25, 38
HBsAg, hepatitis B virus surface antigen	Infectious disease	TPrA, Ru(bpy)$_3^{2+}$	49–51, 84
HAsAg, hepatitis A virus surface antigen	Infectious disease	TPrA, Ru(bpy)$_3^{2+}$	44
HIV-1 p7 antigen	Infectious disease	TPrA, Ru(bpy)$_3^{2+}$	55
IgE, immunoglobulin E	Allergy	TPrA, Ru(bpy)$_3^{2+}$	85
Insulin	Diabetes mellitus	TPrA, Ru(bpy)$_3^{2+}$	86, 87
IL(Interleukin)-18 binding protein	Sepsis	TPrA, Ru(bpy)$_3^{2+}$	88
IL(Interleukin)-2	Immune system	TPrA, Ru(bpy)$_3^{2+}$	89
IL(Interleukin)-4	Immune system	TPrA, Ru(bpy)$_3^{2+}$	89
IL(Interleukin)-6	Immune system	TPrA, Ru(bpy)$_3^{2+}$	24
IL(Interleukin)-8	Immune system	TPrA, Ru(bpy)$_3^{2+}$	24
IL(Interleukin)-10	Immune system	TPrA, Ru(bpy)$_3^{2+}$	89, 90
IL(Interleukin)-10	Immune system	TPrA, Ru(bpy)$_3^{2+}$	88
Interferon-γ	Immune system	TPrA, Ru(bpy)$_3^{2+}$	89

(continued)

Table 1 Continued

Analyte	Application	Detection method	Ref.
LH, lutropin	Reproductive endocrinology	TPrA, Ru(bpy)$_3^{2+}$	36
Osteocalcin	Bone	TPrA, Ru(bpy)$_3^{2+}$	81
Pancreatic phospholipase A2	Pancreatic diseases	Terbium chelate	91
PTH, parathyroid hormone	Ca metabolism	TPrA, Ru(bpy)$_3^{2+}$	92
Prolactin	Reproductive endocrinology	TPrA, Ru(bpy)$_3^{2+}$	36
PSA, prostate specific antigen	Tumor marker	TPrA, Ru(bpy)$_3^{2+}$	25, 29–31
Serum interferon α	Immune system	TPrA, Ru(bpy)$_3^{2+}$	88
T4, thyroxine	Thyroid function	TPrA, Ru(bpy)$_3^{2+}$	25, 40, 41, 43
T3, triiodothyronine	Thyroid function	TPrA, Ru(bpy)$_3^{2+}$	25, 39, 41
TSH, thyroid stimulating hormone	Thyroid function	TPrA, Ru(bpy)$_3^{2+}$	25, 39, 41
TSH	Thyroid function	Terbium chelate	42
Testosterone	Reproductive endocrinology	TPrA, Ru(bpy)$_3^{2+}$	36, 37
Troponin	Myocardial Infarction	TPrA, Ru(bpy)$_3^{2+}$	25, 82, 93–95
Tumor necrosis factor-α	Immune function	TPrA, Ru(bpy)$_3^{2+}$	96
B. Molecular Assays			
Apo 8–100 gene mutation	Metabolism	TPrA, Ru(bpy)$_3^{2+}$	47
Astrovirus	Infectious disease	TPrA, Ru(bpy)$_3^{2+}$	97
Coxsackievirus B3 RNA	Infectious disease	TPrA, Ru(bpy)$_3^{2+}$	65
CMV DNA	Infectious disease	TPrA, Ru(bpy)$_3^{2+}$	57
CMV mRNA	Infectious disease	TPrA, Ru(bpy)$_3^{2+}$	5
Dengue virus RNA	Infectious disease	TPrA, Ru(bpy)$_3^{2+}$	60
Enterovirus	Infectious disease	TPrA, Ru(bpy)$_3^{2+}$	62
Epstein–Barr virus DNA	Infectious disease	TPrA, Ru(bpy)$_3^{2+}$	58
Foot-and-mouth disease	Infectious disease	TPrA, Ru(bpy)$_3^{2+}$	98
HIV-1 RNA	Infectious disease	TPrA, Ru(bpy)$_3^{2+}$	48, 99
HIV DNA	Infectious disease	TPrA, Ru(bpy)$_3^{2+}$	45, 52–54
HPIV1/2/3	Infectious disease	TPrA, Ru(bpy)$_3^{2+}$	100
Influenza virus RNA	Infectious disease	TPrA, Ru(bpy)$_3^{2+}$	63, 64, 101
mRNA	Tumor marker	TPrA, Ru(bpy)$_3^{2+}$	32–35
Prothrombin gene mutation	Venous thromboembolism	TPrA, Ru(bpy)$_3^{2+}$	102

(*continued*)

Table 1 Continued

Analyte	Application	Detection method	Ref.
St. Louis encephalitis	Infectious disease	TPrA, $Ru(bpy)_3^{2+}$	61
Varicella-zoster virus DNA	Infectious disease	TPrA, $Ru(bpy)_3^{2+}$	59
West Nile virus RNA	Infectious disease	TPrA, $Ru(bpy)_3^{2+}$	61
$\Delta F508$ deletion	Cystic fibrosis	TPrA, $Ru(bpy)_3^{2+}$	103
C. Other			
Heavy metals	Toxicology	TPrA, $Ru(bpy)_3^{2+}$	69
Potassium	Electrolytes	TPrA, $Ru(bpy)_3^{2+}$	69
Sodium	Electrolytes	TPrA, $Ru(bpy)_3^{2+}$	68
II. Assays for Coreactants			
Carbon dioxide	Blood gases	NADH,$Ru(bpy)_3^{2+}$	67
Cholesterol	Lipids	oxalate, $Ru(bpy)_3^{2+}$	67
Ethanol	Toxicology	NADH, $Ru(bpy)_3^{2+}$	67, 104
Glucose	Diabetes mellitus	NADH, $Ru(bpy)_3^{2+}$	67, 104
Lactate	Exercise	NADH, $Ru(bpy)_3^{2+}$	104
β-Lactamase	Microbiology	Hydrolyzed β-lactams, $Ru(bpy)_3^{2+}$	

The accepted clinical practice is for all thyroid diagnoses to be confirmed by assays for thyroid markers present in the blood. The series of available tests represent a good example of the flexibility of the ECL technique. Thyroid stimulating hormone (TSH) [39–42], thyroxine (T4) [25,40,41,43], and tri-iodothyronine (T3) [25,39,41] are used to determine whether the thyroid gland is functioning properly and to diagnose possible ailments of the glands. Below-normal concentrations of TSH strongly indicate hypothyroidism—a medical condition affecting millions of individuals. A competent TSH assay must be able to quantify extremely low levels of the hormone, as low as 0.01 μIU/mL, or 50 fM. The ECL-based assay developed for the Elecsys 2010 analyzer is a two-antibody sandwich assay and reports an impressive detection limit of 0.005 μIU/mL, or 25 fM. The T4 assay is generally used with a TSH assay to support the diagnosis in the case of an abnormal TSH result. The molecular weight of free T4 (fT4) is relatively low, which precludes the two-site sandwich assay format. To overcome this problem, the Elecsys ECL-based assay for fT4 is formatted as a direct, one-step competition assay and reports a detection limit of 300 fM. The Elecsys 2010 has a large panel of thyroid markers—TSH, T4, T3, fT4, fT3, T-uptake, thyroglobulin, anti-TPO, and anti-Tg. Other immunoassays available on the Roche Diagnostics instruments are also listed in Table 2 [44]. The menu is very broad, covering a number of disease states.

Table 2 Immunoassays Available on Roche Diagnostics Instruments, the Elecsys 1010 and 2010 and the E-170

Fertility and hormones	Thyroid function	Tumor markers
LH	TSH	AFP
FSH	T3	CEA
Prolactin	T4	Total PSA
Estradiol	Free T3	Free PSA
Progesterone	Free T4	CA 125 II
Testosterone	T-Uptake	CA 15–3
HCG	Thyroglobulin	CA 19–9
Dehydroepiandrosterone sulfate	Anti thyroglobulin	CA 72–4
Cortisol	Anti-thyroid peroxidase	CYFRA 21–1 NSE

Cardiac	Hepatitis	Bone markers
Troponin T	HBsAg	PTH
CKMB	Anti-HBs	β-Crosslaps
Myoglobin	Anti-HBc	Osteocalcin
ProBNP	Anti-HBc IgM	
Digoxin	Anti-HBe	
Digitoxin	Anti-HAV, anti-HAV IgM	

Anemia	Diabetes	Other
Ferritin	Insulin	IgE
Vitamin B_{12}		
Serum folate		
RBC folate		

The use of molecular assays to detect disease states and infectious agents is also an important area of clinical diagnostics. ECL has been successfully employed in a variety of molecular assays in which specific nucleic acid sequences in the sample are hybridized to capture and label DNA or RNA probes [20,45,46], and can be coupled with amplification schemes such as PCR [47] and NASBA [5,48]. Only two molecular assays are commercially available at this time, for application on the NucliSens reader (BioMerieux). They are for the detection and quantification of human cytomegalovirus and HIV-1 mRNA in blood [5,48]. The assays use a $Ru(bpy)_3^{2+}$ label and TPrA coreactant for detection of NASBA RNA amplicons. These are just two of the large number of assays developed for infectious diseases that use ECL detection methods. The only other disease tests cleared by the FDA at this time, however, are a number of immunoassays for the diagnosis of hepatitis A and B infection available for the Roche instruments (Table 2)

[49–51]. The majority of assays for other infectious disease analytes were developed by and for the research community both for the identification of infection and as a research tool for investigating disease models and monitoring therapies. A number of ECL assays have been developed for HIV detection based on molecular probe assays [52–54] and immunoassays [55]. Anticytomegalovirus antibodies have been detected in an ECL assay for monitoring infection and to help in the development of vaccines [56], and direct detection of CMV DNA in serum and plasma is also possible using an ECL probe assay [57].

Molecular probe assays have also been developed for a number of infectious diseases in human samples, including Epstein–Barr virus [58], varicella-zoster virus [59], and dengue virus [60]. Such rapid assays are required for diagnosing disease and for guiding clinical care during recovery. ECL-based NASBA assays have been developed for both West Nile and St. Louis encephalitis viruses and have been used to detect viral RNA from mosquitoes, avian tissues, and human CSF samples [61]. A NASBA assay was also developed for the detection of human enteroviruses in a variety of clinical sample types on the NucliSens reader [62]. H5 [63] and H7 [64] strains of influenza virus and coxsackievirus B3 [65] have been detected in tissue culture and may be useful for patient testing in the near future.

The high sensitivity of the ECL detection method is allowing researchers to identify new disease markers. For example, using the ORIGEN technology, researchers have identified colon cancer cells not normally detected by conventional diagnostic methods [32], and others have shown that elevated serum levels of laminin γ2-chain fragment could indicate the presence of tumor cells [66].

Detection of coreactants by ECL works well for analytes that can be coupled to the enzymatic production of species, such as NADH, that can act as a coreactant [67], or through the specific binding of a coreactant species on or near the electrode surface [18]. Biosensors using this method have been developed for many classical or routine analytes such as glucose and cholesterol [67]. ECL detection has also been used to measure certain electrolytes such as sodium and potassium [68,69]. Assays for the presence of β-lactamase have also been developed based on ECL for the potential detection of bacteria resistant to β-lactam antibiotics [70]. In this case, β-lactam is a poor coreactant, whereas the ring-opened product, formed in the presence of β-lactamase-producing bacteria, can generate ECL in the presence of $Ru(bpy)_3^{2+}$.

V. APPLICATIONS IN LIFE SCIENCES RESEARCH

For more than a decade, ECL has been used in research laboratories to study basic physiological mechanisms and discover the causes and potential cures for disease. A variety of assay formats have been devised, including standard immunoassays to quantify cellular signaling molecules and to study drug metabolism and more elaborate assays such as those to probe enzyme function, ligand–receptor interactions, and protein–DNA binding. In recent years, performing these assays in very high volume and in parallel has become increasingly desirable. High throughput enables screening of compound libraries for drug candidates, and parallel assays such as those that use arrays add extra efficiency by allowing researchers to probe multiple targets simultaneously.

Sandwich immunoassays based on ECL have been used to detect numerous cytokines in serum for clinical applications (see Table 1). Similar assays have been used in pharmaceutical research to detect cytokine responses in cell culture supernatant [105,106], serum [107], and blood [108]. The simultaneous detection of four cytokines from the same sample has been demonstrated using an array of four detection antibodies on a platform designed for drug screening applications (see Fig. 4) [9]. Nucleic acid amplification with ECL detection has been used to identify mRNA coding for cytokines [109].

Assay formats based on ECL have also been developed for the direct measurement of important biological interactions such as receptor–ligand binding and other cellular signaling systems. An example of a direct receptor–ligand binding application is the search for inhibitors of granulocyte colony-stimulating factor (GCSF) using a recombinant human receptor [110]. In this assay, immobilized receptor is incubated with test compounds and labeled GCSF: A decrease in ECL indicates an inhibition of the binding event by the test compound. Researchers have also investigated the interaction of $\beta 1$ integrins with their physiologically relevant ligands by incubating integrins in cell lysates with immobilized ligands and a labeled antibody specific for the integrins [111]. Further examples that demonstrate the flexibility of the ECL technique are available: protein–protein interactions have been probed to study the clearance of amyloid protein [112], an endosome fusion assay was developed to study the mechanisms of membrane fusion [113], and protein–DNA interactions were measured to study heat shock factors [114].

Enzymatic activities have also been investigated using ECL-based measurements. An assay for inhibitors of DNA helicases has been reported that is adaptable for high-throughput screening of compound libraries in the search for antiviral drugs [115]. Assays for other DNA-binding proteins and proteases that affect these proteins are described in a recent patent application [116]. Immunoassays for amyloid peptides have been developed for measuring γ-secretase activity to help understand the mechanism of amyloid peptide formation in Alzheimer's disease [117]. Kinases, enzymes that phosphorylate proteins, are important in the regulation and signaling of many cellular processes and are therefore prime targets for drug development. Kinase activity can be measured by detecting the products of the enzymes: ECL-based assays are formatted by immobilizing the enzyme substrate on a solid support such as an electrode or bead and exposing this substrate to the enzyme (Fig. 7A). Labeled antiphospho-peptide antibodies are added that bind the phosphorylated substrate to afford an ECL signal [4].

Igen International and Meso Scale Discovery have developed and market numerous assays for life science research and drug discovery and have presented these findings at various conferences and industry gatherings [13,14]. Enzyme-based examples include screening assays for HIV reverse transcriptase and integrase, an assay for protease activity using labeled immobilized substrates, high-throughput screening assays for kinase activity, and a number of systems to probe ubiquitylation. Immunoassays have been demonstrated for measuring cyclic-AMP and cyclic-GMP in cell lysates and for detecting various cytokines. Assays are available for probing ligand binding and inhibition using immobilized receptors, some using whole cells or membrane fragments immobilized on the solid support. Epidermal growth factor receptor regulation and phosphorylation of the receptor have also been studied. Other applications include assays to speed antibody development and proteomic applications to search for apoptotic genes and ubiquitylation substrates.

VI. WATER TESTING

Electrogenerated chemiluminescence has also been applied to environmental analysis, including tests for parasites, bacteria, and toxic compounds in environ-mental water sources and drinking water (Table 3). The formats for these assays vary depending on the analyte; however, all reported assays use $Ru(bpy)_3^{2+}$ as the label and TPrA as the coreactant.

Cryptosporidium parvum is a protozoan parasite ubiquitous in water that, if ingested in sufficient amounts, can cause gastrointestinal disease. The infec-tious dose is low, so very sensitive assay methods are required. Immunoassays using an ECL label have been reported for the *Cryptosporidium parvum* oocyst, the form of the parasite typically found in water. Using a sandwich immunoas-say on magnetic beads with ECL detection, oocysts were directly detected in highly turbid water and sewer samples [118]. ECL-based immunoassays that detect only viable organisms have been reported: One detects antigen released upon mechanical disruption of the live oocyst [119], and the other detects mRNA released by the viable oocysts upon heating the sample [120]. These assays have potential use in monitoring the success of water treatment.

Escherichia coli and *Salmonella* sp. can also cause serious enteric disease and are possible contaminants of drinking water. Both *E. coli* O157 and *Salmonella*

Table 3 ECL-Based Assays for Food, Water, and Biological Threat Agents

Analyte	Sample matrix	Assay method[a]	Detection limit	Ref.
Cryptosporidium parvum oocysts	Water	NASBA RNA amplification, ECL probe detection, NucliSens reader	About five viable oocysts per sample	120
	Turbid water	IA, Origen analyzer	50 viable oocysts/mL	119
	Highly turbid water	IA, Origen analyzer	1 oocyst/mL	118
	Karst water samples	IA, Origen analyzer	5 oocysts/mL	125
Escherichia coli O157	Creek water	IA, Origen analyzer	25 cells/mL, 1–2 viable cells/mL after concentration	123
	Drinking water	NASBA mRNA amplification, ECL probe detection, NucliSens reader	40 viable cells/mL	124
	Feces	PATHIgen IA, Origen analyzer	1×10^5 CFU/g	14
	Various food matrices	PATHIgen IA, Origen analyzer	100 cells/sample, more sensitive than culture methods	14
	Ground beef	IA, Origen analyzer	$100 \times$ as sensitive a commercial dipstick	126
	Ground beef, chicken, fish, milk, juices, serum, water	IA, Origen analyzer	1000–2000 cells/mL	121
	Various food and environmental water matrices	IA, Origen analyzer	Not reported	136, 134
Campylobacter	Feces	PATHIgen IA, Origen analyzer	1×10^4 CFU/g	14
	Poultry samples	PATHIgen IA, Origen analyzer	Comparable to culture methods	14
Salmonella	Feces	PATHIgen IA, Origen analyzer	5×10^5 CFU/g	14
	Surface swabs	PATHIgen IA, Origen analyzer	1 CFU/100 cm^2	14

(Continued)

Table 3 Continued

Analyte	Sample matrix	Assay method[a]	Detection limit	Ref.
	Poultry house drag swabs	PATHIgen IA, Origen analyzer	81% positive predictive value vs. culture	14
	Various food matrices	PATHIgen IA, Origen analyzer	Equivalent sensitivity to culture methods	14
	Ground beef, chicken, fish, milk, juices, serum, water	IA, Origen analyzer	1000–2000 cells/mL	121
Listeria monocytogenes	Environmental surface	PATHIgen IA, Origen analyzer	1 CFU/100 cm^2	14
Staphylococcus aureus enterotoxins	Buffer, milk, ground beef, lettuce, potato salad	PATHIgen IA, Origen analyzer	10 pg/mL for SEB	14
	Buffer, various food matrices	IA, Origen analyzer	5–50 ng/mL (SEA, B, C_1, C_2, C_3, D, E)	14
	Serum, tissue, buffer, urine	IA, Origen analyzer	1 pg/mL for SEB	127
	Buffer	IA, Origen analyzer	~0.5 pg/mL	131
Bacillus anthracis	Soil	IA, Origen analyzer	10^5 spores	135, 136
	Buffer	IA, Origen analyzer	100 spores	130
	Buffer	IA, Origen analyzer	1000 CFU/mL	131
	Saliva swab	IA, Origen analyzer	Not reported	134
Botulinus A toxin	Buffer	IA, Origen analyzer	~5 pg/mL	130
	Buffer	IA, Origen analyzer	4 pg/mL	131
Cholera toxin	Buffer	IA, Origen analyzer	~0.5 pg/mL	130
	Buffer	IA, Origen analyzer	2 pg/mL	131
Ricin toxin	Buffer	IA, Origen analyzer	~5 pg/mL	130
	Buffer	IA, Origen analyzer	0.5 pg/mL	131

[a]IA = immunoassay.

typhimurium have been detected with high sensitivity in various environmental water samples and food washings using rapid ECL-based immunoassays [121,122]. Other researchers have observed detection limits of 25 cells/mL in raw water, and one to two viable cells per liter if the sample was first concentrated and enriched by culturing [123]. A highly sensitive and specific assay for viable *E. coli* that does not require a culturing step was also developed for water matrices, based on detection of an mRNA sequence coding for a heat shock protein [124].

VII. FOOD TESTING

Tests for foodborne pathogens are important tools in the food industry in screening products for contamination and in public health use for tracking down sources and causes of disease outbreaks. Although current culture methods can detect the pathogens in the low concentrations required for infectivity, they are typically slow, requiring 2–4 days to complete. ECL-based assays provide an alternative testing method, typically provide results much faster, and are ideal for food testing because of their high tolerance to various matrices (Table 3).

Numerous ECL-based assays have been developed for food pathogens. An example is shown in Figure 10, which shows the results for *E. coli* from a study in which various food matrices were spiked with *E. coli* and *S. typhimurium* and the bacteria were quantified using bead-based ECL immunoassays [121]. The ability of the assays to detect bacteria in nearly all the matrices tested led the authors to recommend the method as a rapid, sensitive, and easy screening method for virulent food pathogens.

Another study investigated the detection of *E. coli* O157:H7 in ground beef, the food most associated with *E. coli* outbreaks [126]. Spike and recovery experiments with ECL immunoassay detection demonstrated that 0.05 colony-forming unit (CFU) of bacteria per gram of beef could be detected using an 18 h enrichment culture. The ECL assay was 100 times as sensitive as a commercially available rapid test kit.

Staphylococcus aureus enterotoxins are also a major cause of food poisoning. These are low molecular weight proteins produced by various strains of the bacteria, and their high toxicity demands a very sensitive detection technique. The toxins also have the potential for use as biological weapons (see Section VIII), and many reports dealing with assays for these toxins discuss applications in both civilian and military arenas. Sensitive and rapid ECL immunoassays for a number of these toxins have been developed using antibodies specific for enterotoxins A, B, C_1, C_2, C_3, D, and E and deemed sensitive enough for food industry use and biowarfare detection [14]. The detection of the B toxin has been demonstrated in a number of biological matrices including serum, urine, tissue, and buffer [127]. Reported sensitivities were 1 pg/mL for all matrices.

Igen International recently commercialized immunoassay kits for detection of pathogens relevant to environmental and food testing. The products, trademarked under the PathIGEN name [2], are magnetic bead–based sandwich immunoassays for use on ORIGEN and newer M-series instruments [14]. Tests currently available detect *E. coli* O157, *Salmonella*, *Listeria*, and *Campylobacter*. Food testing has shown these tests to be much faster than and

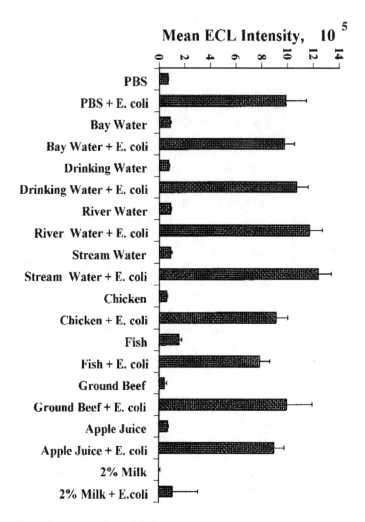

Figure 10 ECL detection of *E. coli* O157:H7 (2000 cells/sample) in diverse sample matrices. Identical amounts of bacteria were spiked into each sample, and the *E. coli* assay was performed on an ORIGEN analyzer. (From Ref. 121.)

at least as sensitive as standard culture methods for pathogen detection. Other applications are possible, such as screening commercial chickenhouses for *Salmonella*, detecting *Listeria* and *Salmonella* in surface swabs, and identifying *E. coli*, *Salmonella*, and *Campylobacter* in fecal samples at sensitivities appropriate for clinical use.

VIII. DETECTION OF BIOLOGICAL THREAT AGENTS

Assays based on ECL have been reported for the detection of numerous threat agents including toxins, bacteria, and viruses using both immunoassay and nucleic acid amplification formats (Table 3). Often, these assays provide a faster and more sensitive result than standard laboratory methods [128,129] such as culturing.

The high sensitivity of ECL-based immunoassays for biowarfare agents has been well established. Early work noted femtogram detection limits for purified biotoxoids (*Staphylococcus aureus* enterotoxin B, botulinus A, cholera, and ricin toxins) [130]. A more recent report added *Bacillus anthracis* and *E. coli* O157:H7 to the list of assays and included a virus (MS2, male-specific coliphage) to simulate the detection of a viral agent such as smallpox, Ebola, or yellow fever virus [131]. Sensitivities are shown in Table 3 and are improved at least 100 times over fluorescence-based plate assays. Recently, DNA aptamers were developed that are specific to cholera toxin, SEB, and anthrax, and a sandwich assay was demonstrated for the latter [132,133].

These reports have established the high sensitivity of ECL assays for agent detection in the research setting; however, field samples come in a variety of matrices. Environmental sampling matrices include water, air filter extracts, soils, surface swabs, and food. In the event of a potential exposure, clinical samples such as saliva, urine, feces, serum, blood, and lesion exudates would require testing to identify those exposed and to monitor therapy. As already discussed, ECL technology is very tolerant of complex matrices and therefore ideal for applications in biological threat detection. Examples of ECL-based agent detection in various matrices include highly sensitive assays for *Staphylcoccus aureus* enterotoxin B in serum, tissue, and urine [127]; detection of anthrax spores in saliva swabs [134]; and an immunoassay for the detection of anthrax spores in soils, a sample matrix important in both military and agricultural testing for this naturally occurring bacterium [135,136].

Further proof of the utility of ECL in real diagnostic arenas is the growing involvement of Igen International with various U.S. government agencies and programs aimed at developing detection systems for homeland security and military defense [14]. These include numerous programs within the Department of Defense for instruments, reagents, and assay development for the detection of biological agents or toxins in environmental and clinical samples. Some of these systems are designed for battlefield use or deployment within mobile laboratories throughout the combat zone. Another cooperative research and development agreement with the U.S. Army Medical Research Institute of Infectious Diseases (USAMRIID) deals with developing tests for food-, water-, and environmentally borne toxins. The intended application is to protect American troops from diseases caused by biological agents, and the tests are designed for use by government agencies, food processors, and analytical laboratories.

IX. CONCLUSION

This chapter has reviewed the use of electrochemiluminescence in the measurement and detection of biologically important compounds. It is clear from the material presented that ECL has made considerable inroads into a variety of disciplines. In terms of sheer volume of assays performed, the field of clinical diagnostics has perhaps benefited most from the introduction of highly sensitive ECL detection methods. Further growth in this area is guaranteed, as diagnostic companies expand their markets and test menus to include newly developed clinical assays. Currently, there is also a great deal of interest in the use of ECL methods for the detection of potential biological warfare agents as well as naturally occurring environmental pathogens. In these particular fields, the insensitivity of ECL techniques to matrix effects provides a clear advantage over other detection methodologies. ECL application in laboratories performing life science research is also considerable, and the recent introduction of high-throughput platforms should greatly enhance the use of ECL in the search for new disease therapeutics. The application of ECL to biological molecule detection in these and related fields will continue to expand, as more researchers and clinicians recognize the benefits of detection systems based on ECL.

ABBREVIATIONS

AFP   α-Fetoprotein

Anti-HBc   Antibody hepatitis B core antigen

Anti-HBs   Antibody hepatitis B surface antigen

Anti-HBe   Antibody hepatitis B e antigen

Anti-HAV   Antibody hepatitis A virus

Anti-Tg   Antithyroglobulin

Anti-TPO   Antithyroid peroxidase

CA   Cancer antigen

CCD   Charge-coupled device

CEA   Carcinoembryonic antigen

CMV   Cytomegalovirus

CYFRA   Cytokeratin fragment

DNA   Deoxyribonucleic acid

dUTP   2′-Deoxyuridine 5′-triphosphate

ECL   Electrogenerated chemiluminescence *or* electrochemiluminescence

ELISA   Enzyme-linked immunosorbent assay

FDA   Food and Drug Administration (U.S.)

FSH   Follicle stimulating hormone (follitropin)

fT3   Free T3

fT4   Free T4

GCSF   Granulocyte colony-stimulating factor

HBeAg   Hepatitis B e antigen

HIV   Human immunodeficiency virus

HbsAg   Hepatitis B viral surface antigen

HasAg   Hepatitis A viral surface antigen

HCG   Human chorionic gonadotropin

HPIV   Human parainfluenza virus

IgE   Immunoglobulin E

IgM   Immunoglobulin M

IA   Immunoassay

IM   Immunomagnetic

IU   International unit

LH   Luteinizing hormone (lutropin)

mRNA   Messenger ribonucleic acid

MS2   Male-specific coliphage

NADH   Nicotinamide adenine dinucleotide

NASBA   Nucleic acid sequence based amplification

NHS   *N*-Hydroxyl succinimide

NSE   Neuron-specific enolase

PCR   Polymerase chain reaction

PMT   Photomultiplier tube

ProBNP   Brain natriuretic peptide

PSA   Prostate-specific antigen

PTH   Parathyroid hormone

RIA   Radioimmunoassay

RNA   Ribonucleic acid

$Ru(bpy)_3^{2+}$ Ruthenium trisbipyridine

SEB   *Staphylococcus aureus* enterotoxin B

STAT   Short turnaround time

T3   Triiodothyronine

T4   Thyroxine

TPrA   Tripropylamine

TSH   Thyroid-stimulating hormone

USAMRIID   US Army Medical Research Institute of Infectious Diseases

REFERENCES

1. Leland, J.K., Powell, M.J. Electrogenerated chemiluminescence. An oxidative-reduction type ECL reaction sequence using tripropylamine. J. Electrochem. Soc. **1990**, *137*, 3127–3131.
2. ORIGEN and PathIGEN are registered trademarks of IGEN International, Inc.
3. Yang, H., Leland, J.K., Yost, D., Massey, R.J. Electrochemiluminescence: a new diagnostic and research tool. Bio/Technology, **1994**, *12*, 193–194.
4. Kibbey, M.C., MacAllan, D., Karaszkiewicz, J.W. Novel electrochemiluminescent assays for drug discovery. J. Assoc. Lab. Automation **2000**, *5*, 45–48.
5. NucliSens and NASBA are trademarks of bioMérieux Inc.
6. Elecsys is a registered trademark of a Member of the Roche Group.
7. Hsueh, Y.T., Collins, S.D., Smith, R.L. DNA quantification with an electrochemiluminescence microcell. Sensors Actuators B: Chem. **1998**, *B49*, 1–4.
8. Sector HTS, Sector PR, Multi-Array, and Multi-Spot are registered trademarks of Meso Scale Discovery, LLC.
9. Glezer, E.N., Johnson, K., Debad, J.D., Tsionsky, M., Jeffrey-Coker, B., Clinton, C., Kishbaugh, A., Leland, J.K., Billadeau, M., Leytner, S., Altunata, S., Sigal, G.B., Wilbur, J.L., Biebuyck, H.A., Wohlstadter, J.N. Electrochemiluminescent microarrays: a new tool for drug discovery and life science research. Abstracts of Papers, 224th ACS Natl. Meeting, Boston, MA, Aug 18–22, 2002, Washington, DC: Am. Chem. Soc. 2002.
10. Miao, W., Choi, J.P., Bard, A.J. Electrogenerated chemiluminescence 69: the tris(2,2′-bipyridine)ruthenium(II), (Ru(bpy)$_3^{2+}$)/tri-n-propylamine (TPrA) system revisited — a new route involving TPrA cation radicals. J. Am. Chem. Soc. **2002**, *124*, 14478–14485.
11. Carter, M.T., Bard, A.J. Investigations of the interaction of metal chelates with DNA. 3. Electrogenerated chemiluminescent investigation of the interaction of tris-(1,10-phenanthroline)ruthenium(II) with DNA. Bioconjugate Chem. **1990**, *2*, 257.
12. Rodriquez, M., Bard, A.J. Electrochemical studies of the interaction of metal

chelates with DNA. 4. Voltammmetric and electrogenerated chemiluminescent studies of the interaction of tris(2,2′-bipyridine)osmium(II) with DNA. Anal. Chem. **1990**, *62*, 2658–2662.

13. Courtesy of Meso Scale Discovery. See also www.meso_scale.com.

14. Courtesy of Igen International, Inc. See also www.igen.com.

15. Keller, G.H., Manak, M.M. DNA Probes. 2nd ed. New York: Stockton Press, 1993, pp. 173–196.

16. Kenten, J.H., Gudibande, S., Link, J., Willey, J.J., Curfman, B., Major, E.O., Massey, R.J. Improved electrochemiluminescent label for DNA probe assays: rapid quantitative assays of HIV-1 polymerase chain reaction products. Clin. Chem. **1992**, *38*, 873–879.

17. PCT WO87/06706 and PCT WO89/04919.

18. Lee, J.G., Yun, K.S., Lee, S.E., Kim, S., Park, J.K. Electrochemiluminescent detection for DNA hybridization using an intercalator. Abstracts of Papers, 222nd ACS Natl. Meeting, Chicago, IL, Aug 26–30, 2001, Washington, DC: Am. Chem. Soc. 2001.

19. Bard, A.J., Debad, J.D., Leland, J.K., Sigal, G.B., Wilbur, J.L., Wohlstadter, J.N. Analytical applications of electrogenerated chemiluminescence. Encycl. Anal. Chem.: Instrum. Appl. **2000**, *11*, 9842–9849.

20. Blackburn, G.F., Shah, H.P., Kenten, J.H., Leland, J.K., Kamin, R.A., Link, J., Peterman, J., Powell, M.J., Shah, A., Talley, D.B., Tyagi, S.K., Wilkins, E., Wu, T.G., Massey, R.J. Electrochemiluminescence detection for development of immunoassays and DNA probe assays for clinical diagnostics. Clin. Chem. **1991**, *37*, 1534–1539.

21. Deaver, D. A new non-isotopic detection system for immunoassays. Nature **1995**, *377*, 758–760.

22. Erler, K. Elecsys immunoassay systems using electrochemiluminescence detection. Wien. Klin. Wochenschr. **1998**, *110(suppl 3)*, 5–10.

23. Namba, Y., Usami, M., Suzuki, O. Development of highly sensitive Ru-chelate based ECL immunoassay. 2. Electrochemical and immunochemical studies on homogeneous and heterogeneous ECL excitation. Anal. Sci. **1999**, *15*, 1087–1089.

24. Yilmaz, N., Erbagci, A.B., Aynacioglu, A.S. Cytochrome P4502C9 genotype in southeast Anatolia and possible relation with some serum tumour markers and cytokines. Acta. Biochim. Polon. **2001**, *48*, 775–782.

25. Stockmann, W., Bablok, W., Luppa, P. Analytical performance of Elecsys 2010. A multicenter evaluation. Wien. Klin. Wochenschr. **1998**, *110(suppl 3)*, 10–21.

26. Sanchez-Carbayo, M., Espasa, A., Chinchilla, V., Herrero, E., Megias, J., Mira, A., Soria, F. New electrochemiluminescent immunoassay for the determination of CYFRA 21-1: analytical evaluation and clinical diagnostic performance in urine samples of patients with bladder cancer. Clin. Chem. **1999**, *45*, 1944–1953.

27. Shimizu, A., Shiraki, K., Ito, T., Sugimoto, K., Sakai, T., Ohmori, S., Murata, K., Takase, K., Tameda, Y., Nakano, T. Sequential fluctuation pattern of serum des-γ-carboxy prothrombin levels detected by high-sensitive electrochemiluminescence

system as an early predictive marker for hepatocellular carcinoma in patients with cirrhosis. Intl. J. Mol. Med. **2002**, *9*, 245–250.

28. Sassa, T., Kumada, T., Nakano, S., Uematsu, T. Clinical utility of simultaneous measurement of serum high-sensitivity des-γ-carboxy prothrombin and *Lens culinaris* agglutinin A-reactive α-fetoprotein in patients with small hepatocellular carcinoma. Eur. J. Gastroenterol. Hepatol. **1999**, *11*, 1387–1392.

29. Xu, X.-H.N., Jeffers, R.B., Gao, J., Logan, B. Novel solution-phase immunoassays for molecular analysis of tumor markers. Analyst. **2001**, *126*, 1285–1292.

30. Butch, A.W., Crary, D., Yee, M. Analytical performance of the Roche total and free PSA assays on the Elecsys 2010 immunoanalyzer. Clin. Biochem. **2002**, *35*, 143–145.

31. Haese, A., Dworschack, R.T., Piccoli, S.P., Sokoll, L.J., Partin, A.W., Chan, D.W. Clinical evaluation of the Elecsys total prostate-specific antigen assay on the Elecsys 1010 and 2010 systems. Clin. Chem. **2002**, *48*, 944–947.

32. Miyashiro, I., Kuo, C., Huynh, K., Iida, A., Morton, D., Bilchik, A., Giuliano, A., Hoon, D.S.B. Molecular strategy for detecting metastatic cancers with use of multiple tumor-specific MAGE-A genes. Clin. Chem. **2001**, *47*, 505–512.

33. O'Connell, C.D., Juhasz, A., Kuo, C., Reeder, D.J., Hoon, D.S. Detection of tyrosinase mRNA in melanoma by reverse transcription-PCR and electrochemiluminescence. Clin. Chem. **1999**, *44*, 1161–1169.

34. Taback, B., Chan, A.D., Kuo, C.T., Bostick, P.J., Wang, H.J., Giuliano, A.E., Hoon, D.S.B. Detection of occult metastatic breast cancer cells in blood by a multimolecular marker assay: correlation with clinical stage of disease. Cancer Res. **2001**, *61*, 8845–8850.

35. Hoon, D.S.B., Kuo, C.T., Wen, S., Wang, H., Metelitsa, L., Reynolds, C.P., Seeger, R.C. Ganglioside GM2/GD2 synthetase mRNA is a marker for detection of infrequent neuroblastoma cells in bone marrow, Am. J. Pathol. **2001**, *159*, 493–500.

36. Gassler, N., Peuschel, T., Pankau, R. Pediatric reference values of estradiol, testosterone, lutropin, follitropin and prolactin. Clin. Lab. **2000**, *46*, 553–560.

37. Sanchez-Carbayo, M., Mauri, M., Alfayate, R., Miralles, C., Soria, F. Elecsys testosterone assay evaluated. Clin. Chem. **1998**, *44*, 1744–1746.

38. Ehrhardt, V., Assmann, G., Baetz, O., Bieglmayer, C., Mueller, C., Neumeier, D., Roth, H.J., Veys, A., Yvert, J.P. Results of the multicenter evaluation of an electrochemiluminescence immunoassay for HCG on Elecsys 2010. Wien. Klin. Wochenschr. **1998**, *110(suppl 3)*, 61–67.

39. Sanchez-Carbayo, M., Mauri, M., Alfayate, R., Miralles, C., Soria, F. Analytical and clinical evaluation of TSH and thyroid hormones by electrochemiluminescent immunoassays. Clin. Biochem. **1999**, *32*, 395–403.

40. Sanchez-Carbayo, M., Mauri, M., Alfayate, R., Miralles, C., Soria, F. Analytical and clinical evaluation of TSH and thyroid hormones by electrochemiluminescent immunoassays. Clin. Biochem. **1999**, *32*, 395–403.

41. Luppa, P.B., Reutemann, S., Huber, U., Hoermann, R., Poertl, S., Kraiss, S., Von Buelow, S., Neumeier, D. Pre-evaluation and system optimization of the Elecsys thyroid electrochemiluminescence immunoassays. Clin. Chem. Lab. Med. **1998**, *36*, 789–796.

42. Kulmala, S., Hakansson, M., Spehar, A.-M., Nyman, A., Kankare, J., Loikas, K., Ala-Kleme, T., Eskola, J. Heterogeneous and homogeneous electrochemiluminoimmunoassays of hTSH at disposable oxide-covered aluminum electrodes. Anal. Chim. Acta. **2002**, *458*, 271–280.

43. Sapin, R., Schlienger, J.-L., Gasser, F., Noel, E., Lioure, B., Grunenberger, F., Goichot, B., Grucker, D. Intermethod discordant free thyroxine measurements in bone marrow-transplanted patients. Clin. Chem. **2000**, *46*, 418–422.

44. See www.rochediagnostics.com

45. Kenten, J.H., Gudibande, S., Link, J., Willey, J.J., Curfman, B., Major, E.O., Massey, R.J. Improved electrochemiluminescent label for DNA probe assays; rapid quantitative assays of HIV-1 polymerase chain reaction products. Clin. Chem. **1992**, *38*, 873–879.

46. Kenten, J.H., Casadei, J., Link, J., Leopold, S., Willey, J., Powell, M., Rees, A., Massey, R. Rapid electrochemiluminescent assay for polymerase chain reaction products. Clin. Chem. **1991**, *37*, 1626–1632.

47. Klingler, K.R., Zech, D., Wielckens, K. Apolipoprotein B-100. Employing the electrochemiluminescence technology of the Elecsys systems for the detection of the point mutation Arg(3500)Gln. Clin. Lab. **2000**, *46*, 41–47.

48. Van Gemen, B., Van Beuningen, R., Nabbe, A., Van Strijp, V., Jurriaans, S., Lens, P., Schoones, R., Kievits, T. The one tube quantitative HIV-1 RNA NASBA nucleic acid amplification assay using electrochemiluminesence labeled probes. J. Virol. Methods **1994**, *49*, 157–168.

49. Kobayashi, Y., Hayakawa, M., Fukumura, Y. Determination of hepatitis B markers in human serum by an electrochemiluminescence immunoassay analyzer, "ECLusys 2010." Igaku Yakugaku **1999**, *42*, 749–756.

50. Kashiwagi, S., Hayashi, J., Asai, T., Nishimura, J., Arai, N., Kanashima, M., Asai, Y. Evaluation of HBc antibody detection kit using electrochemiluminescence immunoassay method. Igaku Yakugaku **1998**, *40*, 119–125.

51. Takahashi, M., Hoshino, H., Ohuchi, Y., Ryan, S., Shimoda, K., Yasuda, K., Tanaka, J., Yoshizawa, K., Hino, K., Iino, S. Fundamental and clinical evaluation of HBcAb measurement using ECL-IA method. Igaku Yakugaku **1998**, *40*, 483–494.

52. Yu, H., Bruno, J.G., Cheng, T.-C., Calomiris, J.J., Goode, M.T., Gatto-Menking, D.L. A comparative study of PCR product detection and quantitation by electrochemiluminescence and fluorescence. J. Biolumin. Chemilumin. **1995**, *10*, 239–245.

53. Schutzbank, T.E., Smith, J. Detection of human immunodeficiency virus type 1 proviral DNA by PCR using an electrochemiluminescence-tagged probe. J. Clin. Microbiol. **1995**, *33*, 2036–2041.

54. Oprandy, J.J., Amemiya, K., Kenten, J.H., Green, R.G., Major, E.O., Massey, R. Electrochemiluminescence-based detection system for the quantitative measurement of antigens and nucleic acids: application to HIV-1 and JC viruses. Technical Advances in AIDS Research in the Human Nervous System [Proc. NIH Symp. on Technical Advances in AIDS Research in the Human Nervous System], Washington, DC, Oct 4–5, 1993 Publisher: Plenum, New York, N.Y.(1995), 281–297.

55. de Baar, M.P., van der Horn, K.H.M., Goudsmit, J., de Ronde, A., de Wolf, F. Detection of human immunodeficiency virus type 1 nucleocapsid protein p7 in vitro and in vivo. J. Clin. Microbiol. **1999**, *37*, 63–67.

56. Ohlin, M., Silvestri, M., Sundqvist, V.-A., Borrebaeck, C.A.K. Cytomegalovirus glycoprotein B-specific antibody analysis using electrochemiluminescence detection-based techniques. Clin. Diagn. Lab. Immunol. **1997**, *4*, 107–111.

57. Boom, R., Sol, C., Weel, J., Gerrits, Y., de Boer, M., van Dillen, P.W. A highly sensitive assay for detection and quantitation of human cytomegalovirus DNA in serum and plasma by PCR and electrochemiluminescence. J. Clin. Microbiol. **1999**, *37*, 1489–1497.

58. Stevens, S.J.C., Vervoort, M.B.H.J., van den Brule, A.J.C., Meenhorst, P.L., Meijer, C.J.L.M., Middeldorp, J.M. Monitoring of Epstein-Barr virus DNA load in peripheral blood by quantitative competitive PCR. J. Clin. Microbiol. **1999**, *37*, 2852–2857.

59. de Jong, M.D., Weel, J.F.L., Schuurman, T., Wertheim-van Dillen, P.M.E., Boom, R. Quantitation of varicella-zoster virus DNA in whole blood, plasma, and serum by PCR and electrochemiluminescence. J. Clin. Microbiol. **2000**, *38*, 2568–2573.

60. Wu, S.-J.L., Lee, E.M., Putvatana, R., Shurtliff, R.N., Porter, K.R., Suharyono, W., Watts, D.M., King, C.-C., Murphy, G.S., Hayes, C.G., Romano, J.W. Detection of dengue viral RNA using a nucleic acid sequence-based amplification assay. J. Clin. Microbiol. **2001**, *39*, 2794–2798.

61. Lanciotti, R.S., Kerst, A.J. Nucleic acid sequence-based amplification assays for rapid detection of West Nile and St. Louis encephalitis viruses. J. Clin. Microbiol. **2001**, *39*, 4506–4513.

62. Fox, D., Han, S., Samuelson, A., Zhang, Y., Neale, M.L., Westmoreland, D. Development and evaluation of nucleic acid sequence based amplification (NASBA) for diagnosis of enterovirus infections using the NucliSens Basic Kit. J. Clin. Virol. **2002**, *24*, 117–130.

63. Collins, R.A., Ko, L.-S., So, K.-L., Ellis, T., Lau, L.-T., Yu, A.C.H. Detection of highly pathogenic and low pathogenic avian influenza subtype H5 (Eurasian lineage) using NASBA. J. Virol. Methods **2002**, *103*, 213–225.

64. Collins, R.A., Ko, L.S., Fung, K.Y., Chang, K.Y., Xing, J., Lau, L.T., Yu, A.C. Rapid and sensitive detection of avian influenza virus subtype H7 using NASBA. Biochem. Biophys. Res. Commun. **2003**, *300*, 507–515.

65. Reetoo, K.N., Osman, S.A., Illavia, S.J., Banatvala, J.E., Muir, P. Development and evaluation of quantitative-competitive PCR for quantitation of coxsackievirus B3 RNA in experimentally infected murine tissues. J. Virol. Methods **1999**, *82*, 145–156.

66. Katayama, M., Sanzen, N., Funakoshi, A., Sekiguchi, K. Laminin γ2-chain fragment in the circulation: a prognostic indicator of epithelial tumor invasion. Cancer Res. **2003**, *63*, 222–229.

67. Jameison, F., Sanchez, R., Dong, L., Leland, J., Yost, D., Martin, M. Electrochemiluminescence-based quantitation of classical clinical chemistry analytes. Anal. Chem. **1996**, *68*, 1298–1302.

68. Lai, R.Y., Chiba, M., Kitamura, N., Bard, A.J. Electrogenerated

chemiluminescence. 68. Detection of sodium ion with a ruthenium(II) complex with crown ether moiety at the 3,3'-positions on the 2,2'-bipyridine ligand. Anal. Chem. **2002**, *74*, 551–553.

69. Muegge, B.D., Richter, M.M. Electrochemiluminescent detection of metal cations using a ruthenium(II) bipyridyl complex containing a crown ether moiety. Anal. Chem. **2002**, *74*, 547–550.

70. Liang, P., Sanchez, R.I., Martin, M.T. Electrochemiluminescence-based detection of β-lactam antibiotics and β-lactamases. Anal. Chem. **1996**, *68*, 2426–2431.

71. Yamaguchi, K., Sawada, T., Yamane, S., Haga, S., Ikeda, K., Igata-Yi, R., Yoshiki, K., Matsuoka, M., Okabe, H., Horii, Y., Nawa, Y., Waltrip, R.W., Carbone, K.M. Synthetic peptide-based electrochemiluminescence immunoassay for anti-Borna disease virus p40 and p24 antibodies in rat and horse serum. Ann. Clin. Biochem. **2001**, *38*, 348–355.

72. Horii, Y., Garcia, J.N.P., Noviana, D., Kono, F., Sawada, T., Naraki, T., Yamaguchi, K. Detection of anti-Borna disease virus antibodies from cats in Asian countries, Japan, Philippines and Indonesia using electrochemiluminescence immunoassay. J. Vet. Med. Sci. **2001**, *63*, 921–923.

73. Fukuda, K., Takahashi, K., Iwata, Y., Mori, N., Gonda, K., Ogawa, T., Osonoe, K., Sato, M., Ogata, S.-I., Horimoto, T., Sawada, T., Tashiro, M., Yamaguchi, K., Niwa, S.-I., Shigeta, S. Immunological and PCR analyses for Borna disease virus in psychiatric patients and blood donors in Japan. J. Clin. Microbiol. **2001**, *39*, 419–429.

74. Khorkova, O.E., Pate, K., Heroux, J., Sahasrabudhe, S. Modulation of amyloid precursor protein processing by compounds with various mechanisms of action: detection by liquid phase electrochemiluminescent system. J. Neurosci. Methods **1998**, *82*, 159–166.

75. Stieber, P., Molina, R., Chan, D.W., Fritsche, H.A., Beyrau, R., Bonfrer, J.M.G., Filella, X., Gornet, T.G., Hoff, T., Inger, W., Van Kamp, G. J, Nagel, D., Peisker, K., Sokoll, L.J., Troalen, F., Untch, M., Domke, I. Evaluation of the analytical and clinical performance of the Elecsys CA 15–3 immunoassay. Clin. Chem. **2001**, *47*, 2162–2164.

76. Filella, X., Friese, S., Roth, H.J., Nussbaum, S., Wehnl, B. Technical performance of the Elecsys CA 72–4 test—development and field study. Anticancer Res. *20*, 5229–5232. 200.

77. Hubl, W., Chan, D.W., Van Ingen, H.E., Miyachi, H., Molina, R., Filella, X., Pitzel, L., Ruibal, A., Rymer, J.C., Bagnard, G., Domke, I. Multicenter evaluation of the Elecsys CA 125 II assay. Anticancer Res. **1999**, *19*, 2727–2733.

78. Ingen, H.E., Chan, D.W., Hubl, W., Miyachi, H., Molina, R., Pitzel, L., Ruibal, A., Rymer, J.C., Domke, I. Analytical and clinical evaluation of an electrochemiluminescence immunoassay for the determination of CA 125. Clin. Chem. **1998**, *44*, 2530–2536.

79. Seck, T., Diel, I., Bismar, H., Ziegler, R., Pfeilschifter, J. Expression of interleukin-6 (IL-6) and IL-6 receptor mRNA in human bone samples from pre- and postmenopausal women. Bone **2002**, *30*, 217–222.

80. Okabe, R., Nakatsuka, K., Inaba, M., Miki, T., Naka, H., Masaki, H., Moriguchi, A., Nishizawa, Y. Clinical evaluation of the Elecsys β-Crosslaps serum assay, a new assay for degradation products of type I collagen C-telopeptides. Clin. Chem. **2001**, *47*, 1410–1414.

81. Scheunert, K., Albrecht, S., Konnegen, V., Wunderlich, G., Distler, W. Chemiluminometric detection of CTX and osteocalcin in cases of postmenopausal women using the Elecsys analyser. Chemiluminescence at the Turn of the Millennium, Albrecht S., Zimmermann T, Brandl H, eds. Schweda-Werbedruch GmbH, Dresdin, **2001**, pp 347–351.

82. Klein, G., Kampmann, M., Baum, H., Rauscher, T., Vukovic, T., Hallermayer, K., Rehner, H., Mueller-Bardorff, M., Katus, H.A. Clinical performance of the new cardiac markers troponin T and CK-MB on the Elecsys 2010. A multicenter evaluation. Wien. Klini. Wochenschr. **1998**, *110(suppl. 3)*, 40–51.

83. Ohlin, M. Cytomegalovirus glycoprotein B-specific antibody analysis using electrochemiluminescence detection based techniques. Clin. Diagn. Lab. **1997**, *4*, 107.

84. Weber, B., Bayer, A., Kirch, P., Schluter, V., Schlieper, D., Melchior, W. Improved detection of hepatitis B virus surface antigen by a new rapid automated assay. J. Clin. Microbiol. **1999**, *37*, 2639–2647.

85. Kobrynski, L., Tanimune, L., Pawlowski, A., Douglas, S.D., Campbell, D.E. A comparison of electrochemiluminescence and flow cytometry for the detection of natural latex-specific human immunoglobulin E. Clin. Diagn. Lab. Immunol. **1996**, *3*, 42–46.

86. Liebert, A., Beier, L., Schneider, E., Kirch, P. Comparison of the Elecsys insulin immunoassay with RIA DPC-, Abbott IMX- and Beckmann ACCESS insulin. Chemiluminescence at the Turn of the Millennium, Albrecht S., Zimmermann T, Brandl H, eds. Schweda-Werbedruch GmbH, Dresdin, **2001**, pp 341–346.

87. Sapin, R., Le Galudec, V., Gasser, F., Pinget, M., Grucker, D. Elecsys insulin assay: free insulin determination and the absence of cross-reactivity with insulin lispro. Clin. Chem. **2001**, *47*, 602–605.

88. Novick, D., Schwartsburd, B., Pinkus, R., Suissa, D., Belzer, I., Sthoeger, Z., Keane, W.F., Chvatchko, Y., Kim, S.H., Fantuzzi, G., Dinarello, C.A., Rubinstein, M. A novel IL-18BP ELISA shows elevated serum IL-18BP in sepsis and extensive decrease of free IL-18. Cytokine **2001**, *14*, 334–342.

89. Sennikov, S.V., Krysov, S.V., Injelevskaya, T.V., Silkov, A.N., Grishina, L.V., Kozlov, V.A. Quantitative analysis of human immunoregulatory cytokines by the electrochemiluminescence method. J. Immunol. Methods **2003**, *275*, 81–88.

90. Swanson, S.J., Jacobs, S.J., Mytych, D., Shah, C., Indelicato, S.R., Bordens, R.W. Applications for the new electrochemiluminescent (ECL) and biosensor technologies. Dev. Biol. Stand. **1999**, *97*, 135–147.

91. Kankare, J., Haapakka, K., Kulmala, S., Nanto, V., Eskola, J., Takalo, H. Immunoassay by time-resolved electrogenerated luminescence. Anal. Chim. Acta **1992**, *266*, 205–212.

92. Hermsen, D., Franzson, L., Hoffmann, J.P., Isaksson, A., Kaufman, J.M., Leary, E., Muller, C., Nakatsuka, K., Nishizawa, Y., Reinauer, H., Riesen, W., Roth, H.J., Steinmuller, T., Troch, T., Bergmann, P. Multicenter evaluation of a new immunoassay for intact PTH measurement on the Elecsys System 2010 and 1010. Clin. Lab. **2002**, *48*, 131–141.

93. Hetland, O., Dickstein, K. Cardiac troponin T by Elecsys system and a rapid ELISA: analytical sensitivity in relation to the TropT (CardiacT) "bedside" test. Clin. Chem. **1998**, *44*, 1348–1350.

94. Ishii, J., Ishikawa, T., Yukitake, J., Nagamura, Y., Ito, M., Wang, J.H., Kato, Y., Hiramitsu, S., Inoue, S., Kondo, T., Morimoto, S., Nomura, M., Watanabe, Y., Hishida, H. Clinical specificity of a second-generation cardiac troponin T assay in patients with chronic renal failure. Clin. Chim. Acta **1998**, *270*, 183–188.

95. Collinson, P.O., Jorgensen, B., Sylven, C., Haass, M., Chwallek, F., Katus, H.A., Muller-Bardorff, M., Derhaschnig, U., Hirschl, M.M., Zerback, R. Recalibration of the point-of-care test for CARDIAC T. Quantitative with Elecsys troponin T 3rd generation. Clin. Chim. Acta **2002**, *307*, 197–203.

96. Moreau, E., Philippe, J., Couvent, S., Leroux-Roels, G. Interference of soluble TNF-α receptors in immunological detection of tumor necrosis factor-α. Clin. Chem. **1996**, *42*, 1450–1453.

97. Tai, J.H., Ewert, M.S., Belliot, G., Glass, R.I., Monroe, S.S. Development of a rapid method using nucleic acid sequence-based amplification for the detection of astrovirus. J. Virol. Methods **2003**, *110*, 119–127.

98. Collins, R.A., Ko, L.S., Fung, K.Y., Lau, L.T., Xing, J., Yu, A.C.H. A method to detect major serotypes of foot-and-mouth disease virus. Biochem. Biophys. Res. Commun. **2002**, *297*, 267–274.

99. Schutzbank, T.E., Smith, J. Detection of human immunodeficiency virus type 1 proviral DNA by PCR using electrochemiluminescence tagged probe. J. Clin. Microbiol. **1995**, *33*, 2036.

100. Hibbitts, S., Rahman, A., John, R., Westmoreland, D., Fox, J.D. Development and evaluation of Nuclisens basic kit NASBA for diagnosis of parainfluenza virus infection with "end-point" and "real-time" detection. J. Virol. Methods **2003**, *108*, 145–155.

101. Shan, S., Ko, L.-S., Collins, R.A., Wu, Z., Chen, J., Chan, K.-Y., Xing, J., Lau, L.-T., Yu, A.C.-H. Comparison of nucleic acid-based detection of avian influenza H5N1 virus isolation. Biochem. Biophys. Res. Commun. **2003**, *302*, 377–383.

102. Gellings, A., Holzem, G., Wielckens, K., Klingler, K.R. Prothrombin mutation: employing the electrochemiluminescence technology of the Elecsys system for the detection of the point mutation at position 20210 in the 3′ untranslated region of the prothrombin gene. Laboratoriumsmedizin **2001**, *25*, 26–30.

103. Stern, H.J., Carlos, R.D., Schutzbank, T.E. Rapid detection of the ΔF508 deletion in cystic fibrosis by allele specific PCR and electrochemiluminescence detection. Clin. Biochem. **1995**, *28*, 470–473.

104. Martin, A.F., Nieman, T.A. Chemiluminescence biosensors using tris(2,2′-bipyridyl)ruthenium(II) and dehydrogenases immobilized in cation exchange polymers. Biosensors Bioelectron. **1997**, *12*, 479–489.

105. Blohm, S., Kadey, S., McAKeon, K., Perkins, S., Sugasawara, R. Use of the ORIGEN electrochemiluminescence detection system for measuring tumor necrosis factor-α in tissue culture media. Biomed. Products, April 1996.

106. Shapiro, L., Heidenreich, K.A., Meintzer, M.K., Dinarello, C.A. Role of p38 mitogen-activated kinase in HIV type 1 production in vitro. Proc. Natl. Acad. Sci. USA **1998**, *95*, 7422–7426.

107. Obenauer-Kutner, L.J., Jacobs, S.J., Kolz, K., Tobias, L.M., Bordens, R.W. A highly sensitive electrochemiluminescence immunoassay for interferon α-2b in human serum. J. Immunol. Methods **1997**, *206*, 25–33.

108. Puren, A.J., Razeghi, P., Fantuzzi, G., Dinarello, C.A. Interleukin-18 enhances lipopolysaccharide-induced interferon-γ production in human whole blood cultures. J. Infecti. Dis. **1998**, *178*, 1830–1834.

109. Motmans, K., Raus, J., Vandevyver, C. Quantification of cytokine messenger RNA in transfected human T cells by RT-PCR and an automated electro-chemiluminescence-based post-PCR detection system. J. Immunol. Methods **1996**, *190*, 107–16.

110. Gopalakrishnan, S.M., Warrior, U., Burns, D., Groebe, D.R. Evaluation of electrochemiluminescent technology for inhibitors of granulocyte colony-stimulating factor receptor binding. J. Biomol. Screening **2000**, *5*, 369–375.

111. Weinreb, P.H., Yang, W.J., Violette, S.M., Couture, M., Kimball, K., Pepinsky, R.B., Lobb, R.R., Josiah, S. A cell-free electrochemiluminescence assay for measuring β1-integrin-ligand interactions. Anal. Biochem. **2002**, *306*, 305–313.

112. Hughes, S.R., Khorkova, O., Goyal, S., Knaeblein, J., Heroux, J., Riedel, N.G., Sahasrabudhe, S. A2-Macroglobulin associates with b-amyloid peptide and prevents fibril formation. Proc. Natl. Acad. Sci. USA **1998**, *95*, 3275–3280.

113. Horiuchi, H., Lippé, R., McBride, H.M., Rubino, M., Woodman, P., Stenmark, H., Rybin, V., Wilm, M., Ashman, K., Mann, M., Zerial, M. A novel Rab5 GDP/GTP exchange factor complexed to Rabaptin-5 links nucleotide exchange to effector recruitment and function. Cell **1997**, *90*, 1149–1159.

114. Mathew, A., Mathur, S.K., Jolly, C., Fox, S.G., Kim, S., Morimoto, R.I. Stress-specific activation and repression of heat shock factors 1 and 2. Mol. Cell. Biol. *21*, 7163–7171.

115. Zhang, L., Schwartz, G., O'Donnell, M., Harrison, R.K. Development of a novel helicase assay using electrochemiluminescence. Anal. Biochem. **2001**, *293*, 31–37.

116. Heroux, J.A., Sigal, G.B., von Borsel, R.W. Assays for measuring nucleic acid damaging activities. US Patent Appl. 2002/0164593A1, Nov. 7, 2002.

117. Zhang, L., Song, L., Terracina, G., Liu, Y., Pramanik, B., Parker, E. Biochemical characterization of γ-secretase activity that produces β-amyloid peptides. Biochemistry **2001**, *40*, 5049–5055.

118. Lee, Y.M., Johnson, P.W., Call, J.L., Arrowood, M.J., Furness, B.W., Pichette, S.C., Grady, K.K., Resh, P., Michell, L., Bergmire-Sweat, D., MacKenzie, W.R., Tsang, V.C.W. Development and application of a quantitative, specific assay for *Cryptosporidium parvum* oocyst detection in high-turbidity environmental water samples. Am. J. Trop. Med. Hyg. **2001**, *65*, 1–9.

119. Call, J.L., Arrowood, M.J., Xie, L.-T., Hancock, K., Tsang, V.C.W. Immunoassay for viable *Cryptosporidium parvum* oocysts in turbid environmental water samples. J. Parasitol. **2001**, *87*, 203–210.

120. Baeumner, A.J., Humiston, M.C., Montagna, R.A., Durst, R.A. Detection of viable oocysts of *Cryptosporidium parvum* following nucleic acid sequence based amplification. Anal. Chem. **2001**, *73*, 1176–1180.

121. Yu, H., Bruno, J.G. Immunomagnetic-electrochemiluminescent detection of *Escherichia coli* O157 and *Salmonella typhimurium* in foods and environmental water samples. Appl. Environ. Microbol. **1996**, *62*, 587–592.

122. Yu, H., Bruno, J. Detection of the coliform bacteria *Escherichia coli* and *Salmonella* sp. in water by a sensitive and rapid immunomagnetic electro-chemiluminescence (ECL) technique. Proce. SPIE—Int. Soc. Opt. Eng. **1995**, *2504(Environmental Monitoring and Hazardous Waste Site Remediation, 1995)*, 241–252.

123. Shelton, D.R., Karns, J.S. Quantitative detection of *Escherichia coli* O157 in surface waters by using immunomagnetic electrochemiluminescence. Appl. Environ. Microbiol. **2001**, *67*, 2908–2915.

124. Min, J., Baeumner, A.J. Highly sensitive and specific detection of viable *Escherichia coli* in drinking water. Anal. Biochem. **2002**, *303*, 186–193.

125. Kuczynska, E., Boyer, D.G., Shelton, D.R. Comparison of immunofluorescence assay and immunomagnetic electrochemiluminescence in detection of *Cryptosporidium parvum* oocysts in karst water samples. J. Microbiol. Methods **2003**, *53*, 17–26.

126. Crawford, C.G., Wijey, C., Fratamico, P., Tu, S.I., Brewster, J. Immunomagnetic-electrochemiluminescent detection of *E. coli* O157:H7 in ground beef. J. Rapid Methods Autom. Microbiol. **2000**, *8*, 249–264.

127. Kijek, T.M., Rossi, C.A., Moss, D., Parker, R.W., Henchal, E.A. Rapid and sensitive immunomagnetic-electrochemiluminescent detection of staphylococcal enterotoxin B. J. Immunol. Methods **2000**, *236*, 9–17.

128. Henchal, E.A., Teska, J.D., Ludwig, G.V., Shoemaker, D.R., Ezzell J.W. Current laboratory methods for biological threat agent detection. Clin. Lab. Med. **2001**, *21*, 661–678.

129. Higgins, J.A., Ibrahim, M.S., Knauert, F.K., Ludwig, G.B., Kijek, T.M., Ezzell, J.W., Courtney, B.C., Henchal, E.A. Sensitive and rapid identification of biological threat agents. Ann. NY. Acad. Sci. **1999**, *894*, 130–148.

130. Gatto-Menking, D.L., Yu, H., Bruno, J.G., Goode, M.T., Miller, M., Zulich, A.W. Sensitive detection of biotoxoids and bacterial spores using an immunomagnetic electrochemiluminescence sensor. Biosensors Bioelectron. **1995**, *10*, 501–507.

131. Yu, H., Raymonda, J.W., McMahon, T.M., Campagnari, A.A. Detection of biological threat agents by immunomagnetic microsphere-based solid phase fluorogenic- and electrochemiluminescence. Biosensors Bioelectron. **2000**, *14*, 829–840.

132. Bruno, J.G., Kiel, J.L. In vitro selection of DNA aptamers to anthrax spores with electrochemiluminescence detection. Biosensors Bioelectron. **1999**, *14*, 457–464.

133. Bruno, J.G., Kiel, J.L. Use of magnetic beads in selection and detection of biotoxin aptamers by electrochemiluminescence and enzymatic methods. BioTechniques **2002**, *32*, 178–183.

134. Yu, H. Enhancing immunoassay possibilities using magnetic carriers in biological fluids. Proce. SPIE—Int. Soc. Opt. Eng. **1997**, *2982(Optical Diagnostics of Biological Fluids and Advanced Techniques in Analytical Cytology)*, 168–179.

135. Bruno, J.G., Yu, H. Immunomagnetic-electrochemiluminescent detection of *Bacillus anthracis* spores in soil matrices. Appl. Environ. Microbiol. **1996**, *62*, 3474–3476.

136. Yu, H., Bruno, J.G. Sensitive bacterial pathogen detection using an immunomagnetic-electrochemiluminescence instrument: potential military and food industry applications. BioMed. Products **1995**, *20*, 20–21.

9

Analytical Applications: Flow Injection, Liquid Chromatography, and Capillary Electrophoresis

Neil D. Danielson
Miami University, Oxford, Ohio, U.S.A.

I. INTRODUCTION

The intent of this chapter is to describe the use of flow injection (FI), liquid chromatography (LC), and capillary electrophoresis (CE) with luminol electro-generated chemiluminescence (ECL) and principally tris(2,2'-bipyridyl)ruthenium(III) $(Ru(bpy)_3^{3+})$ ECL detection. Analytical applications involving other ruthenium complexes such as those with the corresponding phenanthroline ligand $[Ru(phen)_3^{3+}]$ are uncommon but will be mentioned occasionally. Instrumentation involving FI, LC, and CE will be considered; design of sensors will not be covered.

Several quite recent review articles describing analytical applications of ECL have been published, and these will be briefly summarized here. The definition of ECL used in these articles and many other papers is chemiluminescence (CL) produced as the result of an electrochemical reaction or the light produced when molecules, excited as the result of an oxidation–reduction reaction, relax back to the ground state and emit light. ECL mechanisms emphasizing $Ru(bpy)_3^{3+}$ and luminol, analytical applications, limitations such as interfering species and reproducibility, and developments in chemistry and instrumentation are summarized for primarily the last five years in Ref. 1. The ECL of both organic species and metal complexes [mainly luminol and $Ru(bpy)_3^{3+}$, respectively] with strong coverage of biosensors such as those for immunoassay are the primary subjects of Ref. 2. A review of basic principles and applications of $Ru(bpy)_3^{2+}$ ECL including emphasis on dehydrogenase enzyme assays and immunoassays is available [3]. Another article [4] focuses on $Ru(bpy)_3^{3+}$ CL starting with a historical account, followed by a description of different methods of generating $Ru(bpy)_3^{3+}$, and finally presenting a comprehensive discussion of applications. Seven tables summarizing applications subdivided into oxalate and organic acids, amines, amino acids and proteins, pharmaceuticals, other analytes, indirect methods, and immunoassays are included along with text summaries of selected papers. In addition, both proposed reaction mechanisms involving $Ru(bpy)_3^{3+}$ and various analytes are discussed.

The acronym ECL has also been described quite specifically in the literature as CL that results from reactions that occur at an electrode surface in the presence of a photodetector, often a photomultiplier (PMT). Besides the oxidation of $Ru(bpy)_3^{2+}$ to $Ru(bpy)_3^{3+}$ at the electrode, this ECL process in the FI mode likely also involves oxidation of the analyte such as tripropylamine at the electrode to generate a reactive radical reducing intermediate that reacts with $Ru(bpy)_3^{3+}$ to give CL [5]. In this chapter, ECL will most often be used in this context; that is CL produced at an electrode surface.

Because these review articles are quite comprehensive up to their publication date and are readily available, in this chapter I emphasize studies from 1998 and later. Sections II–V discuss FI and LC instrumentation, FI and LC applications, CE instrumentation and applications, and conclusions and possible future developments, respectively. A list of abbreviations is included before the references. Most of the emphasis is on $Ru(bpy)_3^{3+}$ applications, which are better represented than luminol systems in the literature. With respect to applications, the chemical structures of the target compounds and derivatization chemistry are given emphasis. Previous CL review articles did not include chemical structures.

II. FLOW INJECTION AND LIQUID CHROMATOGRAPHIC DETECTOR INSTRUMENTATION

A. Luminol ECL

A schematic diagram for a typical FI instrument for luminol ECL is shown in Figure 1A [6]. The pH (usually very alkaline, pH 12) and ionic strength are adjusted first before mixing with luminol. For the determination of phenolic compounds, the presence of KCl caused a more stable ECL response. After mixing with the injected sample, this solution is pumped into the ECL cell containing a glassy carbon working electrode, an Ag/AgCl reference electrode, and a stainless steel counter electrode. These cells can be fashioned out of a

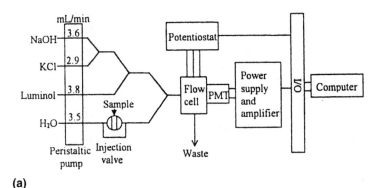

(a)

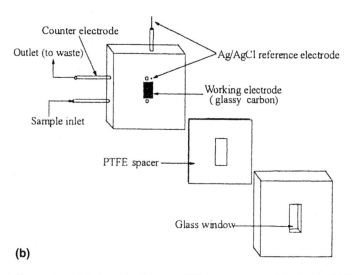

(b)

Figure 1 (a) Schematic diagram of FI instrument used for luminol ECL. (b) Thin-layer flow-through ECL cell. (From Ref. 6.)

solid fluorocarbon polymer block with a PMT detector mounted against the glass window (Fig. 1B) or constructed simply from a glass tube (i.e., 3 mm × 3 cm).

Although most studies of the ECL of luminol have been carried out in alkaline solution, there is one report of the polymerization of luminol on the electrode surface under acidic conditions using electrolysis at 1.2 V vs. SCE [7]. A mechanism of polymerization of luminol analogous to that of aniline to form polyaniline is proposed. Such a polyluminol electrode using FI with an alkaline carrier stream could enhance the ECL of flavin compounds (riboflavin and flavin mononucleotide). It was postulated that the photochemical activity of riboflavin could play a role in the ECL response.

For liquid chromatography, the consideration of flow cell volume is more important. In one study [8], a 37 μL volume flow cell fabricated from PEEK polymer with a gold gauze working electrode, a Pd/PdO reference electrode, and a Pt auxiliary electrode was mounted in a commercially available fluorimeter with the lamp disconnected. In a second study [9], the flow cell fabricated from a Kel-F block with two Pt disc electrodes was only 8 μL in volume. In general, because ECL of luminol is fast, the electrode system is contained within the flow cell.

B. Ru(bpy)$_3^{3+}$ Chemiluminescence

Because Ru(bpy)$_3^{3+}$ is somewhat unstable, it must be generated by oxidation of Ru(bpy)$_3^{2+}$. The three basic approaches for either FI or LC that will be reviewed here are (1) external generation batchwise by either electrolysis or an oxidizing agent and then mixing with the analyte, (2) on-line electrochemical or photochemical generation and then mixing with the analyte, and (3) in situ generation in the presence of the analyte at an electrode positioned in front of the photodetector (ECL). Instrumentally, methods 1 and 2 have the advantage of being compatible with commercially available LC luminescence detectors.

1. Comparison of In Situ Generation (ECL) and External Generation

A comparison of methods 1 and 3 has been made [10] using three approaches: (1) external generation of Ru(bpy)$_3^{3+}$ at an electrode, (2) ECL with a Ru(bpy)$_3^{2+}$ solution, and (3) ECL with immobilized Ru(bpy)$_3^{2+}$. A diagram of the overall instrument is shown in Figure 2, and details of the flow cells are shown in Figure 3. External generation of Ru(bpy)$_3^{3+}$ was done using a glassy carbon sponge electrode, an Ag/AgCl reference electrode, and a Pt counter electrode for 1 h for 200 mL of 2 mM Ru(bpy)$_3^{2+}$. Using an applied potential of 1.3 V, ECL of 2 mM Ru(bpy)$_3^{2+}$ at dual Pt working electrodes and ECL of Ru(bpy)$_3^{2+}$ immobilized on the Pt electrodes coated with a Nafion film were carried out using the same flow cell assembly with a volume of 100 μL. Using proline and oxalate as test compounds for flow injection, the ECL solution mode showed little change in CL as a function of flow rate for oxalate but a sharp decrease in signal with flow rate for proline. This is because the maximum emission intensity profile with time is estimated to be less than 1 s for oxalate but about 3–5 s for proline. The ECL immobilized mode was independent of flow rate as expected.

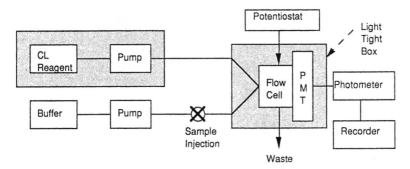

Figure 2 Schematic diagram for experimental comparison of CL modes. CL reagent and pump were used only for external generation mode. Potentiostat was used for the in situ/solution and in situ/immobilized modes. (From Ref. 10.)

The flow rate dependence of the CL reagent and carrier streams in the external generation mode varied for proline and oxalate, with optimum CL values of about 1 and 2 mL/min, respectively. The oxalate and proline working curves for the ECL solution and immobilized modes were linear over four orders of magnitude, but linearity was only three orders of magnitude for the external generation mode. The signal intensities over these linear ranges were consistently ordered ECL immobilized < ECL solution < external generation. The relative intensity ratios (external generation/ECL solution and ECL solution/ECL immobilized) for various compounds are tabulated in Table 1. The ECL solution mode generally

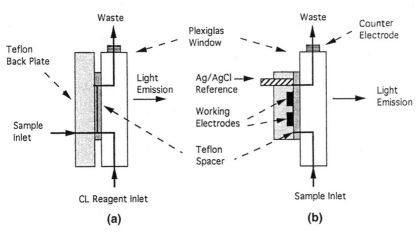

Figure 3 Details of flow cells. (a) External generation mode; (b) in situ/solution and in situ/immobilized modes. (From Ref. 10.)

Table 1 Relative CL Intensity Ratios for Different Modes of Ru(bpy)$_3^{3+}$ Generation

Compound (1 mM)	External generation/ ECL solution	ECL solution/ ECL immobilized
Oxalate	54	8
Proline	226	0.7
NADH	176	0.6
Tripropylamine	105	8
Tryptophan	38	14
5-Hydroxytryptophan	41	21
Histidine	30	87
Serotonin	21	55
Epinephrine	3	5
Gentamicin	348	0.2
Streptomycin	323	0.3
Gly-Pro[a]	nd	nd
Pro-Gly	175	2
Background[b]	25	2

[a]Not determined (nd) because external generation signal was 72 and both ECL signals were 0.
[b]CL intensities for external generation, ECL solution, and ECL immobilized were 40, 1.6, and 0.8, respectively.
Source: Calculated from data in Ref. 10.

produced a stronger signal than that for the immobilized sensor. The external generation mode provided markedly the strongest CL intensities, particularly for aliphatic amines and antibiotics. However, considering time, convenience, and reliability, the ECL solution mode was considered best. Only a short time was needed to polish the electrodes, whereas 2–3 h is required to immobilize the Ru(bpy)$_3^{2+}$, and the sensor must be replaced on a daily basis. The external generation mode involves a 30–60 min oxidation time, and stability of the reagent can be problematic.

A CL system using Ru(bpy)$_3^{2+}$ oxidized by Ce(IV) and an ECL system using an Al working electrode were compared for the determination of morpholine fungicides [11]. Calibration characteristics using dodemorph as the test compound for both methods were similar, showing linearity from about 1×10^{-7} M to 1×10^{-5} M and detection limits at 4×10^{-8} M. Although ECL was deemed the more elegant method, it was less tolerant to methanol, which was used for some of the sample extractions.

2. ECL Modes

Some of the findings in Ref. 10 have been followed up. A more detailed stopped flow analysis of the Ru(bpy)$_3^{3+}$ reaction with oxalate and proline has been

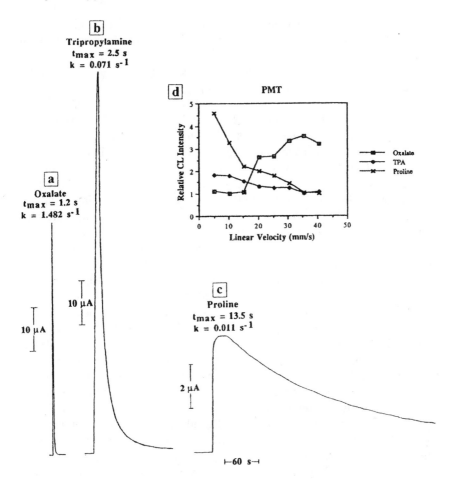

Figure 4 Stopped-flow traces for CL reaction of 1.0 mM Ru(by)$_3^{3+}$ with 0.2 mM (a) oxalate, (b) tripropylamine, and (c) proline. Inset figure (d) displays the CL intensities of each analyte over the range of linear velocities studied by flow injection analysis with the PMT. (From Ref. 12.)

provided using a reaction rate method to differentiate these two compounds in a mixture [12]. Temporal and spatial analyses of Ru(bpy)$_3^{3+}$ ECL in flowing streams were carried out in both stopped flow and FI modes using a charge-coupled device (CCD) for imaging. The maximum signals for the stopped flow CL time profiles for oxalate, tripropylamine, and proline occurred respectively at 1.2 s, 2.5 s, and 13.5 s (Fig. 4). The overall emission profiles covered only a few seconds for oxalate, about 1 min for tripropylamine, and 3–4 min for proline. As expected, the optimum flow rates were inversely related to the time of maximum emission for the three compounds, and this has been an important consideration even for compounds within the same general class such as amines. Using pseudo-first-order decay data, the stopped flow technique could differentiate kinetically between oxalate and proline in a mixture [13].

More information describing the use of $Ru(bpy)_3^{3+}$ immobilized on the fluorinated cation-exchange polymer Nafion and its possibility for regeneration can be found in Ref. 14. Better ECL detection limits and linear ranges were found for tripropylamine and oxalate using a Nafion–silica composite film compared to Nafion alone [15]. In addition, a comparison of ECL in solution and immobilized in a cationic polypyrolle derivative for tris(4,7-diphenyl–1,10-phenanthro-linedisulfonic acid)ruthenium(II) (RuBPS) has been made [16]. In solution, a maximum ECL response occurred with 250 µM of RuBPS, lower than the concentration usually reported for $Ru(bpy)_3^{3+}$. Slower linear velocities gave better ECL for RuBPS and oxalate in contrast to $Ru(bpy)_3^{3+}$ and oxalate; amines such as proline, tripropylamine, and NADH were also studied.

Similar ECL modes for $Ru(bpy)_3^{3+}$ have also been advocated. The generation of $Ru(bpy)_3^{3+}$ using a glassy carbon working electrode at alkaline pH with or without an organic modifier in the carrier stream has been examined [17]. With careful sanding and polishing, the glassy carbon electrode showed good stability in an alkaline mobile phase for at least 40 h. The presence of 10% acetonitrile did reduce the stability of the electrode response. For the determination of underiva-tized amino acids after ion-exchange chromatography, detection limits ranged from 100 fmol for proline to 22 pmol for serine. An ECL flow-through detector using a Pt working electrode and an Ag reference electrode has been fabricated using a silicon photodiode detector [18]. The advantages of the silicon photodiode are primarily the low voltage (12 V) required for the power supply and the potential for miniaturization in the design of a portable instrument. The detector was tested by mixing $Ru(bpy)_3^{2+}$ with tripropylamine as the injected analyte. The limit of detection of 135 ppb was higher than the 4 ppb possible using a PMT detector, but reproducibility of the silicon photodiode detector was better than 2% whereas that for the PMT ranged from 2% to 8%.

New immobilization methods for $Ru(bpy)_3^{2+}$ on films and particles are receiving interest. Recently, the ECL of $Ru(bpy)_3^{2+}$ immobilized in a polystyrene-sulfonate (PSS)–silica–Triton X–100 composite film was reported [19]. The PSS polymer–silica combination immobilized $Ru(bpy)_3^{2+}$ strongly on the electrode, and the Triton X–100 prevented fractures of the film during drying. Flow injection detection limits of oxalate and tripropylamine were 0.1 µM, and linearity covered three orders of magnitude. $Ru(bpy)_3^{2+}$ immobilized on a Dowex–50 cation-exchange resin was developed as an FI chemiluminescent sensor for oxalate, sulfite, and ethanol [20]. A strong CL signal for oxalate and sulfite (10^{-6} and 10^{-7} respective detection limits) and a much weaker one for ethanol (0.5% detection limit) were observed in the presence of either a $KMnO_4$ or Ce(IV) oxidizing agent. The Ru(bpy)$_3{}^{2+}$-immobilized resin was stable for 6 months.

The covalent immobilization of a new CL ruthenium-based dichlorosilyl-bipyridyl ligand complex onto silica particles was of particular interest [21]. The silica particles with the bound CL reagent were packed into a 1.5×15 mm tube that was mounted in front of the PMT. Using sequential FI, zones of sulfuric acid, the oxidizing agent Ce(IV), and the analyte (e.g., codeine) were passed through the detection cell. The CL response was much stronger when codeine was prepared in acetonitrile than in water. After a steady-state response of 100 cycles, a detection limit of 1×10^{-8} M codeine was found to be only somewhat inferior to that for a homogeneous reaction. Standard FI that mixed a sulfuric acid sample carrier stream and a Ce(IV) mobile phase just before the detection cell indicated a detectability for oxalate of 3×10^{-7}, M, which compares favorably to other immobilized Ru complex CL approaches.

3. External Generation

Many commercial CL detectors have large flow cell volumes on the order of 100 µL, which cause excessive band broadening. An LC fluorimeter with an 8 µL flow cell was modified for Ru(bpy)$_3{}^{3+}$ CL [22]. The mobile phase and analyte pass conventionally into the flow cell from the bottom, whereas the Ru(bpy)$_3{}^{3+}$ reagent generated by external batchwise electrolysis enters the detector from the top by means of a fused silica capillary. A stainless steel tube leading out of the top of the flow cell is coaxially fitted with this capillary straight through a tee connector, leaving enough space to allow waste to exit between the tubing walls and out the third end of the tee connector. One advantage of this flow cell design is that the CL reaction takes place directly in the flow cell mounted in front of the PMT. The Ru(bpy)$_3{}^{3+}$ CL reaction is short-lived for tertiary amines, and the optimum light intensity could be missed if the reaction occurs remotely in a mixing tee, away from the PMT. Compared to the standard T-mixing design this flow cell showed enhanced detectability for pheniramine, a pharmaceutical with a tertiary aliphatic amine substituent, at flow rates of less than 200 µL/min. The separation of anticholinergic compounds such as atropine, cyclopentolate, cyclobenzaprine, procyclidine, and dicyclomine under microbore LC conditions using a 1 µL injector and a 2×150 mm reversed-phase C8 column was facilitated by the use of 20% 2-butanone as a mobile phase additive. Recently a 5.5 µL spiral flow cell made with 100 µm × 70 cm Teflon capillary (Fig. 5) was characterized for luminol CL [23]. This detector should be useful for microbore or perhaps capillary LC with Ru(bpy)$_3{}^{3+}$ detection, particularly for mixtures of compounds having different optimum CL emission times.

External batchwise Ru(bpy)$_3{}^{3+}$ production by chemical means with improved long-term reproducibility continues to be researched. Two alternative

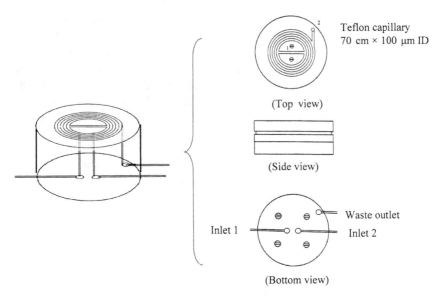

Figure 5   Schematic diagram of the 5.5 µL spiral CL flow cell. (From Ref. 23.)

strategies using lead dioxide were compared by Gerardi et al. [24]. First, $Ru(bpy)_3^{3+}$ was produced using a slurry of lead dioxide at sulfuric acid concentrations of 0.05–2 M. Stability of the CL response of codeine was improved from just a few hours using 0.05 M acid to about 30 h with an acid strength 10 times greater (Fig. 6a). Using a 2 M acid strength, CL stability up to about 240 h was observed. The reason for this improved stability was thought to be the increased difficulty of the oxidation of water at low pH. However, mixing this acidic $Ru(bpy)_3^{3+}$ solution with a high pH carrier to maintain a strong CL signal is problematic, as evidenced by split FI peaks. A second strategy studied was to recirculate the principally $Ru(bpy)_3^{3+}$ solution through lead dioxide particles trapped within a glass wool filter to reoxidize $Ru(bpy)_3^{2+}$ in order to achieve a steady-state concentration. An overall response variation of only 14% was observed over a 90 h period for the reagent prepared in 0.1 M acid (Fig. 6b).

The problem of oxidation of water by $Ru(bpy)_3^{3+}$ that causes its instability was overcome by the synthesis of anhydrous tris(2,2′-bipyridyl)ruthenium(III) perchlorate for dissolution in acetonitrile [25]. It was synthesized by oxidation of $Ru(bpy)_3^{2+}$ with chlorine gas and subsequent precipitation with sodium perchlorate. The stability of this $Ru(bpy)_3^{3+}$ salt in acetonitrile was followed spectrophotometrically at 674 nm, and only a decrease of <6% was noted over a time period of 50 h. Flow injection using codeine to generate the CL response

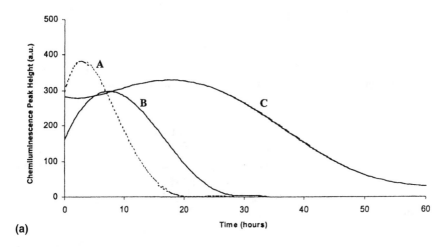

(a)

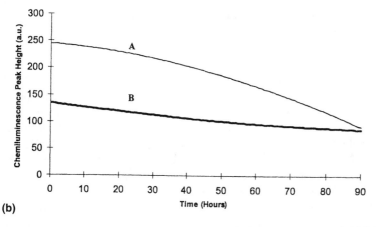

(b)

Figure 6 (a) Reagent response profiles A, B, and C for codeine (5×10^{-6} M) in sodium acetate buffer (50, 100, and 500 mM, respectively, at pH 5.0) using Ru(bipy)$_3^{3+}$ (2.0 mM prepared in 0.05, 0.10, and 0.50 M H$_2$SO$_4$, respectively). (b) Reagent response profiles A and B for codeine (5×10^{-6} M) in sodium acetate buffer (50 and 100 mM, respectively, at pH 5.0) with Ru(bipy)$_3^{3+}$ (2.0 mM prepared in 0.05 and 0.10 M H$_2$SO$_4$, respectively) using the continuous oxidation recirculating system. (From Ref. 24.)

showed stability over a 50 h time period and a decrease in signal of <8%. Besides eliminating the requirement for any prior oxidation step, advantages of this approach are the compatibility of acetonitrile with the carrier stream and the reduced blank response due to the lower buffer concentration required. This Ru(bpy)$_3^{3+}$ perchlorate salt in dry acetonitrile was applied to the FI determination of oxalate in Bayer process samples after passage through a short anion-exchange column [26].

Recently, $Ru(bpy)_3^{2+}$ bound to water-soluble polymers was considered as a potential means of reagent stabilization [27]. A sodium sulfate solution containing PSS bound to $Ru(bpy)_3^{2+}$ generated a lower current profile upon electrolysis than a free $Ru(bpy)_3^{2+}$ solution and a corresponding longer oxidation time (about 2 h for a 100 mL solution). However, the stability of the tripropylamine CL signal at room temperature as measured by at least 90% maintenance of response was markedly longer, up to about 12 h using the polymer bound reagent compared to only about 100 min for the free reagent. The CL signal dropped slowly to the 60% level at 24 h, but CL response was maintained up to 60 h. However, detectability of tripropylamine using the free reagent was about 20 times better than with the polymer-bound reagent. A polyglutamic acid polymer–$Ru(bpy)_3^{3+}$ solution showed better stability than the free reagent but not for as long (16 h) as the PSS–$Ru(bpy)_3^{3+}$ mixture. A possible explanation for the better stability using the PSS polymer is that the more hydrophobic aromatic polymer hinders access of water for oxidation by $Ru(bpy)_3^{3+}$.

4. On-Line Oxidation

An approach using on-line oxidation of $Ru(bpy)_3^{2+}$ in a flowing stream also appears promising. Oxidation of $Ru(bpy)_3^{2+}$ to $Ru(bpy)_3^{3+}$ using a solid-phase reactor packed with 250 μm silica gel particles coated with lead dioxide provided a reproducible FI response for either oxalate [28] or codeine for about 70–80 injections with a detectable signal maintained for 1000 injections [29]. Further improvement in stability was deemed necessary for process analysis. Generation of $Ru(bpy)_3^{3+}$ from a reservoir of 1 mM $Ru(bpy)_3^{2+}$ and 2 mM peroxydisulfate ion by flowing this solution through a coiled PTFE tubing (1 mm × 8m) photoreactor is possible, and this process can be easily interfaced for postcolumn LC derivatization (Fig. 7) [30]. Using a flow rate of 0.5 mL/min, the irradiation time was 12 min. The actual concentration of $Ru(bpy)_3^{3+}$ produced was not determined, but the CL signal-to-noise ratio leveled off at 0.5–1 mM $Ru(bpy)_3^{2+}$. A consistent CL response was found for 3 days when the solution was stored in the refrigerator despite the instability of peroxydisulfate in an aqueous solution. Low detection limits in the 10–100 fmol range for tertiary aliphatic amines were found despite the low reaction pH of 5.5. This photochemical method for $Ru(bpy)_3^{3+}$ production has been used for the FI determination of the pesticides carbofuran and promecarb [31]. Short term repeatability was 1.6% ($n=10$), and reproducibility studied on five different days was excellent at 2%.

On-line electrochemical generation of $Ru(bpy)_3^{3+}$ using a coulometric flow-through reactor has been known for some time [32]. A recent flow diagram of this system is shown in Figure 8 [33]. The guard cell containing the working

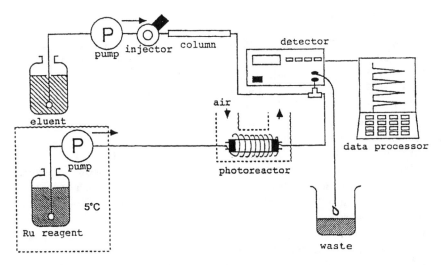

Figure 7 Flow diagram of an instrument combining on-line photochemical generation of Ru(bpy)$_3^{3+}$ with liquid chromatography and chemiluminescence detection. (From Ref. 30.)

electrode where the electrochemical oxidation takes place was porous graphite. Particular care was taken to eliminate dissolved oxygen by using both a He purge and an on-line degasser for all flow streams. In addition, the tube from the mixing tee to the CL detector was metal, not polymeric. Pump pulsation was minimized using the dampers and resistance coils.

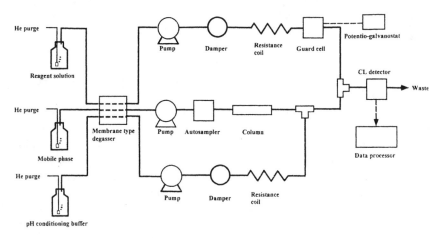

Figure 8 Flow diagram of the on-line electrochemical LC–Ru(bipy)$_3^{3+}$ CL system. (From Ref. 33.)

5. Microfabrication for FI and LC

On-chip generation and detection of $Ru(bpy)_3^{2+}$ ECL has been accomplished in several ways. An interdigitated electrode array (IDA) resting on top of a silicon photodiode has been used [34]. The IDA layout, made from either Au or Pt, consisted of 125 alternate pairs of 1 mm long microbands, with an electrode spacing of 0.8 μm and an electrode width of 3.2 μm. Two array–photodiode assemblies, both contained in only a 5×6 mm² area and incorporated into a 2.25 μL flow channel, were fabricated so the photocurrent produced from the ECL light generated at one assembly could be detected in reference to the other dummy assembly. A calibration curve for $Ru(bpy)_3^{2+}$ from 0.5 to 50 μM showed excellent linearity. This IDA was applied for the determination of codeine at a flow rate of 0.5 mL/min with a 1 min cycle time using a pH 4 acetate buffer containing different 1 mM ruthenium complexes [35]. The bis(2,2′-bipyridyl)(1,10-phenanthroline)ruthenium(II) complex gave a signal about 2.5 times better than that of $Ru(bpy)_3^{2+}$. A detection limit for codeine at 50 μM and a linearity of two to three decades were found. The combination of a microenzymatic reactor and an ECL detector, both fabricated as Pt or carbon IDAs, was incorporated into an FI instrument [36]. The reticulated reactor (2.7 μL) was packed with silica particles with immobilized glucose oxidase. Glucose was defected either by electrochemical oxidation of the enzymatically generated H_2O_2 (2 μM detection limit) or by using the ECL detector in which the oxidation of luminol and H_2O_2 generated light (50 μM detection limit). The ECL detector was also characterized separately using codeine and $Ru(bpy)_3^{3+}$ chemistry, indicating a detection limit of 100 μM. A micro total analytical system (μTAS) moving reagents through 200 μm wide channels on a glass chip by electro-osmotic flow was characterized by using codeine as the analyte and $Ru(bpy)_3^{3+}$ CL detection with light emission enhanced with the nonionic surfactant Triton-X [37]. A similar instrument with negative pressure pumping provided a nanomolar detection limit for atropine upon reaction with chemically oxidized $Ru(bpy)_3^{2+}$ [38].

An electrochemical cell microfabricated from silicon containing gold and optically transparent indium tin oxide (ITO) electrodes was characterized for the determination of DNA amplified by the polymerase chain reaction [39]. A silicon and glass microcell with dimensions of 1.5 cm × 2 cm × 0.2 cm with miocromachined fluid channels and thin-film Pt electrodes was used for the determination of DNA with sample volumes of 150 μL [40]. DNA strands labeled with $Ru(bpy)_3^{2+}$ and bound through a biotin–streptavidin linkage to paramagnetic beads are injected into the microcell shown in Figure 9. They are attracted magnetically to the bottom Pt electrode surface, and the cell is flushed with a tripropylamine solution to remove any unbound DNA. A potential sufficient to oxidize the $Ru(bpy)_3^{2+}$ label is applied, and the resultant ECL is recorded. A detection limit of 40 fmol DNA was possible. Limitations to be addressed are

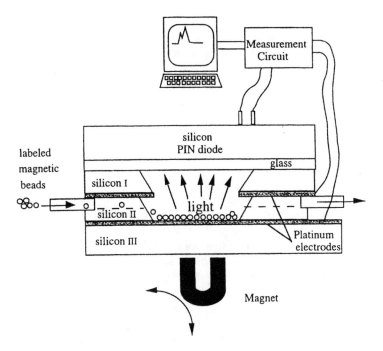

Figure 9 A cross section of the microcell showing the injection of the magnetic beads with attached TBR-labeled DNA strands. TBR = tris (2,2'bipyridyl) ruthenium (From Ref. 40.)

that only part of the bead surface (that within several nanometers of the electrode) is active and the beads are optically opaque possibly obstructing the emitted light.

III. FI AND LC APPLICATIONS OF LUMINOL ECL AND RU(BPY)$_3{}^{3+}$ OR RU(PHEN)$_3{}^{3+}$ CL

The luminol section will be quite brief, dealing mainly with FI of phenolic compounds, FI coupled with oxidase enzyme reactions, and two LC applications. The Ru metal complex section is divided into two main parts: direct CL determination of sample compounds and derivatization chemistry before the CL reaction. Structural considerations related to the Ru(bpy)$_3{}^{3+}$ CL response will be summarized first; then short descriptions of applications of mainly amines, classified when possible by their pharmaceutical activity, follow. It will be shown that derivatization chemistry has allowed for the CL determination of a wider variety of organic compounds. Representative figures of chemical structures to show readers the potential of $Ru(bpy)_3^{3+}$ CL for the determination of compounds of interest are provided. A detailed discussion of coreactants for generation of CL with Ru metal complexes is presented in Chapter 5 of this book.

A. Luminol Applications

Most recent FI applications have involved the determination of aromatic compounds with multiple hydroxyl groups. Pyrogallol [41] and gallic acid [42] were both shown to enhance the ECL of luminol in alkaline solutions. Generation of superoxide radicals at the electrode surface was likely responsible for the greater ECL response. Detection limits of about 2×10^{-8} M with linearity up to 0.1 mM were found for both compounds. Epinephrine was also found to enhance the weak ECL emission of luminol at 425 nm due to the 3-aminophthalate species [43] and showed a detection limit and linearity similar to those described in Ref. 42. A mechanism involving the oxidation of both epinephrine and luminol at the Pt electrode held at 0.6 V, producing respectively a reducing intermediate and the luminol radical, was proposed. Because of the need for dissolved oxygen in the carrier stream, this intermediate can probably react to give the hydroxyl radical, which then reacts with a luminol radical to give enhanced light emission. A cation-exchange column was recommended to eliminate the interference of transition metal ions; however, organic acids were less problematic.

Inhibition of the luminol ECL signal has also been used to analytical advantage. When injected into a luminol carrier stream at pH 12, the polyphenolic compound tannic acid gave a negative peak after oxidation at 0.4 V vs. Ag/AgCl at a glassy carbon electrode [6]. A detection limit of 2×10^{-8} M with linearity up to 1×10^{-5} M was found. Catechol derivatives such as catechol, 3,4-dihydroxybenzoic acid, and chlorogenic acid also quenched the ECL signal when oxidized at 0.8 V vs. Ag/AgCl at pH 12 [44]. A mechanism involving the competition of the catechol derivatives with electrochemically produced reactive oxygen species that can react with luminol to produce light was proposed. Detection limit and linearity data were similar to those described above. Dopamine has also been determined on the basis of its inhibition of luminol ECL [45].

Other luminol ECL FI methods involve lipid hydroperoxides, histidine, hydrazine compounds, and vanadium. The determination of methyl linoleate hydroperoxide in an acetonitrile aqueous solution at 0.45 V was possible owing to the light emission from the reaction of the excited luminol intermediate diazaquinone with the hydroperoxide compound [46]. A detection limit by FI of 0.3 nmol was found. Histidine enhanced the ECL of luminol, and a detection limit of 0.56 μM was indicated [47]. Hydrazine [48] and the antibacterial isoniazid (4-pyridinecarboxylic acid hydrazide) [49] could markedly enhance the ECL of luminol using FI. The detection limit was 6×10^{-9} M for hydrazine and 2.8×10^{-8} for isoniazid. Owing to the electrochemical production of V(II) from vanadate, emission of light was observed upon reaction with luminol [50].

A major focus of ECL with luminol is in conjunction with oxidase enzyme reactions. The basic premise is that light emission results from the oxidation of luminol in the presence of H_2O_2 produced from the enzyme reaction. An optical fiber biosensor based on glucose oxidase immobilized through sol-gel chemistry on a glassy carbon electrode has been used for the FI determination of glucose at concentrations of 50 μM to 10 mM [51]. Using a similar fiber optic biosensor but with either glucose oxidase or lactate oxidase immobilized on polyamide and collagen membranes, FI detection limits for either lactate or glucose were in the 30–60 pmol range [52]. A follow-up study for choline using cholin oxidase immobilized by physical entrapment on a diethylaminoethyl Sepharose–poly(vinyl alcohol) sensing layer indicated an ECL luminol detection limit of 10 pmol [53]. A bilayer electrode based on polymers formed from the anodic polymerization of luminol and tris(5-aminophenanthroline) and an ITO conductive glass was shown to generate light when contacted with H_2O_2 at pH 10 [54]. Further modification of this electrode with glucose oxidase permitted the FI determination of glucose in dilute solutions (0.01–10 mM).

Immunoassays have also been done using luminol ECL. Glucose oxidase was attached to aminosilanized ITO-coated glass wafers, which were then used as working electrodes [55]. Results indicated that ECL enzyme immunoassays would be possible using glucose oxidase as the antibody label. Luminol-labeled antibodies prepared using glutaraldehyde chemistry were used in a 2,4-dichlorophenoxyacetic acid (2,4-D) competitive ECL immunosensor [56]. The 2,4-D as the antigen was immobilized at a glassy carbon surface, where oxidation occurred at 0.5 V in the presence of H_2O_2. Detectability of 0.2 $\mu g/L$ of free 2,4-D by FI was possible.

Only a few papers have reported the use of luminol ECL for liquid chromatography. An early study showed the indirect detection of benzaldehyde, nitrobenzene, and methyl benzoate after separation on a C18 column using a methanol–water mobile phase containing luminol, H_2O_2, and Co^{2+} as the catalyst [9]. Detection limits were 0.01–0.06 μM. The direct LC determination of histamine was achieved by precolumn derivatization with N-(4-aminobutyl)-N-ethylisoluminol isothiocyanate followed by ECL detection [8]. A linear range of 0.5–10 nmol with a detection limit of 1.5 pmol was established.

B. Direct Ru(bpy)$_3{}^{3+}$ CL Determination of Sample Compounds

1. Structural Considerations

Although hydrazine and derivatives [57] as well as some oxygen containing compounds such as oxalate [58,59] and certain ketones [60] can be determined directly by Ru(bpy)$_3{}^{3+}$ CL, most applications involve pharmaceuticals with tertiary or secondary amine groups either present as separate moieties or incorporated in ring structures. Several studies have shown the importance of structural features near the nitrogen atom involved in the oxidation pathway leading to CL. Initially, an inverse linear trend of log CL intensity versus the oxidation potential for the nitrogen compound was established. Tertiary aliphatic amines with the lowest oxidation potentials gave the highest CL responses, followed by secondary aliphatic amines, then primary aliphatic amines [61]. However, the presence of functional groups such as C=O bonded to the nitrogen atom reduced the CL response considerably, as evidenced by the response of the test compound methylacetamide. In addition, the response of proline was significantly better than that of proline peptides [62]. When the hydrogen on the nitrogen of the indole ring was replaced by a hydroxy or methoxy group, the Ru(bpy)$_3{}^{3+}$ CL intensity was markedly weaker, about 50 times less than that for tryptophan [63]. The role of the electron-withdrawing or -donating character of the substituent attached to the α-carbon of amino acids was found to affect the CL response [64]. The relative intensities varied by a factor of 55, with serine giving the lowest response and leucine the highest.

The relationship of structural attributes of tertiary amines to Ru(bpy)$_3{}^{3+}$ CL has been reviewed [65]. It was considered important that the amine group have a hydrogen atom attached to the α-carbon, which after deprotonation will become the radical site. Electron-withdrawing substituents on either the nitrogen or the α-carbon will generally decrease the CL response as noted by the low response of 2-chloroethylamine. Electron-donating substituents such as alkyl groups favor CL. Resonance stabilization present in aromatic amines reduces CL activity. Freedom of the substituents attached to the nitrogen atom to adopt a more planar geometry after oxidation (removal of an electron) will improve CL. An example of this is the comparison of quinuclidine (weak CL), where the nitrogen atom is in a rigid nonplanar geometry, and codeine (strong CL), where the methyl group attached to the nitrogen can rotate freely. Finally, competition from a more easily oxidized moiety such as a phenolic group can limit CL response. The presence of phenolic moieties and aromatic or quaternary nitrogen atoms quenched the CL of certain types of alkaloids [66].

Carbonyl compounds, particularly β-diketones such as 1,3-cyclopentane-dione, can be determined with high sensitivity by Ru(bpy)$_3{}^{3+}$ CL [60]; α-diketones and monoketones are respectively less responsive. A linear correlation was found between CL intensity and the ^{1}H NMR shift of the active methylene proton. A mechanism involving oxidation of the ketone to a radical cation that forms a neutral radical that can react with water to give an alcohol was postulated. A variety of malonic acid derivatives and their esters were CL-active when Ru(bpy)$_3^{3+}$ CL was used, but related cyano compounds such as cyanoacetic acid and malonylnitrile gave the best detection limits in the 5–200 pmol range [67].

2. Specific Applications of Amines

A noncomprehensive listing of amine compound groups with common examples in parentheses determined directly by Ru(bpy)$_3^{3+}$ CL follows: aliphatic amines (ranitidine), alkaloids (codeine), amino acids (proline and tryptophan), antibiotics (erythromycin), antidepressants (amitriptyline), antihistamines (diphenhydramine), aromatic amines (methamphetamine), β-blockers (oxprenolol), and diuretics (hydrochlorothiazide). Structures for all of the compounds in parentheses are given in Figure 10, and a brief discussion of each follows. Newer papers of related compounds will also be discussed here.

Simple aliphatic amines give perhaps the best detection limits, in the femptomole range, and tripropylamine is often used as a test compound for a new instrument. The separation of C$_2$–C$_6$ trialkylamines on a cyano LC column using an aqueous acetate–acetonitrile mobile phase demonstrated the utility of Ru(bpy)$_3^{3+}$ as a postcolumn derivatization reagent [68]. Ion-exchange LC was used to separate glyphosphate (HO$_2$CCH$_2$NHCH$_2$PO$_3$H$_2$) and related compounds (iminodiacetic acid, diethanolamine, hydroxyethylglycine, and glycine) with Ru(bpy)$_3^{2+}$ in the mobile phase and ECL detection [69]. The detection limit of glyphosphate was 0.01 μM, with linearity over five orders of magnitude. The enantiomeric separation of β-amino alcohols was done on a C18 column coated with a N-n-dodecyl-(1R,2S)-norephinephrine stationary phase using a Cu(II) barbital mobile phase and CL detection [70]. The antiulcer drug ranitidine, primarily aliphatic in structure with tertiary and secondary amine groups and only one pair of conjugated double bonds (Fig. 10a), should also be considered a good candidate for Ru(bpy)$_3^{3+}$ CL. Although linearity of response was problematic, a reproducibility of 1.7% and a detection limit of 6×10^{-7} M were determined [71].

Alkaloids having anticholinergic (atropine), analgesic (codeine, see Fig. 10b), and narcotic (cocaine) properties are an important class of compounds that can be detected in the nanomolar range. Flow injection with ECL detection of codeine, heroin, and dextromethorphan indicated linearity from 0.1 to 1 μM with detection limits in the 15–45 nM range [72]. Recently, the alkaloid drugs pethidine, atropine, homatropine, and cocaine were determined by ECL at the sub micromolar level [73]. Solid-phase extraction was used to separate the drugs from urine. The compound sophoridine and related lupin alkaloids such as matrine

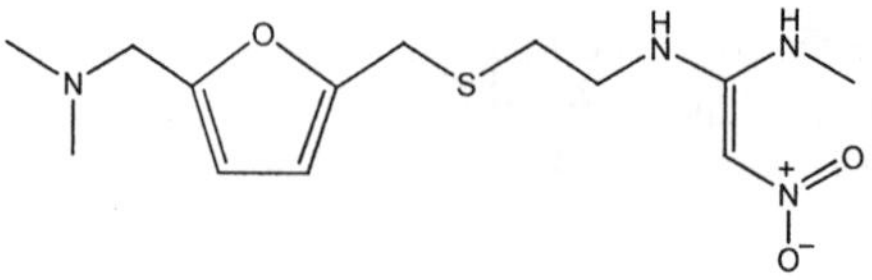

N
O
S
H
N
H
N
N
+
O
O
−

(a)

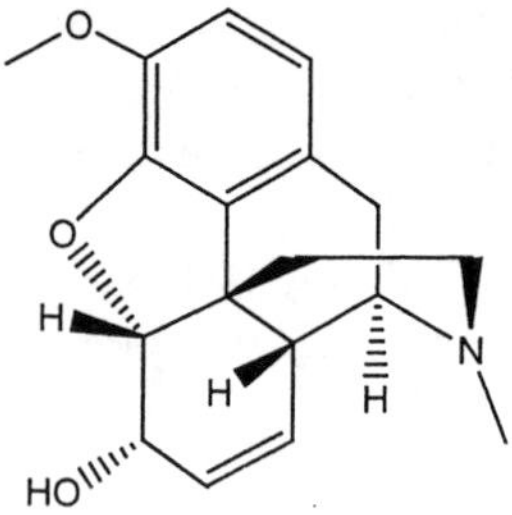

O
O
H
H
H
N
HO

(b)

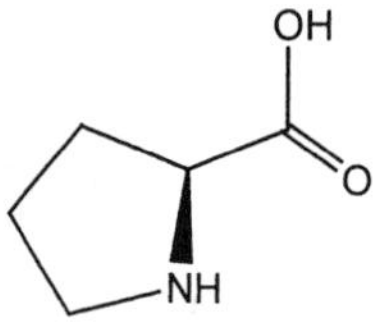

OH
O
NH

(c)

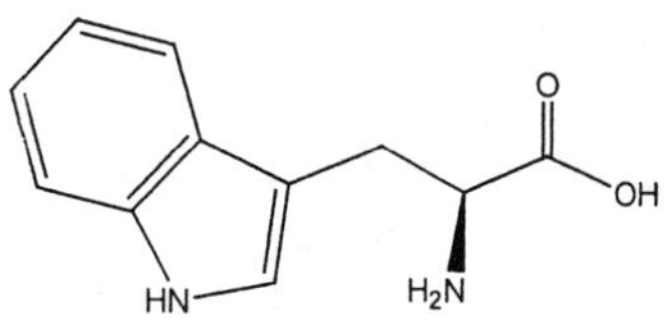

O
OH
HN
H_2N

(d)

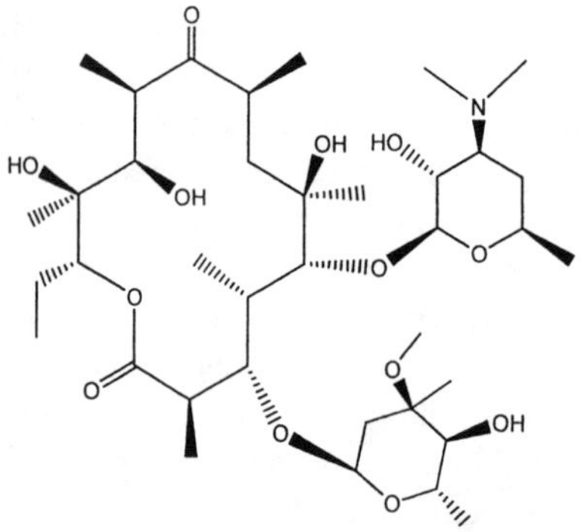

O
OH
HO
OH
HO
N
O
O
O
O
O
O
OH

(e)

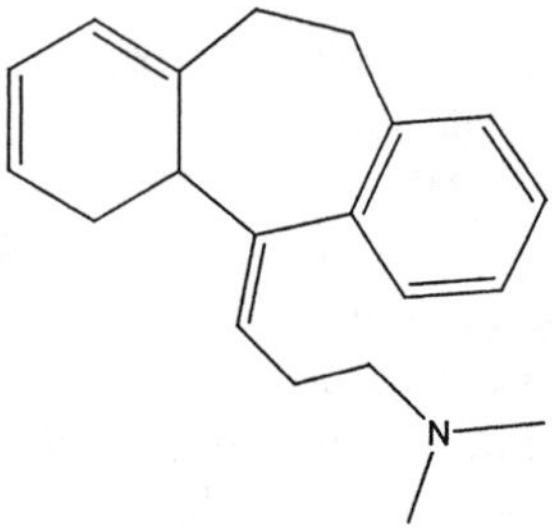

N

(f)

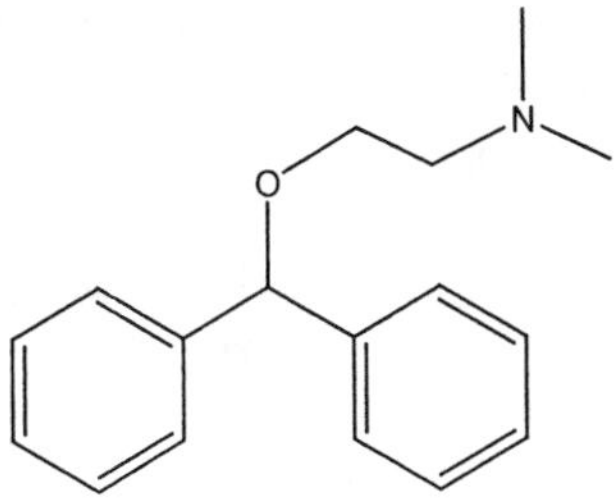

Figure 10 Structure of amines reacting directly with Ru(bipy)$_3^{3+}$ to give CL. (a) Ranitidine; (b) codeine; (c) proline; (d) tryptophan; (e) erythromycin; (f) amitriptyline; (g) diphenhydramine; (h) methamphetamine; (i) oxprenolol; (j) hydrochlorothiazide. See text for more information.

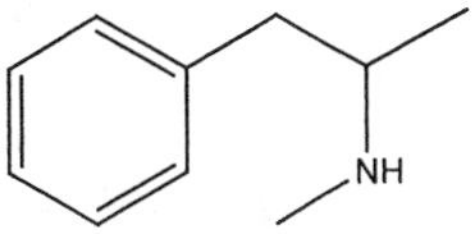

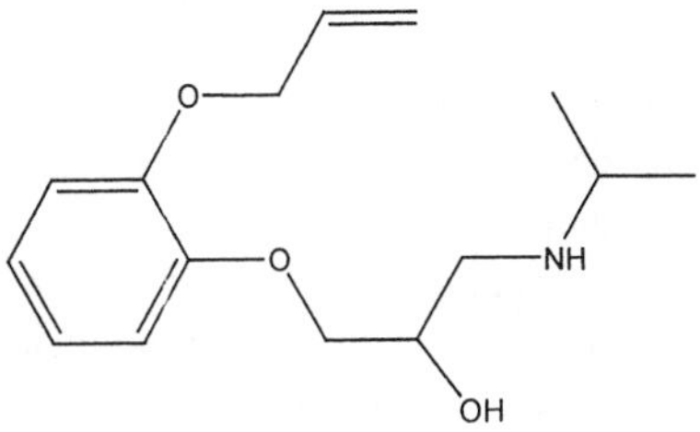

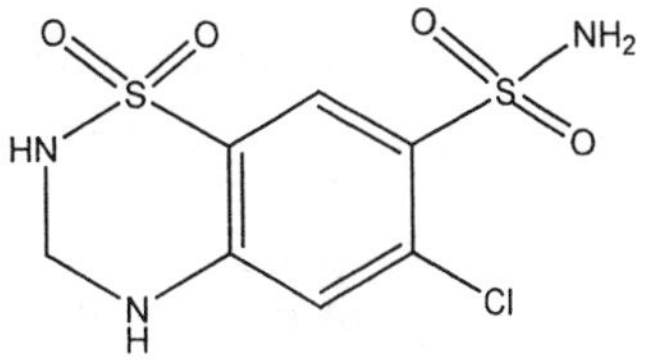

have been determined by LC with Ru(bpy)$_3^{3+}$ ECL detection [74]. Ionization potentials (IPs) defined as IP = $E_{cation} - E_{neutral}$ were calculated for the tertiary nitrogen atom between the two fused rings of the alkaloid structure by a hybrid B3LYP density functional method as implemented in the Gaussian 98 computer package. The alkaloid IP values (7.05, 7.15, and 7.18 eV) were inversely related to the respective ECL signal intensities (46, 38, and 35 mV). In addition, the tertiary, secondary, and primary propylamine calculated IP values (respectively 6.84, 7.54, and 8.30) correlated inversely to the ECL signals of 71, 6, and 2.

The amino acids proline, hydroxyproline, histidine, and tryptophan all have secondary amine substituents, allowing CL detection at generally lower levels than the other amino acids. Hydroxyproline and proline (Fig. 10c) in serum have been determined by ion-pair LC using octanesulfonate with what appears to be a commercial ECL detector, ECR–2000, made by the Comet Co. [75]. The relationship of aging to serum hyroxyproline levels was determined by $Ru(bpy)_3^{3+}$ CL [76]. The separation of D- and L-tryptophan (Fig. 10d) in plasma samples was accomplished using a Cu(II), D-phenylalanine mobile phase with 15% methanol [77]. Flow injection for determination of L-cysteine in pharmaceutical preparations using photogenerated $Ru(bpy)_3^{3+}$ is a recent application [78]. Cystine after reduction to cysteine in a minicolumn packed with Cu-coated Cd particles could also be determined.

The antibiotic erythromycin is an excellent representative of a target compound well suited for $Ru(bpy)_3^{3+}$ CL [79] because (1) it is nonvolatile and cannot be determined, without pyrolysis, by GC; (2) it has no good UV-Vis chromophore such as an aromatic ring but does have a tertiary amine group (Fig. 10e); and (3) there are many other structurally related macrolide antibiotics such as tylosin [80]. The microbore LC/ECL determination of erythromycin in urine and plasma using a 1×150 mm C18 column included $Ru(bpy)_3^{3+}$ in the mobile phase, which did modify the stationary phase [81]. The major peak in the chromatogram was erythromycin, and a detection limit of 0.01 μM or 50 fmol was shown. The antibiotic clindamycin and its phosphate analog both have an N-methylpyrrolidinyl substituent but with no conjugated double bonds in the structure. The $Ru(bpy)_3^{3+}$ CL detection limits for clindamycin and clindamycin-2-phosphate were 100 and 10 ppb, respectively, representing improvement factors of about 100 and 30 compared to UV detection at 214 nm [82].

The cephalosporin antibiotic cefadroxil with a β-lactam (cyclic amide) structure has been determined at concentrations as low as 5×10^{-8} M [83]. Previously, 10 β-lactam antibiotics and their hydrolysis products by corresponding enzymes were determined by $Ru(bpy)_3^{2+}$ ECL [84]. In addition, bacterial broth cultures could be analyzed for the actual β-lactamase enzymes. The antibacterial ofloxacin, which has an N-methyl piperazinyl substituent off a benzoxazine ring, was identified in chicken tissue [85]. Simple aromatic amines with secondary amine substituents such as adrenaline and methamphetamine can be more unpredictable for analysis. Adrenaline was found to strongly inhibit $Ru(bpy)_3^{2+}$ ECL, possibly due to the production of benzoquinone derivatives such as adrenochrome and adrenalinequinone. However, an indirect analytical method was worked out and applied to pharmaceutical samples [86]. Although local anesthetics with tertiary and/or secondary amine substituents such as lidocaine and procaine were determined by FI/ECL at pH 5.5 at the 10^{-8} M level [87], methamphetamine (Fig. 10h) was determined directly by ECL at pH 9 with very low detection limits in the 10^{-11} M range [88]. The effect of surfactants was also considered in this work. Phencyclidine [1-(1-phenylcyclohexyl)piperidine], and its major metabolites were determined in urine after silica solid-phase extraction by reversed-phase LC with $Ru(bpy)_3^{3+}$ CL over a 25–400 ng/L linear range [89]. A similar paper described the determination of disopyramide in serum [90]. The nonsteroidal anti-inflammatory drugs flufenamic acid and mefenamic acid, aromatic carboxylic acids with a secondary amine group, have been determined in pharmaceutical samples by FI using Ru(bpy)$_3^{3+}$ generated on-line with Ce(IV) [91].

The β-blockers atenolol and metoprolol as well as three others similar in structure to oxyprenolol (Fig. 10i) were determined by FI in pharmaceutical preparations at the 8–50 pmol level using Ru(bpy)$_3^{3+}$ ECL detection [92]. This detection mode for β-blockers, as expected, gave much simpler chromatograms for the LC analysis of human urine samples than UV detection. Antihistamines such as diphenhydramine (Fig. 10g) determined by reversed-phase LC with both UV at 214 nm and Ru(bpy)$_3^{3+}$ CL detection indicated comparable detection limits of 5–10 pmol [93]. However, urine samples spiked with low levels of antihistamines such as pheniramine at 0.15 μg/mL were obscured by the large unretained sample matrix peak with UV detection but easily discerned by CL detection with a much smaller unretained peak.

The tricyclic antidepressants have secondary or tertiary aliphatic amine substituents on fused ring structures, as do quite a few important pharmaceuticals such as thiazide diuretics. As an extension of the FI/ECL determination of amitriptyline (Fig. 10f) [94], the compounds imipramine, desipramine, amitriptyline, nortriptyline, and clomipramine have been separated by LC with postcolumn Ru(bpy)$_3^{3+}$ CL detection [95]. Detection limits were 12–105 fmol for a 20 μL injection. The adrenergic blocker yohimbine is structurally based on an indole with a saturated multiring substituent that has tertiary amine, hydroxyl, and ester groups. Recovery from human serum at 50 ng/mL, not quite double the detection limit, was 97% [96]. Using FI with the externally generated Ru(bpy)$_3^{3+}$ CL mode, hydrochlorothiazide (Fig. 10j) was determined in pharmaceutical samples in an acetate/acetic acid carrier stream with detection limits of 1–2 pmol [97].

C. Derivatization Chemistry in Conjunction with Ru(bpy)$_3^{3+}$ CL

Both pre- and online derivatization in conjunction with FI or LC have been reported, and applications seem to be increasing to include non-amine-type compounds. Major compound classes of interest, summarized below, are amines, carboxylic acids and alcohols, sulfur compounds and metal complexes, and enzyme-based reactions. Again, representative chemical structures of derivatives are provided in Figure 11.

1. Amines

Although tertiary amines are usually easily detected by Ru(bpy)$_3^{3+}$ CL, there have been at least two reports of the use of an oxidizing agent, Ce(IV) in sulfuric

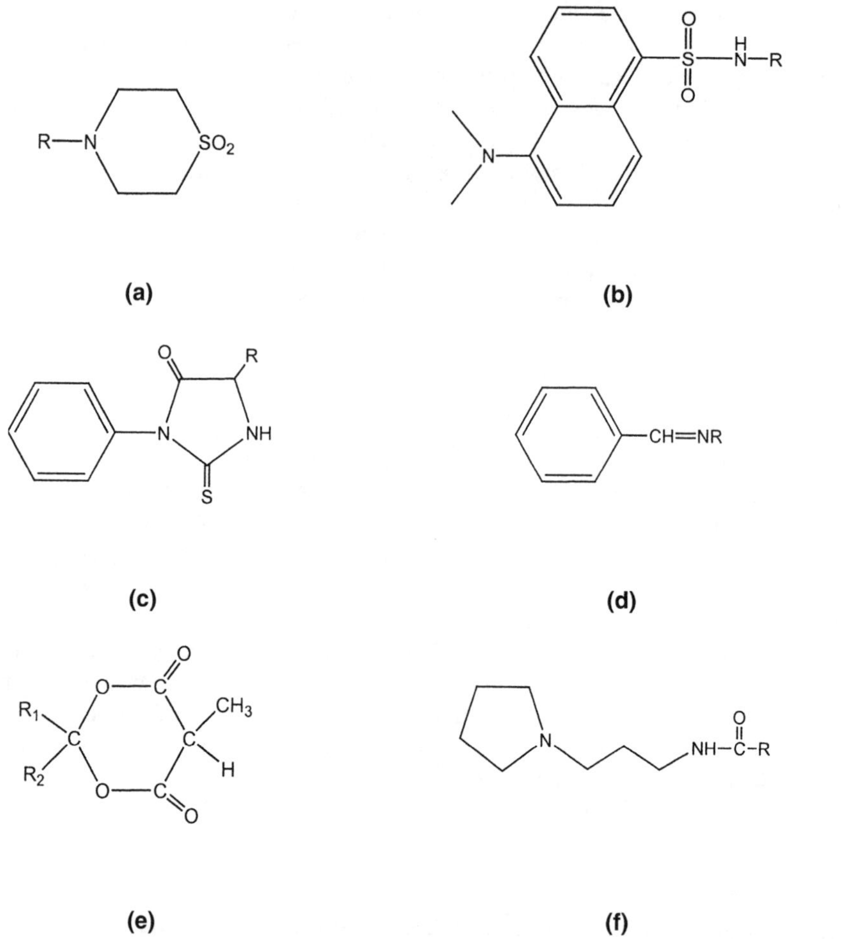

Figure 11   Derivatization products for detection by Ru(bpy)$_3$$^{3+}$ CL. R group(s) from sample compound of interest. (a) 1,4-Thiazane; (b) dansyl; (c) phenylthiohydantoin; (d) Schiff base; (e) β-diketal; (f) NAPP. See text for more information.

acid, to facilitate the reaction by not only producing Ru(bpy)$_3$$^{3+}$ from Ru(bpy)$_3$$^{2+}$ but also generating the analyte radical. Thioxanthene derivatives, similar to phenothiazine in structure except that the ring nitrogen atom is replaced by a double-bonded carbon attached to the alkyl chain with a secondary or tertiary amine, such as zuclopenthixol hydrochloride, flupentixol hydrochloride, and thiothixene, were determined by FI in dosage forms and biological fluids [98]. Fluoroquinolone compounds such as ofloxacin, norfloxacin, and ciprofloxacin that have several tertiary amine moieties (some in fused ring structures) were also determined by FI in pharmaceutical formulations in this way [99]. Detection limits were from 5 to 30 nM with linearity generally from 0.05 to 7 μg/mL. Because the optimum sulfuric acid concentration was found to be 0.1 M, adaptation to LC even using postcolumn mixing might be problematic unless a zirconia-based column with an acidic mobile phase is used.

Because of the importance of improving detection of primary amines such as amino acids, the use of derivatization chemistry to convert these compounds to tertiary amines has been a primary strategy. Primary aliphatic amines such as 4-heptylamine, 3-aminopentane, and cyclopentylamine were allowed to react with divinylsulfone (DVS), $CH_2=CH-SO_2-CH=CH_2$, to give 1,4-thiazane derivatives (Fig. 11a) [100]. An interesting side note is that this reaction was originally published in 1925. Optimum reaction conditions of 40 mM DVS, 50°C, and 15 min at pH 8.0 gave a 98% yield of the DVS-propylamine derivative. The ECL intensity increased in the order straight chain< branched chain <alicyclic type, and the separation of four straight-chain derivatives was shown by LC. Dansyl chloride is a well-known agent for generating fluorescent derivatives of amino acids prior to LC separation. Because of the N,N-dimethylamine substituent on the naphthalene ring (Figure 11b), it was also useful for the determination of amino acids by $Ru(bpy)_3^{3+}$ ECL [101]. Dansyl hydroxide, produced from excess dansyl chloride under alkaline reaction conditions, was found to have a weaker ECL response at high concentrations, and therefore LC resolution from the amino acid derivatives was not problematic as expected.

Preliminary results for phenylisothiocyanate (phenyl$-NH-CH=S$) to yield phenyl thiohydantoin (PTH) derivatives of amino acids for subsequent $Ru(bpy)_3^{3+}$ ECL detection are given in Ref. 102. The PTH derivatives have tertiary amine and secondary amine functionalities, but the ECL signal for PTH-glycine was no better than that for underivatized glycine. This is probably due to the $C=O$ and $C=S$ functional groups, which have electron-withdrawing character, making formation of the radical intermediate more difficult (Fig. 11c). Recently it was shown that PTH derivatives of aliphatic primary and secondary amines can markedly quench the $Ru(bpy)_3^{3+}$ CL response [27]. For example, PTH derivatives of propylamine, ethanolamine, and dihexylamine formed after a 30 min reaction time were found to generate no CL response with $Ru(bpy)_3^{3+}$. The tertiary amines tripropylamine and triethanolamine did not react and could be selectively determined by FI in the presence of the corresponding primary and secondary amine PTH analogs.

A tertiary amine protocol for proteins and peptides was developed using two biotin derivatives with different spacer groups between the biotin moiety and the tertiary amine group [103]. These derivatives were separately incubated with the protein avidin, which has four independent biotin binding sites. A linear CL response using $Ru(bpy)_3^{3+}$ was determined as a function of the molar avidin/derivative ratio from 1:1 to 1:4 and the biotin derivatives could be detected at the 50 nM level.

Conversion of primary amines using an aldehyde to form the corresponding Schiff base or imine compound (Fig. 11d) has been studied using both on-line derivatization and prederivatization modes. Aromatic amines such as benzylamine, phenethylamine, and phenylalanine, upon on-line photolysis using FI, gave a $Ru(bpy)_3^{3+}$ CL response 15–16 times that of the unreacted compounds [104]. Tryptophan and tyrosine as well as the biologically significant aromatic amines L-dopa and tryptamine showed four- to ninefold better CL detectability upon photolysis than the original compounds. CL detection limits of 2–20 pmol were significantly better than those of UV detection at 254 nm. The formation of imine products from photochemically generated benzaldehyde and the aromatic amine was confirmed by gas chromatography coupled with mass spectrometry. Liquid chromatographic separation and CL detection with on-line photolysis was shown. Aromatic amines such as tryptophan, benzylamine, phenylalanine, and tyrosine were reacted with benzaldehyde for 30 min in a 90:10 acetonitrile/water solvent with UV photolysis [27]. Flow injection determination of the reaction products indicated detection limits of 8–20 μg/L in the 1–2 pmol range. The importance of imine formation was shown by a comparison of detectability of 125 μM for aniline and 41 nM for malonaldehydebis(phenylimine), a commercially acquired derivative.

Histamine was reacted with 3-(diethylamino)propionic acid (DEAP) in the presence of a condensation reagent and catalyst to allow the reaction time to be limited to 1 h [33]. Good resolution of the histamine derivative from the starting material was possible in 20 min by ion-pair reversed-phase LC with on-line electrochemical production of $Ru(bpy)_3^{3+}$. Linearity from 2.5 to 250 nM with a detection limit of 70 fmol was given.

The pesticides carbofuran and promecarb, which have a methyl carbamate nitrogen atom, were photochemically converted on-line to methylamine using FI before reaction with peroxydisulfate photogenerated $Ru(bpy)_3^{3+}$ to provide CL detection [31]. Probably because of the carbonyl group neighboring the nitrogen atom, response of the unreacted pesticides to $Ru(bpy)_3^{3+}$ CL was weak. Linearity of response was from about 0.2 to 15 μg/L, and analysis of water, soil, and grains was demonstrated. This analytical approach has been extended to another carbamate pesticide, carbaryl [105].

N-Carboxymethyl derivatives of phenethylamine, cyclohexylamine, and 2-butylamine were separated into enantiomers using a reversed phase silica column coated with *N-n*-dodecyl-L-hydroxyproline [106]. To permit sensitive $Ru(bpy)_3^{3+}$ CL detection, acrylonitrile under basic conditions was mixed postcolumn to generate tertiary amine derivatives.

2. Carboxylic Acids and Alcohols

The use of an auxiliary oxidizing agent to convert organic acids into derivatives that can react with $Ru(bpy)_3^{3+}$ or Ru analogs has been an important approach expanding the applicability of this CL reaction. Chemiluminescence using Ru(phen)$_3$$^{2+}$ was applied to the LC determination of oxalic acid in spinach samples [107]. The reaction chemistry was based on the enhancement of the luminescence by oxalic acid produced from the reaction of Ru(phen)$_3$$^{2+}$ and Ce(IV) in sulfuric acid. Ce(IV) acts as an oxidizing agent to form Ru(phen)$_3$$^{3+}$ as well as the oxalic acid radical, which can then form a radical anion of CO_2. This species then reacts with Ru(phen)$_3$$^{3+}$ to give Ru(phen)$_3$$^{2+*}$, which drops down to the ground state to give light. This analogous reaction using Ru(bpy)$_3$$^{3+}$ was used to distinguish between oxalic acid, which reacts in about 2 s, from tartaric acid, which takes 50 s [108]. Other organic acids with hydroxyl or carbonyl groups such as malonic, pyruvic, lactic, and ascorbic acid react in a similar way. Gluconic acid was oxidized with periodate before CL determination by Ru(bpy)$_3$$^{3+}$ that was generated by Ce(IV) oxidation [109]. Linearity was achieved over about two orders of magnitude from 10^{-5} to 10^{-7} M with a detection limit of 1.5×10^{-8}, and pharmaceutical samples were analyzed. Cinnamic acid, 3-phenyl-2-propenoic acid, was determined similarly using the reaction of $KMnO_4$ with the analyte and Ru(bpy)$_3$$^{3+}$ [110]. The LC determination of cinnamic acid in urine was done with a sensitivity claimed to be two to three orders of magnitude better than previously reported determinations. After separation by LC, amino acids were reacted photochemically with periodate, and the oxalate product was allowed to react with electrogenerated Ru(bpy)$_3$$^{3+}$ [111]. Experimental conditions such as pH, UV irradiation time, and temperature were optimized, and detection limits of 1 pmol for threonine to 40 pmol for glycine were found.

A direct determination of ascorbic acid, a reducing agent, and dehydroascorbic acid by Ru(bpy)$_3$$^{3+}$ ECL at pH 10 was adapted postcolumn for LC [112]. It was shown that actually dehydroascorbic acid was responsible for the ascorbic acid signal. A detection limit of 1×10^{-7} M was found and both analytes could be analyzed in orange juice.

A study of various malonic acid derivatives for derivatization of aliphatic and aromatic ketones to their β-diketone derivatives was undertaken [113]. Methylmalonic acid was found to be the best derivatizing agent using cyclohexanone as a test compound (Fig. 11e). Optimum reaction conditions were determined to be 0.6 M sulfuric acid for 20 min at room temperature. Detection limits for aliphatic ketones such as 2-pentanone were 30–50 fmol. LC/CL detection of methylmalonic acid, an important clinical marker for pernicious anemia, in urine after derivatization should also be possible.

The compounds 2-(2-aminoethyl)-1-methylpyrrolidine and N-(3-aminopropyl)pyrrolidine (NAPP, see Fig. 11f) were found to be effective derivatizing agents for carboxylic acids [33]. Reaction conditions were mild at room temperature for 30 min. The NAPP derivatives for 10 fatty acids were separated by reversed-phase chromatography under isocratic conditions with detection limits in the 70 fmol range. Plasma samples were also analyzed for free fatty acids using this method. New derivatization reagents based on a benzoquino-lizidine structure are 100 times more sensitive for carboxylic acids (0.5–0.6 fmol for myristic and phenylbutylic acids) and have an unusual optimum reaction pH of 1.5–2 [114].

Alcohols have also been determined with an oxidizing agent using a photochemically assisted reaction approach or directly at very alkaline pH. Polyols such as 1,2-diol, 1,3-diol, and saccharides were oxidized in a photochemical reactor to yield oxalate for subsequent $Ru(bpy)_3^{3+}$ CL detection [115]. Detection limits of ethylene glycol and 1,3-propanediol were 38 and 23 pmol, respectively. Both monohydric and polyhydric alcohols in alkaline solutions could be detected through the alkoxide radical by ECL of $Ru(bpy)_3^{3+}$ produced at a glassy carbon electrode [116]. The ECL response decreased with increasing alcohol chain length but, as expected, was proportional to the number of hydroxyl groups. Using Schiff base chemistry with propylamine as the derivatizing agent, glucose could be determined with linearity from the detection limit of 5.2 mg/L to 83 mg/L [27]. The reagent DEAP in the presence of N,N'-dicyclohexylcarbodiimide and 4-dimethylaminopyridine can also functionalize aliphatic alcohols through an ester linkage. Optimum reaction conditions were 2 hr at ambient temperature. The determination of vitamin D_2 by DEAP prederivatization and LC with CL detection can be done with linearity from 3.8 to 750 nM and a detection limit of 40 fmol [117].

Oxygen containing aromatic compounds such as phenols, hydroquinones, catechols, and benzoquinones were shown to efficiently quench $Ru(bpy)_3^{3+}$ ECL [118]. A benzoquinone derivative formed at the electrode surface is proposed to cause energy transfer from the excited state ruthenium complex. The efficiency of the quenching by substituted phenols is not related to the type of functional group but does depend on the substitution pattern, with meta derivatives showing the greatest effect [119].

3.   Sulfur Compounds and Metal Complexes

The sulfhydryl compound 6-mercaptopurine in the presence of $Ru(bpy)_3^{2+}$, hydrogen peroxide, and hydroxide generated an intense light emission with a detection limit of 3×10^{-10} g/L [120]. The important inorganic anion sulfite has been determined by CL using bromate in 5 mM sulfuric acid as a co-oxidant for the analyte and $Ru(bpy)_3^{2+}$ [121]. Sodium dodecylbenzenesulfonate markedly enhanced the CL signal by about a factor of 40. It was proposed that both SO_2^* and $Ru(bpy)_3^{2+*}$ may drop to the ground state and contribute to the CL signal. SO_2 in air could be determined after absorption using a dilute (0.1%) triethanolamine solution. A similar method for sulfite in air using periodate and $Ru(phen)_3^{2+}$ indicated a detection limit of 7×10^{-10} M [122].

The transition metals Cu^{2+} and Co^{2+} after complexation to form emetine dithiocarbamate complexes were separated by LC and detected by $Ru(bpy)_3^{2+}$ ECL [123]. Linear ranges of 1–300 nM for Cu^{2+} and 30–5000 nM for Co^{2+} with detection limits of 650 fg and 17 pg, respectively, were reported.

4. Enzyme-Based Reactions

It appears that there have been no recent papers describing enzyme reactions coupled to $Ru(bpy)_3^{3+}$ CL for FI or LC, and only three representative systems are briefly described here. Ethanol was determined by FI using immobilized alcohol dehydrogenase by means of $Ru(bpy)_3^{3+}$ ECL of the reduced form of nicotinamide adenine dinucleotide (NADH), which has a tertiary nitrogen in the nicotinic ring [124]. The same nitrogen atom is protonated in the substrate NAD^+, which does not react. Glucose was determined in an analogous way using glucose dehydrogenase [125]. Because NADH can be spectrophotometrically determined at 340 nm or by fluorescence, the advantage of $Ru(bpy)_3^{3+}$ ECL detection for dehydrogenase enzymes is not really apparent. The activity of esterase enzymes could be monitored because the hydrolysis product of picolinic acid ethyl ester, picolinic acid, will readily form a mixed ligand complex with $Ru(bpy)_2^{2+}$ that is ECL-active [126].

IV. CAPILLARY ELECTROPHORESIS INSTRUMENTATION AND APPLICATIONS

Use of chemiluminescence detection with capillary electrophoresis (CE) is a promising approach because of the need for simple but sensitive detectors. Spectrophotometry is limited by the short path length of the capillary. Laser based fluorescence detection is generally limited to excitation in the visible wavelength region and is expensive. Several reviews of CL detection with CE have been published [127–131]. These papers describe primarily standard CL reaction chemistry involving luminol, peroxyoxalate, acridium esters, luciferase, and permanganate with some mention of ECL using $Ru(bpy)_3^{3+}$.

Capillary electrophoresis with ECL of luminol and H_2O_2 at either carbon or platinum microelectrodes was the subject of an early paper [132]. The mass detection limits for luminol at carbon and Pt electrodes, respectively, were 92 and 260 amol; however, the carbon electrode provided a more stable response. Amines such as n-octylamine and n-propylamine labeled with a luminol derivative were determined at the femtomole level. However, most CE with ECL detection has focused on $Ru(bpy)_3^{3+}$ chemistry.

A. External Generation

To avoid the problem of decoupling the CE electric field from the voltage applied to the electrode, an instrument (Fig. 12) that has the end of the CE capillary simply immersed in a reservoir of $Ru(bpy)_3^{3+}$ mounted above the PMT has been described [133]. The $Ru(bpy)_3^{3+}$ solution prepared by oxidation of $Ru(bpy)_3^{2+}$ by reaction with lead dioxide in 0.05 M sulfuric acid was found, based on a previous study, to be stable for 6 h. A field amplification injection procedure increased sensitivity by 20-fold and increasing the $Ru(bpy)_3^{3+}$ concentration from 5 to 25 mM improved both peak resolution and sensitivity for the separation of codeine, thebaine, and 6-methoxycodeine. Detection limits of 0.05–0.1 μM with linearity to 500 μM were reported.

B. Electrogenerated Chemiluminescence Mode

Electrogenerated chemiluminescence at the end of the capillary due to $Ru(bpy)_3^{2+}$ present in the run buffer was the earliest published approach for coupling CE to $Ru(bpy)_3^{3+}$ CL detection [134]. A voltage of 1.25 V vs. Ag/AgCl reference electrode was applied to a Pt wire inserted about 3 mm into a 75 μm i.d. capillary. The separated analytes exiting the capillary in a lighttight box could react with $Ru(bpy)_3^{3+}$ to produce light that was directed by a parabolic mirror to the PMT. The CE voltage of 15 or 25 kV was isolated from the voltage applied to the Pt electrode by cracking the capillary 1.5 cm from the end and then encasing the crack with cellulose acetate epoxy, which prevented leakage of the CE run buffer, $Ru(bpy)_3^{2+}$, and analytes. The CE circuit was completed by grounding the buffer filled capsule that was placed over the porous joint. The $Ru(bpy)_3^{3+}$ concentration was optimized with respect to a maximum linear working range of 200 μM for the β-blocker oxprenolol. Because the β-blocker was uncharged at the run buffer pH, neutral surfactants such as Triton X or β-cyclodextrin were added to aid resolution of these compounds. The separation of oxprenolol and acebutanol with an efficiency of about 15,000 plates was shown. The $Ru(bpy)_3^{3+}$ CL detection limits of 2 μM (12 fmol) were comparable to those for CE with UV detection.

A postcolumn introduction method for $Ru(bpy)_3^{2+}$ for subsequent oxidation to $Ru(bpy)_3^{3+}$ at a Pt electrode positioned in an ECL detection cell (actually a Nalgene bottle cap) has been described [135]. The detection cell held the $Ru(bpy)_3^{2+}$ reagent as well as the reference and Pt auxiliary electrodes. The 100 μm separation capillary, which had been etched to form a porous joint, was inserted into the detection cell and positioned as close as possible to and directly

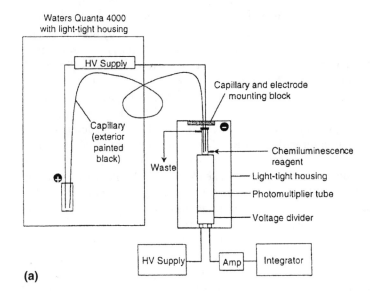

(a)

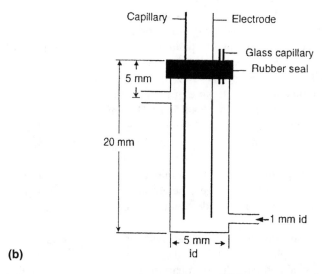

(b)

Figure 12 (a) Schematic diagram of the CE instrument coupled to the CL detector. HV= high voltage (b) Schematic diagram showing details of the glass detection cell, in which a static reservoir of the Ru(bipy)$_3^{3+}$ reagent was maintained. (From Ref. 133.)

opposite the Pt working electrode. The PMT was mounted above the detection cell, with both in a light-tight box. Band broadening indicated by 4000 plates for the proline peak was evident; however, the problem of $Ru(bpy)_3^{2+}$ adsorption to the capillary when introduced on-column as part of the run buffer was eliminated. A limit of detection of 13 μM for proline was noted.

Underivatized amines and amino acids were detected by CE with postcolumn $Ru(bpy)_3^{3+}$ ECL at a 35 μm carbon fiber [136]. The CE electric field was decoupled from the potential applied to the carbon fiber by an on-column fracture covered with a Nafion tube. A spatula acting as the auxiliary electrode held a small pool (100 μL) of electrolyte and $Ru(bpy)_3^{2+}$ in contact with the separation capillary and carbon fiber positioned opposite each other. The PMT was mounted above the spatula end, all in a light-tight box. Replicate separations of triethylamine, proline, valine, and serine were carried out at pH 9.5 in 10 min, with an efficiency of 15,000 plates calculated for serine. The 100 μL reservoir of $Ru(bpy)_3^{2+}$ was adequate for 4 h of operation. Detection limits of 68 nM for proline and 120 nM for triethylamine (3–6 fmol) were, as expected, significantly better than those for amino acids with only primary amine groups such as valine (5.8 μM) and serine (100 μM). When leucine was used in the FI mode with this CE instrument, the detection limit was 330 nM (15 fmol).

An improved CE detection cell for $Ru(bpy)_3^{3+}$ ECL was constructed as shown in Figure 13 [137]. Two major changes can be described. A reaction tube in which the bare CE capillary end and the Pt wire working electrode are inserted was added. A syringe pump set at 10 μL/min replenishes the supply of $Ru(bpy)_3^{2+}$; this alleviates the problem in the previous design of evaporation and dilution from the CE run buffer of the 100 μL reservoir that changed the $Ru(bpy)_3^{2+}$ concentration over time. Because of the low microampere current, a two electrode arrangement was found to be sufficient. Use of Pt electrodes of either 76 or 127 μm o.d., both larger than the 75 μm i.d. CE capillary, provided CL responses independent of alignment. The larger electrode gave a response nine times greater than the smaller electrode because of the higher amount of $Ru(bpy)_3^{3+}$ that could be produced. Detection limits of 0.2–0.7 μM (4–10 fmol) for proline, phenylalanine, and valine were determined with theoretical plate numbers of 74,000–80,000. Masking the detection zone could possibly improve these values.

This instrument has been used for the analysis of peptides and micellar electrokinetic chromatography (MEKC). First, the compositional analysis of two tripeptides was established and reproducible quantification in the 5–10 μg range was possible with detection limits for Gly-Phe-Ala and Val-Pro-Leu of 2.5 pmol and 100 fmol, respectively [138]. The relative luminescence response was dependent on the amino acid R group at the α-carbon as previously known. Second, the ECL mode was found to be compatible for MEKC because the $Ru(bpy)_3^{3+}$ is added postcapillary to avoid precipitation of the anionic surfactant,

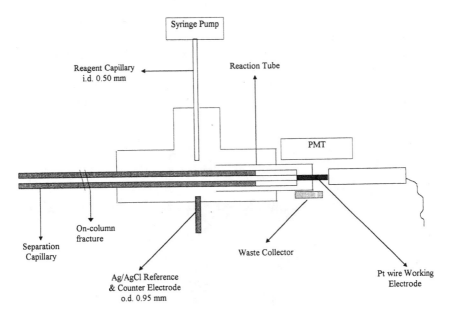

Figure 13 Schematic representation of cell for CE in situ detection of Ru(bpy)$_3^{3+}$ chemiluminescence. (From Ref. 137.)

and the concentration of the surfactant drops below the critical micelle concentration, freeing the analyte for reaction with Ru(bpy)$_3^{3+}$ [139]. The migration time of triethylamine (TEA) as compared to proline was accentuated by the presence of 50 mM sodium dodecylsulfate (SDS) because at the run buffer pH of 9.4, proline is an anion and TEA is a cation. Plate counts of 60,000–70,000 were calculated with a detection limit of 1.5 fmol for TEA. The amines diethylamine, TEA, and N,N'-diisopropylethylamine, which could not be resolved using 5 mM SDS, were baseline separated with 25 mM SDS. A sample of the vapor phase inhibitors morpholine and N,N'-diethanolamine were also analyzed.

An end-column ECL detection method for Ru(bpy)$_3^{2+}$ using a 300 μm o.d. Pt electrode positioned about 220 μm from the 75 μm id CE capillary was recently characterized [140]. No CE electric field decoupler was deemed necessary because of the careful positioning of the electrode and capillary. The required potential applied to the Pt electrode was shifted owing to the high CE voltage. A linear range for tripropylamine from 1×10^{-10} M to 1×10^{-5} M was achieved with an excellent detection limit of 5×10^{-11} M. Lidocaine, an antiarrhythmia drug for the heart, was also determined at low levels because of its tertiary nitrogen moiety. Tramadol and lidocaine were determined in ether extracts of urine in a separate study [141].

Using the same instrument, the antidepressant sulpiride with an N-ethyl-pyrolidinyl substituent, was determined in human plasma or urine after CE separation [142]. The calibration curve was linear from 0.05 to 25 μM with a detection limit of 0.03 μM. Procyclidine, an anticholinergic compound with an N-substituted pyrrolidine group, was detected in urine by Ru(bpy)$_3^{3+}$ CL following separation by CE [143]. Using on-column stacking, a low detection

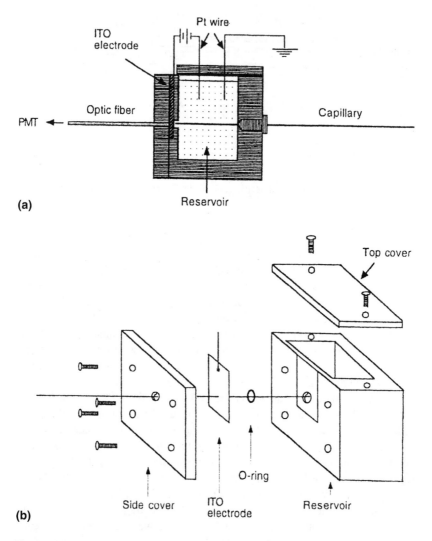

(a)

(b)

Figure 14 Schematic diagram of the indium/tin oxide (ITO) electrode–based ECL detection cell for CE. (a) Overview; (b) exploded view. (From Ref. 144.)

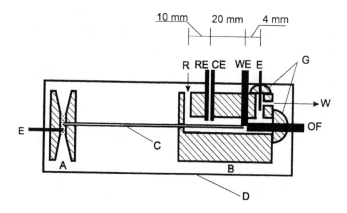

Figure 15 Schematic diagram of the microfluidic device. A, falling drop sample introduction interface; B, reaction and detection cell; C, separation capillary; D, chip baseplate; E, Pt wire electrode; RE, Ag/AgCl reference electrode; CE, counter electrode; WE, Pt working electrode; G, epoxy glue; OF, optical fiber; R, ECL reagent; W, waste. Dimensions not to scale. (From Ref. 148.)

limit of 0.001 μM was achieved. The urine sample was treated by using a cation-exchange cartridge to eliminate matrix effects.

An indium/tin oxide (ITO) working electrode (Fig. 14) was characterized recently for end-column $Ru(bpy)_3^{3+}$ CL detection with CE [144]. Potential control of the ITO electrode versus a Pt pseudo-reference electrode was provided using a battery (1.5 V) without decoupling from the CE electric field. The $Ru(bpy)_3^{2+}$ reservoir held about 8 mL, and the light produced at the ITO-coated glass plate was directed by an optical fiber to the PMT. The ECL reaction of tripropylamine and $Ru(bpy)_3^{3+}$ studied by cyclic voltammetry was found to be similar to previous work at a glassy carbon or Pt electrode. The separation of triethylamine, proline, and hydroxyproline was shown, and a detection limit of 1 μM (12 fmol) was observed for proline. The plate count for proline was only 4000, indicating that perhaps the luminescence zone was too large and the wall jet configuration of the detector could cause turbulence. One limitation of the ITO electrode is that the optimum ECL voltage of 1.5 V can have a corrosive effect [145].

Capillary electrophoresis with a solid-state ECL detector has recently been described [146]. $Ru(bpy)_3^{2+}$ was immobilized in a PSS–silica–poly(vinyl alcohol) film with grafted 4-vinylpyridine. Detection potential, concentration of $Ru(bpy)_3^{3+}$ in the film, and pH of the CE run buffer were all considered. Detection limits of tripropylamine and proline were 0.002 and 2 μM, respectively, similar to that of $Ru(bpy)_3^{2+}$ in solution.

The next step for CE with $Ru(bpy)_3^{3+}$ CL detection is to use a microchip, and this was the subject of a recent presentation [147]. Two recent related publications were found. One described the integration of the separation capillary, a falling drop sample introduction interface, and the ECL detection cell all on a 85 × 15 × 2 mm glass slide platform for both FI and CE [148]. The use of the falling drop interface to isolate the CE voltage allowed sample to be introduced into a reservoir of $Ru(bpy)_3^{2+}$ where ECL occurred at a Pt electrode in front of an optical fiber that transmitted the light to a PMT (Fig. 15). Proline, valine, and phenylalanine were separated by CE in about 1 min using only 1 kV with detection limits of 1–25 μM. In a second paper [149], a microfabricated glass device was characterized for MEKC of $Ru(bpy)_3^{3+}$ and $Ru(phen)_3^{3+}$. In this design the legs of a 'U'-shaped Pt electrode placed across the separation channel act as the working and counter electrodes. What is unusual is that the required potential difference for the ECL reaction is supplied by the electrophoretic electric field. Indirect detection of three amino acids was also reported.

V. CONCLUSIONS AND POSSIBLE FUTURE DEVELOPMENTS

Instruments developed for luminol chemistry have all been based on ECL because the reactive luminol radicals are so short-lived. Applications involving phenols are quite common; extension of the chemistry to other classes of compounds is more limited. More applications involving LC and CE with luminol ECL are needed with real samples to make this technique more widespread.

The development of $Ru(bpy)_3^{3+}$ CL detection methods has been primarily in the direction of ECL because of the advantages of combining oxidation of $Ru(bpy)_3^{2+}$ and possibly the analyte at an electrode together with measurement of light in a flow-through cell. However, new ways of improving the stability of $Ru(bpy)_3^{3+}$ generated externally batchwise have been published, and both on-line electrochemical or photochemical production of $Ru(bpy)_3^{3+}$ are quite easy to implement and are also compatible with existing CL detectors. Photochemical reactivity can also be used to expand the variety of compounds detectable by $Ru(bpy)_3^{3+}$ CL. Commercial development of both CL and ECL detectors with low-volume flow cells would speed applications into microbore and capillary LC. In addition, to the best of my knowledge, CL detection is not commercially available for CE. Published designs indicate that application of $Ru(bpy)_3^{3+}$ CL detection to capillary electrokinetic chromatography should be relatively straightforward. In the future, microchip CE with $Ru(bpy)_3^{3+}$ CL detection will definitely be further developed. Undoubtedly, some of the microfabrication sytems described in Section II.B.5 could be adapted to capillary electrophoresis.

More applications involving organic compounds without an amine group will likely also be possible either directly or through derivatization chemistry.

Use of computer programs to run theoretical calculations for ionization constants of organic substituents should become an important way to predict the $Ru(bpy)_3^{3+}$ CL response of many organic compounds.

ABBREVIATIONS

C18	Octadecyl
CCD	Charge-coupled device
CE	Capillary electrophoresis
CL	Chemiluminescence
2,4-D	2,4-Dichloro phenoxy acetic acid
DEAP	3-(Diethylamino)propionic acid
DVS	Divinylsulfone
ECL	Electrogenerated chemiluminescence, or CL at an electrode
FI	Flow injection
GC/MS	Gas chromatography coupled with mass spectrometry
IDA	Interdigitated array
IP	Ionization potential
ITO	Indium/tin oxide
Kel-F	Polychlorotrifluoroethylene
MEKC	Micellar electrokinetic chromatography
μTAS	Micro total analytical system
NAD+	Nicotinamide adenine dinucleotide (oxidized form)
NADH	Nicotinamide adenine dinucleotide (reduced form)
NAPP	N-(3-Aminopropyl)pyrrolidine
PEEK	Poly(ether ether ketone)
PMT	Photomultiplier
PSS	Polystyrenesulfonate
PTH	Phenylthiohydantoin
RuBPS	tris(4,7-diphenyl-1,10-phenanthroline-sulfonic acid) ruthenium (II)
SCE	Saturated calomel electrode
SDS	Sodium dodecylsulfate
TEA	Triethylamine
UV	Ultraviolet

REFERENCES

1. Knight, A.W. A review of recent trends in analytical applications of electrogenerated chemiluminescence. Trends Anal. Chem. **1999**, *18*, 47–62.
2. Fahnrich, K.A., Pravda, M., Guilbault, G.G. Recent applications of electrogenerated chemiluminescence in chemical analysis. Talanta. **2001**, *54*, 531–559.
3. Lee, W.Y. Tris(2,2′-bipyridyl)ruthenium(II) electrogenerated chemiluminescence in analytical science. Mikrochim. Acta **1997**, *127*, 19–39.
4. Gerardi, R.D., Barnett, N.W., Lewis, S.W. Analytical applications of tris(2,2′-bipyridyl)ruthenium(III) as a chemiluminescent reagent. Anal. Chim. Acta **1999**, *378*, 1–41.
5. Leland, J.K., Powell, M.J. Electrogenerated chemiluminescence: an oxidative-reduction type ECL reaction sequence using tripropylamine. J. Electrochem. Soc. **1990**, *137*, 3127–3131.
6. Sun, Y.-G., Cui, H., Li, Y.-H., Zhao, H.-Z., Lin, X.-Q. Flow injection analysis of tannic acid with inhibited electrochemiluminescent detection. Anal. Lett. **2000**, *33*, 2281–2291.
7. Zhang, G.-F., Chen, H.-Y. Studies of polyluminol modified electrode and its application in electrochemiluminescence analysis with flow system. Anal. Chim. Acta **2000**, *419*, 25–31.
8. Steiger, O.M., Kamminga, D.A., Brumelhuis, A., Lingeman, H. Liquid chromatography with luminol-based electrochemiluminescence detection— determination of histamine. J. Chromatogr. A. **1998**, *799*, 57–66.
9. Wang, Y., Yeung, E.S. Indirect detection for liquid chromatography based on electrogenerated luminol chemiluminescence. Anal. Chim. Acta **1992**, *266*, 285–300.
10. Lee, W.-Y., Nieman, T.A. Evaluation of use of tris(2,2′-bipyridyl)ruthenium(III) as a chemiluminescent reagent for quantitation in flowing streams. Anal. Chem. **1995**, *67*, 1789–1796.
11. Gonzalez, J.M., Greenway, G.M., McCreedy, T., Qijun, S. Determination of morpholine fungicides using the tris(2,2′-bipyridine)ruthenium(II) chemiluminescence reaction. Analyst. **2000**, *125*, 765–769.
12. Shultz, L.L., Stoyanoff, J.S., Nieman, T.A. Temporal and spatial analysis of electrogenerated Ru(bpy)$_3^{3+}$ chemiluminescent reactions in flowing streams. Anal. Chem. **1996**, *68*, 349–354.
13. Shultz, L.L., Nieman, T.A. Stopped-flow analysis of Ru(bpy)$_3^{3+}$ chemiluminescent reactions. J. Biolumi. Chemilum. **1998**, *13*, 85–90.
14. Downey, T.N., Nieman, T.A. Chemiluminescent detection with a regenerable reagent by using tris(2,2′-bipyridyl)ruthenium(II) immobilized in Nafion. Anal. Chem. **1992**, *64*, 261–268.
15. Khramov, A.N., Collinson, M.M. Electrogenerated chemiluminescence of tris(2,2′-bipyridyl)ruthenium(II) ion-exchanged in Nafion-silica composite films. Anal. Chem. **2000**, *72*, 2943–2948.
16. Blanchard, R.M., Martin, A.F., Nieman, T.A., Guerrero, D.J., Ferraris, J.P. Electrogenerated chemiluminescence using solution phase and immobilized

tris(4,7-diphenyl-1,10-phenanthrolinedisulfonic acid)ruthenium(II). Mikrochim. Acta **1998**, *130*, 55–62.

17. Jackson, W.A., Bobbitt, D.R. Chemiluminescent detection of amino acids using in situ generated Ru(bpy)$_3^{3+}$. Anal. Chim. Acta **1994**, *285*, 309–320.

18. Knight, A.W., Greenway, G.M., Chesmore, E.D. Development of a silicon photodiode, electrogenerated chemiluminescence, flow-through detector. Anal. Proc. **1995**, *32*, 125–127.

19. Wang, H., Xu, G., Dong, S. Electrochemiluminescence of tris(2,2'-bipyridine)ruthenium(II) immobilized in poly(p-styrenesulfonate)–silica–Triton X-100 composite thin films. Analyst **2001**, *127*, 1095–1099.

20. Lin, J.M., Qu, F., Yamada, M. Chemiluminescent investigation of tris(2,2'-bipyridyl)ruthenium(II) immobilized on a cationic ion-exchange resin and its application to analysis. Anal. Bioanal. Chem. **2002**, *374*, 1159–1164.

21. Barnett, N.W., Bos, R., Brand, H., Jones, P., Lim, K.F., Purcell, S.D., Russell, R.A. Synthesis and preliminary analytical evaluation of the chemiluminescence from (4-[4-(dichloromethylsilanyl)-butyl]-4'-methyl-2,2'-bipyridyl)bis(2,2'-bipyridyl)ruthenium(II) covalently bonded to silica particles. Analyst **2002**, *127*, 455–458.

22. Holeman, J.A., Danielson, N.D. Microbore liquid chromatography of tertiary amine anticholinergic pharmaceuticals with tris(2,2'-bipyridine)ruthenium(III) chemiluminescence detection. J. Chromatogr. Sci. **1995**, *33*, 297–302.

23. Zhang, W., Danielson, N.D. "Characterization of a Micro Spiral Flow Cell for Chemiluminescence Detection", Microchem. J. **2003**, *75*, 255–264.

24. Gerardi, R.D., Barnett, N.W., Jones, P. Two chemical approaches for the production of stable solutions of tris(2,2'-bipyridyl)ruthenium(III) for analytical chemiluminescence. Anal. Chim. Acta **1999**, *388*, 1–10.

25. Barnett, N.W., Hindson, B.J., Lewis, S.W., Purcell, S.D., Jones, P. Preparation and preliminary evaluation of anhydrous tris(2,2'-bipyridyl)ruthenium(III) perchlorate as a temporally stable reagent for analytical chemiluminescence. Anal. Chim. Acta **2000**, *421*, 1–6.

26. Barnett, N.W., Lewis, S.W., Purcell, S.D., Jones, P. Determination of sodium oxalate in Bayer liquor using flow-analysis incorporating an anion exchange column and tris(2,2'-bipyridyl)ruthenium(II) chemiluminescence detection. Anal. Chim. Acta **2002**, *458*, 291–296.

27. Bolden, M.E. Investigation of the tris(2,2'-bipyridine)ruthenium(III) chemiluminescence reaction for the determination of aromatic amines and aldehydes by flow injection or liquid chromatography. PhD Dissertation, Miami University, Oxford, OH, 2002.

28. Barnett, N.W., Bowser, T.A., Russell, R.A. Determination of oxalate in alumina process liquors by ion chromatography with post column chemiluminescence detection. Anal. Proc. **1995**, *32*, 57–59.

29. Barnett, N.W., Bowser, T.A., Gerardi, R.D., Smith, B. Determination of codeine in process streams using flow injection analysis with chemiluminescence detection. Anal. Chim. Acta **1996**, *318*, 309–317.

30. Yamazaki, S., Shinozaki, T., Tanimura, T. Detection of tertiary amines with chemiluminescent reaction using tris(2,2'-bipyridine)ruthenium(III) prepared by

on-line photochemical oxidation. J. High Resolut Chromatogr. **1998**, *21*, 315–316.

31.   Perez-Ruiz, T., Martinez-Lozano, C., Tomas, V., Martin, J. Chemiluminescence determination of carbofuran and promecarb by flow injection analysis using two photochemical reactions. Analyst **2002**, *127*, 1526–1530.

32.   Uchikura, K., Kirisawa, M. Generation of chemiluminescence upon reaction of alicyclic tertiary amines with tris(2,2′-bipyridine)ruthenium(III) using flow electrochemical reactor. Anal. Sci. **1991**, *7*, 803–804.

33.   Morita, H., Konishi, M. Electrogenerated chemiluminescence derivatization reagents for carboxylic acids and amines in high performance liquid chromatography using tris(2,2′-bipyridine)ruthenium(II). Anal.Chem. **2002**, *74*, 1584–1589.

34.   Fiaccabrino, G.C., de Rooij, N.F., Koudelka-Hep, M. On-chip generation and detection of electrochemiluminescence. Anal. Chim. Acta **1998**, *359*, 262–267.

35.   Michel, P.E., Fiaccabrino, G.C., de Rooij, N.F., Koudelka-Hep, M. Integrated sensor for continuous flow electrochemiluminescent measurements of codeine with different ruthenium complexes. Anal. Chim. Acta **1999**, *392*, 95–103.

36.   L'Hostis, E., Michel, P.E., Fiaccabrino, G.C., Strike, D.J., de Rooij, N.F., Koudelka-Hep, M. Microreactor and electrochemical detectors fabricated using Si and EPON SU-8. Sensors Actuators B **2000**, *64*, 156–162.

37.   Greenway, G.M., Nelstrop, L.J., Port, S.N. Tris(2,2′-bipyridyl)ruthenium(II) chemiluminescence in a microflow injection system for codeine determination. Anal. Chim. Acta **2000**, *405*, 43–50.

38.   Greenwood, P.A., Merrin, C., McCreedy, T. Greenway, G.M. Chemiluminescence μTAS for the determination of atropine and pethidine. Talanta **2002**, *56*, 539–545.

39.   Hsueh, Y.-T., Smith, R.L., Northrup, M.A. A microfabricated, electro-chemiluminescence cell for the detection of amplified DNA. Sensors Actuators B: Chem. **1996**, *33*, 110–114.

40.   Hsueh, Y.-T., Collins, S.D., Smith, R.L. DNA quantification with an electrochemiluminescence microcell. Sensors Actuators B **1998**, *49*, 1–4.

41.   Sun, Y.G., Cui, H., Lin, X.Q., Li, Y.H., Zhao, H.Z. Flow injection analysis of pyrogallol with enhanced electrochemiluminescent detection. Anal. Chim. Acta **2000**, *423*, 247–253.

42.   Sun, Y.G., Cui, H., Li, Y.H., Li, S.F., Lin, X.Q. Determination of gallic acid by flow injection with electrochemiluminescent detection. Anal. Lett. **2000**, *33*, 3239–3252.

43.   Zheng, X., Guo, Z., Zhang, Z. Flow-injection electrogenerated chemiluminescence determination of epinephrine using luminol. Anal. Chim. Acta **2001**, *441*, 81–86.

44.   Sun, Y.-G., Cui, H., Li, Y.-H., Lin, X.-Q. Determination of some catechol derivatives by a flow injection electrochemiluminescent inhibition method. Talanta **2000**, *53*, 661–666.

45.   Zhu, L.D., Li, Y.X., Zhu, G.Y. Flow injection determination of dopamine based on inhibited electrochemiluminescence of luminol. Anal. Lett. **2002**, *35*, 2527–2537.

46.   Sakura, S., Terao, J. Determination of lipid hydroperoxides by electrochem-iluminescence. Anal. Chim. Acta **1992**, *262*, 59–65.

47.   Zhu, L.D., Li, Y.X., Zhu, G.Y. Flow injection analysis of histidine with enhanced

electrogenerated chemiluminescence of luminol. Chin. Chem. Lett. **2002**, *13*, 1093–1096.

48. Zheng, X.W., Zhang, Z.J., Guo, Z.H., Wang, Q. Flow-injection electrogenerated chemiluminescence detection of hydrazine based on its in-situ electrochemical modification at a pre-anodized platinum electrode. Analyst **2002**, *127*, 1375–1379.

49. Zheng, X.W., Guo, Z.H., Zhang, Z.J. Flow-injection electrogenerated chemiluminescence determination of isoniazid using luminol. Anal. Sci. **2001**, *17*, 1095–1099.

50. Li, J.J., Du, J.X., Lu, J.R. Flow-injection electrogenerated chemiluminescence determination of vanadium and its application to environmental water samples. Talanta **2002**, *57*, 53–57.

51. Zhu, L., Li, Y.X., Zhu, G.Y. A novel flow through optical fiber biosensor for glucose based on luminol electrochemiluminescence. Sensors Actuators B Chem. **2002**, *86*, 209–214.

52. Marquette, C.A., Blum, L.J. Luminol electrochemiluminescence-based fibre optic biosensors for flow injection analysis of glucose and lactate in natural samples. Anal. Chim. Acta **1999**, *381*, 1–10.

53. Tsafack, V.C., Marquette, C.A., Leca, B., Blum, L.J. An electrochemiluminescence-based fibre optic biosensor for choline flow injection analysis. Analyst **1999**, *125*, 151–155.

54. Wang, C.H., Chen, S.M., Wang, C.M. Co-immobilization of polymeric luminol, iron(II)tris(5-aminophenanthroline) and glucose oxidase at an electrode surface, and its application as a glucose optrode. Analyst **2002**, *127*, 1507–1511.

55. Wilson, R., Kremeskotter, J., Schiffrin, D.J., Wilkinson, J.S. Electrochemiluminescence detection of glucose oxidase as a model for flow injection immunoassays. Biosensors Bioelectron **1996**, *11*, 805–810.

56. Marquette, C.A., Blum, L.J. Electrochemiluminescence of luminol for 2,4-D optical immunosensing in a flow injection analysis system. Sensors Actuators B Chem. **1998**, *51*, 100–106.

57. Nonidez, W.K., Leyden, D.E. Effect of reducing agents of pharmacological importance on the chemiluminescence of tris-(2,2′-bipyridine)ruthenium(II). Anal. Chim. Acta **1978**, *96*, 401–404.

58. Skotty, D.R., Nieman, T.A. Determination of oxalate in urine and plasma using reversed-phase ion-pair high performance liquid chromatography with tris(2,2′-bipyridyl)ruthenium(II)-electrogenerated chemiluminescence detection. J. Chromatogr. B **1995**, *665*, 27–36.

59. Wu, F.W., He, Z., Luo, Q.Y., Zeng, Y. High-performance liquid chromatographic determination of oxalic acid in tea using tris(1,10-phenanthroline)ruthenium(II) chemiluminescence. Anal. Sci. **1998**, *14*, 971–973.

60. Uchikura, K. Electrochemiluminescence reaction by mixing tris(2,2′-bipyridine)ruthenium(III) with ketones. Anal. Sci. **1999**, *15*, 1049–1050.

61. Noffsinger, J.B., Danielson, N.D. Generation of chemiluminescence upon reaction of aliphatic amines with tris(2,2′-bipyridine)ruthenium(III). Anal. Chem. **1987**, *59*, 865–868.

62. He, L., Cox, K.A., Danielson, N.D. Chemiluminescence detection of amino acids,

peptides, and proteins using tris(2,2′-bipyridyl)ruthenium(III). Anal. Lett. **1990**, *23*, 195–210.

63. Uchikura, K., Kirisawa M. Chemiluminescence of tryptophan with electro-generated tris(2,2′-bipyridine)ruthenium(III). Chem. Lett. **1991**, *8*, 1373–1376.

64. Brune, S.N., Bobbitt, D.R. Role of electron-donating withdrawing character, pH, and stoichiometry on the chemiluminescent reaction of tris(2,2′-bipyridyl)ruthenium(III) with amino acids. Anal. Chem. **1992**, *64*, 166–170.

65. Knight, A.W., Greenway, G.M. Relationship between structural attributes and observed electrogenerated chemiluminescence (ECL) activity of tertiary amines as potential analytes for the tris(2,2′-bipyridine)ruthenium(II) ECL reaction—a review. Analyst **1996**, *121*, 101R–106R.

66. Barnett, N.W., Gerardi, R.D., Hampson, D.L., Russell, R.A. Some observations on the chemiluminescent reactions of tris(2,2′-bipyridyl)ruthenium(III) with certain *Papaver somniferum* alkaloids and their derivatives. Anal. Commun. **1996**, *33*, 255–260.

67. Saito, K., Murakami, S., Yamazaki, S., Muromatsu, A., Hirano, S., Takahashi, T., Yokota, K., Nojiri, T. Detection of active methylene compounds having cyano and carbonyl groups with chemiluminescent reaction using tris(2,2′-bipyridine) ruthenium(III) prepared by on-line electrochemical oxidation. Anal. Chim. Acta **1999**, *378*, 43–46.

68. Noffsinger, J.B., Danielson, N.D. Liquid chromatography of aliphatic trialkylamines with post-column chemiluminescent detection using tris(2,2′-bipyridine)ruthenium(III). J. Chromatogr. **1987**, *387*, 520–524.

69. Ridlen, J.S., Klopf, G.J., Nieman, T.A. Determination of glyphosphate and related compounds using HPLC with tris(2,2′-bipyridyl)ruthenium(II) electrogenerated chemiluminescence detection. Anal. Chim. Acta **1997**, *341*, 195–204.

70. Yamazaki, S., Tanimura, T., Uchikura, K. Direct enantiomeric separation of β-alcohols with a tertiary amine moiety by ligand exchange chromatography with chemiluminescence detection. Chromatography **1998**, *19*, 201–205.

71. Barnett, N.W., Hindson, B.J., Lewis, S.W. Determination of ranitidine and salbutamol by flow injection analysis with chemiluminescence detection. Anal. Chim. Acta **1999**, *384*, 151–158.

72. Greenway, G.M., Knight, A.W., Knight, P.J. Electrogenerated chemiluminescent determination of codeine and related alkaloids and pharmaceuticals with tris(2,2′-bipyridine)ruthenium(II). Analyst **1995**, *120*, 2549–2552.

73. Song, Q.J., Greenway, G.M., McCreedy, T. Tris(2,2′-bipyridine)ruthenium(II) electrogenerated chemiluminescence of alkaloid type drugs with solid phase extraction preparation. Analyst **2001**, *126*, 37–40.

74. Chen, X., Yi, C.Q., Li, M.J., Lu, X., Li, Z., Li, P.W., Wang, XR. Determination of sophoridine and related lupin alkaloids using tris(2,2′-bipyridine)ruthenium electrogenerated chemiluminescence. Anal. Chim. Acta **2002**, *466*, 79–86.

75. Uchikura, K., Sakurada, K., Tezuka, K., Koike, K. Determination of free hydroxyproline and proline in serum by HPLC with electrogenerated chemiluminescence detection using tris(2,2′-bipyridine)ruthenium(II). Bunseki Kagaku **2002**, *51*, 953–957.

76. Koike, K., Li, Y.M., Seo, M., Sakurada, I., Tezuka, K., Uchikawa, K. Free 4-hydroxyproline content in serum of bedridden aged people is elevated due to fracture. Biol. Pharm. Bull. **2000**, *23*, 101–103.

77. Uchikura, K., Kirisawa, M. Electrochemiluminescence determination of D-, L-tryptophan using ligand exchange high performance liquid chromatography. Anal. Sci. **1991**, *7*, 971–973.

78. Perez-Ruiz, T., Martinez-Lozano, C., Tomas, V., Martin, J. Flow injection chemiluminescent method for the successive determination of l-cysteine and l-cystine using photogenerated tris(2,2′-bipyridyl)ruthenium(III). Talanta **2002**, *58*, 987–994.

79. Danielson, N.D., He, L., Noffsinger, J.B., Trelli, L. Determination of erythromycin in tablets and capsules using flow injection analysis with chemiluminescent detection. J. Pharm. Biomed. Anal. **1989**, *7*, 1281–1285.

80. Danielson, N.D., Holeman, J.H., Bristol, D.C., Kirzner, D.H. Simple methods for the qualitative identification and quantitative determination of macrolide antibiotics. J. Pharm. Biomed. Anal. **1993**, *11*, 121–130.

81. Ridlen, J.S., Skotty, D.R., Kissinger, P.T., Nieman, T.A. Determination of erythromycin in urine and plasma using microbore liquid chromatography with tris(2,2′-bipyridyl)ruthenium(II) electrogenerated chemiluminescence detection. J. Chromatogr. B **1997**, *994*, 393–400.

82. Targove, M.A., Danielson, N.D. High performance liquid chromatography of clindamycin antibiotics using tris(bipyridine)ruthenium(III) chemiluminescence detection. J. Chromatogr. Sci. **1990**, *28*, 505–509.

83. Tomita, I.N., Bulhoes, O.S.L. Electrogenerated chemiluminescence determination of cefadroxil antibiotic. Anal. Chim. Acta **2001**, *442*, 201–206.

84. Liang, P., Sanchez, R.I., Martin M.T., Electrochemiluminescence-based detection of β-lactam antibiotics and β-lactamases. Anal. Chem. **1996**, *68*, 2426–2431.

85. Suzuki, K., Aoki, Y., Kido, Y., Tsuji, A., Ito, K., Maeda, M. Development of electrogenerated chemiluminescence HPLC for ofloxacin using tris(2,2′-bipyridine)ruthenium(II) and its application to residual analysis in chicken tissues. J Food Hyg Soc Jpn **1999**, *40*, 23–28.

86. Li, F., Cui, H., Lin, X.-Q. Determination of adrenaline by using inhibited Ru(bpy)$_3^{2+}$ electrochemiluminescence. Anal. Chim. Acta **2002**, *471*, 187–194.

87. Chen, X., Li, M., Li, Z., Yi, C., Wang, X., Wang, B. Study of electro-chemiluminescence based on tris(2,2′-bipyridyl)ruthenium(II) and some amine drugs using flow injection analysis. Fenxi Huaxue **2002**, *30*, 513–517.

88. Knight, A.W., Greenway, G.M. Electrogenerated chemiluminescence determination of some local anaesthetics. Anal. Commun. **1996**, *33*, 171–174.

89. Chiba, R., Fukushi, M., Tanaka, A. Simultaneous determination of phenylcyclidine and its metabolites in rat urine by high performance liquid chromatography. Anal. Sci. **1998**, *14*, 979–982.

90. Chiba, R., Yamamoto, N., Tanaka, A. High performance liquid chromatography with chemiluminescence detection of disopyramide in human serum. Anal. Sci. **1998**, *14*, 1153–1155.

91. Aly, F.A., Al-Tamimi, S.A., Alwarthan, A.A. Determination of flufenamic acid and mefenamic acid in pharmaceutical preparations and biological fluids using flow

injection analysis with tris(2,2'-bipyridyl)ruthenium(II) chemiluminescence detection. Anal. Chim. Acta **2000**, *416*, 87–96.

92. Park, Y.J., Lee, D.W., Lee, W.Y. Determination of beta-blockers in pharmaceutical preparations and human urine by high performance liquid chromatography with tris(2,2'-bipyridyl)ruthenium(II) electrogenerated chemiluminescence detection. Anal. Chim. Acta **2002**, *471*, 51–59.

93. Holeman, J.A., Danielson, N.D. Liquid chromatography of antihistamines using post-column tris(2,2'-bipyridine)ruthenium(III) chemiluminescence detection. J. Chromatogr. A **1994**, *679*, 277–284.

94. Dolman, S.J.L., Greenway, G.M. Determination of amitriptyline using electrogenerated chemiluminescence. Anal. Commun. **1996**, *33*, 139–141.

95. Yoshida, H., Hidaka, K., Ishida, J., Yoshikuni, K., Nohta, H., Yamaguchi, M. Highly selective and sensitive determination of tricyclic antidepressants in human plasma using high performance liquid chromatography with post-column tris (2,2'-bipyridyl)ruthenium(III) chemiluminescence detection. Anal. Chim. Acta. **2000**, *413*, 137–145.

96. Chiba, R., Shinriki, M., Ishii, Y., Tanaka, A. High performance liquid chromatographic determination of yohimbine in serum by chemiluminescence detection. Anal. Sci. **1998**, *14*, 975–978.

97. Holeman, J.A., Danielson, N.D. Chemiluminescence detection of thiazide compounds with tris(2,2'-bipyridine)ruthenium(III). Anal. Chim. Acta **1993**, *277*, 55–60.

98. Aly, F.A., Al-Tamimi, S.A., Alwarthan. A.A. Flow injection chemiluminometric determination of some thioxanthene derivatives in pharmaceutical formulations and biological fluids using the [Ru(dipy)$_3$$^{2+}$]-Ce(IV) system. Anal. Sci. **2001**, *17*, 1257–1261.

99. Aly, F.A., Al-Tamimi, S.A., Alwarthan, A.A. Chemiluminescence determination of some fluoroquinolone derivatives in pharmaceutical formulations and biological fluids using [Ru(bipy)$_3$$^{2+}$]-Ce(IV) system. Talanta **2001**, *53*, 885–893.

100. Uchikura, K., Kirisawa, M., Sugii, A. Electrochemiluminescence detection of primary amines using tris(bipyridine)ruthenium(III) after derivatization with divinylsulfone. Anal. Sci. **1993**, *9*, 121–123.

101. Skotty, D.R., Lee, W.-Y., Nieman, T.A. Determination of dansyl amino acids and oxalate by HPLC with electrogenerated chemiluminescence detection using tris(2,2'-pyridyl)ruthenium(II) in the mobile phase. Anal. Chem. **1996**, *68*, 1530–1535.

102. Jackson, W.A., Bobbitt, D.R. Chemiluminescent detection of amino acids using in situ generated Ru(bpy)$_3$$^{3+}$. Anal. Chim. Acta **1994**, *285*, 309–320.

103. Waguespack, B.L., Lillquist, A., Townley, J.C., Bobbitt, D.R. Evaluation of a tertiary amine labeling protocol for peptides and proteins using Ru(bpy)$_3$$^{3+}$-based chemiluminescence detection. Anal. Chim. Acta **2001**, *441*, 231–241.

104. Bolden, M.E., Danielson, N.D. Liquid chromatography of aromatic amines with photochemical derivatization and tris(bipyridine)ruthenium(III) chemilumine-scence detection. J. Chromatogr. A **1998**, *828*, 421–430.

105. Perez-Ruiz, T., Martinez-Lozano, C., Tomas, V., Martin, J. Chemiluminescence

determination of carbaryl using photolytic decomposition and photogenerated tris(2,2′-bipyridyl)ruthenium(III). Anal. Chim. Acta **2003**, *476*, 141–148.

106. Yamazaki, S., Ban'i, K., Tanimura, T. Chiral separation of carboxymethyl derivatives of amines by liquid chromatography on reversed-phase silica gel coated with N-n-dodecyl-L-hydroxyproline. J. High Resolut. Chromatogr. **1999**, *22*, 487–489.

107. Wu, F., He, Z., Luo, Q., Zeng, Y. HPLC determination of oxalic acid using tris(1,10-phenanthroline)ruthenium(II) chemiluminescence—application to the analysis of spinach. Food Chem. **1999**, *65*, 543–546.

108. He, Z., Gao, H., Yuan, L., Luo, Q., Zeng, Y. Simultaneous determination of oxalic acid and tartaric acid with chemiluminescence detection. Analyst **1997**, *122*, 1343–1345.

109. Han, H.Y., He, Z.K., Han, H.Y., He, Z.K., Li, X.Y., Zeng, Y. Chemiluminescence determination of gluconic acid in pharmaceutical formulations using Ru(bipy)$_3{}^{2+}$-KIO$_4$-Ce(IV) system. Anal. Lett. **1999**, *32*, 2297–2310.

110. Han, H.Y., He, Z.K., Zeng, Y.E. Pulse injection analysis with chemiluminescence detection: determination of cinnamic acid using the Ru(bipy)$_3{}^{2+}$-KMnO$_4$ system. Anal. Chim. Acta **1999**, *402*, 113–118.

111. Yokota, K., Saito, K., Yamazaki, S., Muromatsu, A. New detection method of alpha-, beta-, and gamma-amino acids coupled with on-line photochemical oxidation and tris(2,2′-bipyridine)ruthenium(III) chemiluminescence. Anal. Lett. **2002**, *35*, 185–194.

112. Zorzi, M., Pastore, P., Magno, F. A single calibration graph for the direct determination of ascorbic and dehydroascorbic acids by electrogenerated luminescence based on Ru(bpy)$_3{}^{2+}$ in aqueous solution. Anal. Chem. **2000**, *72*, 4934–4939.

113. Uchikura, K. Tris(2,2′-bipyridine)ruthenium(III) chemiluminescence detection of carbonyl compounds with methylmalonic acid. Anal. Sci. **2000**, *16*, 453–454.

114. Morita, H., Konishi, M. Electrogenerated chemiluminescence derivatization reagent, 3-isobutyl-9,10-dimethoxy-1,3,4,6,7,11b-hexahydro-2H-pyrido[2,1-a]isoquinolin-2-ylamine, for carboxylic acids in high performance liquid chromatography using tris(2,2′-bipyridine)ruthenium(II). Anal. Chem. **2003**, *75*, 940–946.

115. Yamazaki, S., Hara, K., Yokota, K., Ikegami, T., Konari, D., Saito, K. Detection of polyols by HPLC coupled with an on-line photochemical oxidation reactor and chemiluminescence detection. HRC—J. High Resolut. Chromatogr. **2000**, *23*, 127–130.

116. Chen, X., Sato, M., Lin, Y.J. Study of the electrochemiluminescence based on tris(2,2′-bipyridine)ruthenium(II) and alcohols in a flow injection system. Microchem. J. **1998**, *58*, 13–20.

117. Morita, H., Konishi, M. A new electrogenerated chemiluminescence derivatization reagent, 3-(diethylamino)propionic acid, for alcohol in HPLC using tris(2,2′-bipyridine)ruthenium(II). J. Liquid Chromatogr. Rel. Tech. **2002**, *25*, 2413–2423.

118. McCall, J., Alexander, C., Richter, M.M. Quenching of electrogenerated

chemiluminescence by phenols, hydroquinones, catechols, and benzoquinones. Anal. Chem. **1999**, *71*, 2523–2527.

119. McCall, J., Richter, M.M. Phenol substituent effects on electrogenerated chemiluminescence quenching. Analyst **2000**, *125*, 545–548.

120. He, Z., Liu, X., Luo, Q., Yu, X., Zeng, Y. Chemiluminescent determination of 6-mercaptopurine in pharmaceutical preparation. Anal. Sci. **1995**, *11*, 415–417.

121. Wu, F., He, Z., Meng, H., Zeng, Y. Determination of sulfite in sugar and sulfur dioxide in air by chemiluminescence using the Ru(bipy)$_3$$^{2+}$-KBrO$_3$ system. Analyst **1998**, *123*, 2109–2112.

122. He, Z., Wu, F., Meng, H., Yuan, L., Luo, Q., Zeng, Y. Chemiluminescence determination of sulfur dioxide in air using tris(1,10-phenanthroline)ruthenium-KIO$_4$ system. Anal. Lett. **1999**, *32*, 401–410.

123. Tsukagoshi, K., Miyamoto, K., Nakajima, R., Ouchiyam, N. Sensitive determination of metal ions by liquid chromatography with tris(2,2′-bipyridine)ruthenium(II) complex electrogenerated chemiluminescence detection. J. Chromatogr. A **2001**, *919*, 331–337.

124. Yokoyama, K., Sasaki, S., Ikebukuro, K., Takeuchi, T., Karube, I., Tokitsu, Y., Masuda, Y. Biosensing based on NADH detection coupled to electrogenerated chemiluminescence from ruthenium tris(2,2′-bipyridine). Talanta **1994**, *41*, 1035–1040.

125. Martin, A.F., Nieman, T.A. Glucose quantitation using an immobilized glucose dehydrogenase enzyme reactor and a tris(2,2′-bipyridyl)ruthenium(II) chemiluminescent sensor. Anal. Chim. Acta **1993**, *281*, 475–481.

126. Dong, L., Martin, M.T. Enzyme-triggered formation of electrochemiluminescent ruthenium complexes. Anal. Biochem. **1996**, *236*, 344–347.

127. Baeyens, W.R.G., Ling, B.L., Imai, K., Calokerinos, A.C., Schulman, S.G. Chemiluminescence detection in capillary electrophoresis. J. Microcolumn Sep. **1994**, *6*, 195–206.

128. Garcia Campana, A.M., Baeyens, W.R.G., Zhao, Y. Chemiluminescence detection in capillary electrophoresis. Anal. Chem. **1997**, *69*, 83A–88A.

129. Staller, T.D., Sepaniak, M.J. Chemiluminescence detection in capillary electrophoresis. Electrophoresis **1997**, *18*, 2291–2296.

130. Huang, X.-J., Fang, Z.-L. Chemiluminescence detection in capillary electrophoresis. Anal. Chim. Acta **2000**, *414*, 1–14.

131. Kuyper, C., Milofsky, R. Recent developments in chemiluminescence and photochemical reaction detection for capillary electrophoresis. TRAC—Trends Anal. Chem. **2001**, *20*, 232–240.

132. Gilman, S.D., Silverman, C.E., Ewing, A.G. Electrogenerated chemiluminescence detection for capillary electrophoresis. J. Microcolumn Sep. **1994**, *6*, 97–106.

133. Barnett, N.W., Hindson, B.J., Lewis, S.W., Purcell, S.D. Determination of codeine, 6-methoxycodeine, and thebaine using capillary electrophoresis with tris(2,2′-bipyridyl)ruthenium(II) chemiluminescence detection. Anal. Commun. **1998**, *35*, 321–324.

134. Forbes, G.A., Nieman, T.A., Sweedler, J.V. On-line electrogenerated Ru(bpy)$_3$$^{3+}$

chemiluminescent detection of β-blockers separated with capillary electrophoresis. Anal. Chim. Acta **1997**, *347*, 289–293.

135. Dickson, J.A., Ferris, M.M., Milofsky, R.E. Tris(2,2′-bipyridyl)ruthenium(III) as a chemiluminescent reagent for detection in capillary electrophoresis. HRC—J. High Resolut. Chromatogr. **1997**, *20*, 643–646.

136. Bobbitt, D.R., Jackson, W.A., Hendrickson, H.P. Chemiluminescent detection of amines and amino acids using in situ generated $Ru(bpy)_3^{3+}$ following separation by capillary electrophoresis. Talanta **1998**, *46*, 565–572.

137. Wang, X., Bobbitt, D.R. In situ cell for electrochemically generated $Ru(bpy)_3^{3+}$-based chemiluminescence detection in capillary electrophoresis. Anal. Chim. Acta **1999**, *383*, 213–220.

138. Hendrickson, H.P., Anderson, P., Wang, X., Pittman, Z., Bobbitt, D.R. Compositional analysis of small peptides using capillary electrophoresis and $Ru(bpy)_3^{3+}$-based chemiluminescence detection. Microchem. J. **2000**, *65*, 189–195.

139. Wang, X., Bobbitt, D.R. Electrochemically generated $Ru(bpy)_3^{3+}$-based chemiluminescence detection in micellar electrokinetic chromatography. Talanta **2000**, *53*, 337–345.

140. Cao, W., Liu, J., Yang, X., Wang, E. New technique for capillary electrophoresis directly coupled with end-column electrochemiluminescence detection. Electrophoresis **2002**, *23*, 3683–3691.

141. Cao, W., Liu, J., Qiu, H., Yang, X., Wang, E. Simultaneous determination of tramadol and lidocaine in urine by end-column capillary electrophoresis with electrochemiluminescence detection. Electroanalysis **2002**, *14*, 1571–1576.

142. Liu, J., Cao, W., Qiu, H., Sun, X., Yang, X., Wang, E. Determination of sulpiride by capillary electrophoresis with end-column electrogenerated chemiluminescence detection. Clin. Chem. **2002**, *48*, 1049–1058.

143. Sun, X.H., Liu, J.F., Cao, W.D., Yang, X.R., Wang, E.K., Fung, Y.S. Capillary electrophoresis with electrochemiluminescence detection of procyclidine in human urine pretreated by ion-exchange cartridge. Anal. Chim. Acta **2002**, *470*, 137–145.

144. Chiang, M.T., Whang, C.W. Tris(2,2′-bipyridyl)ruthenium(III)-based electrochemiluminescence detector with indium/tin oxide working electrode for capillary electrophoresis. J. Chromatogr. A **2002**, *934*, 59–66.

145. Wilson, R., Akhavan-Tafti, H., DeSilva, R., Schaap, A.P. Comparison between acridan ester, luminol, and ruthenium chelate electrochemiluminescence. Electroanalysis **2001**, *13*, 1083–1092.

146. Cao, W., Jia, J., Yang, X., Dong, S., Wang, E. Capillary electrophoresis with solid-state electrochemiluminescence detector. Electroanalysis **2002**, *23*, 3692–3698.

147. Waguespack, B.L., Bobbitt, D. Integration of $Ru(bpy)_3^{3+}$-based chemiluminescence with microfabricated CE devices for the analysis of amines, amino acids, and proteins. 223rd ACS Natl Meeting, Orlando, FL, **2002**, Apr 7–22.

148. Huang, X.-J., Wang, S.-L., Fang, Z.-L. Combination of flow injection with capillary electrophoresis.8. Miniaturized capillary electrophoresis system with flow injection sample introduction and electrogenerated chemiluminescence detection. Anal. Chim. Acta **2002**, *456*, 167–175.

149. Arora, A., Eijkel, J.C.T., Morf, W.E., Manz, A. A wireless electrochem-
      iluminescence detector applied to direct and indirect detection for electrophor-
      esis on a microfabricated glass device. Anal. Chem. **2001**, *73*, 3282–3288.

10

ECL Polymers and Devices

Mihai Buda*

University of Texas at Austin, Austin, Texas, U.S.A.

I. INTRODUCTION

The field of organic light-emitting devices, for which the name "molecular light-emitting devices" might be better suited because it actually includes many inorganic active compounds [such as the highly popular tris(8-hydroxyquinolinato) aluminum], has grown immensely in the last decade. Numerous studies are published every year, many of them reporting new substances with improved emission properties, stability, and lifetimes. Moreover, displays made with polymer light-emitting devices are currently being commercialized. These devices are no longer only toys for scientists around the world; their study has evolved into a new, growing technology.

Given the vast literature, the topic of organic light-emitting diodes (OLEDs) would have to be covered in a book by itself, and a rather large book even without going into detail. Because excellent reviews and books are available, the reader wanting to find out more about organic electroluminescence in general and OLEDs in particular is directed to Refs. 1–10. Here we will focus on a single class of light-emitting devices, the so-called light-emitting electrochemical cells (LECs). Their properties are much more similar to those of a typical electrochemical system than to those of a classical amorphous semiconductor, and therefore it is more appropriate to discuss them in an electrochemistry book. There is also another reason for discussing only LECs: They are much less pop-

*Present address: Department of Applied Physical Chemistry & Electrochemistry, Univ. Politehnica Bucharest, Calea Grivitei 132, ZIP: 010737, Bucharest, Romania. E-mail: mihai@catedra.chfiz.pub.ro

ular in the organic light-emitting literature, and they deserve a more detailed treatment. The history of solution electrogenerated chemiluminescence (ECL), with which these devices have much in common, is more than 30 years old and is now a well-established field in electrochemistry.

Because LECs involve light generation in a solid phase, we find it appropriate to give here an insight, without going into detail, into solid-state ECL, with emphasis on its application to light-emitting devices. Thus, we shall examine a few examples of solid-state ECL, even when not directly related to light-emitting devices (LEDs), examples that we believe will help the reader to have a broader view. Not all solid-state ECL research has been focused on LEDs, so we shall mention, where necessary, some other potential applications of solid-state ECL (e.g., in analytical chemistry).

The material found in this chapter is divided into two major parts: three-electrode systems, studied in a classical electrochemical experimental setup, and two-electrode systems, typical for LEDs, each part comprising several subsections. In the first part, we present ECL data obtained using modified electrodes. The second part deals with LECs and their features, as well as with some theoretical considerations. We sincerely hope that the reader interested in learning about OLEDs in general and LECs in particular will find this material useful and will benefit from both the presentation and the cited references.

II. THREE-ELECTRODE SETUP FOR ECL WITH MODIFIED ELECTRODES

In this section we review and discuss briefly ECL on electrodes modified with pure emitters or emitters dispersed in an inert matrix and adsorbed monolayers (Langmuir–Blodgett films).

The study of luminescent materials immobilized onto an electrode in a three-electrode setup allows each electrochemical process involved in the light generation to be studied separately. Thus, one can discuss only the reduction or the oxidation of the active material, processes that occur together in an LEC and are impossible to separate in a two-electrode cell. By studying the oxidation and reduction processes separately, one can also obtain the redox potentials for these processes, potentials that can be used to estimate highest occupied molecular orbital (HOMO) and lowest unoccupied molecular orbital (LUMO) energies [11]. The HOMO/LUMO energy values are in turn useful for calculating the injection barriers in light-emitting devices. In many cases the use of such modified electrodes is the easiest way to obtain such values, especially when the polymer is completely insoluble in polar solvents, its electrochemical behavior being thus inaccessible in solution.

Even though the electrochemistry of such modified electrodes is somwhat complicated by ion insertion and deinsertion processes (required to maintain electroneutrality in the film), these studies can provide valuable information about the charge transport properties, material stability, and energetics of the redox reactions, with a simple and easy-to-use experimental setup [12].

As already hinted in the first paragraph, several approaches exist for obtaining modified electrodes with an active luminescent material. The first, and probably easiest, way is to polymerize or deposit (e.g., by spin-coating) the active material or to immobilize the active material in a solid matrix onto the electrode. After coating the electrode, one can immerse it in an inert electrolyte (in which the deposited substance is insoluble) and study its electrochemical and photochemical properties. Another approach is to form a monolayer of active material by adsorbing it onto a metal or semiconductor electrode. In the first two cases, thick films are usually obtained (micrometer or submicrometer range), although in principle thinner films can be obtained; in the latter case, very thin layers (nanometers or even less) of active material result.

A. ECL from Electrodes Modified with Pure Emitters

There have been only a few studies dealing with ECL from pure active materials deposited as solid films onto electrodes. An attractive candidate for such modified electrode ECL is tris(4-vinyl-4'-methyl-2,2'-bipyridyl)ruthenium(II) [Ru(vbpy)$_3^{2+}$]. Modified electrodes can be easily obtained by electropolymerization of the monomer ruthenium complex in acetonitrile, either at constant current or by potential cycling [13].

The thickness of the polymer layer can be adjusted by controlling the charge passed during electropolymerization and is typically between ~40 Å and 1 μm [14].

Ru(vbpy)$_3^{2+}$ PVDPA

When the potential is pulsed between +1.5 and −1.5 V (vs. SSCE) in an acetonitrile electrolyte, the modified electrodes exhibit an orange luminescence (λ_{max} ~650 nm); the ECL has a quite modest lifetime, only ~20 min.

Another attractive electrochemiluminescent polymer is poly(vinyl-9, 10-diphenylanthracene) (PVDPA). The polymer, with an average molecular weight of about 48,000, can be deposited as a film by spin-coating on different substrates (SnO_2, glassy carbon, and platinum) from benzene solutions. The film exhibits reversible oxidation and reduction (1.4 and −1.88 V vs. SCE, respectively) (Fig. 1) in nonaqueous solvents (propylene carbonate, acetonitrile, and tetrahydrofuran), with oxidation/reduction potentials close to those of 9,10-diphenylanthracene in solution, indicating that the interaction along the polymer chain is weak. When the potential is pulsed between 1.6 and −2.0 V (SCE), the modified electrode shows blue electroluminescence (λ_{max} ~440 nm) (Fig. 1), which quickly (5–10 min) fades as the radical anion and radical cation forms of the polymer generated during oxidation and reduction dissolve in the electrolyte [15].

An interesting polymer widely used for solid-state electroluminescent devices is 4-methoxy-(2-ethylhexoxyl)-2,5-polyphenylenevinylene (MEH-PPV). Richter et al. [16] reported the ECL of a Pt electrode coated with a layer (~100 nm) of MEH-PPV. The modified electrode shows fairly good reversible oxidation and reduction (+0.4 and −1.9 V vs. Fc/Fc$^+$) (Fig. 2a) in acetonitrile. The electrode is reasonably stable in solution; some dissolution of the oxidized form is observed during repetitive scans. When the potential is stepped from +0.4 to −2.35 V, an orange luminescence is observed (λ_{max} ~620 nm), the ECL spectrum being

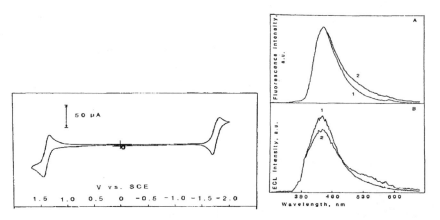

Figure 1 Left: Cyclic voltammogram of a PVDPA film (~400 nm) on SnO_2 in propylene carbonate (0.2 M TBABF$_4$, 0.1 V/s). Right: Fluorescence (A) and ECL (B) spectra of a PVDPA film in propylene carbonate. Potential stepped between 1.6 and −2.0 V (SCE); 100 ms pulse width. (From Ref. 15. Copyright 1985 by Elsevier Science.)

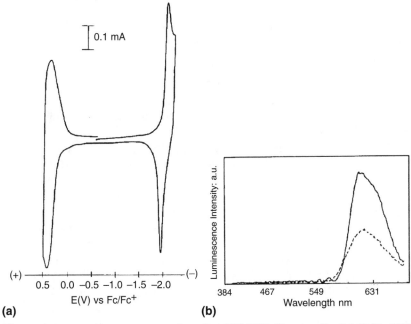

MEH-PPV   MDO-PPV   PAT6

essentially the same as the photoluminescence spectrum (Fig. 2b). Again, the ECL lifetime is very short (1–2 min), the film changing color on oxidation from red to deep blue. Janakiraman et al. [17] observed the same ECL of MEH-PPV coated on a Pt electrode by slow evaporation of chloroform solution and simulated the ECL transients using a diffusion model coupled with the second-order light-generating annihilation reaction. The simulated results agree reasonably well with

Figure 2 (a) Cyclic voltammetry of an (b) MEH-PPV film onto Pt (0.5 M Bu$_4$NBF$_4$, acetonitrile, 0.1 V/s). (b) Room-temperature photoluminescence spectrum (solid lines) and ECL spectrum (dashed line) for the same MEH-PPV film. ECL was generated by stepping the potential between 0.4 V (vs. Fc/Fc$^+$) for 1 s and –2.35 V for 0.5 s. (From Ref. 16. Copyright 1994 by Elsevier Science.)

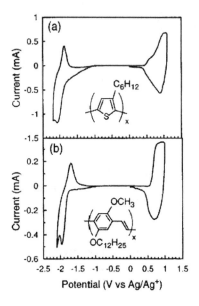

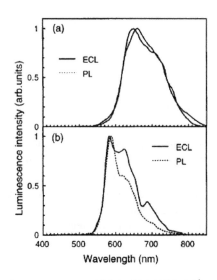

Figure 3 Cyclic voltammograms (left) and luminescence spectra (right) of (a) ITO/PAT6- and (b) ITO/MDOPPV-modified electrodes in acetonitrile. (From Ref. 18.)

the experimental transients, giving an apparent diffusion coefficient (for the slowest species) in the film of about 5×10^{-13} cm^2/s and a bimolecular annihilation rate of 3×10^3 L/(mol s).

Nambu and coworkers [18,19] reported the ECL of poly(3-hexylthiophene) (PAT6) and poly(2-methoxy-5-dodecyloxy-p-phenylenevinylene) (MDO-PPV) films formed onto ITO by spin-coating from chloroform solutions. PAT6 (~50 nm thick) shows almost reversible reduction (−1.2 V vs. Ag/Ag$^+$) in acetonitrile, but the oxidation peak (1.1 V) is rather broad and not very well defined (Fig. 3a). A linear scan rate dependence of the peak currents was found, typical for thin-layer films. The MDO-PPV film electrode (~320 nm thick) shows similar behavior, with a broad oxidation peak (1.0 V) and a more reversible and well-defined reduction peak (−1.9 V). PAT6 films show orange ECL (λ_{max} ~640 nm) when the potential is pulsed between 1.1 and −2.2 V (vs. Ag wire quasi-reference), whereas MDO-PPV shows red ECL (λ_{max} ~580 nm) when the potential is pulsed between 1.0 and −1.9 V (Fig. 3b). For both polymers the light emission is not identical during both pulses, the authors calculating that 17% of the monomer units are oxidized and only 10% are reduced.

Buda et al. [20] studied the ECL of ITO electrodes modified with Ru(bpy)$_2$(Es-bpy-Es), containing a long alkyl chain bipyridine derivative, which

Ru(bpy)₂(Es-bpy-Es)

gives a water-insoluble ruthenium chelate. The films (deposited by spin-coating) show strong ECL in water when pulsed between −1.0 and 1.3 V (vs. Ag/AgCl). These films are reasonably stable in water at open circuit (up to 6–8 h), but their properties tend to degrade rather quickly on reduction. This is not surprising, though, because it is well known that the reduced species of $Ru(bpy)_3^{2+}$ and derivatives are not very stable in the presence of water.

Although such modified electrodes are interesting because they allow the reduction and oxidation processes to be studied independently, they are subject to several limitations:

1. The film-forming process, whether by electrochemical polymerization or casting from solution, leads to rather porous films [16,20,21]. Because in many cases the oxidation/reduction of these films occurs at rather extreme potentials, unwanted solvent-related electrochemical processes that may occur in the pores can lead to side reactions that limit the electrochemical stability of these films and consequently shorten the ECL lifetime. Such processes can be also a source of errors when trying to estimate some properties of the films.
2. Even if the polymer film itself is perfectly insoluble when resting in a solvent, either the oxidized or reduced form (or both) may be much more soluble, leading to material loss in solution. In some cases the dissolution process can be also related to degradation products that accumulate during continuous cycling of the film.
3. During both oxidation and reduction of the film, ions from the electrolyte penetrate the film to preserve the electroneutrality; along with the ions, some solvent also enters the film [22]. Ion and solvent insertion not only leads to changes in film thickness by swelling [23] but may also cause ion pairing [24] and even more subtle structural changes. When the film is cycled continuously between the oxidized and reduced forms, the film's structure changes and films may eventually be delaminated. Also, when the potential is pulsed between the values needed to generate the reduced and oxidized

forms, it is possible that the ions inserted in the previous process are not totally expelled from the film when the potential is changed. As a result, more ions of the opposite charge are required to enter the film in the subsequent step; this process is likely to have a negative influence on the long-time behavior of these films.

4. It is often difficult to obtain pure polymers, and unwanted impurities and oligomers may adversely influence the behavior of the deposited films.

In all cases the ECL lifetime, of the order of minutes, is very modest, but, as pointed out above, this is not surprising. Even though such modified electrodes may not be used directly in a practical device, their importance lies not in the device itself but rather in their ability to give valuable information about the phenomena that occur in solid-state ECL.

B. ECL from Monolayer Modified Electrodes

An alternative way to modify the electrode surface with a pure emitter is to form an adsorbed monolayer. For such a layer to be obtained, the molecule must be able to bind strongly on some solid surface (usually metals or carbon, but semiconductors can also be used). Such modified electrodes have almost exclusively analytical applications: the amount of material deposited on the electrode is very small and will lead to very low ECL intensity, impractical for a light-emitting device. Still, monolayer ECL can provide useful information for the scientist interested in light-emitting devices. For example, because in an adsorbed monolayer the excited state is generated in very close proximity to the electrode, one can obtain information about the phenomenon of luminescence quenching by the metal substrate [25,26], which is not yet fully understood.

Because the practical application of monolayer-modified electrodes is mostly as sensors for various analytes, the reported ECL for such systems involves a coreactant, usually oxalate ions. The ECL generated with coreactants is treated in more detail in Chapter 5, so here we discuss it only briefly.

Many studies in the literature deal with potentially luminescent monolayers, most of them involving adsorbed derivatives of tris(2,2′-bipyridine)ruthenium [27–32] and only a few reporting ECL experiments. The first report of ECL from an adsorbed monolayer of a tris(2,2′-bipyridine)ruthenium(II) derivative was published in 1988 [33]. Zhang and Bard found that the ECL for a monolayer of a suitable derivatized tris(2,2′-bipyridine)ruthenium(II) [Ru(bpy)$_2$(bpy-C19)] using oxalate ions as coreactant was three orders of magnitude lower on Pt and Au electrodes than on In-doped SnO$_2$, showing that the metal electrode quenching of the ECL is very important.

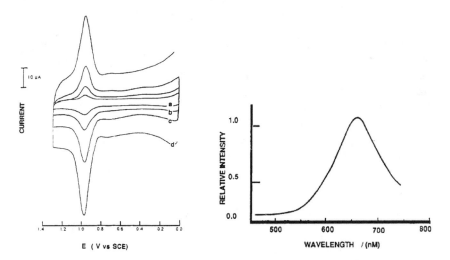

Ru(bpy)$_2$(bpy-C19)

In a more recent study, Obeng and Bard [34] employed a more stable tris(bipyridyl)ruthenium(II) derivative [Ru(bpy)$_2$(bpy-SH)$^{2+}$] that spontaneously adsorbs onto Au and ITO electrodes. The Ru(bpy)$_2$(bpy-SH)$^{2+}$ monolayer gives intense ECL (Fig. 4) when oxidized in the presence of oxalate ions in solution, again the ECL signal being stronger for monolayers formed on ITO. However, the authors noticed that the surface coverage on Au electrodes is much lower, and the electrochemistry of the monolayer is complicated by gold oxidation. Sato and Uosaki [35] studied the ECL of a closely related thiol derivative, Ru(bpy)$_2$(HS-bpy-SH)$^{2+}$, in the presence of oxalate ions. Their observations are quite similar, showing better electrochemistry and ECL for monolayers formed on ITO.

Figure 4 Left: Cyclic voltammetry of a Ru(bpy)$_2$(bpy-SH)$^{2+}$ monolayer on ITO in 0.5 M H$_2$SO$_4$ at various scan rates. Right: ECL spectrum of a Ru(bpy)$_2$(bpy-SH)$^{2+}$ monolayer on polycrystalline gold in 0.1 M Na$_2$C$_2$O$_4$ and 0.4 M Na$_2$SO$_4$ (pH 4.7) biased to 1.25 V vs. SCE. (From Ref. 34. Copyright 1991 by the American Chemical Society.)

Ru(bpy)$_2$(bpy-SH)$^{2+}$ Ru(dp-bpy)$_3$$^{2+}$ Ru(bpy)$_2$(HS-bpy-SH)$^{2+}$

Miller et al. [36] studied the same Ru(bpy)$_2$(bpy-C19)$^{2+}$ and Ru(dp-bpy)$_3$$^{2+}$ as potential probes for ECL imaging. HOPG and ITO electrodes were used for monolayer formation, and ECL was generated by oxidation in the presence of oxalate ions in solution. The strong ECL signal allows the monolayer to be imaged with a CCD camera. The luminescent tris(bipyridyl)ruthenium(II) derivatives can also be dispersed in a nonluminescent monolayer and the resulting film imaged using the same technique. Because no light excitation is needed, in contrast with fluorescent probes, the experimental design is much simpler and allows better sensitivities.

In 1994, Xu and Bard [37] found that even the underivatized Ru(bpy)$_3$$^{2+}$ interacts strongly with HOPG, Au, and Pt surfaces, forming adsorbed monolayers. The ECL mechanism using tripropylamine as coreactant is somewhat complicated by the desorption of the complex adsorbed onto metal electrodes. As a result, ECL from the dissolved ruthenium chelate is also observed. The ruthenium chelate appears to be more strongly bound to HOPG, monolayers obtained on these electrodes being more stable and showing longer ECL lifetimes.

C. ECL from Emitters Immobilized in an Inert Matrix

The insertion or immobilization of the active material in an inert matrix has several attractive features. First, one can tune the system's properties by choosing different host materials with different properties (stability, binding groups, porosity, etc.). Second, the concentration of the emitter can be varied easily, thus reducing the triplet–triplet annihilation reaction that is usually observed when the concentration of the excited state is high [38]. Third, one can photochemically excite the emitter molecule and then study the quenching effect of the reduced/oxidized species by generating them electrochemically.

Like adsorbed monolayer electrodes, these types of modified electrodes have been studied mainly for their potential applications in analytical chemistry, but useful mechanistic information is also available from these studies. All the studies discussed here involve as emitters $Ru(bpy)_3^{2+}$ or derivatives immobilized in various matrices.

1. ECL from $Ru(bpy)_3^{2+}$ Immobilized in Nafion-Modified Electrodes

Nafion is an inert cationic ion-exchange polymer, and it is thus ideal for immobilizing cationic emitters; it has been widely employed to immobilize $Ru(bpy)_3^{2+}$ and derivatives. Nafion-modified electrodes are easily obtained by simply dipping the electrode in a Nafion solution (e.g., ethanol) and then drying it in air. Immobilization of the cationic emitter, such as $Ru(bpy)_3^{2+}$, is then attained by ion-exchange after immersion of the Nafion-modified electrode in a solution containing $Ru(bpy)_3^{2+}$. The whole procedure is very simple, producing good quality films. The amount of immobilized emitter can be calculated by determining the charge needed for oxidation (or reduction) of the modified electrode.

Rubinstein and Bard first reported in 1980 [39] the ECL of $Ru(bpy)_3^{2+}$ immobilized in a Nafion matrix with emitter concentrations in the range of $2{\times}10^{-6}$ to $4{\times}10^{-6}$ mol/cm^2. The modified electrode is very stable (at open circuit) in water and aqueous solutions, reproducible cyclic voltammograms being obtained after 3 weeks of immersion. When the immobilized emitter is oxidized in aqueous solution in the presence of oxalate, strong ECL is observed. In aqueous solution containing 20% acetonitrile, annihilation-type ECL is indicated by the alternating electrochemical generation of Ru(I) and Ru(III) species. The ECL, however, decays fast in this case, mainly due to the dissolution of the Nafion in the solvent. In a more detailed study [51], a mechanism was proposed for the ECL generated from Nafion-immobilized $Ru(bpy)_3^{2+}$ using oxalate as coreactant. It was shown that for such thick films the process is controlled by mass transport; the apparent diffusion coefficient for the attached ruthenium chelate was found to be $\sim 10^{-9}$ cm^2/s. The complex reaction mechanism, which includes a step of quenching by $Ru(bpy)_3^{3+}$, was checked by digital simulation of both electrochemical and ECL experimental data, the simulation showing good agreement with the experiment.

2. ECL from $Ru(bpy)_3^{2+}$ Immobilized in Silica-Based Modified Electrodes

Electrodes modified with inorganic materials, including silica-modified electrodes obtained by sol-gel techniques, are now widely used because they

$[PS\text{-}CH_2CH_2NHCO\text{-}(Ru^{II})_{18}]^{36+}$

have several attractive features. It is not the purpose of this chapter to discuss the details of this procedure, so we encourage the reader to check the original references as well as reviews treating such modified electrodes in detail [40].

Sykora and Meyer [41] reported the ECL of $[PS\text{-}CH_2CH_2NHCO\text{-}(Ru^{II})_{18}]^{36+}$ immobilized in a silica sol-gel matrix in the presence of oxalate ions. They also reported intense orange annihilation-type ECL when the reduced and oxidized species of $Ru(bpy)_3^{2+}$ were generated alternately by pulsing the potential between -1.24 and $+1.26$ V ($vs.$ SSCE), but the annihilation ECL lifetime was short, with half-lives of 3–5 min (10 Hz pulsing frequency). One reason for the short lifetime is the presence of chloride counter ions, which can be oxidized at the potential values needed for Ru(III) generation.

In a series of papers [42–45], Collinson and coworkers reported the ECL of silica-based modified electrodes containing immobilized $Ru(bpy)_3^{2+}$. The ECL, in all cases of the coreactant type using either oxalate or tripropylamine, showed reasonable lifetimes when a modified microelectrode was used, with a drop of only a few percent in 2 h or 10–20% in 24 h when tripropylamine was used as coreactant (Fig. 5). The authors note that the optimization of the procedure for obtaining the silica-based modified electrodes should help to increase both the lifetime and the emission intensity. Although neither the coreactant type of ECL nor the small electrode size is practical for a light-emitting device, the good results reported with these modified electrodes are promising.

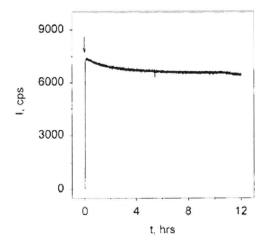

Figure 5 Intensity–time transient for gel-encapsulated Ru(bpy)$_3^{2+}$ (10 mM) and tripropylamine (5 mM) at 1.15 V (13 μm Pt ultramicroelectrode, gel prepared with 0.1M phosphate buffer, pH 6.2). (From Ref. 43. Copyright 1999 by the American Chemical Society.)

3. ECL from Ru(bpy)$_3^{2+}$ Immobilized in Polymer-Modified Electrodes

Ghosh and Bard [46] studied the electrochemistry and ECL in aqueous solutions of modified electrodes of poly(tris(bipyrazine)ruthenium(II)) containing immobilized Ru(bpy)$_3^{2+}$. The polymer film is formed from either an aged solution ($\geq$7 days) of 1 mM Ru(bpz)$_3$Cl$_2$ and 7.5–8.0 M H$_2$SO$_4$ by scanning the potential between 2.4 and 0.6 V (vs. SSCE), or from a 1 mM bipyrazine solution in 6 M H$_2$SO$_4$ by scanning the potential between 2.1 and 1.5 V. The exact nature of the polymer film is not known, and it is unclear whether the oxidized protonated bipyrazine or the hydroxyl ions are responsible for polymerization initiation. To incorporate Ru(bpy)$_3^{2+}$, the modified electrodes are then cycled in an aqueous solution containing 0.14 mM Ru(bpy)$_3^{2+}$. The interactions between Ru(bpy)$_3^{2+}$ and the polymer matrix are weak, the loss of Ru(bpy)$_3^{2+}$ being observed during continuous oxidation potential scans of the electrode in a 0.05 M KNO$_3$ aqueous solution. If, however, as little as 0.05 mM Ru(bpy)$_3^{2+}$ is added to the supporting electrolyte, the loss becomes negligible. ECL is observed when the modified electrode containing Ru(bpy)$_3^{2+}$ is oxidized in aqueous oxalate solutions, but with low emission intensities. The authors relate the weak emission to a large concentration of Ru(bpy)$_3^{3+}$ formed in the film during oxidation, which is known as an efficient quencher for the Ru(bpy)$_3^{2+}$ excited state (see Table 1). The charge transport through these films appears to occur by both electron hopping and physical diffusion, an apparent diffusion coefficient of ~2.2$\times$10^{-9} cm^2/s being obtained for the overall process.

Table 1   Quenching of the Excited State of Ru(bpy)$_3{}^{2+}$ Immobilized in an Inert Matrix

Matrix	Ru(bpy)$_3{}^{3+}$ quenching rate constant (M^{-1} s^{-1})	Ref.
Polystyrenesulfonate	1.2×10^8	47
Nafion	$1-5 \times 10^7$	50
Nafion	5×10^7	51

In an interesting study, even though not reporting ECL, Majda and Faulkner [47] studied the quenching of the excited state of trisbipyridylruthenium(II) (generated photochemically) in a polystyrenesulfonate matrix by electrogenerated Ru(bpy)$_3{}^{3+}$, showing that the quenching process is a simple bimolecular reaction between the excited state and Ru(bpy)$_3{}^{3+}$. Several other authors have estimated the quenching rate constant for this process (see Table 1). The quenching of the excited state by the very species that led to its formation is an important process that limits the quantum efficiency of light-emitting devices; it is known that, once formed by the annihilation reaction, the exciton can hop to nearby sites [38] and thus migrate in regions where the concentration of the reduced/oxidized species is greater. Thus, it is useful to have quantitative data about this quenching mechanism. Table 1 shows some data for the quenching of the excited state of Ru(bpy)$_3{}^{2+}$, immobilized in various matrices, by electrogenerated Ru(bpy)$_3{}^{3+}$. Bearing in mind that the results are likely to depend on various parameters such as the film thickness as well as emitter concentration and the matrix in which it is immobilized, the results can be considered to be in reasonable agreement.

Apart from the useful information that such modified electrodes can provide, we shall only mention here that modified electrodes containing Ru(bpy)$_3{}^{2+}$ have been successfully employed as sensors for various analytes, such as oxalate, alkylamines, and NADH [48,49].

III. TWO-ELECTRODE LIGHT-EMITTING DEVICES—SOLID-STATE LECs

A. Introduction

Light is emitted in LECs from an excited molecule formed by the annihilation of electrons (or radical anions in solution) with holes (or radical cations). The light emission process is very similar with fluorescence, except that the excited state is generated through an electrochemical mechanism rather than photochemically.

The idea of using electrochemical cells as tools for studying light-emitting processes is by no means new; it was suggested in 1965 [52]. After the mechanism of annihilation-type ECL was proposed [53], the idea of using this phenomenon for light-emitting devices was immediately acknowledged. In the following years several attempts were made to make practical devices (see Section III.B). However, the efforts did not lead to a real practical device, mainly because of two major drawbacks. First, their operation lifetimes remained short, far below the requirements for, e.g., a color-active display. Second, the use of liquid solvents, in some cases expensive solvents, highly pure and very dry, made them less attractive for practical devices. Nevertheless, with all their shortcomings, solution LECs (SLECs) pioneered what is now the very rich and dynamic field of organic electroluminescence.

Since Tang and VanSlyke published in 1987 [54] their report on solid-state electrogenerated luminescence with high quantum efficiency (~1%) at moderate voltages (<10 V) from an "organic" material, tris(8-hydroxyquinolinato)aluminum (AlQ$_3$), the interest in these devices has grown progressively, with many new active materials being reported. At the same time, the interest in LECs, this time solid-state LECs (SSLECs), was renewed [55–57,199]. Solid-state LECs are more appealing as practical devices, and they did indeed benefit from the discovery of new materials such as solid electrolytes. The main difference between solid-state and solution electroluminescence lies in the transport properties of the medium. While in solution the oxidized and reduced species move *physically* (by diffusion and/or migration); in condensed phase the charge transport occurs almost exclusively through electron hopping between fixed active sites (see Section IV.A).

When talking about organic light-emitting devices, it is useful to distinguish between organic light-emitting diodes (OLEDs) and light-emitting electrochemical cells (LECs). They are related to each other, the major difference between them being the presence in LECs of a (relatively) large concentration of mobile ions. The presence of mobile ions leads to some important differences between LECs and OLEDs [58]. The "turn-on" voltage of LECs is close to the optical bandgap and depends weakly on the film thickness, they have improved

AlQ$_3$

quantum efficiencies (due to the balanced charge injection; see below), and their electrical and luminescence characteristics are symmetrical and usually not related to the type of electrode (depending on the mobile ion concentration). All of these features are desirable in a practical device. Symmetrical characteristics allow the LECs to be driven in alternating current mode, and the weak dependence of their properties on the nature of the electrode eliminates the need to use low-work-function metals such as calcium, which are quite reactive and difficult to work with.

Even though we will discuss LECs from an electrochemist's point of view, in the following pages we shall use the terminology of solid-state physics rather than that of electrochemistry, referring to the reduced species as "electrons" and to oxidized species as "holes."

The general requirements for efficient organic light-emitting devices [6] are related to some important basic features:

1. Charge Balance.   For an efficient annihilation, light-generating, reaction to take place and to maximize the excited state generation, the concentration of electrons must exactly match the concentration of holes. In a LEC this is generally true even without special precautions (such as the use of low-work-function electrodes or hole and electron transport layers). LECs usually have a large concentration of mobile ions, and therefore, being unable to support large internal electric fields, they must be (almost) electroneutral. As a result, the charges injected at each electrode must exactly balance each other. This is not always true, however, and the balanced charged injection depends on the concentration of mobile ions. Moreover, in some cases parallel, usually unwanted, electrochemical processes may also be responsible for charge injection. For example, when $Ru(bpy)_3^{2+}$ is used as emitter, a second or even third reduction (electron-generating) of the metal chelate may occur, whereas only one oxidation process (hole-generating) is available. The injected positive and negative charges are still balanced, but the species involved in light generation become unbalanced, and therefore the quantum efficiency is lowered.

2. Singlet-to-Triplet Ratio.   The annihilation between electrons and holes results in the formation of either a singlet or triplet excited state (see Chapter 4 for more details). Only the singlet is allowed to decay radiatively with organic molecules; for the triplet state this process is forbidden (except for molecules with heavy metal atoms). For statistical reasons, the ratio between the singlet and triplet states is 1:3, and therefore, at least theoretically, the maximum (internal) quantum efficiency for organic molecules is limited to 25%. Some studies, both experimental and theoretical, have shown, however, that in principle 100% internal quantum efficiency can be achieved [59,60].

3. Quantum Efficiency.   For maximizing the efficiency of such devices, materials with very high luminescence quantum efficiency should be used: the higher the quantum efficiency, the more efficient is the conversion of electric energy into light. However, using a material with a high luminescence quantum

efficiency does not necessarily ensure a high quantum efficiency in an organic LED. Even when the charge injection is balanced, there are several other factors that can affect the quantum (and power) efficiency. Unwanted impurities, either present from the beginning or formed during operation, can act as luminescence quenchers, lowering the quantum efficiency. Also, if the excited state is generated close to a metal electrode, effective quenching by the metal can lead to a drastic decrease in quantum efficiency of the light-emitting devices.

4. External Light Output. The light passes through several layers of material before reaching the detector (or the human eye), such as the emitter layer itself, the transparent electrode (typically ITO), the substrate onto which the transparent electrode is deposited (glass or plastic), and air. Optical phenomena such as interference, absorption, and internal reflection will lower the total amount of light that reaches the detector. Therefore, the effectively measured quantum efficiency, the so-called external quantum efficiency, can be as low as 20–25% from the internal quantum efficiency. In the following pages, whenever we talk about the quantum efficiency, we shall assume it to be the measurable external quantum efficiency.

5. Response Time. The response time can be defined as the time needed to reach a certain brightness level after the bias is applied. A good device should respond rapidly when applying a bias, long delays being impractical. Moreover, a LEC with fast response times can be operated under a.c. voltage, resulting in improved operating lifetimes and sometimes higher efficiencies [232]. In many cases, due to the limited ionic mobility, LECs have rather slow response times (sometimes of the order of tens of seconds) (Fig. 7), which is one of their major drawbacks. The response time can be improved in several ways: by increasing the ionic conductivity, by using devices with a "frozen junction" (see Section III.F), or by applying a special voltage program [228].

6. Lifetime. Because the lifetime during operation depends on the (initial) brightness level, it is usually referred to as the time after which the brightness level reaches half of an initial brightness of 100 cd/m^2 (a typical value for a desktop monitor). The best available OLEDs (with both polymeric and small-molecule emitters) have operation lifetimes exceeding 10,000 h and in some cases even more than 50,000 h [61]. At present, solid-state LECs still suffer from relatively short lifetimes, the best LECs available having lifetimes of the order of 1000 h, but further progress in this field is expected to improve their lifetimes.

B. Solution ECL Devices

Although the purpose of this chapter is to present solid-state LECs (SSLECs) we think it appropriate to show here some of the results, both old and new,

Rubrene

about solution LECs (SLECs). We will analyze a few data dealing with degradation of ECL in liquid electrochemical cells, because this is the main reason for their failure as practical devices. It is also instructive to understand the failure mode of these devices, because many of the factors that affect their lifetimes are relevant for current SSLECS as well. We should also mention that, although not strictly related to ECL experiments, when the luminescent polymers are soluble in polar solvents, solution electrochemistry can provide useful estimates of the HOMO-LUMO energies, such measurements being now quite common [11,62,72,216].

Device failure and degradation for solution LECs have been reported mostly for $Ru(bpy)_3^{2+}$ and rubrene ECL cells, two emitters that have been widely employed in SLECs. As early as 1975, Laser and Bard [63] discussed in detail the operation of ECL devices and the factors affecting their lifetime and failure. They pointed out and discussed several causes for device failure:

1. Gradual loss of electroactive substances in either solvent- or impurity-related reactions
2. Side chemical or electrochemical reactions of the electrogenerated species, solvent, supporting electrolyte, and/or impurities, which decrease their life or generate quenchers
3. Electrode blocking by film formed during operation

Analyzing the lifetime of solution light-emitting devices based on $Ru(bpy)_3^{2+}$ in acetonitrile, they underlined the necessity of precise potential control, in order to avoid the second and third reduction of the metal chelate. If these processes are allowed to occur, irreversible precipitation onto the Pt electrode occurs, blocking the electrode surface and destroying the active substance (Fig. 6, left). This requirement can be fulfilled in a three-electrode setup, but it is not as easily attained in a two-electrode, practical device. They also showed that it is important to use a potential program for generating ECL and that a simple square-wave signal may not be appropriate for efficient light generation. For thin-layer electrochemical cells as two-electrode devices, with either rubrene (in benzonitrile) or $Ru(bpy)_3^{2+}$

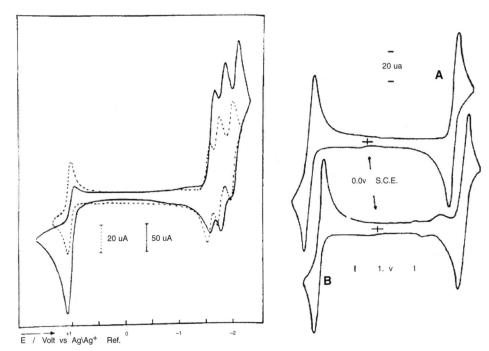

Figure 6 Left: Cyclic voltammetry of a 0.6 mM Ru(bpy)$_3$(ClO$_4$)$_2$ solution (0.1 M tetrabutylammonium perchlorate/acetonitrile, Pt electrode, 100 mV/s), before (dotted line) and after (solid line) 5 h of ECL (pulsing between +1.2 and −1.9 V vs. Ag/Ag$^+$ at 100 Hz). (From Ref. 63.) Right: Cyclic voltammetry of a 0.4 mM rubrene solution (0.05 M tetrabutylammonium perchlorate/acetonitrile, Pt electrode, 100 mV/s) before (curve A) and after (curve B) 4 h of ECL generation in a thin layer cell. (From Ref. 65.)

(in acetonitrile) as emitters, satisfactory lifetimes were obtained only if the supporting electrolyte was absent, thus proving that either the impurities present in tetrabutylammonium perchlorate (TBAP) or the salt itself drastically affect the operational lifetime. The typical half-life in the absence of supporting electrolyte is about 10 min, whereas for cells containing 0.1 M TBAP it is less than 3 min.

Dunnett and Voinov [64] studied the ECL decay of rubrene in benzonitrile, showing that the simultaneous reduction of benzonitrile, which occurs together with rubrene reduction, is likely to critically affect the ECL lifetime. They suggested that the radical anions generated in benzonitrile reduction can be involved in the annihilation of rubrene radical cations, thus being partially responsible for the observed ECL. Impurities were also detected in the benzonitrile, such as phenol and benzyl alcohol, which can protonate the rubrene radical anions or react with the electrophilic radical cations, further affecting the ECL lifetime. When the rubrene thin-layer cell (26 μm thick) is operated at 2.8 V, the ECL decays with a half-life of only 2 min. The rubrene oxidation and reduction waves disappear after prolonged operation, new waves being observed on the cyclic voltammogram. Even when the cell is operated at 2 V, below the ECL threshold of 2.35 V, the degradation is still noticeable, with apparently the same mechanism. Brilmyer and Bard [65] reached similar conclusions (Fig. 6, right), showing that the failure of rubrene-based solution LECs is due, at least in part, to the degradation of the rubrene molecule, isomers of dihydrorubrene being detected in spent thin-layer SLECs. Thus, degradation of the emitter molecule proves to be an important factor in the failure mechanism of these devices.

Schaper et al. [66] investigated the rubrene ECL in thin-layer cells, using low polarity solvents (1,2-dimethoxyethane). They showed that such cells work well even without the addition of supporting electrolyte: the amount of ionic impurities is still large enough to provide a minimum ionic conductivity ($\sim 10^{-7}\ \Omega^{-1}cm^{-1}$).

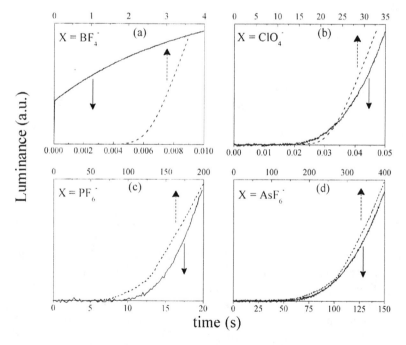

Figure 7 Luminance transients for an ITO/Ru(bpy)$_3$(X$^-$)$_2$/GaSn LEC (100 nm, 2.5 V) device in air (solid lines) and in a dry box (dotted lines). X$^-$ = (a) BF$_4^-$; (b) ClO$_4^-$; (c) PF$_6^-$; (d) AsF$_6^-$. (From Ref. 231. Copyright 2002 by the American Chemical Society.)

The typical half-life of such a device is about 50 h, with brightness levels still in the visible range after 500 h. They also reported that the current and onset of luminescence do not depend on the electrode separation, partly because of free convection of the liquid electrolyte. Indeed, the ECL phenomenon in liquid systems is linked to the onset of electrohydrodynamic convection, the ECL being observed macroscopically as structured, honeycomblike, patterns. The mass transport mechanism is thus not only diffusional; convection contributes also, a feature that contributes to the constancy of the light and current onset on cell thickness. The interesting phenomenon of patterned ECL due to electrohydrodynamic convection was given attention in later, more detailed, studies [67–69].

Chang et al. [70] reported ECL from polymer solutions. BDOH-PF solutions (see Table 2 for structure) in dichlorobenzene, prepared without adding supporting electrolyte, show blue electroluminescence (Fig. 8), with quantum efficiencies of 1% and lifetimes of ~30 min at 10 V. Because the main reason for the short lifetime is solvent evaporation, sealed devices have longer lifetimes. Similar devices were reported with concentrated, gel-like, MEH-PPV solution (8% w/w) in either dichlorobenzene or cyclohexanone [71]. They show rather poor performance that improves upon addition of a surfactant (dibenzo-18-crown-6). The role of the added surfactant is not clear; it was suggested that it prevents the phase separation of MEH-PPV and dichlorobenzene and improves the contact between the electrode and the polymer gel solution. The mechanism of these polymer SLECs is not a true annihilation ECL, apparently involving solvent radicals [72]. The ECL is always observed near the cathode for BDOH-PF and near the anode for MEH-PPV; presumably the dichlorobenzene or cyclohexanone is oxidized at the anode (or reduced at the cathode), generating radicals that move fast, and generates light by annihilation with essentially immobile polymer radicals formed near the other electrode. The light intensity for BDOH-PF is higher in dichlorobenzene, which has a higher oxidation potential than cyclohexanone, chlorobenzene radicals favoring the formation of the polymeric excited state. MEH-PPV SLECs show better performance in cyclohexanone, because the dichlorobenzene radical anion appears to be less stable. Surprisingly, even in these viscous polymer solutions convection is still observed [70,72].

More inert solvents, such as toluene, were employed in solution ECL of poly(9,9′-dioctylfluorene) [73]. As expected, the device is highly resistive;

Poly(9,9′-dioctylfluorene)

Table 2 LECs That Have Polymeric Fluorene Derivatives as Emitters

Emitter	Structure	LEC type[a]	Emission	Brightness	Max. efficiency[b]	Ref.
BDOH-PF		ITO/P + LiTf/Al ITO/P + PEO + LiTf/Al	Sky blue White	190 cd/m² at 3.1 V 1000 cd/m² at 3.5 V 400 cd/m² at 4 V	4% (12 lm/W) 8 lm/W 2.4%	81,83
DHF-*co*-DHF-*co*-BTOHF		ITO/P + PEO + LiTf/Al	Blue-green	1 μW/cm² at 11–12 V	—	82
BDOHF-ANT		ITO/P + LiTf/Al	Blue to green-blue	100 cd/m² at 4.0–4.5 V	—	197
BDOHF/ANT-ANT						
BDOHF-FLUO						

(*Continued*)

Table 2 Continued

Emitter	Structure	LEC type[a]	Emission	Brightness	Max. efficiency[b]	Ref.
PDHF	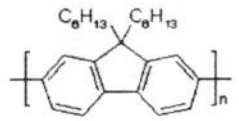	ITO/P + THA-TSFI/Al[c]	—	~400 cd/cm² at 6 V	—	198

[a]P = polymer emitter; LiTf = lithium triflate; PEO = poly(ethylene oxide).
[b]Expressed as quantum efficiency (%) or luminous efficiency (lm/W).

[c]THA-TSFI =

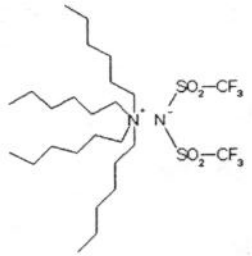

Figure 8 Picture of the ECL from a polymer solution LEC with BDOH-PF as emitter. (From Ref. 70.)

charge is injected only above 20 V, and the current reaches 100 A/m² at around 100 V. The ECL mechanism of this device is unclear, the authors suggesting that the charge injection could be similar to that of a normal OLED. However, the current–voltage curve cannot be described by simple thermionic emission; it is likely that this device has much in common with the rubrene-dimethoxyethane system reported by Schaper et al. [66].

Polymer SLECs are still a rather new field of study, and it remains to be seen whether true annihilation-type polymer SLECs can be obtained, because the polymer molecules are too bulky to diffuse fast enough in solution.

C. Solid-State LECs: Some Practical Issues

A solid-state LEC (SSLEC) comprises several key components: an emitting substance, an electrolyte, an ionic conductor (usually a salt), and two electrodes, one of them transparent (Fig. 9). Not all the components need to be added separately. In some cases the polymeric emitter can also act as a solid electrolyte (see Section III.D), and in others the emitter itself is a salt with a large enough ionic conductivity (see Section III.E).

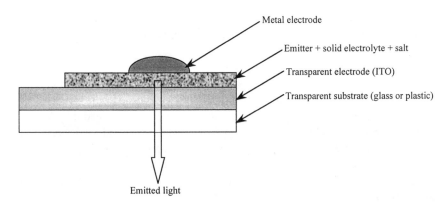

Figure 9 Schematic picture of a solid-state light-emitting electrochemical cell.

The emitters used in solid-state LECs are quite numerous, either polymeric materials or "small" molecules. Widely used emitters are PPV and its derivatives, polymeric fluorenes, and Ru(bpy)$_3^{2+}$-type metal chelates, and they are discussed in more detail in the following sections.

The active film, containing the emitter, electrolyte, and salt, is usually deposited by spin-coating from a mixed solution in a suitable solvent (the most common being acetonitrile, chloroform, cyclohexanone, or THF). Spin-coating is a cheap and easy method for film formation, usually giving good quality films. The film's thickness depends on both the solution concentration and the spinning rate, and it can be adjusted by varying these parameters; typical values range from 100 to 500 nm. When insoluble polymers are to be deposited as films, the (soluble) monomer is spin-coated instead, and the obtained film is then polymerized (by thermal or photochemical initiation) [224].

The polymer electrolyte most commonly used is poly(ethylene oxide) (PEO), which has good solvent properties for ionic substances. Unfortunately, because in many cases the polymer emitters used are rather nonpolar, mixing them with PEO can result in phase separation due to limited miscibility between polymers; adding a salt to the system increases this chance. This is one motivation for developing polymeric emitters with polar side groups that are capable of dissolving salts without the need to add a solid electrolyte. Crown ethers are also used as solid electrolytes [206,237] because of their good ability to solvate alkaline cations such as Li$^+$ and their wider miscibility range with nonpolar polymers; improved ionic dissociation was found in such electrolytes [219], but they are usually more expensive.

A widely used salt is lithium trifluoromethane sulfonate (LiCF$_3$SO$_3$), commonly known as lithium triflate (LiTf). It is soluble in a wide variety of polar and moderately polar organic solvents, being reasonably stable. In some cases, though, it may lead to the formation of ion pairs and/or ion aggregates, lowering the total ionic concentration [74]. If needed, other lithium or tetraalkylammonium salts can be used as well [206]; in some cases even room temperature molten salts, which can serve as both electrolyte and source of mobile ions, have been used with good results [198].

In most cases, the transparent electrode used for preparing LECs is indium-tin oxide (ITO). Thin, semitransparent metal films deposited onto glass can be used also, but these films are usually highly resistive and may lead to unwanted heating during operation. ITO-covered glass (or plastic) is preferred, because it has good conductivity and optical properties. ITO is also commercially available in a wide variety of types and on various substrates. Several alternatives to ITO have been proposed, that have somewhat better properties, but they have been used only occasionally. Alternatives to ITO include other doped semiconducting oxides, such as $Ga_{0.12}In_{1.88}O_3$, $Ga_{0.08}In_{1.28}Sn_{0.64}O_3$, and $Zn_{0.5}In_{1.5}O_3$ [75,76], which have better optical properties and higher work functions. However, because the electrode work function is not so important for LECs, ITO is probably still the best option with respect to both price and availability and optical and electrical characteristics. When employed as an electrode, the ITO surface must be cleaned before use, various procedures being available. Wet cleaning procedures, such as sonication in organic solvents (acetone, isopropanol, ethyl alcohol, water–ethanolamine mixture), are an easy and relatively efficient way of obtaining good surfaces. Some surface contamination with organic substances results when this method is used, but this is not usually a major concern for LECs. Other, rather expensive, methods include O_2 plasma cleaning and inert gas (Ne, Ar) sputtering [77,78], which give very clean, organics-free surfaces [79] with higher work functions. Because for LECs we are not so concerned about the electrode work function, wet-cleaning procedures are likely to be preferred.

The second contact is usually a metal deposited by vacuum evaporation. Aluminum is widely used, but noble metals such as Au and Ag can be employed with almost the same results. The use of liquid alloys (Ga-In and Ga-Sn eutectics) and metals (Hg) was also reported [229–231], contacts being made by printing the liquid metal with a syringe. Although when using liquid metal contacts the contact area is not very well defined and is far from being perfect, this method produces easily short-free contacts. In some cases, the second contact can also be ITO, giving a totally symmetrical LEC [80].

D. Polymer LECs

We have grouped polymer SSLECs into three major categories, depending on the emitter type: fluorene polymers, PPV and derivatives, and other polymers. The

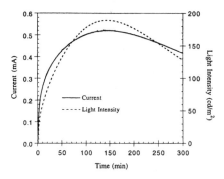 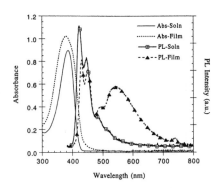

Figure 10 Left: Current and light intensity as a function of time of an ITO/BDOH-PF + LiCF$_3$SO$_3$/Al LEC biased at 3.1 V. Right: Optical absorption and PL spectra of BDOH-PF in a dilute solution in THF (1 mg/100 mL) and in a thin film spin-cast from a solution in THF (20 mg/mL). (From Ref. 81. Copyright 1996 by the American Chemical Society.)

results reported in the literature are summarized in three tables, with only the most interesting ones being analyzed in more detail.

1. LECs with Polymeric Fluorene Derivatives

Polyfluorenes show high solid-state fluorescence quantum yields, in some cases close to 70%, and are therefore attractive candidates fo. polymer LECs. Results from the literature for LECs with polyfluorene derivatives are summarized in Table 2; a typical result is shown in Figure 10. Polyfluorene itself is nonpolar, and therefore phase separation occurs when it is mixed with PEO. Improved derivatives have been synthesized that have ethylene oxide side groups and better properties [81,82]. Interestingly, Yang and Pei reported [83] that one of these derivatives, BDOH-PF (see Table 2), emits white light when blended with PEO instead of the blue-green color expected from such an emitter. The emission shift appears to be related to the strong phase separation that occurs when the polymer is mixed with PEO.

The lifetimes of LECs with polyfluorene emitters are still quite modest, not exceeding 10–20 h.

2. LECs with Polyphenylenevinylene and Derivatives

Polyphenylenevinylene (PPV) and its many derivatives are widely employed for both LECs and OLEDs because of their good photoluminescent quantum yields and stability. After Heeger and coworkers [199] reported the first SSLEC with PPV as the emitter, interest in this type of LEC gradually increased; the mostimportant results are shown in Table 3. Figure 11 shows a typical result for this type of cell.

Table 3 LECs with PPV and Derivatives as Emitters

Emitter	Emitter Structure	LEC type[a]	Emission	Brightness	Max. efficiency[b]	Ref.
MEH-PPV		ITO/P + PEO + LiTf/Al	Orange	>100 cd/m² at ±4 V	1%	199
PPV		ITO/P + PEO + LiTf/Al	Green		0.1–0.2%	199
PPV		ITO/P + PEO +LiTf/Al	Green	100 cd/m² at 4 V; 8 cd/m² at 3 V	2%	200, 201
PPV+MEH-PPV		ITO/P1 + PEO + LiTf/P2 + PEO + LiTf/Al	Red-orange or green	—	—	202
PPV		ITO/P + PEO + LiTf/Al	Green	—	0.0045%	203
		ITO/P+PEO+LiTf/LiPSS +PEO/Al	Green	—	0.02%	203
		ITO/P+PEO+LiTf/LiPSS +PEO/P+PEO+LiTf/Al[c]	Green	—	0.029%	203
MEH-PPV		ITO/P + PEO + LiTf/Al;	Orange	10–20 cd/m² at 3 V	1–2.5%	204
		ITO/P+PEO+OCA +LiTf/Al[c]		1000 cd/m² at 3 V	1–2.5%	204
MEH-PPV		ITO/P + PUI/Al[c]	White			205
MEH-PPV		ITO/P + DCH-18Cr6 + LiTf/Al;	Orange	90.3 cd/m² at 3 V	0.55%	206
MEH-PPV		ITO/P + DCH-18Cr6 + LiM/Al[c]	Orange	389 cd/m² at 3 V	0.59%	206
MEH-PPV		ITO/P/PEO+TBABF₄/Al[c]	Orange	—	0.43%	207
MEH-PPV		ITO/P/SCC/Al	Orange	—	0.86%	208
MEH-PPV		ITO/P/SAC/Al[c]	Orange	172 cd/m²	0.74%	208

(Continued)

Table 3 Continued

Emitter	Emitter Structure	LEC type[a]	Emission	Brightness	Max. efficiency[b]	Ref.
MEH-PPV		ITO/P + THA-TSFI/Al[c]	Orange	0.5 μW/cm²	2.6×10^{-4d}	198
BuEH-PPV	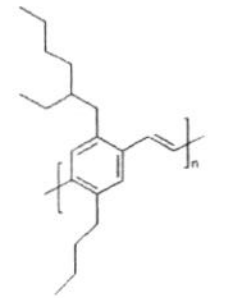	ITO/P + LiTf + OCA/Al[c]	Green	—	3% (4.4 lm/W at 6 V)	209
		ITO/P + LiTf/Al	Green	21,400 cd/m² at 8.0 V	2.5% (5.6 lm/W)	209
BCHA-PPV	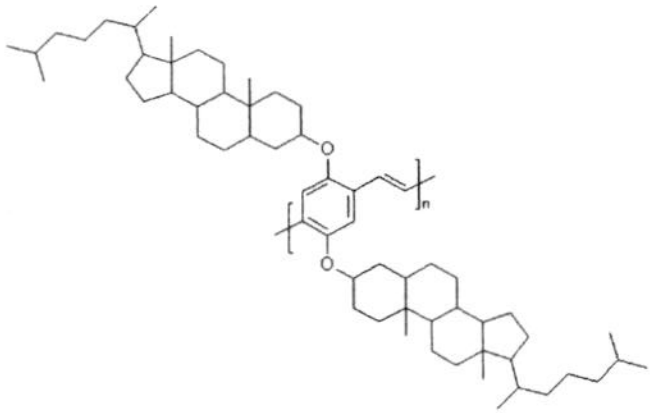	ITO/P+LiTf/Al	Yellow	—	3% at 3 V	209
BTEM-PPV	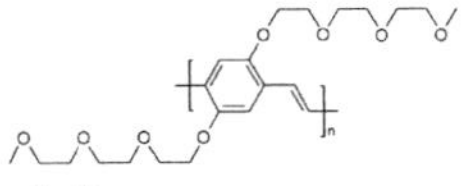	ITO/P + LiTf/Al	Orange	35 cd/m² at 3.0 V	0.35%	210, 211
		ITO/P + LiTf/Au	Orange	—	—	
		ITO/P + LiTf/Ag	Orange	—	—	
15C5-DMOS-PPV	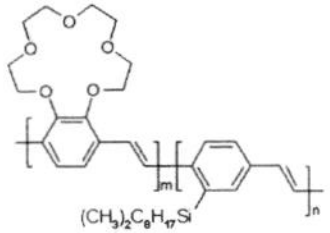	ITO/P + PEO + LiTf/Al	Yellow-green	—	1.9–2.3 %	212
DB-alt-BTEM-PPV	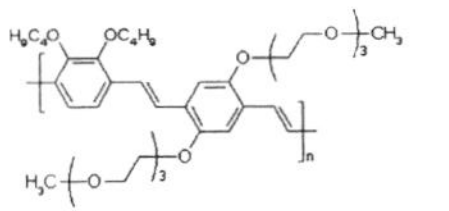	ITO/P + LiTf/Al	—	—	0.9 cd/A	213
		ITO/P+ PEO + LiTf/Al	—	—	1.5 cd/A	213
BDMOS-co-BTEM-PPV	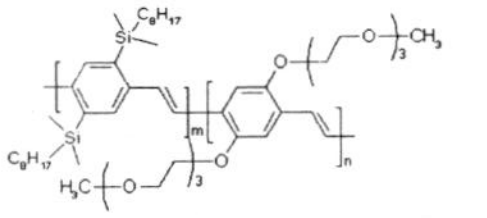	ITO/P + LiTf/Al	Orange	0.15 cd/m² at 3 V	0.03 cd/A	213–215
		ITO/P + PEO + LiTf/Al	Orange	—	0.49 cd/A	

(Continued)

Table 3 Continued

Emitter	Emitter Structure	LEC type[a]	Emission	Brightness	Max. efficiency[b]	Ref.
DNVB-TEO	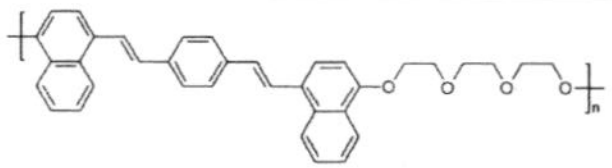	ITO/P + PEO + LiTf/Al	Blue-green	1.3 cd/m² at 3.6 V	0.174 lm/W	216
MDNVB-TEO	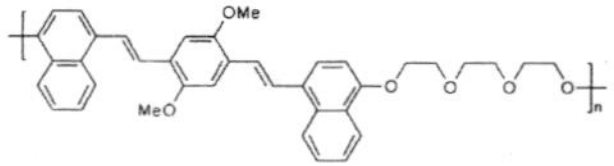	ITO/P + PEO + LiTf/Al	Blue-green	1.7 cd/m² at 3.3 V	0.042 lm/W	216
PPV-EO	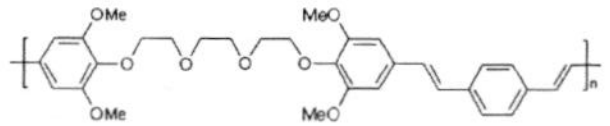	ITO/P + LiTf/Al	Blue	—	—	217

[a]P = polymer emitter; LiTf = lithium triflate; PEO = poly(ethylene oxide).
[b]Expressed as quantum efficiency (%) or luminous efficiency (lm/W or cd/A).
[c]LiPSS = poly(styrene-sulfonic acid) lithium salt; OCA = octylcyanoacetate;

PUI =

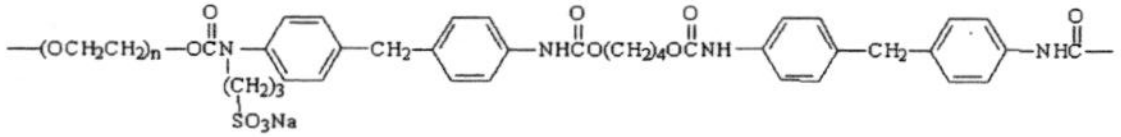

DCH-18Cr6 = 2,3,11,12-dicyclohexano-1,4,7,10,13,16-hexaoxacyclooctadecane; LiIM = lithium imide; TBABF$_4$ = tetrabutylam-monium tetrafluoroborate;

SCC =

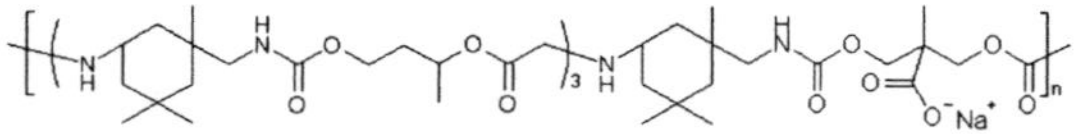

SAC =

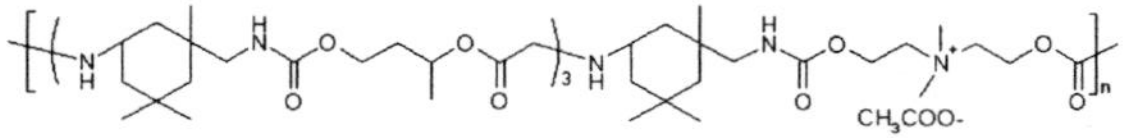

THA-TSFI =

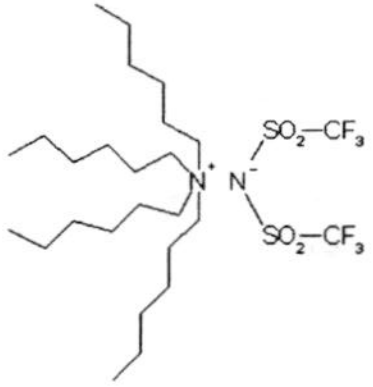

[d]Power efficiency.

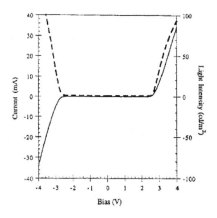 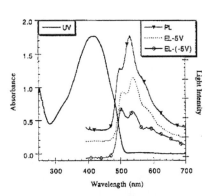

Figure 11 Left: Typical current (solid line) and light intensity (dashed line) versus voltage for an ITO/ PPV + PEO + LiCF₃SO₃/Al LEC. The voltage scans from 0 to 4 V and from 0 to –4 V, respectively. Right: Absorption, photoluminescence (PL), and electroluminescence (EL) spectra from the same LEC. (From Ref. 200. Copyright 1996 by Elsevier Science.)

As with other polymeric emitters used in LECs, phase separation occurs when PPV and nonpolar derivatives are used. Adding to the blend compounds that help prevent phase separation leads to an increase in device performance [204]. Several derivatives have been synthesized, combining the fluorescence of the PPV moiety with ion-solvating properties of ethylene glycol or crown ether groups (see Table 3). In some cases, changes in the absorption spectrum of the emitter are observed when a lithium salt is added [84,210]. The presence of the Li⁺ ion near the polymer backbone induces changes in the electron density, thus resulting in changes in the optical bandgap. This is an interesting phenomenon, because it allows the emitted light to be tuned simply by adding a different salt, without the need to synthesize new materials as emitters. The shift is small, however (~10–15 nm in absorption and <5 nm in emission); moreover, if the alkaline ions are too strongly complexed by the polar groups on the polymer backbone, the ionic conductivity decreases, leading to poor performance.

Tunable emission can be achieved by using two different emitters spin-coated separately [202] (Fig. 12). Because of asymmetrical charge injection and very different transport properties for electrons and holes, the annihilation maximum is close to one electrode (the cathode). If the LEC is operated under forward and reverse bias, different light colors are observed; with the MEH-PPV side negative, the emission is red-orange (MEH-PPV acting as emitter), whereas with the PPV side negative, green light is observed. Tasch et al. [211] reported that tunable color LECs can also be obtained by blending two different emitters,

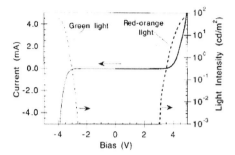
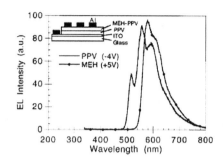

Figure 12 Left: Current–voltage and light–voltage characteristics of a bilayer MEH-PPV/PPV LEC under forward and reverse bias conditions. When forward-biased, the LEC emits red-orange light from the MEH-PPV layer. When reverse-biased, the LEC emits green light from the PPV layer. Right: Electroluminescent spectra from the bilayer LEC under forward and reverse bias conditions. Inset shows a schematic diagram of the bilayer device structure. (From Ref. 202.)

MEH-PPV and m-LPP (see Table 4 for structure) with PEO and lithium triflate. The emission color of the resulting LEC can be adjusted by simply varying the applied bias and shifts from orange (<4 V, MEH-PPV emission dominant) to green (>6 V, m-LPP emission dominant). Owing to the relatively large differences in the optical bandgap for the two polymers and very limited interaction between them, the emission occurs as if two different LECs with different turn-on voltages were operated in parallel.

Several studies [85–87,204] have shown that films made with MEH-PPV-type polymers mixed with PEO have a three-dimensional network structure (Fig. 13) with domains, due to the limited miscibility of MEH-PPV and PEO. The two phases, one containing the solid electrolyte and the other the MEH-PPV-type emitter, are responsible for ion transport and light emission, respectively. It is unclear as to how the phase separation affects the operating lifetime of these LECs.

The best available LECs with PPV derivatives show reasonably long lifetimes: less than 20% luminescence loss, for an initial brightness of 300 cd/m², after 100 h of operation was reported [204]. In an effort to improve the performances of PPV-type LECs, a number of studies [88,198,206] have focused on using different electrolytes and salts instead of the classical PEO and lithium triflate. The results are promising, showing that improvement is still expected in this area.

3.   LECs with Other Polymers

Various other polymers have been used as emitters for LECs, ranging from ladder-type poly(p-phenylene) to Ru(bpy)$_3^{2+}$-based polymers. These are summa-

Table 4 LECs with Various Polymers as Emitters

Emitter	Emitter structure	LEC type[a]	Emission	Brightness	Max. efficiency[b]	Ref.
DOHO-PPP		ITO/P + PEO + LiTf/Al	Blue	—	2%	199
m-LPPP		ITO/P + PEO + LiTf/Al; ITO/P + PEO + LiClO₄/Al; ITO/P + PEO + LiTf/Ag; ITO/P + PEO + LiTf/Au	Blue-green	250 cd/m² at 10 V	0.3%	211, 218, 219
						219
		ITO/P + DCH18C6 + LiTf/Al[c]	Blue-green	1500 cd/m² at 6 V	—	211, 218, 219
TOD-PC		ITO/P + PEO + LiTf/Ca	Blue	—	0.02%	220
DT-PQX		ITO/P + PEO + LiTf/Ca	Blue-green	—	<0.02%	220
TOD-PC + DT-PQX		ITO/P + PEO + LiTf/Ca	Orange-yellow	—	0.01%	220
Ru-polyester		ITO/P/Al	Orange-red	80–300 cd/m²	0.01–0.08%	221

(Continued)

Table 4 Continued

Emitter	Emitter structure	LEC type[a]	Emission	Brightness	Max. efficiency[b]	Ref.
Ru-polyester		ITO/P/Al	Orange-red	200–300 cd/m^2 at 7.5 V	0.1–0.2%	222
Ru-polyester		ITO/P + PAA/Al[d]	Orange-red	6–30 cd/m^2	0.2–3%	222
		ITO/P + SPS/Al	Orange-red	12.8 µW/cm^2	0.035%	223
		ITO/P + S-PPP/Al	Orange-red	6.2 µW/cm^2	0.023%	223
		ITO/P + PMA/Al	Orange-red	0.65 µW/cm^2	0.014%	223
		ITO/P + PAA/Al[d]	Orange-red	2.5 µW/cm^2	2.6%	223
Ru-poly		ITO/P/Au	Orange-red	—	0.92%	224
P3OT		ITO/P + PEO + TBATf/Al[e]	Orange	—	0.01%	225

[a]P-polymer emitter; LiTf-lithium triflate; PEO-poly(ethylene oxide).
[b]Expressed as quantum efficiency (%).
[c]DCH18C6 = dicyclohexano-18-crown-6.
[d]Films prepared using layer-by-layer technique. PAA = poly(acrylic acid) sodium salt; SPS = poly(styrene sulfonic acid) sodium salt; S-PPP = sulfonated poly(p-phenylene); PMA = poly(methacrylic acid) sodium salt; PAA = poly(acrylic acid) sodium salt.
[e]TBATf = tetrabutylammonium triflate.

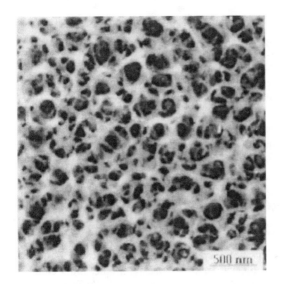

Figure 13 Transmission electron micrograph of a MEH-PPV + PEO + LiCF$_3$SO$_3$ + octylcyanoacetate composite. PEO was dissolved out of the blends by immersing the films in water for several hours. The extracted films were shadowed by Pt/Pd from the bottom side. (From Ref. 204.)

rized in Table 4. As with the others polymer LECs, we shall detail below only the most interesting results reported in the literature.

The ladder-type poly(p-phenylene) m-LPP shows rather peculiar behavior when mixed with PEO, with nonsymmetrical current–voltage characteristics and bias-dependent emission [89,218]. The nonsymmetrical characteristics have been attributed to a very restricted ionic mobility of the Li$^+$ ions [218], but other causes, such as purity of the materials used for LEC preparation, cannot be ruled out. LECs made with crown ether solid electrolyte show the usual symmetrical current–voltage characteristics, and IR spectroscopic studies prove that the degree of ion pairing is much lower in this case [219].

An interesting type of LEC was reported with a mixture of TOD-PC and DT-PQX (see Table 4 for structures) [220]. TOD-PC and DT-PQX show intense blue and blue-green fluorescence, respectively. Charge injection in TOD-PC devices is mainly p-type, because TOD-PC can be easily oxidized and very difficult to reduce. In the case of DT-PQX LECs, very low currents were obtained, probably because of the high ionization potential of DT-PQX. LECs made with a mixture of the two polymers show, surprisingly, a bright yellow-orange emission in both fluorescence and electroluminescence. The main reason for this unexpected behavior, which can prove useful in manufacturing color-tunable LECs, is apparently the mixed recombination of electrons injected in DT-PQX with holes injected in TOD-PC.

In a detailed analysis of $Ru(bpy)_3^{2+}$-type polymer emitters, Rubner's group [221–223], reported the production of $Ru(bpy)_3^{2+}$-polyester LECs using a layer-by-layer deposition technique with various polyanions [223]. The layer-by-layer deposition is performed by alternate dipping of the ITO substrate in the corresponding polyion solution. The resulting film has a high degree of molecular interpenetration between the alternate layers of the $Ru(bpy)_3^{2+}$ polyester (the polycation) and the polyanion; the best performances are achieved when the polyanion is the sodium salt of poly(acrylic acid). The film composition can be adjusted by varying the pH of the dipping solutions, thus allowing control of the distance between the Ru(II) sites. When the desired multilayer structure has been built, Al is evaporated on top to serve as the second contact. The quantum efficiency of these devices increases as the total amount of $Ru(bpy)_3^{2+}$ polyester decreases, proving that self-quenching effects are important. Other factors, such as changes in charge injection and transport, may also be responsible for improved quantum efficiencies. However, dilution of the emitter results in a decrease in device brightness as well as higher driving voltages due to increased resistance of the film, so a compromise between high quantum efficiency and reasonable driving voltage and brightness has to be made in a practical device. For example, devices with 46% $Ru(bpy)_3^{2+}$ polyester show 3% quantum efficiency, with a reasonable brightness of 25 cd/m^2 at 12 V. If the $Ru(bpy)_3^{2+}$ polyester concentration is increased to 58%, the brightness reaches 30 cd/m^2 at 12 V, but the quantum efficiency drops to 2%. On the other hand, devices with pure $Ru(bpy)_3^{2+}$ polyester spin-coated films have a quantum efficiency of only 0.2% but 265 cd/m^2 brightness at only 8 V [222].

It is unclear which species are responsible for the ionic conductivity of these devices, because the polyions themselves are too bulky to be mobile and provide reasonable conductivity. It is likely that protons present in the poly(acrylic acid) play this role, because Fourier transform infrared (FTIR) studies have shown that only about half of the acidic groups are used to construct the film, the remainder being protonated [222]. These devices show poorer performance in reverse bias (Al wired positive) due to oxidation of the aluminum contact, but it is still possible to drive them in a.c. mode (Fig. 14). Overall, the layer-by-layer technique seems to give better results than spin-coating, allowing better control on the amount of emitter introduced in these highly ordered films and consequently increasing the quantum efficiency. Elliott et al. [224] studied related $Ru(bpy)_3^{2+}$-based polymer spin-coated films using ITO and Au electrodes that showed good quantum efficiencies, a little less than 1%.

E. Small-Molecule LECs

The only small-molecule devices for which LEC-type behavior has been reported are the ones based on ionic metal chelates such as $Ru(bpy)_3^{2+}$ and derivatives.

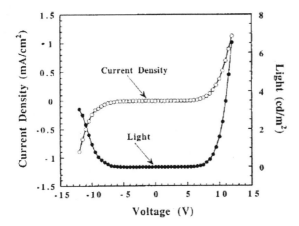

Figure 14 Current–voltage and light–voltage characteristics of a device composed of 20 bilayers of the 46% $Ru(bpy)_3^{2+}$ polyester + poly(acrylic acid) sodium salt system. (From Ref. 222. Copyright 1999 by the American Chemical Society.)

$Ru(bpy)_3^{2+}$ and its derivatives have been extensively employed in electrogenerated chemiluminescence studies in liquid solutions; their good quantum efficiency and electrochemical behavior make them excellent subjects for such studies (see Chapter 7). LEC-type devices with small-molecule emitters are presented in Table 5.

Spin-coated films of $Ru(bpy)_3^{2+}$ (and derivatives) are amorphous and do not tend to crystallize even during prolonged heating at moderate temperatures (~120°C). These devices show LEC characteristics, the emitter film having a reasonable ionic conductivity (especially for small counter ions such as BF_4^- and ClO_4^- [231,232]) even without the addition of any solvent or salts. The reason for this relatively high ionic conductivity is not known; it was speculated that traces of water from the atmospheric moisture penetrate the film and help solvate the ions [231], but there is still a need for direct evidence to confirm this hypothesis. Crystalline films of $Ru(bpy)_3(ClO_4)_2$, which are less likely to contain traces of water (because the perchlorate salt does not contain any crystallization water), show even greater ionic conductivity. Single-crystal films ~1.5 µm thick grown between two ITO electrodes show relatively large currents and brightness levels, with an external quantum efficiency of 3.4% [80].

The first report of an SSLEC with a $Ru(phen)_3^{2+}$ derivative (see Table 5 for structure) was published in 1996 [226], but the device performed poorly, with a quantum efficiency of only about 0.005%. Later studies [227] showed that blending the $Ru(phen)_3^{2+}$ derivative with PEO and lithium triflate significantly improves the performance, but the maximum quantum efficiency achieved was still low (0.2%). Other studies have confirmed that $Ru(phen)_3^{2+}$-type SSLECS have rather poor quantum efficiency [230].

Table 5 LECs with Small Molecule Emitters

Acronym	Emitter structure	LEC type[a]	Emission	Brightness	Max. efficiency[b]	Ref.
Ru(Sphen)$_3$²⁺		ITO/S/Al	Orange-red	30–35 cd/m²	0.005%	226
Ru(bpy)$_3$²⁺		ITO/S/Al ITO/S + PEO + LiTf/Al ITO/S/Al	Orange-red Orange-red Red-orange	50 cd/m² 100 cd/m² —	0.01% 0.02% 0.3–0.4%	227 227 228
Ru(bpy)$_2$(HO-bpy-OH)²⁺		ITO/S/Al	Red-orange	600–1000 cd/m²	1%	228
Ru(bpy)$_2$(Es-bpy-Es)²⁺		ITO/S/Al	Red	50–600 cd/m² at 3–5 V	0.1–0.4 %	228
Ru(Es-bpy-Es)$_3$²⁺		ITO/S/Al	Red	50–600 cd/m² at 3–5 V	0.1–0.4%	228

(continued)

Table 5 Continued

Acronym	Emitter structure	LEC type[a]	Emission	Brightness	Max. efficiency[b]	Ref.
Ru(bpy)$_3$$^{2+}$		ITO/S/Ga-Sn	Red-orange	500 cd/m² at 3.0 V; 2000–3000 cd/m² at 4.0 V	1.4–2%	229–231
Ru(bpy)$_3$$^{2+}$		ITO/S + PMMA/Ag[c]	Red-orange	200–350 cd/m² at 5 V (a.c.-driven)	2.7% (3.3 lm/W)	232, 233
Ru(Tbu-bpy-Tbu)$_3$$^{2+}$		ITO/S + PMMA/Ag[c]	Red-orange	—	4.1% (4.9 lm/W)	232
Rubpy-A	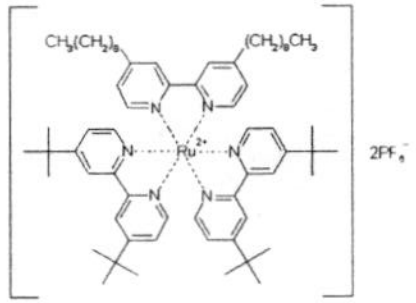	ITO/S + PMMA/Ag[c]	Red-orange	15 cd/m² at 5 V (a.c.-driven)	4.8% (5.6 lm/W)	232
Rubpy-3	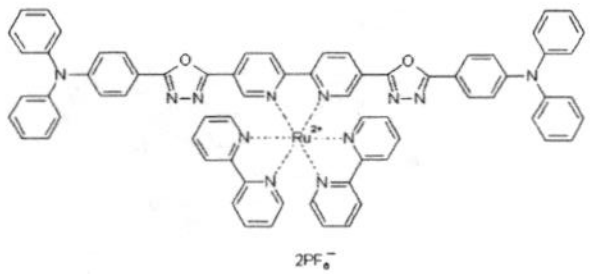	ITO/S + PVOH/Al[c]	Red	730 cd/m²	0.1%	234
Rebpy-3	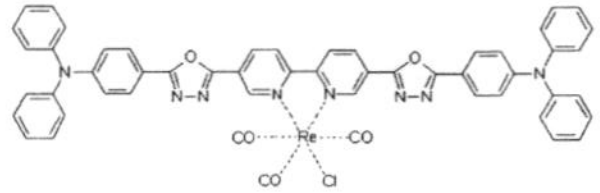	ITO/S + PC/Al[c]	Red	—	0.1%	234

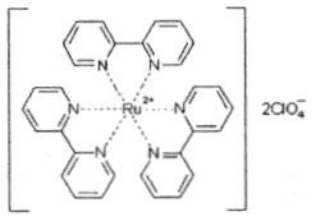

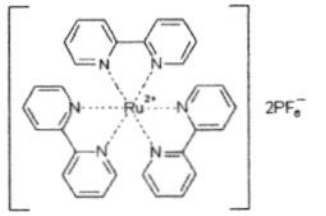

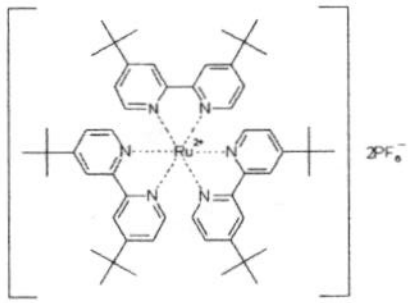

(continued)

Table 5 Continued

Acronym	Emitter structure	LEC type[a]	Emission	Brightness	Max. efficiency[b]	Ref.
Rubpy-1a	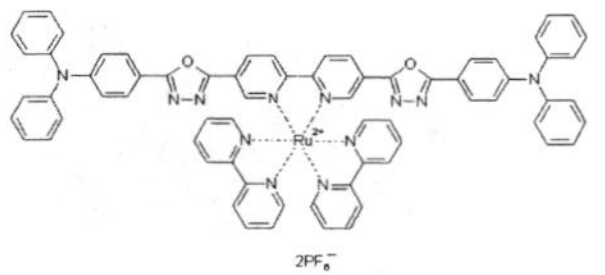	ITO/S + PVOH/Al[c]	Red-orange	650 cd/m^2	0.1%	235
Rubpy-1b	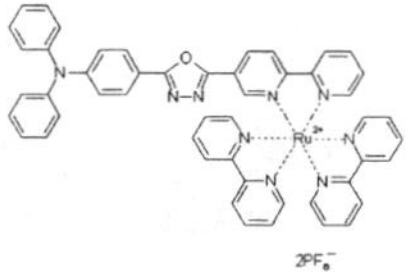	ITO/S + PVOH/Al[c]	Red	130 cd/m^2	0.07%	235
Rubpy-2a	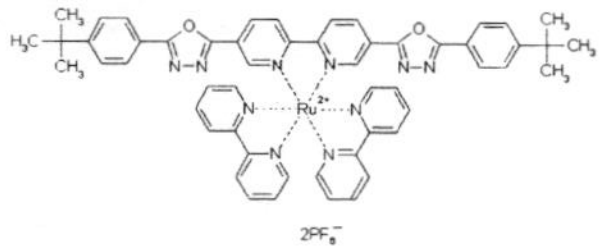	ITO/S + PVOH/Al[c]	Red	160 cd/m^2	0.04%	235
Rubpy-2b	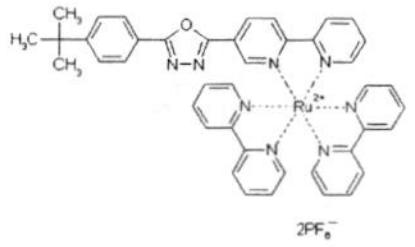	ITO/S + PVOH/Al[c]	Red	190 cd/m^2	0.03%	235
Rebpy-1a	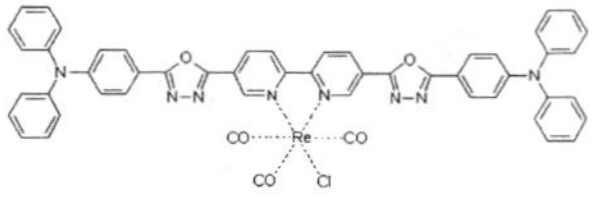	ITO/S + PC/Al[c]	Red-orange	730 cd/m^2	0.1%	235
Rebpy-1b	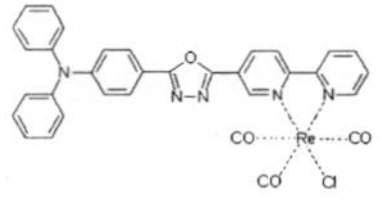	ITO/S + PC/Al[c]	Red-orange	350 cd/m^2	0.08%	235

(continued)

Table 5 Continued

Acronym	Emitter structure	LEC type[a]	Emission	Brightness	Max. efficiency[b]	Ref.
Rebpy-2a		ITO/S + PC/Al[c]	Red	175 cd/m²	0.06%	235
Rebpy-2b		ITO/S + PC/Al[c]	Red	180 cd/m²	0.05%	235
Os(bpy)$_2$(PhP)$^{2+}$		ITO/S/Au	Red-orange	6000 cd/m² at 6 V; 330 cd/m² at 3 V	1%	236
(bpy)$_2$-Ru-L$_6$-Ru(bpy)$_2$$^{4+}$		ITO/S + 18C6 + LiTf/Al[c]	Orange	5000 cd/m² at 8.2 V	1.4×10^{-3}%[d]	237
		ITO/PEDOT-PSS/S+18C6 +LiTf/Al[c]	Orange	5000 cd/m² at 9.6 V	0.76×10^{-3}%[d]	
		ITO/PEDOT-PSS+18C6 +Li-TFSI/Al[c]	Orange	5000 cd/m² at 4.3 V	4.3×10^{-3}%[d]	

[a]S = small-molecule emitter; LiTf = lithium triflate; PEO = poly(ethylene oxide).
[b]Expressed as quantum efficiency (%) or luminous efficiency (lm/W).
[c]PMMA = poly(methyl methacrylate); PVOH = poly(vinyl alcohol); PC = polycarbonate; 18C6 = 18-crown-6; PEDOT-PSS = poly(3,4-ethylene dioxythiophene-2,5-diyl) doped with poly(styrene sulfonate); Li-TFSI = Li[(CF$_3$SO$_2$)$_2$N$^-$].
[d]Power efficiency.

Figure 15 Photograph showing emission from two contacts of an ITO/Ru(bpy)$_3$(ClO$_4$)$_2$/ Ga-In LEC. (From Ref. 229. Copyright 2000 by the American Chemical Society.)

Much better results were obtained with Ru(bpy)$_3^{2+}$ (Fig. 15) and derivatives as emitters [228–231], quantum efficiencies as high as 2% being observed. Devices made using PF$_6^-$ counter ions show slow response times [228,231] (Figs. 7 and 16); the response time can be decreased by using smaller counter ions [231,232] or by applying a high-voltage pulse for a short time, to build up the ionic charge profiles, followed by a steady, low operating voltage [228]. It was reported that Ru(bpy)$_3$(PF$_6$)$_2$ LECs containing a thin electron transport layer between the emitter layer and the Al contact also show much faster response times [90], but it is not at all clear how the electron transport layer influences their behavior.

Blending the Ru(bpy)$_3^{2+}$ emitter with inert polymers (such as PMMA) increases the quantum efficiency but also requires higher operating voltages. If Ag cathodes are used instead of Al cathodes, which tend to degrade during storage [91], the quantum efficiency for devices blended with PMMA can reach 2.5–3% [233]. Alternating current operation of such devices leads to an increase in both quantum efficiency and operation lifetime, >1000 h half-life being obtained at 5 V and 1 kHz (50% duty cycle), with brightness levels exceeding 100 cd/m^2 [233] (Fig. 18). Recent studies employing various Ru(bpy)$_3^{2+}$ derivatives have shown that better performance can be achieved by tuning the emitter's structure, quantum efficiencies up to 4.8% being reported [232]. Other studies have employed various related metal chelates, such as tris(2,2'-bipyridyl)osmium(II) and related derivatives [230,236] and other ruthenium(II) and even rhenium(I) chelates [234,235].

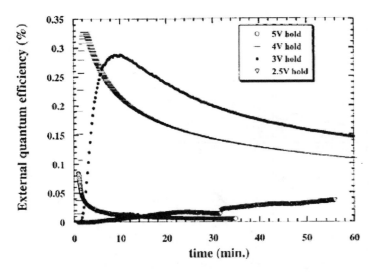

Figure 16 Device efficiency (%, external) as a function of time and applied voltage for an ITO/Ru(bpy)$_3$(PF$_6$)$_2$/Al LEC. (From Ref. 228. Copyright 1999 by the American Chemical Society.)

The best available LECs with small-molecule emitters still appear to be the ones based on Ru(bpy)$_3^{2+}$ and its derivatives, which have reasonably long lifetimes (>1000 h), with good quantum efficiencies and brightness levels.

F. "Frozen Junction" LECs

"Frozen junction" LECs are a special case of LECs where the concentration (ionic, hole, and electron) gradients are formed by applying a bias at a high temperature and then freezing the ionic concentration gradient by cooling the device to a temperature below the glass transition. Electron hopping still occurs at these low temperatures, but the ion movement is essentially stopped. As a result, the LEC-type behavior changes into a diode-like one, showing rectification and fast transients, due to the still fast electron hopping charge transport mechanism.

Maness et al. [92] first employed this technique to study the ECL of poly[Ru(vbpy)$_3$](PF$_6$)$_2$ films prepared by electropolymerization onto gold interdigitated array electrodes (IDAs). Concentration gradients of the oxidized and reduced species of poly[Ru(vbpy)$_3$](PF$_6$)$_2$ were generated by controlling separately the potentials of the two IDA finger electrodes, using a bipotentiostat, until a steady state was reached. The film-covered IDA was then dried and cooled

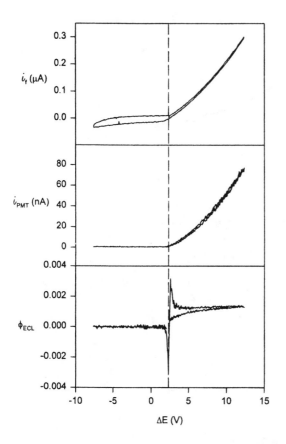

Figure 17 Current (top), light emission (middle), and ECL efficiency (bottom) of a [Ru(bpy)$_2$(bpy(CO$_2$MePEG350)$_2$)](ClO$_4$)$_2$ film containing concentration gradients prepared at room temperature with a ΔE = 2.4 V bias and then frozen at $-20\,^\circ$C. The applied voltage was swept ± 10 V at 10 V/s and centered at ΔE = 2.4 V. (From Ref. 93. Copyright 1997 by the American Chemical Society.)

at $0\,^\circ$C while the bias (+2.6 V) was maintained to preserve the concentration gradients. The resulting film shows typical diode-like behavior: current flows and light is generated at +3.6 V, and very low currents are observed if the bias is reversed (-3.6 V). The frozen concentration gradients slowly relax back to an equilibrium state if left at zero bias; the ion movement still occurs, even if very slowly. If the film is kept in reverse bias for a long time, the relaxation is accelerated. In principle, the concentration gradients can be preserved longer either by using less mobile ions or by freezing them at lower temperatures.

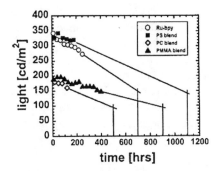

Figure 18 Light emission vs. time of pristine $Ru(bpy)_3(PF_6)_2$ devices and blend devices with poly(methyl methacrylate) (PMMA), polycarbonate (PC), and polystyrene (PS). All blend devices contain ~25 vol% polymer. Devices were operated at 5 V and 50% duty cycle at 1 kHz. The solid lines are linear extrapolations of the light emission to determine the half-life of each device. (From Ref. 233.)

In a similar study [93], a room-temperature molten salt ($[Ru(bpy)_2(bpy$ $(CO_2MePEG350)_2)](ClO_4)_2$) was employed as the emitter. The molten salt is very viscous at room temperature ($\eta \sim 10^7$ cP) and has an apparent glass transition at $-5°C$. The ionic conductivity of this molten salt falls from $10^{-8} \ \Omega^{-1} \ cm^{-1}$ at room temperature to $\sim 10^{-11} \ \Omega^{-1} \ cm^{-1}$ at the temperature of the glass transition. The molten salt is cast onto Pt IDA, after which the frozen concentration gradients are built in as described earlier [92]. The frozen concentration LEC shows rectification (Fig. 17) and fast rise in both current and light during a potential step, but the quantum efficiency is quite low (0.0013% at $-20°C$). As with the poly[Ru(vbpy)_3](PF_6)_2 films, the frozen junction is degraded under prolonged reverse biasing.

Gao et al. [94,95] used the same technique for MEH-PPV+PEO LECs, showing that response times in the microsecond range can be obtained with junctions

$[Ru(bpy)_2(bpy(CO_2MePEG350)_2)]-(ClO_4)_2$

frozen at 150 K. The frozen junction was found to be very stable at temperatures below 200 K, which roughly corresponds to the glass transition temperature of PEO.

Room temperature frozen junction LECs, much more attractive as practical devices based on PPV-type polymers, have also been reported [209], with very high brightness (>20,000 cd/m^2 at 8 V) and quantum efficiency (2.4%). The response time (~30 μs) is also much faster than for regular LECs. In conclusion, it appears that frozen junction LECs are attractive candidates for practical devices, because they combine the balanced charge injection of LECs with the fast response of OLEDs.

G. Lifetime and Device Degradation of LECs

The operating lifetime of LECs is still the critical issue with this type of electroluminescent device. Even though lifetimes exceeding 1000 h have been reported [233], the value is still well below the 10,000 h minimum requirement for displays.

In contrast to the OLED literature [61,96–100], there are not many studies that discuss in detail the factors affecting the operating lifetime of LECs; such a study is quite difficult to perform, because in many cases LECs consist of complex mixtures of three or four different materials, each with its own properties. Therefore, stability (thermal, electrochemical) issues have to be addressed for each component, with a considerable experimental effort, and it is sometimes difficult to ascribe the failure of these devices to one component or another. However, there are at least a few common stability issues between solution and solid-state LECs, also common to OLEDs, mostly related to overoxidation or overreduction of the emitter, side electrochemical reactions of the solid electrolyte and/or salt (see also Section III.B), and quencher generation during operation [96,100,103]. Atmospheric moisture and O$_2$ are also likely to shorten the lifetime of LECs, either by reacting with the electrogenerated radical species or by being involved directly in electrochemical reactions [101].

Kervella et al. [102] have investigated in detail the failure mode of a LEC based on DHF-*co*-BTOHF (see Table 2 for structure). Because the quantum efficiency of such LECs remains almost constant during operation, albeit both current and brightness decay sharply (Fig. 19), they suggested that the electrochemical degradation of the polymer, due to overoxidation or overreduction, is responsible for this behavior. The salt used as ionic conductor can also be electrochemically degraded; the total amount of light generated was found to be four times lower, with a faster decay of the quantum efficiency, if Li[(CF$_3$SO$_2$)$_2$N] was used instead of lithium triflate. As Kervella et al. [102] pointed out, degradation of the anions of salts used as ion conductors in solid electrolytes is a major issue in lithium batteries, and further progress in this field may lead to improvement of LECs too.

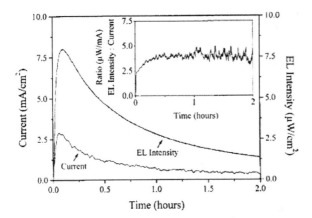

Figure 19 Evolution as a function of time of the current–voltage and electroluminescence intensity–voltage curves of a fresh ITO/DHF-*co*-BTOHF + LiCF₃SO₃/Al LEC biased at 15 V. Inset: The corresponding EL intensity/current ratio. (From Ref. 102. Copyright 2001, The Electrochemical Society.)

In the case of Ru(bpy)$_3^{2+}$-based LECs, it was found [103] that device degradation is likely to be caused by a quencher, Ru(bpy)$_2$(H$_2$O)$_2^{2+}$, formed during device operation through the reaction of the excited state with traces of water, devices operated under dry, inert atmosphere showing longer lifetimes. Changes in the absorption spectrum were also observed after device failure [230]; some degradation of the Ru(bpy)$_3^{2+}$ emitter seemingly occurs too, being also suggested as a reason for device degradation in other related studies [104]. For LECs with a binuclear Ru(bpy)$_3^{2+}$-type emitter [237], device degradation was also related to the degradation of the Ru(bpy)$_3^+$ species due to traces of water.

There is still the need for a more detailed analysis of LEC failure; the available experimental data do not allow one to draw a firm conclusion, but it appears that degradation of the emitter due to overoxidation and/or overreduction and degradation products that act as luminescence quenchers are likely to be two of the major reasons for LEC degradation.

IV. CHARGE TRANSPORT IN SSLECs—ELECTRON HOPPING

A. Electron Hopping Between Fixed Sites

The difference between the charge transport in a fluid medium and in a solid is that in the first case the transport is due to physical movement (or displacement),

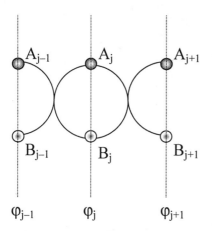

Figure 20 Schematic picture of electron hopping between adjacent donor (A) and acceptor (B) redox sites.

whereas in the second case it is mainly due to electron hopping (Fig. 20). The contribution of electron hopping to charge transport was acknowledged long ago [105,106], and it was shown that in solution this contribution is very small [107]. Interest in the electron hopping mechanism grew after the discovery of redox polymers, and theoretical and experimental analysis of this phenomenon was developed as a consequence [108–114].

Savéant [115] derived a transport equation for electron hopping between fixed sites that was later generalized by Baldy et al. [116] and Mohan and Sangaranrayanan [117]:

$$\frac{\partial C_A}{\partial t} = D_E \frac{\partial}{\partial x}\left[\frac{\partial C_A}{\partial x} + \frac{nF}{RT}C_A\left(1-\frac{C_A}{C^0}\right)\frac{\partial \varphi}{\partial x}\right] \tag{1}$$

$$\frac{\partial C_B}{\partial t} = D_E \frac{\partial}{\partial x}\left[\frac{\partial C_B}{\partial x} + \frac{nF}{RT}C_B\left(1-\frac{C_B}{C^0}\right)\frac{\partial \varphi}{\partial x}\right] \tag{1'}$$

where D_E is the equivalent diffusion coefficient for electron hopping; C_A, C_B are the concentrations of donor and acceptor sites (C^0 being the total concentration of redox sites, i.e., $C_A + C_B = C^0$); and φ is the electrostatic potential. Thus, electron hopping is a second-order process, characterized by two concentrations: that of the acceptor and that of the donor. The equivalent diffusion coefficient of electrons can be expressed in terms of the exchange rate of the electron hopping between the two fixed sites [118,119],

$$D_E = k_{et}\delta^2/6 \tag{2}$$

where k_{et} is the (first-order) electron transfer rate constant (which can be also viewed as the jump frequency between the two fixed sites) and δ is the average nearest-neighbor separation. k_{et} depends also on the separation between the fixed sites, because the probability of the hopping event depends on the energy barrier, and can be written as [119,120]

$$k_{et} = Ae^{-(\delta - r)/\gamma} \tag{3}$$

where A is a constant, r is the contact distance (the molecular diameter), and γ is a distance related to the electronic coupling in the system, depending on the height of the energy barrier (thus depending also on the environment of the redox sites). Typical γ values lie in the range 0.5–1 Å [120]. Electron hopping has many similarities with the Marcus theory of electron transfer reactions; indeed, if the electron exchange can be considered outer-sphere, then the Marcus formalism can be used to calculate the rate of electron hopping [116,119,121]. In some cases, when extensive ion pairing occurs (which is not uncommon for LECs), a more complicated treatment is necessary [122,123].

The average distance between the redox sites can be calculated by assuming various packings for the active sites [119]. The simplest estimation can be made by taking the redox site to be a point (with zero dimensions), in which case [118]

$$\delta = 0.554c^{-1/3} \tag{4}$$

where c is the (number) concentration of the redox sites. Of course, this is rather a gross estimate, and attempts have been made to calculate this distance using more realistic assumptions. For example, one can assume that the redox sites are hard spheres with a radius r, in which case one obtains a rather complicated expression for the average neighbor distance [119]:

$$\delta = \left(\frac{3}{4\pi c}\right)^{1/3} e^{\xi} \left[\Gamma\left(\frac{4}{3}\right) - \sum_{n=0}^{\infty} \frac{(-1)^n \xi^{(n+4/3)}}{n!(n+4/3)}\right] \tag{5}$$

where $\xi = (4/3)\pi r^3 c$ is the (dimensionless) excluded volume, i.e., the total volume occupied by the redox sites themselves. The electron hopping diffusion coefficient thus proves to be a complex function of active site concentration, because δ is found in both Eqs. (2) and (3), but useful estimates can still be made using Eq. (4) at constant concentration.

When physical movement of the charged species is important too, the total diffusion coefficient is given by

$$D_{app} = D_{phys} + D_{hop} \tag{6}$$

where D_{phys} is the "true" diffusion coefficient of the charged species (due to physical movement) and D_{hop} is the contribution from electron hopping [Eq. (2)].

More detailed theoretical analyses of coupled physical diffusion and electron hopping have recently been given [124,125]. We estimate, however, that for SSLECs, and especially for polymer ones, the charge transport occurs almost exclusively by electron hopping.

Because electron hopping is the preferred charge transport mechanism in SSLECs, the above formalism may be useful for a quantitative analysis of charge transport properties of such devices (especially for the ones employing fixed-site emitters, such as small-molecule emitters and $Ru(bpy)_3^{2+}$-based polymers); such an analysis is still lacking, however. It should be pointed out, though, that Eqs. (1) and (1') are based on several assumptions [115,116]. The most important is the condition $\Delta\varphi \ll kT/e$, where $\Delta\varphi$ is the potential difference across the distance δ, under which the equations can be linearized [115]. In other words, if these equations are to describe the charge transport of an SSLEC, then the electric field must be reasonably small at every point. If we take the hopping distance as ~1 nm [119,126], the condition becomes $E \ll$ ~2.5×10^5 V/cm; accepting 10% of this value as the upper limit for the validity of these equations, then E ~2.5×10^4 V/cm. The value of the electric field in an SSLEC can be considerably larger (~10^6 V/cm) near the electrodes, where a space charge region exists due to the accumulation of the ions. Large electric fields may also be encountered in the annihilation region; the electrons and holes annihilate each other, with neutral species resulting, and therefore the electric field increases sharply as the local ion concentration may still be large enough [68] (Fig. 28). The second major assumption is that the activity coefficients of all species are constant. This is true only for small changes in the system's properties; however, LECs usually have rather large concentration gradients of electrons and holes. Activity effects have been previously reported for other types of polymers [127], but it is difficult to consider them experimentally in a quantitative manner.

There have been numerous studies, both experimental and theoretical, about electron hopping between fixed sites, and it is beyond our purpose to give a detailed account here. More details can be found in the cited references, as well as other useful reviews [128–133].

Typical values for the electron hopping equivalent diffusion coefficient range from 10^{-11} to 10^{-7} cm^2/s, corresponding to (first-order) electron exchange rates of ~10^3 to ~10^8 s^{-1} [126]. For example, solvent (acetonitrile) swollen films of poly[Ru(vbpy)$_3$] [92] show a diffusion coefficient of 1.9×10^{-8} cm^2/s for hole hopping [between Ru(II) and Ru(III) sites] and 7.7×10^{-8} cm^2/s for electron hopping [between Ru(I) and Ru(II) sites]. It is likely that in such systems the slow, rate-determining, step for electron hopping is the hopping event between sites on different polymer chains rather than hopping between sites on the same chain. For small-molecule films, which are more tightly packed, equivalent diffusion coefficients for electron hopping as high as 6.6×10^{-6} cm^2/s were reported [134]. In this case, the small distance between the redox sites leads

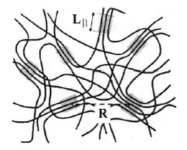

Figure 21 Schematic view on the structure of conjugated polymers. The lines represent polymer chains, and the shaded squares mark the regions with crystalline order. (From Ref. 135. Copyright 2002 by Elsevier Science.)

to a very high rate of electron exchange and thus to a large equivalent diffusion coefficient.

B. Electron Hopping in Conjugated Polymers

In conjugated polymers, electrons move freely along the polymer backbone, where there is some degree of delocalization, hopping only between the polymer chains. Conducting polymers are composed primarily of ordered "crystalline" regions, with close packing of the chains, and "amorphous" regions, with almost no order (Fig. 21). The charge transport mechanism is therefore somewhat intermediate between band transport (metal-like) and localized transport; electrons are not completely delocalized, as in a metal, but they are not confined to a fixed position either, as in redox polymers [135]. From a quantum point of view, inside the "metallic" regions the wave function is extended over the entire region (electron delocalization), whereas in the "amorphous" regions the wave function tends to be localized on one chain. However, even in this case, where electrons are not necessarily confined at fixed points, the formalism presented above may still be applied, albeit probably resulting in even more approximate results. The rate of electron exchange between sites can be written in the same way [135]:

$$D_E = WR^2 \tag{7}$$

where W is a hopping frequency and R is an average distance between the "metallic" grains. The hopping frequency contains a tunneling probability that decreases exponentially with the distance. Again, Marcus theory of electron transfer can be used to describe theoretically the electron hopping in conjugated polymers [136–138].

The structure of such a polymer can thus be viewed as a network of conducting regions separated by nonconducting regions, and because the transport occurs easily within the conducting region, the overall charge transport will be limited by electron hopping *between* such regions. Structural order is actually a major issue for obtaining polymeric materials with very high mobilities that can be used in molecular electronic devices [139]. In a LEC, however, the structure is even more complicated: The phase separation that commonly occurs in such devices results in a three-dimensional network structure (see Section III.D.2) in which the polymeric emitter regions are separated by solid electrolyte regions. Therefore, a theoretical, quantitative description of electron hopping in such a system is quite complicated, because the two phases now have different compositions also. In some cases, electron hopping may occur from both fixed sites and delocalized regions. Polymers that have fixed redox sites along conjugated polymer chains, very interesting for LEC emitters, have been synthesized [140,141].

In a LEC, electron hopping is coupled with ion movement; ionic conductivity in solid electrolytes is a very dynamic and rich field with many practical applications, e.g., in solid-state batteries [142]. Many studies have dealting with ionic conduction in polymer electrolytes; an interesting review that also discusses LECs can be found in Ref. 143. In LECs, because of the limited ionic conductivity as a result of both ion pairing [74] and slow ion movement, the transient behavior is almost always limited by the slow ionic mobility. For example, the Li^+ diffusion coefficients have been measured for MEH-PPV-modified electrodes, showing values of 1.6×10^{-8} cm^2/s when MEH-PPV is blended with PEO and 1.7×10^{-9} cm^2/s for pure MEH-PPV [144]. These values, although not very small, are still small enough to significantly slow the transient response of such devices.

V. METAL/ORGANIC INTERFACES IN SSLECs

A. Interface Structure

Metal/organic interfaces have been intensely studied because understanding them is very important for both OLEDs and SSLECs. Almost all studies deal with such interfaces for OLEDs, which lack high concentrations of mobile ions; however, some similarities still exist. We give here only a brief introduction and outline some important consequences for SSLECs; such a complex problem cannot be treated adequately in a few pages.

The metal/organic interface in an OLED is usually described within the framework of the Mott–Schottky model. In the Mott–Schottky limit, the barriers for charge injection are given by the differences in the work functions of the two materials. The theory was originally developed for metal/inorganic semiconductor interfaces, where the semiconductor can be described by band theory. In a disordered solid, however, there are no conduction and valence bands; instead, electrons and holes are promoted to the lowest unoccupied molecular orbital (LUMO) and highest occupied molecular orbital (HOMO), respectively, of the molecules adjacent to the interface, and from there they hop to available neighboring sites.

When defining the vacuum level for a given material, generally it is not possible to refer the work function to the vacuum level of an electron at rest at infinite distance from the metal surface. Instead, it is referred to the vacuum level of an electron just outside the solid surface. As a result, the work function of any material will depend on the surface dipoles and the influence of these dipoles on the electron's energy just outside the solid. This is why the measured work function of metals depends on the surface orientation, because the Fermi level is a bulk property of the metal. The difference between the energies of the electrons at rest just outside the solid and those at infinite distance is related to the surface dipole layer formed by electronic orbitals "spilling" outside the metal surface. Therefore, the work function of any solid will be different if its surface is changed, for example, by adsorbing a thin layer of an organic substance. The bulk properties of the solid do not change (e.g., the Fermi level), but the environment in which the electron just outside the solid phase sits does change. For the same reason, the work function of any material will depend on temperature. Briefly, one can say that the dependence of the work function on the surface type reflects the differences in spillage of the electron cloud outside the surface. This effect is frequently noticed and is well known in solid-state physics [145,146] and electrochemistry [147,148].

The interaction of the metal (or ITO) contact with the organic substance will lead to deviations from the Mott–Schottky model (Fig. 22), and in some cases these deviations are quite large. As a matter of fact, many experimental data show that the applicability of the Mott–Schottky model for calculating the charge injection barriers for metal/organic interfaces is the exception rather than the rule. For SSLECs, which usually contain a complex mixture of substances, such studies have not been attempted; results exist only for classical OLEDs [146,149–157]. A more detailed discussion of this topic is not intended here, because it is less relevant for SSLECs; more information can be found in the references given above and in several reviews [158–161].

The situation would be qualitatively similar for SSLECs, but the interactions are likely to be much stronger because now we are dealing with large concentrations of freely moving ions, not only with surface dipoles. Also, very polar impurities, such as traces of water (likely to exist even in dry PEO), will increase such interactions. It is well known that water adsorbs on metal

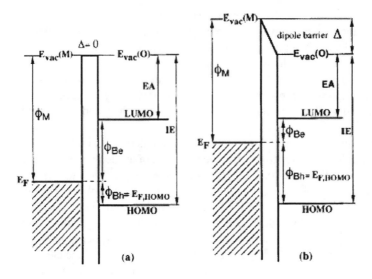

Figure 22 Schematic of an organic/metal interface energy diagram (a) without and (b) with an interface dipole Δ. φ_{Be} and φ_{Bh} are the electron and hole barriers, and $E_{vac}(O)$ and $E_{vac}(M)$ are the organic and metal vacuum levels, respectively. (From Ref. 149.)

electrodes, leading to strong influences on the double-layer structure in aqueous solutions [162–164]; moreover, even organic substances and free ions themselves can adsorb onto electrodes and thus interact specifically with them [165,166]. As a rule of thumb, the more polar the organic substance, the stronger the interaction effect, but, of course, the interaction strength will also depend on the metal.

The existence of mobile ions will lead to charge redistribution at the interface and the development of an electrical double layer. Actually, a sound theoretical description of the metal/organic interface for SSLECs needs a good description of the electrical double layer in solid electrolytes, and several attempts have been made in this respect [167,168], but we are still far from understanding all the underlying phenomena and their consequences. Experimental probing of double-layer properties and structure in SSLEC-type systems is not an easy task, and little has been done in this respect [169,180,181]. Phase separation and surface roughness of the electrodes are likely to make such attempts even more difficult.

The main difference between the electrical double layer in a fluid medium and in a condensed one is that mobile ions can move freely and may occupy (almost) any position in a fluid medium, whereas in the latter case the condensed nature of the medium and the polymer structure will usually allow the free ions to exist in only some regions [168]. Ion transport usually occurs through "segmental motion" [143]: A polymer segment carrying ions moves to another location where ions can hop to another neighboring segment; anions and cations may have very different mobilities in such a medium [143,170].

There is another effect in SSLEC-type systems in addition to the strong interfacial interactions and the development of an electrical double layer. Fermi-Dirac statistics (and the Boltzmann statistics, a limiting case of the Fermi-Dirac one), which were used to describe LECs [186] in order to give the distribution of electrons over the energy levels in the organic phase, is strictly valid only for noninteracting particles. If the particles interact strongly, as is the case for ions at relatively large concentrations, then Fermi–Dirac statistics is no longer appropriate. The density of states will depend on the number of particles (i.e., on concentration) [147]. In other words, from a thermodynamic point of view, activities have to be used instead of concentrations in order to be able to use the same equations. In a classical semiconductor, with small concentrations of dopant, this may not be a problem, but in a LEC, which usually contains large concentrations of mobile ions as well as of electrons and holes, one has to pay attention to this effect.

The metal/organic interface in SSLECs is thus very complicated, and its theoretical treatment is still quite limited owing to the great complexity of the system. Moreover, the complex mixture of materials usually found in SSLECs makes such attempts even less likely to give easy answers. We still rely on qualitative treatments, based sometimes on unrealistic assumptions. Hopefully, modern double-layer theories [171,172], applied eventually for solid-state electrolytes, will help us understand better the structure and the role of the interfaces in SSLECs.

B. Charge Injection in SSLECs

Solid-state LECs are electrochemical cells, and therefore if some charge is injected at one electrode, the same charge, but of opposite sign, must be injected at the other electrode. This is a general feature of electrochemical cells, which are systems with a high concentration of free ions and are therefore essentially electroneutral. The electroneutrality assumption, however, is always an approximation, although a very good one for systems with a large concentration of free ions (see Ref. 173 for more details). On the other hand, if the concentration of mobile ions is small, then the electroneutrality condition may no longer apply; the electric field in the bulk of the cell may be significant, and therefore the charge injection need not be symmetrical. Therefore, we can say that the charge injection in LECs is expected to be symmetrical only if the concentration of mobile ions is high. Even so, a large concentration of mobile ions does not necessarily mean that the device remains electroneutral in all cases. For example, if we have a LEC with a 0.1 M concentration of mobile ions but we expect to inject 0.5 M electrons at the cathode, then the device may not be electroneutral simply because we inject much more charge than existed before. So we can say that the device will remain electroneutral as long as the charge injected at the electrodes remains much smaller than the total charge of mobile ions (in other words, we should not change appreciably the ionic strength during the charge injection process). This is not necessarily true for LECs, especially when the concentration of mobile ions is small, but in many cases we can safely assume that this condition is fulfilled.

There are currently two mechanisms for charge injection in SSLECs. In the first model, supported by the (former) UNIAX group [199], the doping process (i.e., oxidation and reduction) is facilitated by the presence of the electrolyte and mobile ions compared to OLEDs, due to changes in the tunneling mechanism, as well as by changes in conformation and lattice fluctuation [174]. When a bias (larger than the emitter's "bandgap") is applied, the emitter starts to be reduced (*n*-doped) at the cathode and oxidized (*p*-doped) at the anode (the *p-n* junction model). The highly doped states are very conductive, and therefore the contacts become Ohmic, with a very thin tunneling barrier. At the same time, the mobile ions move to preserve the local electroneutrality (Figs. 23, 24). In this case, the electric field develops mainly in the bulk region [175,199,201] (Fig. 24); the charge injection process is somewhat similar to that of an electron-tunneling mechanism

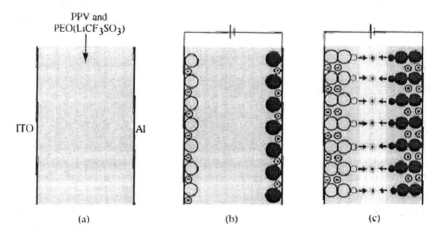

(a)　　　　　　　　(b)　　　　　　　　(c)

Figure 23 Schematic diagram of the electrochemical processes in a solid-state light-emitting electrochemical cell in the *p-n* junction model. (O) An oxidized molecule; (●) a reduced molecule; (⊖) an anion; (⊕) a cation; (o) a hole; (•) an electron; (*) a photon. (From Ref. 201. Copyright 1996 by the American Chemical Society.)

(through a very thin barrier), which is treated in the OLED literature using the Fowler–Nordheim formalism [176].

Because the concept of doping is somewhat misleading, another model was proposed, supported by the Cavendish group [58], in which the presence of mobile ions facilitates the charge injection through a space-charge effect. When the bias is applied, the mobile ions accumulate at the electrodes, thus increasing the local electric field until it reaches a value large enough for the charge to be injected in the HOMO/LUMO of the emitter (Fig. 25). The charge injection still occurs through a tunneling mechanism, but now it is assisted by the very large electric field at the interface as the ions accumulate. In this case, the electric field is built mainly at the interfaces, just as in a typical electrochemical cell, while the bulk remains essentially electroneutral (the "field-free" model) [58,177]. If we assume that there are no interactions between the solvent, emitter, and mobile ions (which is very unlikely), we can think that the electron and hole injection barriers are similar as in OLEDs, but now the charge injection occurs more easily, without the need to apply large voltages; the ions increase the electric field exactly where is needed, at the contacts. Of course, this process depends on the concentration of mobile ions; if the concentration is small, then the space-charge region will lead to lower electric fields near the electrodes; at the limit, where no mobile ions are present, we recover the OLED-type behavior, which is highly dependent on the nature of the metal contact. Experimental results show that such a dependence exists for SSLECs too, albeit it is much less pronounced, provided that the ionic concentration is relatively small [178].

De Mello et al. [179] tried to unify the two treatments, showing that the *p-n* junction model and the field-free one can both be used to describe SSLEC behavior but under different biasing conditions. The field-free model is adequate for devices operated under normal conditions, and the *p-n* junction model is better employed when devices are cooled under a steady bias (thus "freezing" the ions) and then operated at larger voltages.

Experimental evidence exists for both mechanisms, and it is difficult to distinguish between them [174,175,179,208]. The field-free model, however, is typical for an electrochemical cell, and if one considers that the only difference between a SLEC and an SSLEC is the charge transport mechanism (physical movement *vs.* electron hopping), then the field-free model would seem to be better suited to describe the operation of electrochemical cells. SIMS depth profiling [180,181] has shown that ions do redistribute at zero bias, owing to the built-in potential (which in turn is due to the different work functions of materials used as contacts), so one expects that this will also happen when a bias is applied. Nevertheless, it is also possible that both mechanisms are responsible for charge injection at the electrodes, with various contributions to the overall process. Adding a salt and a polymer electrolyte would certainly change the interfacial structure and therefore change the n and p doping processes, but at the

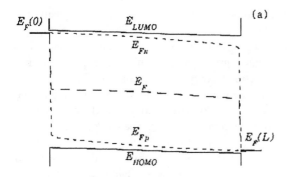

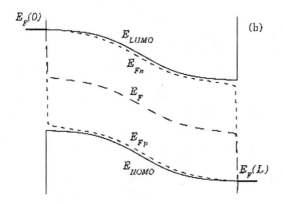

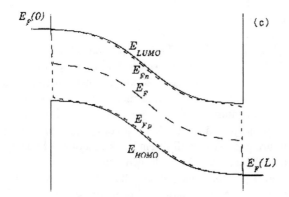

Figure 24 Energy diagrams in the *p-n* junction model calculated for various applied biases. (From Ref. 186. Copyright 1998 by the American Chemical Society.)

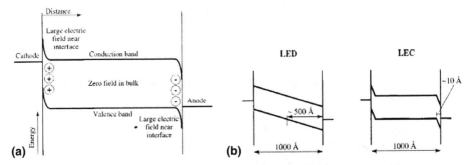

Figure 25 Left: Field-free model of the band diagram for an LEC device operating in forward bias. Positive ionic space charge accumulates close to the cathode. Negative ionic space charge accumulates close to the anode. This redistributes the electric field away from the bulk of the device towards the interfaces. The steady-state shape of the barriers for electron and hole injection are set by solutions to Poisson's equation and Boltzmann statistics. Typically the barrier widths are less than 10 Å in a device of thickness 1000 Å with an ionic charge density in excess of $10^{20} cm^{-3}$. Right: Field free model for the band diagrams showing the comparison between an LED and an LEC with high ionic concentration ($>10^{-30}$ cm) weakly biased in the forward direction. The LED resembles a capacitor with plate spacing of 1000 Å. The LEC resembles two capacitors in series each having a typical plate spacing of less than 10 Å. Reprinted with permission from reference [177].

same time the local electric field at the interface, as the ions accumulate, would facilitate the charge injection [182]. Actually, the only major difference between the two mechanisms is the way in which the electric field distributes along the LEC. In both cases charge is injected and must be transported toward the middle of the device by an electron hopping mechanism. It must be pointed out that in *both* models a concentration gradient of both holes and electrons exists, and simply observing that reduced and oxidized species exist in the film does not allow a clear distinction between the two mechanisms. Also, in both models the type of contact is not significant for device behavior. In the *p-n* junction model the charge injection is due entirely to the doping mechanism and therefore depends only on the emitter's bandgap, whereas in the field-free model, even if there were a weak dependence on the nature of the contact, it would be obscured by the very large electric field at the interface (in the high ionic concentration limit), which controls the overall process of charge injection. As a result, the contacts can be considered Ohmic in both cases.

We should also mention that modern techniques, using STM tips as contacts on molecularly sharp boundaries (adsorbed monolayers), which allows the study of electron transfer at molecular levels, may help in better understanding the charge injection phenomenon in organic molecules [183,184].

VI. GENERAL THEORY OF LECs AND DEVICE MODELING

A general theory of SSLECs should address the following issues:

1. Metal/organic and organic/organic interfaces—ion redistribution and formation of the electrical double layer
2. Charge transport—electron hopping for electrons and holes and ion movement
3. Charge injection
4. Electron–hole annihilation and light emission

A theoretical model would thus include a charge injection mechanism, coupled with transport equations (electron hopping for electrons and holes and Nernst–Planck equations for ions) and the annihilation reaction between electrons and holes. Depending on the concentration of mobile ions, either the electroneutrality condition or the Poisson equation would be included in the model.

A detailed theoretical description for steady-state LECs was proposed by Smith [185] for polymer LECs, based on the *p-n* junction model, where the salt generating the mobile ions is different from the emitting substance and the ions are considered to be inert. This model involves symmetrical devices with identical electrodes for easy calculations, but the ionic concentration is considered to be relatively low, because the value for the free energy of ion dissociation is taken to be as high as 0.6 eV. A somewhat similar treatment is due to Manzanares et al [186], who solve, at steady state, the drift–diffusion (or Nernst–Planck) equations

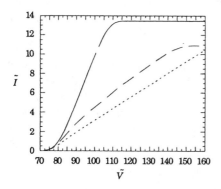

Figure 26 Simulated steady-state current-voltage curves in the *p-n* junction model for various salt concentrations and annihilation rates. Reprinted with permission from reference [186]. Copyright 1998 American Chemical Society.

for electron, hole, and ion transport, coupled with the electroneutrality condition throughout the device (Fig. 26). It is not clear, however, whether this condition is appropriate or not, because the concentration of the injected charges is comparable to the ionic charge. Also in this mechanism, Eq. (1) might be better employed to describe electron and hole movement rather than the classical Nernst–Planck equation. Several studies were focused on the a.c. behavior of SSLECs and the difference between the impedance of SSLECs and OLEDs, using the same *p-n* junction model [187–190]; similar treatments for polymer SLECs have also been reported [191].

A detailed description, based on the field-free model [177], outlines the possibility of having both unipolar and bipolar injection in SSLECs. Debye–Hückel theory is employed to calculate the steady-state distribution of ions, because it allows one to calculate such distributions analytically. However, the theory is too approximate, especially for large ionic concentrations (typical of SSLECs), and therefore a more accurate treatment, using transport equations coupled with the Poisson equation instead of the electroneutrality assumption, was attempted; the injection current is assumed to be of the Fowler–Nordheim type. Transients are discussed also; in the field-free model the current is not injected instantaneously when the bias is applied. First the electric field must be built at the interfaces, and, because this is a relatively slow process controlled by ion movement, some delay is observed [177]. A more detailed non-steady-state analysis for SSLECs, based on the same field-free model, was attempted for small-molecule, $Ru(bpy)_3^{2+}$-type, SSLECs [231]. The Nernst–Planck transport equations for electrons, holes, and ions, together with the Poisson equation, are solved numerically to give the current–voltage and light–voltage characteristics of such SSLECs. The difficult numerical solution did not allow the use of Eq. (1) for electron and hole transport, but the use of the Nernst–Planck equation seems to be a reasonable approximation. Charge injection barriers similar to OLEDs are considered at the electrode; qualitative agreement between simulation and experiment is good (Fig. 27), one interesting feature being that charge injection is asymmetrical, because only the small counter ions are mobile, the bulky $Ru(bpy)_3^{2+}$ being essentially fixed.

Similar attempts have been carried out not only for SSLECs, but also for SLECs. An interesting theoretical analysis of rubrene ECL in a thin-layer electrochemical cell [68] was carried out in essentially the same way as the field-free model. Here, because we are dealing with a liquid cell, there is no doubt that the mechanism is entirely electrochemical. Charge transport equations, coupled with a combination of Poisson equation and the electroneutrality condition (because the concentration of the supporting electrolyte is low) are solved numerically, and the ionic, radical anion, and radical cation profiles are calculated (outside the double-layer regions developed near the electrodes). It is shown also that when the concentrations of radical anions and cations become large, some potential

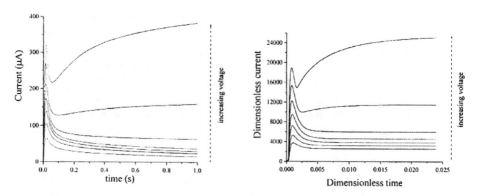

Figure 27 Simulated (left) and Experimental (right) current transients for different bias values for an ITO/Ru(bpy)$_3$(ClO$_4$)$_2$/GaSn LEC. (From Ref. 231. Copyright 2002 by the American Chemical Society.)

drop is seen in the middle of the cell too, similar to a *p-n* junction (Fig. 28). This is a direct consequence of the fact that when the charge injected becomes comparable to the ionic charge, the electrochemical cell is, in fact, no longer electroneutral. Similar analytical and numerical calculations carried out for SLECs with no supporting electrolyte show the same type of behavior and prove that ionic concentrations as low as 10^{-10} mol/cm^3 can still provide an electric field at interfaces large enough to promote an electrochemical reaction [192].

An interesting electrochemical treatment of SSLECs was given by Riess and Cahen [193], but because the authors assumed that the charge transfer coefficient is 1, a quite unusual value for electrochemical reactions, when calculating the current contribution due to electrochemical processes, the results are somewhat questionable.

Finally, a key issue in SSLEC modeling has to do with whether the device can be considered electroneutral at all times, i.e., for any applied bias, or whether significant deviations from electroneutrality occur. In the latter case, one must use the Poisson equation to calculate the charge distribution, significantly complicating the problem. Moreover, the steady-state Poisson–Boltzmann distribution, often used when dealing with charged species at steady state, is only approximately valid in such systems; it is strictly valid for point charges: holes and electrons are small particles and can be considered point charges, but ions are much larger, and their finite size will lead to errors in the Poisson–Boltzmann ion distribution. Finite ion size and the Poisson–Boltzmann distribution of ions is a very well known issue in electrical double-layer theories, and many efforts have been made to include such effects without overly complicating the theoretical formalism [194–196].

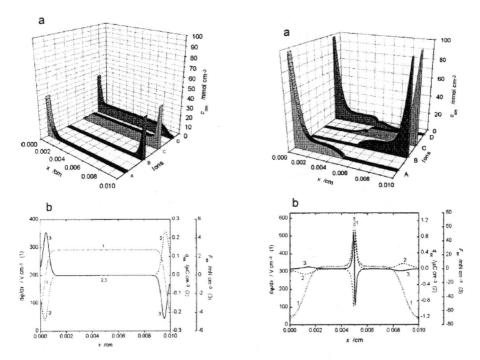

Figure 28 Simulation for a thin-layer ECL cell at constant current. Interelectrode distribution of (a) ions (A) R^-, (B) R^+, (C) An, and (D) Cat and (b) (1) gradient of electric potential, (2) local uncompensated charge q_{ex}, (3) driving force density $F_{ex} = -q_{ex}(\partial\varphi/\partial x)$. Equal diffusion coefficients of all species: 6.2×10^{-6} cm^2/s were assumed; $I = 0.135$ mA. Left: Time $t = 10$ ms after beginning of the electrolysis, solution resistance $R = 20.19$ kΩ. Right: $t = 60$ ms after beginning of the electrolysis, solution resistance $R = 17.54$ kΩ. (From Ref. 68. Copyright 1998 by the American Chemical Society.)

VII. CONCLUSIONS

The main purpose of this chapter was to outline and discuss practical, as well as theoretical, features of LECs. Solid-state light-emitting electrochemical cells are attractive for light-emitting devices, but there are still many issues that need to be solved before they are actually wed in commercial devices. Even though important progress has been achieved, the operating lifetime of SSLECs still needs to be improved for them to be competitive with the best available OLEDs. The device degradation mechanism of SSLECs has yet to be clarified. The synthesis of new materials is likely to lead to better performance in terms of efficiency and lifetime, but a better understanding of their operation and degradation mechanisms will help us look for the materials we need. Also, there is a need for better theories to describe device behavior, ones capable of both prediction and explanation of device characteristics. We can only hope that future efforts will address these issues and answer these questions.

REFERENCES

1. Kalinowski, J. Electroluminescence in organics. J. Phys. D: Appl. Phys. **1999**, *32*, R179–R250.
2. Mitschke, U., Bäuerle, P. The electroluminescence of organic materials. J. Mater. Chem. **2000**, *10*, 1471–1507.
3. Armstrong, N.R., Wightman, R.M., Gross, E.M. Light-emitting electrochemical processes. Annu. Rev. Phys. Chem. **2001**, *52*, 391–422.
4. Greenham, N.C., Friend, R.H. Semiconductor device physics of conjugated polymers. Solid State Phys. **1995**, *49*, 1–149.
5. Brütting, W., Berleb, S., Mückl, A.G. Device physics of organic light-emitting diodes based on molecular materials. Org. Electron. **2001**, *2*, 1–36.
6. Patel, N.K., Cina, S., Burroughes, J.H. High-efficiency organic light-emitting diodes. IEEE J. Quantum Electron. **2002**, *8*, 346–361.
7. Chen, C.H., Shi, J. Metal chelates as emitting materials for organic electroluminescence. Coord. Chem. Rev. **1998**, *171*, 161–174.
8. Kraft, A., Grimsdale, A.C., Holmes, A.B. Electroluminescent conjugated polymers—seeing polymers in a new light. Angew. Chem. Int. Ed. **1998**, *37*, 402–428.
9. Forrest, S., Burrows, P., Thompson, M. The dawn of organic electronics. IEEE Spectrum, **2000**, *37*, 29–34.
10. Sheats, J.R., Antoniadis, H., Hueschen, M., Leonard, W., Miller, J., Moon, R., Roitman, D., Stocking, A. Organic electroluminescent devices. Science, **1996**, *273*, 884–888.
11. Li, Y., Cao, Y., Gao, J., Wang, D., Yu, G., Heeger, A.J. Electrochemical properties of luminescent polymers and polymer light-emitting electrochemical cells. Synth. Met. **1999**, *99*, 243–248.
12. Murray, R.W. Chemically modified electrodes. Acc. Chem. Res. **1980**, *13*, 135–141.
13. Abruña, H.D., Denisevich, P., Umaña, M., Meyer, T.J., Murray, R.W. Rectifying interfaces using two-layer films of electrochemically polymerized vinylpyridine and vinylbipyridine complexes of ruthenium and iron on electrodes. J. Am. Chem. Soc. **1981**, *103*, 1–5.
14. Abruna, H.D., Bard, A.J. Electrogenerated chemiluminescence. 40. A chemiluminescent polymer based on the tris(4-vinyl-4'-methyl-2,2'-bipyridyl)Ru(II) system. J. Am. Chem. Soc. **1982**, *104*, 2641–2642.

15. Fan, F.-R-F., Mau, A., Bard, A.J. Electrogenerated chemiluminescence. A chemiluminescent polymer based on poly(vinyl-9,10-diphenylanthracene). Chem. Phys. Lett. **1985**, *116*, 400–404.

16. Richter, M.M., Fan, F-R.F., Klavetter, F., Heeger, A.J., Bard, A.J. Electrochemistry and electrogenerated chemiluminescence of films of the conjugated polymer 4-methoxy-(2-ethylhexoxyl)-2,5-polyphenylenevinylene. Chem. Phys. Lett. **1994**, *226*, 115–120.

17. Janakiraman, U., Dini, D., Preusser, A., Holmes, A.B., Martin, R.E., Doblhofer, K. Electrochemiluminescence of conjugated polymer. Synth. Met. **2001**, *121*, 1685–1686.

18. Nambu, H., Hamaguchi, M., Yoshino, K. A comparative study of electrogenerated chemiluminescence in poly(3-hexylthiophene) and poly(2-methoxy-5-dodecyloxy-p-phenylenevinylene). J. Appl. Phys. **1997**, *82*, 1847–1852.

19. Hamaguchi, M., Nambu, H., Yoshino, K. Electrogenerated chemiluminescence from poly(3-hexylthiophene). Jpn. J. Appl. Phys. Part 2. **1997**, *36*, L124–L126.

20. Buda, M., Gao, F.G., Pile, D., Bard, A.J., unpublished results.

21. Gurunathan, K., Murugan, A.V., Marimuthu, R., Mulik, U.P., Amalnerkar, D.P. Electrochemically synthesised conducting polymeric materials for applications towards technology in electronics, optoelectronics and energy storage devices. Mater. Chem. Phys. **1999**, *61*, 173–191.

22. Jureviciute, I., Bruckenstein, S., Hillman, A.R. Counter-ion specific effects on charge and solvent trapping in poly(vinylferrocene) films. J. Electroanal. Chem. **2000**, *488*, 73–81.

23. Mathias, M.F., Haas, O. Effect of counterion type on charge transport at redox polymer-modified electrodes. J. Phys. Chem. **1993**, *97*, 9217–9225.

24. Forster, R.J., Vos, J.G. Ionic interactions and charge transport properties of metallopolymer films on electrodes. Langmuir, **1994**, *10*, 4330–4338.

25. Zhang, X., Bard, A.J. Electrogenerated chemiluminescent emission from an organized (L-B) monolayer of a Ru(bpy)$_3^{2+}$-based surfactant on semiconductor and metal electrodes. J. Phys. Chem. **1988**, *92*, 5566–5569.

26. Burin, A.L., Ratner, M.A. Exciton migration and cathode quenching in organic light emitting diodes. J. Phys. Chem. A **2000**, *104*, 4704–4710.

27. Gaines Jr., G.L., Behnken, P.E., Valenty, S.J. Monolayer films of surfactant ester derivatives of tris(2,2′-bipyridine)ruthenium(II)$^{2+}$. J. Am. Chem. Soc. **1978**, *100*, 6549–6559.

28. Valenty, S.J., Behnken D.E., Gaines Jr., G.L. Preparation and monolayer properties of surfactant tris (2,2′-bipyridine)ruthenium(II) derivatives. Inorg. Chem. **1979**, *18*, 2160–2164.

29. Sprintschnik, G., Sprintschnik, H.W., Kirsch, P.P., Whitten, D.G. Preparation and photochemical reactivity of surfactant ruthenium(II) complexes in monolayer assemblies and at water-solid interfaces. J. Am. Chem. Soc. **1977**, *99*, 4947–4954.

30. Ghosh, P.K., Spiro, T.G. Photoelectrochemistry of tris(bipyridyl) ruthenium(II) covalently attached to n-type SnO$_2$. J. Am. Chem. Soc. **1980**, *102*, 5543–5549.

31. Murakata, T., Miyashita, T., Matsuda, M. Formation of monolayer and multilayer

containing a ruthenium complex with no alkyl chain substituent. J. Phys. Chem. **1988**, *92*, 6040–6043.

32. Kalyanasundaram, K. Photophysics, photochemistry and solar energy conversion with tris(bipyridyl) ruthenium(II) and its analogues. Coord. Chem. Rev. **1982**, *46*, 159–244.

33. Zhang, X., Bard, A.J. Electrogenerated chemiluminescent emission from an organized (L-B) monolayer of a Ru(bpy)$_3$$^{2+}$-based surfactant on semiconductor and metal electrodes. J. Phys. Chem. **1988**, *92*, 5566–5569.

34. Obeng, Y.S., Bard, A.J. Electrogenerated chemiluminescence. 53. Electrochemistry and emission from adsorbed monolayers of a tris(bipyridyl)ruthenium(II)-based surfactant on gold and tin oxide electrodes. Langmuir, **1991**, *7*, 195–201.

35. Sato, Y., Uosaki, K. Electrochemical and electrogenerated chemiluminescence properties of tris(2,2′-bipyridine)ruthenium(II)-tridecanethiol derivative on ITO and gold electrodes. J. Electroanal. Chem. **1995**, *384*, 57–66.

36. Miller, C.J., McCord, P., Bard, A.J. Study of Langmuir monolayers of ruthenium complexes and their aggregation by electrogenerated chemiluminescence. Langmuir, **1991**, *7*, 2781–2787.

37. Xu, X.-H., Bard, A.J. Electrogenerated chemiluminescence. 55. Emission from adsorbed Ru(bpy)$_3$$^{2+}$ on graphite, platinum, and gold. Langmuir, **1994**, *10*, 2409–2414.

38. Ikeda, N., Yoshimura, A., Tsushima, M., Ohno, T. Hopping and annihilation of ^{3}MLCT in the crystalline solid of [Ru(bpy)$_3$]X$_2$ (X = Cl$^-$, ClO$_4^-$ and PF$_6^-$). J. Phys. Chem. A. **2000**, *104*, 6158–6164.

39. Rubinstein, I., Bard, A.J. Polymer films on electrodes. 4. Nafion-coated electrodes and electrogenerated chemiluminescence of surface-attached Ru(bpy)$_3$$^{2+}$. J. Am. Chem. Soc. **1980**, *102*, 6641–6642.

40. Walcarius, A. Electrochemical applications of silica-based organic-inorganic hybrid materials. Chem. Mater. **2001**, *13*, 3351–3372.

41. Sykora, M., Meyer, T.J. Electrogenerated chemiluminescence in SiO$_2$ sol-gel polymer composites. Chem. Mater. **1999**, *11*, 1186–1189.

42. Collinson, M.M., Martin, S.A. Solid-state electrogenerated chemiluminescence in sol-gel derived monoliths. Chem. Commun. **1999**, 899–900.

43. Collinson, M.M., Taussig, J., Martin, S.A. Solid-state electrogenerated chemiluminescence from gel-entrapped ruthenium(II) tris(bipyridine) and tripropylamine. Chem. Mater. **1999**, *11*, 2594–2599.

44. Khramov, A.N., Collinson, M.M. Electrogenerated chemiluminescence of tris(2,2′-bipyridyl)ruthenium(II) ion-exchanged in Nafion-silica composite films. Anal. Chem. **2000**, *72*, 2943–2948.

45. Collinson, M.M., Novak, B., Martin, S.A., Taussig, J.S. Electrochemiluminescence of ruthenium(II) tris(bipyridine) encapsulated in sol-gel glasses. Anal. Chem. **2000**, *72*, 2914–2918.

46. Ghosh, P.K., Bard, A.J. Polymer films on electrodes. Part XV. The incorporation of Ru(bpy)$_3$$^{2+}$ into polymeric films generated by electrochemical polymerization of 2,2′-bipyrazine (bpz) and Ru(bpz)$_3$$^{2+}$. J. Electroanal. Chem. **1984**, *169*, 113–128.

47. Majda, M., Faulkner, L.R. Luminescence as an indicator of electron exchange. Dynamics in poly(styrenesulfonate) films containing tris(2,2′-bipyridine)ruthenium complexes. J. Electroanal. Chem. **1984**, *169*, 97–112.

48. Downey, T.M., Nieman, T.A. Chemiluminescence detection using regenerable tris(2,2′-bipyridyl)ruthenium(II) immobilized in Nafion. Anal. Chem. **1992**, *64*, 261–268.

49. Zhao, C.-Z., Egashira, N., Kurauchi, Y., Oga, K. Electrochemiluminescence sensor having a Pt electrode coated with a Ru(bpy)$_3$$^{2+}$-modified chitosan/silica gel membrane. Anal. Sci. **1998**, *14*, 439–441.

50. Buttry, D.A., Anson, F.C. Electrochemical control of the luminescent lifetime of Ru(bpy)$_3$$^{2+*}$ incorporated in Nafion films on graphite electrodes. J. Am. Chem. Soc. **1982**, *104*, 4824–4829.

51. Rubinstein, I., Bard, A.J. Polymer films on electrodes. 5. Electrochemistry and chemiluminescence at Nafion-coated electrodes. J. Am. Chem. Soc. **1981**, *103*, 5007–5013.

52. Anderson, L.B., Reilley, C.N. Thin-layer electrochemistry: use of twin working electrodes for the study of chemical kinetics. J. Electroanal. Chem. **1965**, *10*, 538–552.

53. Faulkner, L.R., Bard, A.J. Electrogenerated chemiluminescence. I. Mechanism of anthracene chemiluminescence in N,N-dimethylformamide solution. J. Am. Chem. Soc. **1968**, *90*, 6284–6290.

54. Tang, C.W., VanSlyke, S.A. Organic electroluminescent diodes. Appl. Phys. Lett. **1987**, *51*, 913–915.

55. Yu, G., Heeger, A.J. High efficiency photonic devices made with semiconducting polymers. Synth. Met. **1997**, *85*, 1183–1186.

56. Heeger, A.J. Light emission from semiconducting polymers: light-emitting diodes, light-emitting electrochemical cells, lasers and white light for the future. Solid State Commun. **1998**, *107*, 673–679.

57. Heeger, A.J., Diaz-Garcia, M.A. Semiconducting polymers as materials for photonic devices. Curr. Opin. Solid State Mater. Sci. **1998**, *3*, 16–22.

58. deMello, J.C., Tessler, N., Graham, S.C., Li, X., Holmes, A.B., Friend, R.H. Ionic space-charge assisted current injection in organic light emitting diodes. Synth. Met. **1997**, *85*, 1277–1278.

59. Adachi, C., Baldo, M.A., Thompson, M.E., Forrest, S.R. Nearly 100% internal phosphorescence efficiency in an organic light emitting device. J. Appl. Phys. **2001**, *90*, 5048–5051.

60. Hong, T.-M., Meng, H.-F. Spin-dependent recombination and electroluminescence quantum yield in conjugated polymers. Phys. Rev. B **2001**, *63*, 075206–1–075206–4.

61. Popovic, Z.D., Aziz, H. Reliability and degradation of small molecule-based organic light-emitting devices (OLEDs). IEEE J. Selected Top. Quant. **2002**, *8*, 362–371.

62. Yang, C., He, G., Wang, R., Li, Y. Solid-state electrochemical investigation of poly[2-methoxy, 5-(2′-ethylhexyloxy)-1,4-phenylenevinylene], J. Electroanal Chem. **1999**, *471*, 32–36.

63. Laser, D., Bard, A.J. Electrogenerated chemiluminescence XXIII. On the operation and lifetime of ECL devices. J. Electrochem. Soc. **1975**, *122*, 632–640.

64. Dunnett, J.S., Voinov, M. Reasons for the decay of the electrochemical luminescence of rubrene dissolved in benzonitrile. J. Electroanal. Chem. **1978**, *89*, 181–189.

65. Brilmyer, G.H., Bard, A.J. Electrogenerated chemiluminescence XXXVI. The production of direct current ECL in thin layer and flow cells. J. Electrochem. Soc. **1980**, *127*, 104–110.

66. Schaper, H., Köstlin, H., Schnedler, E. New aspects of D-C electrochemiluminescence. J. Electrochem. Soc. **1982**, *129*, 1289–1294.

67. Orlik, M., Rosenmund, J., Doblhofer, K., Ertl, G. Electrochemical formation of luminescent convective patterns in thin-layer cells. J. Phys. Chem. B **1998**, *102*, 1397–1403.

68. Orlik, M., Doblhofer, K., Ertl, G. On the mechanism of electrohydrodynamic convection in thin-layer electrolytic cells. J. Phys. Chem. B **1998**, *102*, 6367–6374.

69. Orlik, M. On the onset of self-organizing electrohydrodynamic convection in the thin-layer electrolytic cells. J. Phys. Chem. B **1999**, *103*, 6629–6642.

70. Chang, S.-C., Yang, Y., Pei, Q. Polymer solution light-emitting devices. Appl. Phys. Lett. **1999**, *74*, 2081–2083.

71. Chang, S.-C., Yang, Y. Polymer gel light-emitting devices. Appl. Phys. Lett. **1999**, *75*, 2713–2715.

72. Chang, S.-C., Li, Y., Yang, Y. Electrogenerated chemiluminescence mechanism of polymer solution light-emitting devices. J. Phys. Chem. B **2000**, *104*, 11650–11655.

73. Edel, J.B., deMello, A.J., deMello, J.C. Solution-phase electroluminescence. Chem. Commun. **2002**, 1954–1955.

74. Wenzl, F.P., Holzer, L., Tasch, S., Scherf, U., Müllen, K., Winkler, B., Mau, A.W.H., Dai, L., Leising, G. Turn on behavior of light emitting electrochemical cells. Synth. Met. **1999**, *102*, 1138–1139.

75. Cui, J., Wang, A., Edleman, N.L., Ni, J., Lee, P. Armstrong, N.R., Marks, T.J. Indium tin oxide alternatives—high work function transparent conducting oxides as anodes for organic light-emitting diodes. Adv. Mater. **2001**, *13*, 1476–1480.

76. Marks, T.J., Veinot, J.G.C., Cui, J., Yan, H., Wang, A., Edleman, N.L., Ni, J., Huang, Q., Lee, P., Armstrong, N.R. Progress in high work function TCO OLED anode alternatives and OLED nanopixelation. Synth. Met. **2002**, *127*, 29–35.

77. Schlaf, R., Murata, H., Kafafi, Z.H. Work function measurements on indium tin oxide films. J. Electron Spectrosc. Relat. Phenom. **2001**, *120*, 149–154.

78. Kugler, T., Johansson, Å., Dalsegg, I., Gelius, U., Salaneck, W.R. Electronic and chemical structure of conjugated polymer surfaces and interfaces: applications in polymer-based light-emitting devices. Synth. Met. **1997**, *91*, 143–146.

79. Liau, Y.-H., Scherer, N.F., Rhodes, K. Nanoscale electrical conductivity and surface spectroscopic studies of indium-tin oxide. J. Phys. Chem. B **2001**, *105*, 3282–3288.

80. Liu, C., Bard, A.J. Individually addressable submicron scale light-emitting devices based on electroluminescence of solid $Ru(bpy)_3(ClO_4)_2$ films. J. Am. Chem. Soc. **2002**, *124*, 4190–4191.

81. Pei, Q., Yang, Y. Efficient photoluminescence and electroluminescence from a soluble polyfluorene. J. Am. Chem. Soc. **1996**, *118*, 7416–7417.

82. Stephan, O., Collomb, V., Vial, J-C., Armand, M. Blue-green light-emitting diodes and electrochemical cells based on a copolymer derived from fluorene. Synth. Met. **2000**, *113*, 257–262.

83. Yang, Y., Pei, Q. Efficient blue-green and white light-emitting electrochemical cells based on poly[9,9-bis(3,6-dioxaheptyl)-fluorene–2,7-diyl]. J. Appl. Phys. **1997**, *81*, 3294–3298.

84. Holzer, L., Winkler, B., Wenzl, F.P., Tasch, S., Dai, L., Mau, A.W.H., Leising, G. Light-emitting electrochemical cells and light-emitting diodes based on ionic conductive poly phenylene vinylene: a new chemical sensor system. Synth. Met. **1999**, *100*, 71–77.

85. Carvalho, L.M., Santos, L.F., Guimarães, F.E.G., Gonçalves, D., Gomes, A.S., Faria, R.F. Morphology of 2,5-disubstituted poly(p-phenylene-vinylene) with oligo(ethylene oxide) side chains/PEO-salt blends. Synth. Met. **2001**, *119*, 361–362.

86. Santos, L.F., Carvalho, L.M., Guimarães, F.E.G., Gonçalves, D., Faria, R.F. Electrical and optical properties of light emitting electrochemical cells using MEH-PPV/PEO:lithium salt blends. Synth. Met. **2001**, *121*, 1697–1698.

87. Yang, C.Y., Hide, F., Heeger, A.J., Cao, Y. Nanostructured polymer blends: novel materials with enhanced optical and electrical characteristics. Synth. Met. **1997**, *84*, 895–896.

88. Yang, C., He, G., Sun, Q., Li, Y. Polymer light-emitting electrochemical cells with 1-methoxy-4-(2-ethylhexyloxy)benzene as salt carrier. Synth. Met. **2001**, *124*, 449–453.

89. Wenzla, F.P., Pachlera, P., Lista, E.J.W., Somitsch, D., Knoll, P., Patil, S., Guentner, R., Scherf, U., Leising, G. Self-absorption effects in a LEC with low Stokes shift. Physica E **2002**, *13*, 1251–1254.

90. Shiratori, S. A quick operating organic EL device using Ru complex light-emitting layer with 2,5-bis(4-tert-butyl-2-benzoxazolyl)thiophene electron transport layer. Mater Sci. Eng. B **2001**, *85*, 149–153.

91. Rudmann, H., Rubner, M.F. Single layer light-emitting devices with high efficiency and long lifetime based on tris(2,2'-bipyridyl)ruthenium(II) hexafluorophosphate. J. Appl. Phys. **2001**, *90*, 4338–4345.

92. Maness, K.M., Terrill, R.H., Meyer, T.J., Murray, R.W., Wightman, R.M. Solid-state diode-like chemiluminescence based on serial, immobilized concentration gradients in mixed-valent poly[Ru(vbpy)$_3$](PF$_6$)$_2$ films. J. Am. Chem. Soc. **1996**, *118*, 10609–10616.

93. Maness, K.M., Masui, H., Wightman, R.M., Murray, R.W. Solid state electrochemically generated luminescence based on serial frozen concentration gradients of Ru$^{III/II}$ and Ru$^{II/I}$ couples in a molten ruthenium 2,2'-bipyridine complex. J. Am. Chem. Soc. **1997**, *119*, 3987–3993.

94. Gao, J., Yu, G., Heeger, A.J. Polymer light-emitting electrochemical cells with frozen p-i-n junction. Appl. Phys. Lett. **1977**, *71*, 1293–1295.

95. Gao, J., Li, Y., Yu, G., Heeger, A.J. Polymer light-emitting electrochemical cells with frozen junctions. J. Appl. Phys. **1999**, *86*, 4594–4599.

96. Aziz, H., Popovic, Z.D., Hu, N.-X., Hor, A.-M., Gu Xu. Degradation mechanism of small molecule-based organic light-emitting devices. Science. **1999**, *283*, 1900–1902.

97. Sheats, J.R., Roitman, D.B. Failure modes in polymer-based light-emitting diodes. Synth. Met. **1998**, *95*, 79–85.

98. Sato, Y., Ichinosawa, S., Kanai, H. Operation characteristics and degradation of organic electroluminescent devices. IEEE J. Sel. Top. Quant. **1998**, *4*, 40–48.

99. Shen, J., Wang, D., Langlois, E., Barrow, W.A., Green, P.J., Tang, C.W., Shi, J. Degradation mechanisms in organic light emitting diodes. Synth. Met. **2000**, *111–112*, 233–236.

100. Popovic, Z.D., Aziz, H., Ioannidis, A., Hu, N.-X., dos Anjos, P.N.M. Time-resolved fluorescence studies of degradation in tris(8-hydroxyquinoline)aluminum (AlQ_3)-based organic light emitting devices (OLEDs). Synth. Met. **2001**, *123*, 179–181.

101. Schaer, M., Nüesch, F., Berner, D., Leo, W., Zuppiroli, L. Water vapor and oxygen degradation mechanisms in organic light emitting diodes. Adv. Funct. Mater. **2001**, *11*, 116–121.

102. Kervella, Y., Armand, M., Stéphan, O. Organic light-emitting electrochemical cells based on polyfluorene. Investigation of the failure modes. J. Electrochem. Soc. **2001**, *148*, H155–H160.

103. Kalyuzhny, G., Buda, M., McNeill, J., Barbara, P., Bard, A.J. Stability of thin-film solid-state electroluminescent devices based on tris(2,2′-bipyridine)ruthenium(II) complexes. J. Am. Chem. Soc. **2003**, *125*, 6272–6283,.

104. Bernhard, S., Barron, J.A., Houston, P.L., Abruña, H.D., Ruglovksy, J.L., Gao, X., Malliaras, G.G. Electroluminescence in ruthenium(II) complexes. J. Am. Chem. Soc. **2002**, *124*, 13624–13628.

105. Dahms, H. Electronic conduction in aqueous solution. J. Phys. Chem. **1968**, *72*, 362–364.

106. Ruff, I., Friedrich, V.J. Transfer diffusion. I. Theoretical. J. Phys. Chem. **1971**, *75*, 3297–3302.

107. Ruff, I., Friedrich, V.J., Demeter, K., Csillag, K. Transfer diffusion. II. Kinetics of electron exchange reaction between ferrocene and ferricinium ion in alcohols. J. Phys. Chem. **1971**, *75*, 3303–3309.

108. Buttry, D.A., Anson, F.C. Electron hopping vs. molecular diffusion as charge transfer mechanisms in redox polymer films. J. Electroanal. Chem. **1981**, *130*, 333–338.

109. Pickup, P.G., Kutner, W., Leidner, C.R., Murray, R.W. Redox conduction in single and bilayer films of redox polymer. J. Am. Chem. Soc. **1984**, *106*, 1991–1998.

110. Laviron, E. A multilayer model for the study of space distributed redox modified electrodes. Part I. Description and discussion of the model. J. Electroanal Chem. **1980**, *112*, 1–9.

111. Andrieux, C.P., Savéant, J.M. Electron transfer through redox polymer films. J. Electroanal Chem. **1980**, *lll*, 377–381.

112. Laviron, E., Roullier, L., Degrand, C. A multilayer model for the study of space distributed redox modified electrodes. Part II. Theory and application of linear potential sweep voltammetry for a simple reaction. J. Electroanal Chem. **1980**, *112*, 11–23.

113. Chidsey, C.E.D., Murray, R.W. Electroactive polymers and macromolecular electronics. Science. **1986**, *231*, 25–31.

114. Chidsey, C.E.D., Murray, R.W. Redox capacity and direct current electron conductivity in electroactive materials. J. Phys. Chem. **1986**, *90*, 1479–1484.

115. Savéant, J.M. Electron hopping between fixed sites. Equivalent diffusion and migration laws. J. Electroanal. Chem. **1986**, *201*, 211–213.

116. Baldy, C.J., Elliott C.M., Feldberg, S.W. Electron hopping in immobilized polyvalent and/or multicouple redox systems. J. Electroanal. Chem. **1990**, *283*, 53–65.

117. Mohan L.S., Sangaranarayanan, M.V. A generalised diffusion-migration equation for long distance electron hopping between redox centres. J. Electroanal. Chem. **1992**, *323*, 375–379.

118. Chandrasekar, S. Stochastic problems in physics and astronomy. Rev. Mod Phys. **1943**, *15*, 1–89.

119. Fritsch-Faules, I., Faulkner, L.R. A microscopic model for diffusion of electrons by successive hopping among redox centers in networks. J. Electroanal. Chem. **1989**, *263*, 237–255.

120. Blauch, D.N., Savéant, J.-M. Effects of long-range electron transfer on charge transport in static assemblies of redox centers. J. Phys. Chem. **1993**, *97*, 6444–6448.

121. Masui, H., Murray, R.W. Room-temperature molten salts of ruthenium tris(bipyridine). Inorg. Chem. **1997**, *36*, 5118–5126.

122. Savéant, J.-M. Electron hopping between localized sites. Coupling with electroinactive counterion transport. J. Phys. Chem. **1988**, *92*, 1011–1013.

123. Savéant, J.-M. Effect of ion pairing on the mechanism and rate of electron transfer. Electrochemical aspects. J. Phys. Chem. B. **2001**, *105*, 8995–9001.

124. Umamaheswari, J., Sangaranarayanan, M.V. Charge transport through chemically modified electrodes: a general analysis for ion exchange and covalently attached redox polymers. J. Phys. Chem. B **1999**, *103*, 5687–5697.

125. Umamaheswari, J., Sangaranarayanan, M.V. Nonequilibrium thermodynamics formalism for charge transport in redox polymer electrodes. J. Phys. Chem. B, **2001**, *105*, 2465–2473.

126. Terrill, R.H., Murray, R.W. Electron hopping transport in electrochemically active, molecular mixed valent materials. Mol. Electron. **1997**, 215–239.

127. Daum P., Murray, R.W. Charge-transfer diffusion rates and activity relationships during oxidation and reduction of plasma-polymerized vinylferrocene films. J. Phys. Chem. **1981**, *85*, 389–396.

128. Dalton, E.F., Surridge, N.A., Jernigan, J.C., Wilbourn, K.O., Facci, J.S.,

Murray, R.W. Charge transport in electroactive polymers consisting of fixed molecular redox sites. Chem. Phys. **1990**, *141*, 143–157.

129. Kaneko, M. Charge transport in solid polymer matrixes with redox centers. Prog. Polym. Sci. **2000**, *26*, 1101–1137.

130. Blauch, D.N., Savéant, J.-M. Dynamics of electron hopping in assemblies of redox centers. Percolation and diffusion. J. Am. Chem. Soc. **1992**, *114*, 3323–3332.

131. Savéant, J.-M. Electron hopping between localized sites. Effect of ion pairing on diffusion and migration. General rate laws and steady-state responses. J. Phys. Chem. **1988**, *92*, 4526–4532.

132. Savéant, J.-M. Electron hopping between fixed sites. "Diffusion" and "migration" in counter-ion conservative redox membranes at steady state. J. Electroanal. Chem. **1988**, *242*, 1–21.

133. Buck, R.P. Coupled electron hopping-anion displacement in plane sheet fixed-site polymer membranes. J. Electroanal. Chem. **1989**, *258*, 1–12.

134. Forster, R.J., Keyes, T.E., Majda, M. Homogeneous and heterogeneous electron transfer dynamics of osmium-containing monolayers at the air/water interface. J. Phys. Chem. B, **2000**, *104*, 4425–4432.

135. Prigodin, V.N., Epstein, A.J. Nature of insulator-metal transition and novel mechanism of charge transport in the metallic state of highly doped electronic polymers. Synth. Met. **2002**, *125*, 43–53.

136. Baldo, M.A., Forrest, S.R. Interface-limited injection in amorphous organic semiconductors. Phys. Rev. B, **2001**, *64*, 085201-1–085201-17.

137. Cornil, J., Beljonne, D., Calbert, J.-P., Brédas, J.-L. Interchain interactions in π-conjugated materials: impact on electronic structure, optical response and charge transport. Adv. Mater. **2001**, *13*, 1053–1067.

138. Soos, Z.G., Bao, S., Sin, J.M., Hayden, G.W. Compensation temperature in molecularly doped polymers. Chem. Phys. Lett. **2000**, *319*, 631–638.

139. Dimitrakopoulos, C.D., Malenfant, P.R.L. Organic thin film transistors for large area electronics. Adv. Mater. **2002**, *14*, 99–117.

140. Zhu, S.S., Kingsborough R.P., Swager, T.M. Conducting redox polymers: investigations of polythiophene–$Ru(bpy)_3^{n+}$ hybrid materials. J. Mater. Chem. **1999**, *9*, 2123–2131, and references therein.

141. Walters, K.A., Trouillet, L., Guillerez S., Schanze, K.S. Photophysics and electron transfer in poly(3-octylthiophene) alternating with Ru(II)- and Os(II)-bipyridine complexes. Inorg. Chem. **2000**, *39*, 5496–5509.

142. Tarascon, J.-M., Armand, M. Issues and challenges facing rechargeable lithium batteries. Nature, **2001**, *414*, 359–367.

143. Riess, I. Polymeric mixed ionic electronic conductors. Solid State Ionics, **2000**, *136–137*, 1119–1130.

144. Yang, C., He, G., Wang, R., Li, Y. Solid-state electrochemical investigation of poly[2-methoxy,5-(2'-ethyl-hexyloxy)–1,4-phenylene vinylene]. J. Electroanal. Chem. **1999**, *471*, 32–36.

145. Herring, C., Nichols, M.H. Thermionic emission. Rev. Mod Phys. **1949**, *21*, 185–271.

146. Ishii, H., Sugiyama, K., Ito, E., Seki, K. Energy level alignment and interfacial electronic structures at organic/metal and organic/organic interfaces. Adv. Mater. **1999**, *11*, 605–625.

147. Reiss, H. The Fermi level and the redox potential. J. Electrochem. Soc. **1985**, *89*, 3783–3791.

148. Reiss, H. The absolute electrode potential. Tying the loose ends. J. Electrochem. Soc. **1988**, *135*, 247C–258C.

149. Hill, I.G., Rajagopal, A., Kahn, A. Molecular level alignment at organic semiconductor-metal interfaces. Appl. Phys. Lett. **1998**, *73*, 662–664.

150. Rajagopal, A., Wu, C.I., Kahn, A. Energy level offset at organic semiconductor heterojunctions. J. Appl. Phys. **1998**, *83*, 2649–2655.

151. Berleb, S., Brütting, W., Paasch, G. Interfacial charges in organic hetero-layer light emitting diodes probed by capacitance-voltage measurements. Synth. Met. **2001**, *122*, 37–39.

152. Seki, K., Ito, E., Ishii, H. Energy level alignment at organic/metal interfaces studied by UV photoemission. Synth. Met. **1997**, *91*, 137–142.

153. Ishii, H., Seki, K. Energy level alignment at organic/metal interfaces studied by UV photoemission: breakdown of traditional assumption of a common vacuum level at the interface. IEEE Trans. Electron Devices, **1997**, *44*, 1295–1301.

154. Hill, I.G., Milliron, D., Schwartz, J., Kahn, A. Organic semiconductor interfaces: electronic structure and transport properties. Appl. Surf. Sci. **2000**, *166*, 354–362.

155. Lee, S.T., Hou, X.Y., Mason, M.G., Tang, C.W. Energy level alignment at Alq/metal interfaces. Appl. Phys. Lett. **1998**, *72*, 1593–1595.

156. Lee, S.T., Wang, Y.M. Hou, X.Y., Tang, C.W. Interfacial electronic structures in an organic light-emitting diode. Appl. Phys. Lett. **1999**, *74*, 670–672.

157. Gao, Y. Surface analytical studies of interface formation in organic light-emitting devices. Acc. Chem. Res. **1999**, *32*, 247–255.

158. Bardeen, J. Surface states and rectification at a metal semiconductor contact. Phys. Rev. **1947**, *71*, 717–727.

159. Tung, R.T. Schottky barrier height. J. Vac. Sci. Technol. B, **1993**, *11*, 1546–1552.

160. Tung, R.T. Formation of an electric dipole at metal-semiconductor interfaces. Phys. Rev. B, **2001**, *64*, 205310-1–205310-15.

161. Tung, R.T. Recent advances in Schottky barrier concepts. Mater. Sci. Eng. Rep. **2001**, *35*, 1–238.

162. O'M, J., Bockris, M., Devanathan, A., Müller, K. On the structure of charged interfaces. Proc. Roy. Soc. Lond. A, **1963**, *274*, 55–79.

163. Trasatti, S. Physical, chemical and structural aspects of the electrode/solution interface. Electrochim. Acta, **1983**, *28*, 1083–1093.

164. Guidelli, R., Schmickler, W. Recent developments in models for the interface between a metal and an aqueous solution. Electrochim. Acta, **2000**, *45*, 2317–2338.

165. Ritchie, I.M., Bailey, S., Woods, R. The metal-solution interface. Adv. Colloid Interface Sci. **1999**, *80*, 183–231.

166. Zimmermann, R., Dukhin, S., Werner, C. Electrokinetic measurements reveal interfacial charge at polymer films caused by simple electrolyte ions. J. Phys. Chem. B, **2001**, *105*, 8544–8549.

167. Armstrong, R.D., Horrocks, B.R. The double layer structure at the metal-solid electrolyte interface. Solid State Ionics, **1997**, *94*, 181–187.

168. Horrocks B.R., Armstrong, R.D. Discreteness of charge effects on the double layer structure at the metal/solid electrolyte interface. J. Phys. Chem. B, **1999**, *103*, 11332–11338.

169. Ouisse, T., Stéphan, O., Armand, M., Leprêtre, J.-C. Double-layer formation in organic light-emitting electrochemical cells. J. Appl. Phys. **2002**, *92*, 2795–2802.

170. Vincent, C.A. Ion transport in polymer electrolytes. Electrochim. Acta, **1995**, *40*, 2035–2040.

171. Schmickler, W. Electronic effects in the electric double layer. Chem. Rev., **1996**, *96*, 3177–3200.

172. Schmickler, W. Recent progress in theoretical electrochemistry. Annu. Rep. Prog. Chem. C, **1999**, *95*, 117–161.

173. Feldberg, S.W. On the dilemma of the use of the electroneutrality constraint in electrochemical calculations. Electrochem. Commun. **2000**, *2*, 453–456.

174. Johansson, T., Mammo, W., Andersson, M.R., Inganäs, O. Light-emitting electrochemical cells from oligo(ethylene oxide)-substituted polythiophenes: evidence for in situ doping. Chem. Mater. **1999**, *11*, 3133–3139.

175. Gao, J., Heeger, A.J., Campbell, I.H., Smith, D.L. Direct observation of junction formation in polymer light-emitting electrochemical cells. Phys. Rev. B, **1999**, *59*, R2482–R2485.

176. Fowler, R.H., Nordheim, L. Electron emission in intense electric fields. Proc. Roy. Soc. Lord A, **1928**, *119*, 173–181.

177. deMello, J.C., Tessler, N., Graham, S.C., Friend, R.H. Ionic space-charge effects in polymer light-emitting diodes. Phys. Rev. B, **1998**, *57*, 12951–12963.

178. Buda, M., Bard, A.J. Charge injection into light emitting devices based on tris(2,2′-bipyridine)ruthenium(II) and derivatives. to be published.

179. deMello, J.C., Halls, J.J.M., Graham, S.C., Tessler N., Friend, R.H. Electric field distribution in polymer light-emitting electrochemical cells. Phys. Rev. Lett. **2000**, *85*, 421–424.

180. Holzer, L., Wenzl, F.P., Sotgiu, R., Gritsch, M., Tasch, S., Hutter, H., Sampietro, M., Leising, G. Charge distribution in light emitting electrochemical cells. Synth. Met. **1999**, *102*, 1022–1023.

181. Gritsch, M., Hutter, H., Holzer, L., Tasch, S. Local ion distribution inhomogeneities in polymer based light emitting cells. Mikrochim. Acta, **2000**, *135*, 131–137.

182. Moderegger, E., Wenzl, F.P., Tasch, S., Leising, G., Scherf, U., Annan, K.O. Comparison of the internal field distribution in light-emitting diodes and light-emitting electrochemical cells. Adv. Mater. **2002**, *12*, 825–827.

183. Bumm, L.A., Arnold, J.J., Dunbar, T.D., Allara D.L., Weiss, P.S. Electron transfer through organic molecules. J. Phys. Chem. B, **1999**, *103*, 8122–8127.

184. Fan, F.-R.F., Yang, J., Cai, L., Price Jr., D.W., Dirk, S.M., Kosynkin, D.V., Yao, Y., Rawlett, A.M. Tour, J.M., Bard, A.J. Charge transport through self-assembled monolayers of compounds of interest in molecular electronics. J. Am. Chem. Soc. **2002**, *124*, 5550–5560.

185. Smith, D.L. Steady state model for polymer light-emitting electrochemical cells. J. Appl. Phys. **1997**, *81*, 2869–2880.

186. Manzanares, J.A. Reiss, H., Heeger, A.J. Polymer light-emitting electrochemical cells: a theoretical study of junction formation under steady-state conditions. J. Phys. Chem. B, **1998**, *102*, 4327–4336.

187. Campbell, I.H., Smith, D.L., Neef, C.J., Ferraris, J.P. Capacitance measurements of junction formation and structure in polymer light-emitting electrochemical cells. Appl. Phys. Lett. **1998**, *72*, 2565–2567.

188. Yu, G., Cao, Y., Zhang, C., Li, Y., Gao, J., Heeger, A.J. Complex admittance measurements of polymer light-emitting electrochemical cells: ionic and electronic contributions. Appl. Phys. Lett. **1998**, *73*, 111–113.

189. Li, Y., Gao, J., Yu, G., Cao, Y., Heeger, A.J. AC impedance of polymer light-emitting electrochemical cells and light-emitting diodes: a comparative study. Chem. Phys. Lett. **1998**, *287*, 83–88.

190. Li, Y., Gao, J., Wang, D., Yu, G., Cao, Y., Heeger. A.J. A.C. impedance of frozen junction polymer light-emitting electrochemical cells. Synth. Met. **1998**, 191–194.

191. Chang, S.-C., Yang, Y., Wudl, F., He, G., Li, Y. AC impedance characteristics and modeling of polymer solution light-emitting devices. J. Phys. Chem. B, **2001**, *105*, 11419–11423.

192. Schaper, H., Schnedler, E. On the charge distribution in thin-layer under electrostatic conditions. Application to electrolyte-free electrochemiluminescence. J. Electroanal. Chem. **1982**, *137*, 39–49.

193. Riess, I., Cahen, D. Analysis of light emitting polymer electrochemical cells. J. Appl. Phys. **1997**, *82*, 3147–3151.

194. Fawcett, W.R., Henderson, D.J. A simple model for the diffuse double layer based on a generalized mean spherical approximation. J. Phys. Chem. B, **2000**, *104*, 6837–6842.

195. Kuo, Y.-C., Hsu, J.-P. Double-layer properties of an ion-penetrable charged membrane: effect of sizes of charged species. J. Phys. Chem. B, **1999**, *103*, 9743–9748.

196. Borukhov I., Andelman, D. Steric effects in electrolytes: a modified Poisson-Boltzmann equation. Phys. Rev. Lett. **1997**, *79*, 435–438.

197. Lee, J.-Ik, Hwang, D.-H., Park, H., Do, L.-M., Chu, H.Y., Zyung, T., Miller, R.D. Light-emitting electrochemical cells based on poly(9,9-bis(3,6-dioxaheptyl)-fluorene-2,7-diyl). Synth. Met. **2000**, *111*, 195–197.

198. Panozzo, S., Armand, M., Stéphan, O. Light-emitting electrochemical cells using a molten delocalized salt. Appl. Phys. Lett. **2002**, *80*, 679–681.

199. Pei, Q., Yu, G., Zhang, C., Yang, Y., Heeger, A.J. Polymer light-emitting electrochemical cells. Science, **1995**, *269*, 1086–1088.

200. Pei, Q., Yang, Y. Solid-state polymer light-emitting electrochemical cells. Synth. Met. **1996**, *80*, 131–136.

201. Pei, Q., Yang, Y., Yu, G., Zhang, C., Heeger, A.J. Polymer light-emitting electrochemical cells: in situ formation of a light-emitting p-n junction. J. Am. Chem. Soc. **1996**, *118*, 3922–3929.

202. Yang, Y., Pei, Q. Voltage controlled two color light-emitting electrochemical cells. Appl. Phys. Lett. **1996**, *68*, 2708–2710.

203. Majima, Y., Hiraoka, T., Takami, N., Hayase, S. Novel approach for hole-blocking in light-emitting electrochemical cells. Synth. Met. **1997**, *91*, 87–89.

204. Cao, Y., Yu, G., Heeger, A.J., Yang, C.Y. Efficient, fast response light-emitting electrochemical cells: electroluminescent and solid electrolyte polymers with interpenetrating network morphology. Appl. Phys. Lett. **1996**, *68*, 3218–3220.

205. Yin, C., Zhao, Y., Yang, C., Zhang, S. Single-ion transport light-emitting electrochemical cells: designation and analysis of the fast transient light-emitting responses. Chem. Mater. **2000**, *12*, 1853–1856.

206. Cao, Y., Pei, Q., Andersson, M.R., Yu, G., Heeger, A.J. Light emitting electrochemical cells with crown ether as solid electrolyte. J. Electrochem. Soc. **1997**, *144*, L317-L320.

207. Lee, T.-W., Lee, H.-C., Ok Park, O. High-efficiency polymer light-emitting devices using organic salts: a multilayer structure to improve light-emitting electrochemical cells. Appl. Phys. Lett. **2002**, *81*, 214–216.

208. Lee, T.-W., Ok Park, O. Polymer light-emitting energy-well devices using single-ion conductors. Adv. Mater. **2001**, *13*, 1274–1278.

209. Yu, G., Cao, Y., Andersson, M., Gao, J., Heeger, A.J. Polymer light-emitting electrochemical cells with frozen p-i-n junction at room temperature. Adv. Mater, **1998**, *10*, 385–388.

210. Holzer, L., Wenzl, F.P., Tasch, S., Leising, G., Winkler, B., Dai, L., Mau, A.W.H. Ionochromism in a light-emitting electrochemical cell with low response time based on an ionic conductive poly-phenylene vinylene. Appl. Phys. Lett. **1999**, *75*, 2014–2016.

211. Tasch, S., Holzer, L., Wenzl, F.P., Gao, J., Winkler, B., Dai, L., Mau, A.W.H., Sotgiu, R., Sampietro, M., Scherf, U., Müllen, K., Heeger, A.J., Leising, G. Light-emitting electrochemical cells with microsecond response times based on PPPs and novel PPVs. Synth. Met. **1999**, *102*, 1046–1049.

212. Morgado, J., Cacialli, F., Friend, R.H., Chuah, B.S., Moratti, S.C., Holmes, A.B. Luminescence properties of PPV-based copolymers with crown ether substituents. Synth. Met. **2000**, *111–112*, 449–452.

213. Morgado, J., Friend, R.H., Cacialli, F., Chuah, B.S., Rost, H., Moratti, S.C., Holmes, A.B. Light-emitting electrochemical cells based on poly(p-phenylene vinylene) copolymers with ion-transporting side groups. Synth. Met. **2001**, *122*, 111–213.

214. Morgado, J., Cacialli, F., Friend, R.H., Chuah, B.S., Rost, H., Holmes, A.B. Light-emitting devices based on a poly(p-phenylenevinylene) statistical copolymer with oligo(ethylene oxide) side groups. Macromolecules **2001**, *34*, 3094–3099.

215. Morgado, J., Cacialli, F., Friend, R.H., Chuah, B.S., Rost, H., Moratti, S.C., Holmes, A.B. Luminescence properties of a PPV-based statistical copolymer with glyme-like side groups. Synth. Met. **2001**, *119*, 595–596.

216. Sun, Q., Wang, H., Yang, C., He, G., Li, Y. Blue-green light-emission LECs based on block copolymers containing di(α-naphthalene vinylene)benzene chromophores and tri(ethylene oxide) spacers. Synth. Met. **2002**, *128*, 161–165.

217. Wang, H., Wang, X., Liu, D. Design and synthesis of novel luminescent copolymers containing ionic conductive blocks on the skeletons. Synth. Met. **2002**, *126*, 219–223.

218. Tasch, S., Gao, J., Wenzl, F.P., Holzer, L., Leising, G., Heeger, A.J., Scherf, U., Müllen, K. Blue and green light-emitting electrochemical cells with microsecond response times. Electrochem. Solid State Lett. **1999**, *2*, 303–305.

219. Wenzl, F.P., Pachler, P., Tasch, S., Somitsch, D., Knoll, P., Scherf, U., Annan, K.O., Leising, G. Ion dissociation in crown ether based wide gap LECs. Synth. Met. **2001**, *121*, 1735–1736.

220. Yang, Y., Pei, Q. Light-emitting electrochemical cells from a blend of *p* and *n*-type luminescent conjugated polymers. Appl. Phys. Lett. **1997**, *70*, 1926–1928.

221. Lee, J.-K., Yoo, D., Rubner, M.F. Synthesis and characterization of an electroluminescent polyester containing the Ru(II) complex. Chem. Mater. **1997**, *9*, 1710–1712.

222. Wu, A., Yoo, D., Lee, J.-K., Rubner, M.F. Solid-state light-emitting devices based on the tris-chelated ruthenium(II) complex: 3. High efficiency devices via a layer-by-layer molecular-level blending approach. J. Am. Chem. Soc. **1999**, *121*, 4883–4891.

223. Wu, A., Lee, J., Rubner, M.F. Light emitting electrochemical devices from sequentially adsorbed multilayers of a polymeric ruthenium(II) complex and various polyanions. Thin Solid Films **1998**, *327–329*, 663–667.

224. Elliott, C.M., Pichot, F., Bloom, C.J., Rider, L.S. Highly efficient solid-state electrochemically generated chemiluminescence from ester-substituted trisbipyridineruthenium(II)-based polymers. J. Am. Chem. Soc. **1998**, *120*, 6781–6784.

225. Greenwald, Y., Hide, F., Heeger, A.J. The electrochemistry of poly(3-octylthiophene) based light emitting electrochemical cells. J. Electrochem. Soc. **1997**, *144*, L241–L243.

226. Lee, J.-K., Yoo, D.S., Handy, E.S., Rubner, M.F. Thin film light emitting devices from an electroluminescent ruthenium complex. Appl. Phys. Lett. **1996**, *69*, 1686–1688.

227. Lyons, C.H., Abbas, E.D., Lee, J.-K., Rubner, M.F. Solid-state light-emitting devices based on the trischelated ruthenium(II) complex. 1. Thin film blends with poly(ethylene oxide). J. Am. Chem. Soc. **1998**, *120*, 12100–12107.

228. Handy, E.S., Pal, A.J., Rubner, M.F. Solid-state light-emitting devices based on the tris-chelated ruthenium(II) complex. 2. Tris(bipyridyl)ruthenium(II) as a high-brightness emitter. J. Am. Chem. Soc. **1999**, *121*, 3525–3528.

229. Gao, F.G., Bard, A.J. Solid-state organic light-emitting diodes based on tris(2,2′-bipyridine)ruthenium(II) complexes. J. Am. Chem. Soc. **2000**, *122*, 7426–7427.

230. Gao, F.G., Bard, A.J. High-brightness and low-voltage light-emitting devices based on trischelated ruthenium(II) and tris(2,2′-bipyridine)osmium(II) emitter layers and low melting point alloy cathode contacts. Chem. Mater. **2002**, *14*, 3465–3470.

231. Buda, M., Kalyuzhny, G., Bard, A.J. Thin-film solid-state electroluminescent devices based on tris(2,2′-bipyridine)ruthenium(II) complexes. J. Am. Chem. Soc. **2002**, *124*, 6090–6098.

232. Rudmann, H., Shimada, S., Rubner, M.F. Solid-state light-emitting devices based

on the tris-chelated ruthenium(II) complex. 4. High-efficiency light-emitting devices based on derivatives of the tris(2,2′-bipyridyl)ruthenium(II) complex. J. Am. Chem. Soc. **2002**, *124*, 4918–4921.

233. Rudmann, H., Rubner, M.F. Single layer light-emitting devices with high efficiency and long lifetime based on tris(2,2′ bipyridyl)ruthenium(II) hexafluorophosphate. J. Appl. Phys. **2001**, *90*, 4338–4345.

234. Gong, X., Ng, P.K., Chan, W.K. Trifunctional light-emitting molecules based on rhenium and ruthenium bipyridine complexes. Adv. Mater. **1998**, *10*, 1337–1340.

235. Chan, W.K., Ng, Po K., Gong, X., Hou, S. Light-emitting multifunctional rhenium(I) and ruthenium(II) 2,2′-bipyridyl complexes with bipolar character. Appl. Phys. Lett. **1999**, *75*, 3920–3922.

236. Bernhard, S., Gao, X., Malliaras, G.G., Abruña, H.D. Efficient electroluminescent devices based on a chelated osmium(II) complex. Adv. Mater. **2002**, *14*, 433–436.

237. Leprêtre, J.-C., Deronzier, A., Stéphan, O. Light-emitting electrochemical cells based on ruthenium(II) using a crown ether as solid electrolyte. Synth. Met. **2002**, *131*, 175–183.

11
Miscellaneous Topics and Conclusions

Allen J. Bard
The University of Texas at Austin, Austin, Texas, U.S.A.

I. INTRODUCTION

This chapter deals briefly with several topics related to ECL and its applications that were not covered in the previous chapters. It concludes with some ideas about future work in ECL.

II. LOW LEVEL EMISSION AND INVERSE PHOTOEMISSION

There are numerous examples of cases where very low levels of light are recorded during an ECL experiment, when the conditions, such as the applied potential range, seem inconsistent with accepted models for ECL. Such is the case for so-called preannihilation ECL mentioned in Chapter 1. One must be aware, however, that there are a number of processes that can produce low-level emission and that current technology allows the detection of light at exceedingly low levels. Thus trace impurities, either in the solution originally or formed in a decomposition reaction of an electrogenerated species, can sometimes act as coreactants. Weak chemiluminescence results from reactions of many organic compounds with oxygen [1]. In addition, there are a number of artifacts, such as the accumulation in solution of reaction products from the counter electrode or insufficient shielding of the counter electrode to prevent its possible light emission from being detected by the photomultiplier or CCD, that can lead to reports of ECL at a working electrode under conditions that are difficult to explain. In general, one should be very cautious about experiments that produce apparent ECL at very low levels.

Low-level (background) emission does establish the sensitivity limits in many ECL analytical procedures. In addition to the sources mentioned above, a background emission caused by inverse photoemission can occur. Inverse photoemission (sometimes also called bremsstrahlung spectroscopy) occurs when an electron impinging on a metal causes the emission of a photon. Although this phenomenon is usually found with metals in vacuum and has been used to map the band structures of metals [2], a number of studies have demonstrated that it can also occur in solution in electrochemical cells. The first experiments of this kind were performed by McIntyre and Sass [3], who coined the term charge-transfer-reaction inverse-photoemission spectroscopy (CTRIPS) to describe these kinds of studies. The basic idea of CTRIPS is shown in Figure 1. If a reduced species, e.g., an anion radical A^- in MeCN, is electrogenerated and then oxidized at a very positive potential (but a potentials at which no cation radical can be formed), the electron is injected into the metal at a level characteristic of the A/A^- couple. However, the Fermi level of the metal is controlled by the applied potential. Most of the electrons relax to the Fermi level by a radiationless process, but a few photons are emitted and can be detected (Fig. 1a). The analogous process can be carried out with hole injection, during which a cation radical is generated and then reduced at a fairly negative potential (Fig. 1b). In choosing reactants for these types of processes, one generally selects species such as benzophenone (anion radical) or thianthrene (cation radical) that do not produce annihilation ECL. A number of other papers have described

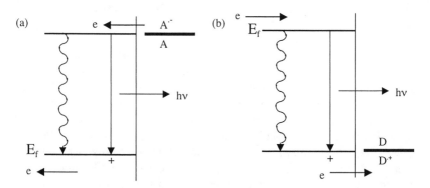

Figure 1 Schematic representation of inverse photoemission process at metal/electrolyte interface for (a) electron injection into positive metal and (b) hole injection into negative metal.

this phenomenon, almost all of the processes being carried out in nonaqueous solutions in the 1980s and early 1990s [4–6], Experiments in aqueous solutions were also carried out [7]. Recently, Gosavi and Marcus [8] elaborated a theoretical model for CTRIPS.

III. HOT ELECTRON INJECTION AND UP-CONVERSION AT SEMICONDUCTORS

As discussed in Chapter 1 (see Fig. 1), electroluminescence (EL) occurs at a semiconductor/solution interface. Note the similarities of this EL process and the inverse photoemission process discussed in the previous section. The difference is that EL at semiconductors is usually much more efficient (by many orders of magnitude), because the radiationless processes are less rapid. This has been seen with many semiconductors and is useful in mapping surface and interface states within the region between the conduction and valence bands, for example for GaN and $Ga_xIn_{1-x}N$ [9].

A somewhat related example is the direct formation of excited states at semiconductor or insulator surfaces by an electron from the electrode reacting with a cation radical in solution to form an emitting species (Fig. 2). The direct reduction of thianthrene cation radical ($TH^{\cdot+}$) to form an excited state

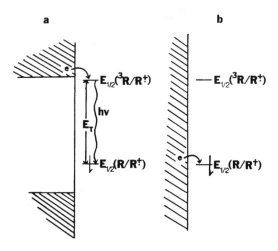

Figure 2 Electron transfer from electrodes to oxidized solution species (a) on a semiconductor, (b) on a metal electrode.

does not occur at a metal electrode (e.g., Pt), because any TH* formed is quenched by energy and electron transfer at the metal electrode [10]. However, this quenching is not as efficient at a semiconductor electrode, and direct formation of excited states at electrodes such as *n*-ZnO and *n*-CdS has been reported [11].

Direct formation of excited states is also possible by tunneling of hot electrons through a thin insulating film, such as Ta_2O_5 film on a Ta electrode, as shown in Figure 3 [12]. A comparison of the ECL obtained by the reduction of TH$^+$ at Pt and the oxide-coated electrode is shown in Figure 4. Similar processes occur in aqueous solutions at an Al electrode coated with an oxide film [13]. However, the Al_2O_3 film is less stable in aqueous solution during cathodization.

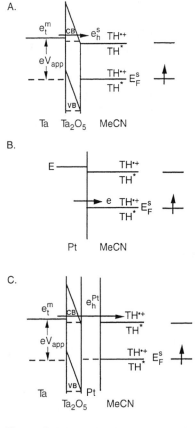

Figure 3 Schematic of electron transfer from electrodes to oxidized TH$^+$ species in solution at (a) Ta/Ta$_2$O$_5$, (b) Pt, and (c) Ta/Ta$_2$O$_5$/Pt electrodes. (From Ref. 12b.)

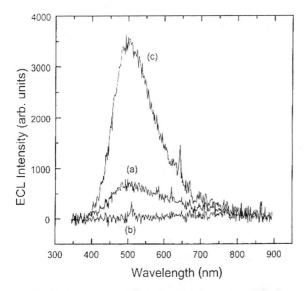

Figure 4 ECL spectra of 7 mM $TH^{+}/0.1$ M TBAP/MeCN solution at (a) Ta/2.5 nm Ta_2O_5, (b) Pt, and (c) Ta/2.5 nm Ta_2O_5/40 nm Pt electrodes. The exposure time was 5 min. The potential was pulsed between 0 and -2.7 V vs. SCE at 0.1 s intervals. (From Ref. 12b.)

IV. ELECTROGENERATED CHEMILUMINESCENCE IMAGING

Electrogenerated chemiluminescence provides an excellent way of imaging electrodes and solution flows such as those at an RRDE (Chapter 3, Fig. 17). A number of other papers have used ECL of, e.g., the $Ru(bpy)_3^{2+}$ system, to image electrodes, including in more recent times ultramicroelectrodes and arrays [14].

At semiconductor electrodes, as described in the preceding section, ECL combined with the concept of photoelectrochemistry at semiconductors can be used for the up-conversion of radiation. Up-conversion is the conversion of longer wavelength light (e.g., infrared) to shorter wavelength light (e.g., visible). It is used for imaging (e.g., night vision), usually employing solid-state devices and high voltages. The principle is based on using a semiconductor electrode (e.g., n-GaAs or n-InP) as the working electrode in an ECL cell with precursors A and D (e.g., A, DPA; D, TMPD) [15]. In the dark, the n-type semiconductor can carry out reductions but not oxidations, so cycling the potential between the regions where A is reduced and D is oxidized on Pt results in only A^{-} formation and no ECL emission. When the semiconductor is irradiated with light of energy greater than the bandgap, i.e., red light for the above semiconductors, then oxidation is promoted (because holes are photogenerated in the valence band of

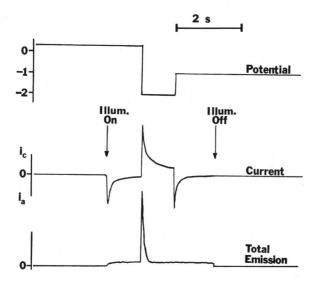

Figure 5 Potential–, current–, and emission–time profiless illustrating the photoinduced oxidation of 10-methylphenothiazine and the reduction of fluoranthene at an n-type InP electrode yielding ECL upon illumination of the n-InP with a 3 mW He-Ne laser. (From Ref. 15.)

the semiconductor). Thus the annihilation reaction between $D^{.+}$ and $A^{.-}$ is possible and ECL emission from A^* (for DPA blue-violet emission) is observed. Thus an up-conversion from red light to blue-violet light occurs (Fig. 5). This scheme was proposed for actual imaging but never reduced to practice [16].

V. SCANNING OPTICAL MICROSCOPY WITH ECL

Light sources based on ECL are discussed in Chapter 10. A unique light source is based on ECL at an ultramicroelectrode (UME) and uses this for scanning optical microscopy. The principle is shown in Figure 6. ECL is generated at the UME in the usual way, using either annihilation in an aprotic solvent or a coreactant system in water, as the tip is scanned in the XY plane above the sample to be imaged. The light is collected by a detector, e.g., a photomultiplier tube, beneath the sample, and an image is obtained by plotting the detector response as a function of XY position. The resolution obtainable by this technique depends on the UME radius a and the distance d between the UME and the sample. With the first experiments using this technique [17], electrode radii were ~1–10 μm, with

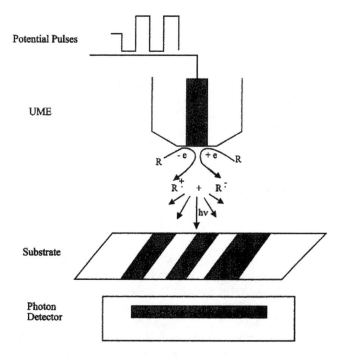

Figure 6 Schematic diagram illustrating the operating principles of ECL generation at an SECM in the alternating potential pulse mode. ECL is generated by the annihilation scheme $R^{+} + R^{-} \rightarrow 2R + h\nu$ and is detected by a PMT after attenuation by the substrate.

similar d values, so the resolution of the images was not very high. In later experiments [18,19], much smaller electrodes were employed and higher resolution was obtained. Various approaches were employed. In one [18], etched Pt tips coated with electrophoretic paint with effective diameters of 60–100 nm were attached to a tuning fork to maintain d at a very small value. ECL was generated with a constant potential in an aqueous $Ru(bpy)_3^{2+}/TPrA$ system. A resolution of the order of 200 nm was reported. In another [19], the electrode was a flame-etched carbon fiber that had no insulating coating near the tip end. The ECL solution was 25 mM DPA in benzonitrile (BN), and emission was generated by an annihilation reaction (DPA^{+} and BN^{-}). By generating the reactants with a square-wave frequency of 20 kHz, the reaction was confined to the end of the tip, and a resolution of about 600 nm was reported without any feedback control of d. The estimated power of the ECL tip source was 1.8 pW, which was sufficient to allow transmittance spectroscopy of the sample.

The results obtained so far with an ECL-based tip light source are quite promising, especially for optical imaging of samples immersed in solution.

However, to be widely useful, true near-field scanning optical microscopic (NSOM) resolution levels must be obtained routinely and rapidly. This probably will require feedback loop control with a computer and electronics dedicated just to this function as carried out with commercial NSOM instruments. Moreover, the ECL solutions may be inconvenient in certain applications, e.g., with biological samples. An alternative, not yet explored, is the use of a solid-state light-emitting electrode as described in Chapter 10, although making this of the required small size will be very challenging.

VI. USING ECL TO OBTAIN THERMODYAMIC AND MECHANISTIC INFORMATION

The extensive studies of ECL, particularly of organic species in aprotic media, have clearly demonstrated that the observation of emission as a function of reactant concentration, solution conditions, and potential can provide very useful insight into the existence and energetics of often very unstable intermediates and help elucidate complex mechanisms. An example is the mechanism of the reaction in the $Ru(bpy)_3^{2+}$ and TPrA system discussed in Chapter 5. A close examination of the mechaism of ECL in this system established the existence of the TPrA cation radical $Pr_2NCH_2CH_2CH_3^{+\cdot}$ and allowed its lifetime to be estimated. In addition, the ECL process in this case provided evidence for the complex mechanism of aliphatic amine oxidation and the intermediacy of the free radical $Pr_2NC\cdot HEt$ following deprotonation.

The free energies of these species can also be estimated from ECL experiments, based on the energetic requirements for the observation of ECL, i.e., the energy of the electron transfer reaction that produces ECL must be larger than the energy of the excited state, $-\Delta H^0 = E^O(D^+\cdot,D) - E^O(A,A^-) - T\Delta S^0 > E_S$. For example, it is possible to examine the behavior of TPrA as a coreactant with the electrogenerated cation radicals of a number of aromatic hydrocarbons whose oxidation spans a wide range of potentials in MeCN/benzene [20]. By observing which hydrocarbons (R) produce ECL and which do not, and knowing the potentials for oxidation of R to $R^{+\cdot}$ and the singlet state energies, one can estimate the potential for the reduction of $Pr_2NC\cdot HEt$ as about -1.7 V vs. SCE in this solvent. Moreover, by studying the quenching of the fluorescence of R by TPrA and assuming that it proceeds by electron transfer quenching, one can also estimate the potential of the $TPrA^{+\cdot}/TPrA$ couple as 0.9 V vs. SCE. Although ECL has not found wide application in the physical-organic chemistry community, it should be possible to design ECL experiments that probe many mechanisms of interest.

VII. CONCLUSIONS

Electrogenerated chemiluminescence has come a long way in the last 40 years in terms of both understanding the phenomena associated with it and finding applications. The immunoassay systems based on ECL had sales in excess of $400 million in 2002. However, there are still new reactions to investigate and understand, new ECL phenomena to probe, and new applications to develop. There are continued reports in the literature of new organic and inorganic compounds that produce ECL, and the combined electrochemical, spectroscopic, and ECL data lead to useful insights into the structures and interactions in these. Advances in computational chemistry, which already provides important information about ECL, will continue to play a role.

In the analytical field, improvement of sensitivity is always of interest, especially in detection of biological species. The "Holy Grail" in such studies is the ECL detection of single molecules, and it is only a matter of time until this is accomplished by utilization of various amplification schemes, because single-molecule detection by fluorescence measurements is well established. Continued progress in light-emitting electrochemical cells based on polymer or solid-state systems could lead to light sources and displays.

REFERENCES

1. Lee D.C.-S., Wilson, T. In Cormier, M.J., Hercules, D.M., Lee, J. eds. Chemiluminescence and Bioluminescence, Plenum Press, New York, **1973**, p. 265.
2. Dose, V. Prog. Surf. Sci. **1983**, *13*, 225–284.
3. McIntyre, R., Sass, J.K. J. Electroanal. Chem. **1985**, *196*, 199; Phys. Rev. Lett. **1986**, *56*, 651.
4. (a) McIntyre, R., Roe, D.K., Sass, J.K., Gerischer, H. Ber. Bunsenges. Phys. Chem. **1987**, *91*, 488; (b) McIntyre, R., Roe, D.K., Sass, J.K., Storck, W. J. Electroanal. Chem. **1987**, *228*, 293.
5. (a) Ouyang, J., Bard, A.J. J. Phys. Chem. **1987**, *91*, 4058; (b) **1988**, *92*, 5201.
6. (a) Uosaki, K., Murakoshi, K., Kita, H. J. Phys. Chem. **1991**, *95*, 779; (b) Murakoshi, K., Uosaki, K. J. Phys. Chem. **1992**, *96*, 4593; (c) Murakoshi, K., Uosaki, K. Phys. Rev. B. **1993**, *47*, 2278; (d) Uosaki, K., Murakoshi, K., Kita, H. Chem. Lett. **1990**, 1159.
7. McCord P., Bard, A.J., unpublished experiments; McCord, P.A., Ph.D. Dissertation, The University of Texas at Austin, 1992.
8. Gosavi, S., Marcus, R.A. Electrochim. Acta, **2003**, *49*, 3.
9. Hung, C.-J., Halaoui, L.I., Bard, A.J., Grudowski, P., Dupuis, R.D., Molstad, J. DeSalvo, F.J. Electrochem. Solid State Letts. **1998**, *1*, 142 and references therein.

10. Chandross, E.A., Visco, R.E. J. Phys. Chem. **1968**, *72*, 378.

11. (a) Yeh, L.R., Bard, A.J. Chem. Phys. Lett. **1976**, *44*, 339; (b) Luttmer, J.D., Bard, A.J. J. Electrochem. Soc. **1978**, *125*, 1423; (c) Gleria, M., Memming, R. Z. Phys. Chem. **1976**, *101*, 171; (d) Grabner, E.W. Electrochim. Acta **1975**, *20*, 7.

12. (a) Sung, Y.-E., Gaillard, F., Bard, A.J. J. Phys. Chem. B. **1998**, *102*, 9797. (b) Sung, Y.-E., Bard, A.J. J. Phys. Chem. B. **1998**, *102*, 9806; (c) Gaillard, F., Sung, Y.-E., Bard, A.J. J. Phys. Chem. B. **1999**, *103*, 667.

13. Kulmala, S., Ala-Kleme, T., Heikkila, L., Väre, L. J. Chem. Soc. Faraday Trans. **1997**, *93*, 3107 and references therein.

14. (a) Szunerits, S., Tam, J.M., Thouin, L., Amatore, C., D.R. Walt, Anal. Chem. **2003**, *75*, 4382; (b) Engstrom, R.C., Johnson, K.W., Desjarlais, S. Anal. Chem. **1987**, *59*, 670; (c) Hopper, P., Kuhr, W.G. Anal. Chem. **1994**, *66*, 1966 and references therein.

15. Luttmer, J.D., Bard, A.J. J. Electrochem. Soc. **1979**, *126*, 414.

16. Nicholson, M.M. J. Electrochem. Soc. **1972**, *119*, 461.

17. (a) Fan, F.-R.F. Cliffel, D., Bard, A.J. Anal. Chem. **1998**, *70*, 2941 (b) Maus, R.G., McDonald, E.M., Wightman, R.M. Anal. Chem. **1999**, *71*, 4944.

18. Zu, Y., Ding, Z., Zhou, J., Lee, Y., Bard, A.J. Anal. Chem. **2001**, *73*, 2153.

19. Maus, R.G., Wightman, R.M. Anal. Chem. **2001**, *73*, 3993.

20. Lai, R.Y., Bard, A.J. J. Phys. Chem. B. **2003**, *107*, 3335.

Index